CONSERVATION IN THE UNITED STATES

CONSERVATION IN THE UNITED STATES

Second Edition

by

RICHARD M. HIGHSMITH, JR.

J. GRANVILLE JENSEN

ROBERT D. RUDD

Oregon State University

GREENWOOD PRESS, PUBLISHERS
WESTPORT, CONNECTICUT

Library of Congress Cataloging in Publication Data

Highsmith, Richard Morgan, 1920-
Conversation in the United States.

Reprint of the ed. published by Rand McNally, Chicago, in series: Rand McNally geography series.
Includes bibliographical references and index.
1. Natural resources--United States.
2. Conservation of natural resources--United States.
I. Jensen, John Granville, 1911- joint author.
II. Rudd, Robert D., 1924- joint author.
III. Title.
[HC103.7.H5 1978] 333.7'2 78-6195
ISBN 0-313-20511-6

Reprinted with the permission of Rand McNally College Publishing Company

Reprinted in 1978 by Greenwood Press, Inc.
51 Riverside Avenue, Westport, CT. 06880

Printed in the United States of America

10 9 8 7 6 5 4 3 2 1

Dedicated to our children.

May they intelligently accept their responsibilities in the nation's future.

PREFACE TO THE SECOND EDITION

Americans are truly fortunate people. Their heritage includes the richest and most varied resource base of any nation in the world. Through application of a high order of science and technology to this base, they have developed an economy that provides them with a higher standard of living and more material benefits than are enjoyed by any other people.

In this achievement the record in resource management and use has been clouded by ideas of abundance and inexhaustibility and largely guided by a desire for immediate individual profit. Undue despoliation and misuse have been common. Although improvements have been made in recent decades, the rate of environmental change is increasing rapidly with advancement of technology, but adjustments still are inadequate if citizens of today are to pass along to future generations the bases for maintaining the American way of life. In the face of increasing population, rising standards of living, and expanding world responsibilities, resource requirements will increase. The obligation to lift the level of stewardship, therefore, is great.

It is the conviction of the authors that resource adequacy, management, and use should be the concern of every citizen of this great democracy. Whether he be farmer, forester, scientist, office worker, or teacher, an individual's livelihood and material well-being depend upon the wealth of the resource base—and so does that of his progeny. Everyone has the responsibility to promote practices and regulations which assure the future, for all policies, programs, and practices stem from the majority opinion of the body politic.

The authors firmly believe that conservation must become a national way of life, and that its precepts should be inculcated into the educational process. This book is presented in that interest. It contains a systematic and reasoned appraisal of the extent and character of the nation's resources, the evolution of their use and misuse, and present conservation practices in the light of their adequacy for the future.

Professor Highsmith was responsible for the general organization of the book and the preparation of Chapters 1 through 6, 18, and 20. Chapters 7 through 10, 14 through 16, and 19 were prepared by Professor Jensen. Chapters 11, 12, 13, and 17 were prepared by Professor Rudd. The entire manuscript was read and critiques were done several times by all. The finished product represents a composite of our thinking.

Acknowledgments

Many persons, through their published works or through personal correspondence, contributed significant ideas for this book. Most statistical data and much other information were drawn from governmental sources. These greatly facilitated the preparation and are acknowledged in footnotes within the text. All interpretations, however, are the complete responsibility of the authors.

Most of the maps and other drawings, based primarily upon Federal government sources, were prepared by our former colleague, Mr. Jon Leverenz, for the first edition. The standards set by his quality work have been maintained by the Company in this second edition. Pictures were supplied by a variety of agencies and companies. Credit is given within the text, but here we acknowledge special appreciation.

Richard M. Highsmith, Jr.
J. Granville Jensen
Robert D. Rudd

Corvallis, Oregon

CONTENTS

List of Maps

List of Tables

CONSERVATION
IN THE UNITED STATES

1

A Perspective on Resources and Conservation

CONSERVATION HAS BECOME A POPULAR TERM in the American vocabulary. It is used to describe resource management techniques, to identify efficient practices, to promote resource planning and development programs, to encourage the protection of wildlife, to popularize the preservation of wild land, and so on. It is used by resource managers, government agencies, planners, special interest groups, teachers, politicians, and journalists. Generally, by identification with conservation each group desires to imply some quality of "good" in the relations between its programs, interests, or philosophies and natural resources. Aside from this common aim, there are obvious differences in the conceptual meanings specifically assigned to the term by different groups.

In the broadest sense, conservation is concerned with the maintenance and improvement of the fundamental capital of a society—its total resource base. In a real sense, conservation is an endless program of stewardship. It has as its major goal the perpetual maintenance of a society, and it requires a variety of decisions and actions *today* to assure that there will be resources to fulfill the societal needs of *tomorrow*. By the very nature of its goal, conservation includes a dynamic system of techniques, practices, use efficiencies, plans, policies, research, cooperative efforts, restraints, and so on. These, in turn, sweep across the interests of many individuals, groups, and agencies. And the goal and means of conservation should be among the interests of the citizenry at large because everyone has a stake in the perpetuation of the society!

Resource Perception and Use

Before proceeding further with a discussion of conservation, it will be of value to consider some fundamental characteristics of natural resources, including the variables that influence their identity, development, and durability. Natural resources derive from the earth's physical-biological system (natural environment) as the result of a human group's perception of a functional utility or value of some element or complex of elements. In any given circumstance any natural feature or combination of features—rock, sand, mineral, sunshine, canyon, water, vegetation, cropland, physical site, and so forth—may be identified as a resource. The processes of bringing natural environmental elements into use may be called resource development.

Resource identity, then, is influenced by variables inherent in human societies as well as in the earth's physical-biological system. The characteristics of the physical-biological system vary from place to place, and the characteristics of populations and their cultures (which influence the perception of environmental utility) vary in space and time. Thus, because of the nature of the influential variables and the possibilities and probabilities of change in any one, resources may be said to have dynamic qualities.

RESOURCE VARIABLES

The major variables involved in the complex of physical (or natural) and human relationships which result in the perception of resources and the development of resource use systems can be grouped into five interdependent classes or sets. These are identified as the physical-biological, the technological, the social, the economic, and the formal-institutional variables.[1]

1. The physical-biological variables set the outside limits of the natural permissibility of a given environmental system for every stage of development. They are dependent upon ecological conditions and/or physical presence. For particular resources and uses, some spatial units offer greater natural possibilities than others.
2. The technological variables are related to the kinds and levels of resource-converting and space-adjusting techniques possessed by an occupying society.[2] Resource-converting techniques include the sciences, arts, tools, materials, and so forth that provide the means for bringing elements, materials, or features of the natural environment into human use or service. Space-adjusting techniques, similarly, are those that may be employed either to shorten the effective distance of travel and transportation (which stimulate and permit spatial interaction) or to permit intensification of space employment above that allowed by a natural land surface. When man gains a high level of technology, he attains a considerable capacity for "ecological dominancy."
3. The social variables include all kinds of public attitudes and pressures that bear upon decision makers, potential developers, and workers, as well as users. Thus they influence the adoptability and acceptability of particular resources and resource use systems to the social structure of an occupying group.
4. The economic variables include those features that have bearing upon the profits derived from an input mix required to bring a resource into use. Profit is, perhaps, the strongest motive for resource development as well as the strongest test of the viability of an established resource use. (But what is profitable to a business firm with relatively short-range interests may not be profitable to a society with long-range interests.) The major economic variables include the costs associated with resources, technology, labor, transportation, and marketing, and the size, dispersion, and purchasing power of the consumers.
5. The formal-institutional variables are comprised of the various organizational forms of a social order that are supported by law or strong traditions. These tend to circumscribe the possible actions of resource developers and divide individual from public responsibilities. Especially important in the resource context are those forms that have to do with real property ownership, the rights of owners, public policies that permit or encourage resource actions, and policies that restrain resource actions.

In summation, natural elements become natural resources when they are adoptable and useful to a human society, which occurs when favorable relationships exist between physical, technical, social, economic, and institutional permissibilities, and need or demand and developmental-managerial ability.

RESOURCE STABILITY

The complexities and dynamics of the resource concept are indicated by the fact that resources result from a cultural appraisal of environment utility. What is useful to one group may not be useful to another group; or what is useful to one group at a given time may not be useful, at least for the same purpose, at another time. As suggested earlier, man is a dynamic agent who varies through space and time in numbers, wants, abilities, attitudes, and social organization. Especially because man's number, ingenuity, and capability are not static, resources are neither finite nor stable. Advancement in technology, growth in population, and desire for higher standards of living have brought continual refinements in resource use and broadened perception of environmental utility. Thus, the resource base of this nation today is not the same as it was fifty years ago, nor will it be the same fifty years from now.

The improvement of technology has not only reduced many former environmental restraints but has also changed attitudes respecting the role of the environment in human support. Many Americans today seem to be willing to place full reliance on the capabilities of technology to meet the material needs of the future; this is understandable in the light of the significant advances in technology and the shifts in population from rural to urban living that have occurred in recent decades. There is little doubt that the technological capabilities will continue to increase. To be sure, these advances have improved and will continue to improve resource use efficiencies. But technology has not reduced societal dependence upon the natural environment for basic capital. In fact, the quantity and variety of natural resources required today are greater than ever. Moreover, the demands for natural resources as well as for technology will continue to increase.

Growth in population, and the national and world-wide desire for improved living standards, leave little room for complacency. Yet, as Zimmerman has appropriately stated, "a business civilization is apt to neglect the care of its basic assets."[3] In such a society there is a tendency to think in terms of the present, whereas rational, harmonious use and adequate care of the basic assets require a long-range view and continual adjustments. The human record on the earth is a history of environmental change and modification. Some modifications have produced lasting social benefits; others have yielded long-enduring social strains. Some resources are depleted and others are altered by the processes of use. Moreover, because of the interrelations of the subsystems of the environment, uses made of one subsystem nearly always have an impact upon the use potentials of another. Thus there is much evidence to support the concept of total environment conservation —or, perhaps more appropriately, total environment management.

A Resource Classification

Because of the possible and probable changes in the variables that influence their identity and use, resources can be appraised only in synoptic terms. Today's farmland may become urban land in a few years; dry grazing land may become irrigated cropland; an open pit mine may become a recreation lake; and so on. Nevertheless, it is possible to classify elements of the physical-biological system according to their potentials as resources from the point of view of today. The classification that follows is based upon a grouping of resources according to their similarities in nature, their durability when in use, and their requirements in management practice and policies for conservation. It consists of three major groups, two of which have two subdivisions, making a total of five categories.

1. Resources which through the actions of man are not only sustainable, but are improvable. This category includes the agricultural lands for both crop and animal production, forest lands for timber and other purposes, and other types of land used for biotic production. Other examples are surface waters, fish, and wildlife.
2. Resources which are nonimprovable, and are depleted with use. This group

includes mainly the minerals, of which two categories are recognized:

   a. Those that are exhaustible, particularly the fuels, coal, oil, and gas; also minerals that have a low ratio of recovery for re-use such as sulphur, potassium, phosphate, tin, and boron.
   b. Those that are depletable, but which once in use have a high ratio of re-use, such as copper, iron, lead, gold, and silver.

3. Resources which are not improvable or increasable and are not depletable, but are, for the most part, continuous or endless in time. Two categories are recognized:
   a. Those that are essentially beyond influence and damage by man, such as solar energy, macroclimate, oceans (except along coasts or if used as dumps for radio-active material), and space in the sense of area.
   b. Those that are endless in time, but are susceptible to damage by man, such as air, site, and scenery.

These are the categories of natural assets that are the base for the primary support of people and economies. At any given time or in any given place, of course, the individual resources will not be of equal significance. Physical, technical, societal, economic, or legal influences may limit some, and may enhance others. In the establishment of its occupance, however, a society must have concern for the total environment and the interrelations among the elements, as well as for the characteristics of durability and limits of use possible for each—those that are currently useful as well as those that may be only potential resources. In the long run this is simply good business, for neutral or even negative qualities of the environment today may be assets tomorrow.

A Conservation Philosophy

No nation can attain affluence or remain great without managing its resource base in a manner that will provide future supplies. This is well documented in the annals of world history (Figure 1 – 1).[4] Resources form the essential foundations for human existence, providing the living, working, and recreational space and the basis for food, clothing, shelter, equipment, and all other material aspects of civilization. The relative wealth of resources in the possession of a people has direct bearing on their level of living, and this, in turn, has much influence on the societal elements of their civilization. Because of this basic dependence, it is clear that every generation of every society has an ethical responsibility to pass along to the next a legacy of natural as well as scientific and cultural wealth—the means for maintaining or improving the established way of life.

The United States has been singularly blessed in the wealth and variety of its resources. The demand on this foundation, however, is increasing and there is no longer a rich land frontier into which the nation can expand to help nourish continuing economic growth. The frontiers of the present and the future lie mainly in the possibilities of improving and intensifying man-resource relations in the area now contained within the national boundaries. Although shortages exist in some resources and are imminent in others, the total foundation remains strong and diverse.

The major barrier to maximization of enduring social benefit from the nation's natural endowments is the social and economic attitude toward resources; this problem has a corollary in the lack of definitive societal goals. As products of, and participants in a business society supported by relative abundance, Americans have developed a readily observable tendency to sacrifice basic assets for immediate profits, expecting that advancing technology will be the implement for meeting future needs. Inherent in the pervasive societal philosophy are the concepts of private property, private enterprise, private gain, and protection of private rights. Not so firmly fixed, however, is an essential

1 – 1 *Land degradation.* Imprudent use and poor management result in the reduction in the productive capacity of land. This is the case on the upper coastal plain of Mississippi. Although the total area of severe land deterioration in the United States is not large, this is by no means an isolated example. (Department of Agriculture)

corollary concept, private responsibility; with rights go responsibilities to the society. Private rights and the rights of each individual, city, and state to make mistakes have been carried to the extreme. As a result, remedial actions are requiring a significant share of the national budget today to eliminate slums, relocate highways, purify water, combat air pollution, and so forth. To be sure, the nation has advanced significantly in resource management technology. We can, for example, largely control the problem of land erosion. But one may wonder if that problem has not been replaced with one more ominous, the "cultural erosion" of the landscape quality. In a private enterprise democracy, decisions respecting resource matters are distributed among many private and public owners, and governmental policies and restraints are responsive to the voice of the body politic; therefore, the future of the society must depend in large measure upon an enlightened, knowledgeable, ethically responsible citizenry.

It would seem axiomatic that any society growing within finite space and having concern for quality of life must come to terms with the natural environment and plan the orderly advance of its occupancy. Resource use systems must be viable in the context of the natural environment as well as of the culture. There must be minimization of practices and use systems that produce undesirable, irreversible results for the sake of short-term benefits, and even a minimization of those that produce results that are very expensive to correct. It would seem, moreover, that every society will have to exercise increasing caution and restraint to achieve, in the face of growing numbers of people, a harmonious perpetual adjustment to its resource base. Such would appear to be the only insurance policy for a secure future; this is the essence of a conservation philosophy (Figure 1 – 2).

A dynamic state of harmony between man and nature cannot be achieved by any simple program. It requires consideration of the total environment, of technological capability, of societal needs, goals, and institutions, and of the potential impacts of human choices. It must be self-correcting, and couple moral responsibility with economic responsibility.[5] Especially, it requires cooperation among all levels of resource managers, and between a concerned citizenry and a responsive government. It would seem that the following considerations are essential parts of a conservation philosophy.

1) The development of a sense of public responsibility supported by at least the major share of the citizenry in order that desirable legislation and social and economic pressures be brought to bear on the resource problems to assure viable use systems.
2) The consideration of the total resource base—the total environment—to insure that economic exigencies in one given time will not result in unnecessary damage to potential resources or other resources of lesser economic value.
3) The continual improvement of the productive capacities of those resources that are susceptible to improvement.
4) The restriction to their higher orders of use of those resources that are limited in stock and exhaustible.

1 – 2 *Conservational management of land.* This presents a sharp contrast to the scene shown in Figure 1 – 1. In this Wisconsin scene a cultural balance has been substituted for the natural balance. Hills and steeper slopes are in woods and grass, waterways are grassed, and croplands are protected by contour and strip-crop farming practices. (Department of Agriculture)

5) The continual search through science and technology for new resources to take pressure off those that are diminishing.
6) The continual development and use of substitutes based upon resources of more generous supply for those resources in use that are limited in stock.
7) The continual striving to maximize use and minimize waste to derive the utmost value from each resource.
8) The continual striving to achieve optimum multiple use for resource complexes where appropriate, thus expanding total value.
9) The continual inventory of the character, quantity, and quality of resources available.
10) The continual projection of resource needs, as far as possible, to avoid sudden appearance of shortages or other problems.
11) The balancing and harmonizing of the rights of present citizens against those of future generations, and the harmonizing of the rights of private property interests and the rights of public interests.
12) The continual planning and programming of all development, whether private or public, with consideration given to all aspects of resource availability and human welfare in the interest of achieving the greatest possible maintainable economic and social benefits.
13) Cooperation on all levels, including the allocation of duties, powers, and responsibility to private, state, and Federal interests in the cause of the common good of the nation.

Many would suggest that some of the facets of such a multi-faceted philosophy are not acceptable to the American society—that the relinquishment of some long-standing private property rights is too great a price to pay. Others would suggest that a harmony or balance between the American society and its environment is impossible—that only in a stable farming or pastoral economy and culture can such a harmony be achieved. By its very nature an industrial society alters its natural environment. American experience would support the conclusion that nothing has been so constant as change. But "man's own wisdom is his premier resource—the key resource that unlocks the Universe."[6] Men of wisdom certainly must come to recognize a greater responsibility to their society and their progeny, and thus to apply the principles of conservation to environmental management to minimize disruptive changes, to maximize the use of nonrenewable and scarce resources, and in general, to advance the occupance and use of their earth habitat in an orderly and rational manner.

FOOTNOTES

[1] See also Walter Firey, *Man, Mind, and Land, A Theory of Resource Use* (Glencoe, Illinois: The Free Press, 1960), 256 pp. Firey assesses resource use systems according to the theories of ecology, enthnology, and economics, testing whether they are possible, adoptable, or gainful.

[2] These terms are borrowed from Edward A. Ackerman, "Geography as a Fundamental Research Discipline," *Research Paper No. 53* (Chicago: Department of Geography, Univ. of Chicago, June, 1958), p. 26.

[3] Erich Zimmerman, *World Resources and Industries* (New York: Harper and Row, Publishers, Inc., 1951), p. 4.

[4] For example see W. C. Lowdermilk, "Conquest of the Land through Seven Thousand Years," *Agriculture Information Bulletin No. 99* (Washington: Soil Conservation Service, Department of Agriculture, August, 1953), 30 pp.

[5] For a more complete statement on the idea of ethical responsibility see Aldo Leopold, *A Sand County Almanac* (New York: Oxford Univ. Press, 1949), pp. 201–26.

[6] Zimmerman, *op. cit.* p. 7.

2

The Conservation Movement

During the period of the rapid rise of the nation to the eminent position as the leading producer of the world, the record of resource management was not all good. Considered from the perspective of today, it is clear that growth from the beginning of settlement until the 1930's was supported by large expenditures of resource capital. Farmlands commonly were mined of their fertility and much topsoil was lost. Forests, in general, were cut without adequate provision for regeneration. Mineral deposits were skimmed for their highest quality ores. Streams were stained with effluvia from mills and muddied with sediments from careless land use. Wildlife was slaughtered and some species nearly exterminated. Indeed, extravagant resource spending characterized almost every facet of the nation's economic development.

In fairness to our forefathers, however, it must be stressed that the despoilment of resources that occurred was not waste in a deliberate sense, nor was it economic waste in the time setting. It was, rather, a natural reaction of a small population to the development of a land which seemed so abundantly endowed with resources that stocks could never be exhausted. It was, furthermore, a natural result of the prevailing economic conditions and philosophy, the ease of gaining property, the equipment available, and the technological limitations of the times. Under these conditions only quality timber and minerals could pay their way to market; and it was often more economical to abandon "worn out" farm land than to cope with declining fertility. Resource abundance resulted in low economic valuation. The exuberant exploitations of resource wealth that characterized the early decades may be regarded as part of the growing pains of the nation.

With the growth of the nation to maturity —accompanied by increasing knowledge of resource character and requirements; improvements in science, technology (including both resource-converting and space-adjusting techniques), and equipment; and changing economic conditions—resource management has improved considerably. Happily, the original endowments were large. Thus, if wisdom born of experience is used to bring forethought to development activity in the years ahead, the potentials remain liberal and the future relatively bright.

An appreciation of the factors involved in shaping the attitudes and decisions of the American people in resource matters is essential. The evolution of the conservation movement, therefore, is traced from the colonial period to the present in the remaining pages of this chapter.[1]

The Colonial Period

Although their influence for three centuries was small, conservation leaders have never been lacking in America. Several of the founding fathers of the colonies recognized the need for establishing a proper relationship with resources. For example, in 1626 Plymouth Colony passed an ordinance prohibiting the cutting of timber on colony lands without official consent. William Penn's ordinance in 1681 required that when clearing land in the Pennsylvania Colony one acre be left in trees for every five acres cleared.[2] Many of the early colonists had some background in soil and forest management, having come from western Europe where such care already was necessary. None, however, had knowledge of how to live in a forest wilderness or dreamed that the time would come when population would increase to the point where every acre of land had to be counted. Land hunger could be readily satisfied; work was the major requirement. Utilizing methods brought from Europe or learned from the Indians, and by trial and error, the settlers carved out their farms from the forest. If one was not satisfied there were new lands and rich prospects farther west.

During the Colonial period every man was a farmer—even doctors, teachers, and preachers tilled the land at least part time. As settlement crystallized along the Atlantic seaboard, a few of the more learned individuals became aware of increasing soil erosion and diminishing fertility and carried out experimental work to control these problems. In 1748 Jared Elliot, a minister, doctor, part-time farmer of Killingsworth, Connecticut, published the first American book on agriculture. A large part of this work was devoted to English practices, but many of his own observations and experimentations were included.[3] By 1800 a score of such publications appeared, but distribution was limited and interest remained small. In the meantime, the forest was regarded more as an enemy than a friend; it had to be cleared to make way for farms, and it harbored the Indians.

The time was not ripe for the development of a conservational relationship with the environment; it was neither economically nor so-

TABLE 2–1
ACQUISITION OF THE PUBLIC DOMAIN, 1781–1867*

Acquisition	Area			Cost
	Land	Water	Total	
	Acres	Acres	Acres	
State cessions (1781–1802)	233,415,680	3,409,920	236,825,600[1]	$ 6,200,000
Louisiana Purchase (1803)	523,446,400	6,465,280	529,911,680	$23,213,568
Red River Basin[2]	29,066,880	535,040	29,601,920	
Cession from Spain (1819)	43,342,720	2,801,920	46,144,640	$ 6,674,057
Oregon Compromise (1846)	180,644,480	2,741,760	183,386,240	———
Mexican cession (1848)	334,479,360	4,201,600	338,680,960	$16,295,149
Purchase from Texas (1850)	78,842,880	83,840	78,926,720	$15,496,448
Gadsden Purchase (1853)	18,961,920	26,880	18,988,800	$10,000,000
Alaska Purchase (1867)	365,481,600	9,814,400	375,296,000	$ 7,200,000
TOTAL PUBLIC DOMAIN	1,807,681,920	30,080,640	1,837,762,560	$85,079,222

* *Public Land Statistics, 1964* (Washington: Department of the Interior, Bureau of Land Management, 1965), p. 4.

[1] Georgia Cession (56,689,920 acres) 1802.

[2] Red River Basin of the North, south of the 49th parallel. There is disagreement as to the method and exact date of acquisition. Some hold that it was part of the Louisiana Purchase, others maintain that it was acquired from Britain.

cially adoptable. On the contrary, abundance of resources everywhere engendered a spendthrift attitude that was to prevail for some time to come.

The Young Republic, 1781 – 1870

During the first century of national existence the entire land area of the forty-eight contiguous states and Alaska was acquired, and settlement spread from coast to coast (Figure 2 – 1). Between 1781 and 1802 the original states ceded most of their western land claims to the Federal government, creating a public domain of more than two hundred million acres. Other additions were made for a grand total of more than 1.8 billion acres by the end of the period. The amounts, manner of acquisitions, and distribution are shown in Table 2–1 and Figure 2 – 2.

Indian claims caused relatively few problems in the settlement of the West. A policy was established in 1783 "prohibiting persons from settling on, purchasing, or receiving lands claimed by Indians without the express authority of the Congress."[4] Nevertheless, as the frontier moved west the Indians gradually lost their land, treaties were signed (more than 350 in number), and the Indians were placed on reservations as wards of the Federal government under the administration of the Bureau of Indian Affairs, established in 1824. This procedure gradually resulted in cession of native rights and the opening of the lion's share of Indian claims to settlement (Figure 2 – 3).

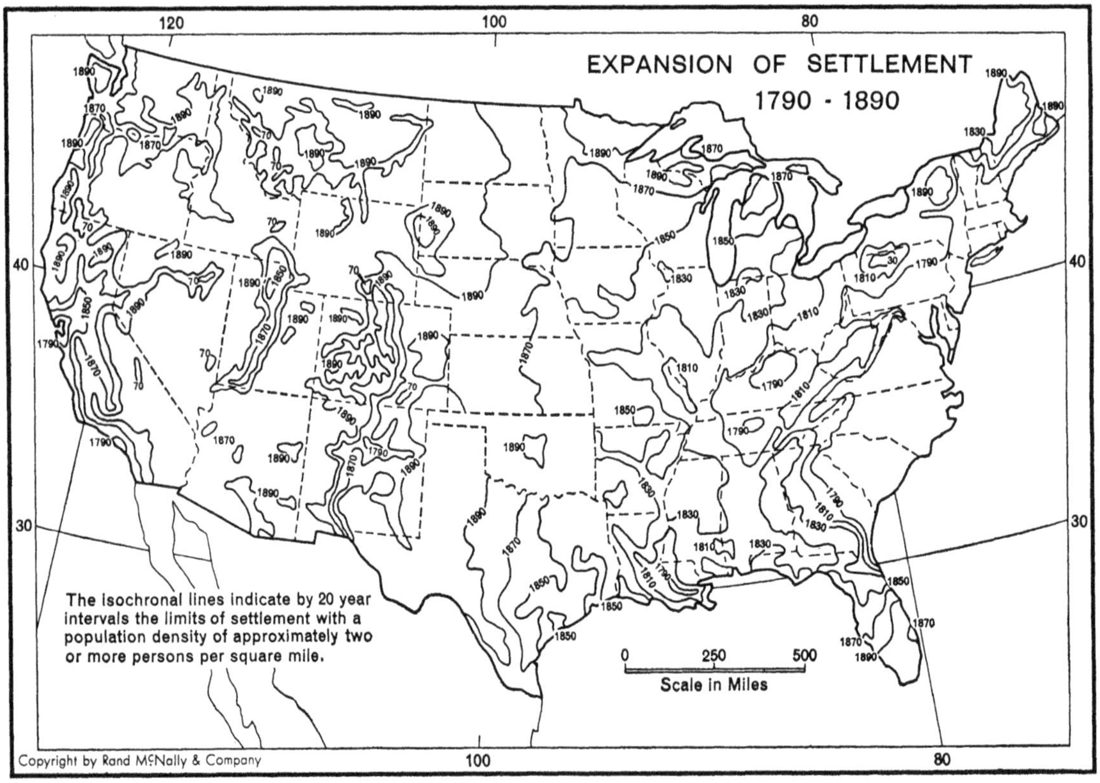

2 – 1 ***Expansion of settlement, 1790–1890.*** **This map shows the westward expansion of the frontier. It should be noted that the frontier was not a terminal line, but a ragged fringe. The part of the Great Plains that is now Oklahoma was not opened to settlement until 1889. In the West, settlement was confined to places with readily available water supplies for many years. (Adapted from Agricultural Handbook No. 153, p. 29)**

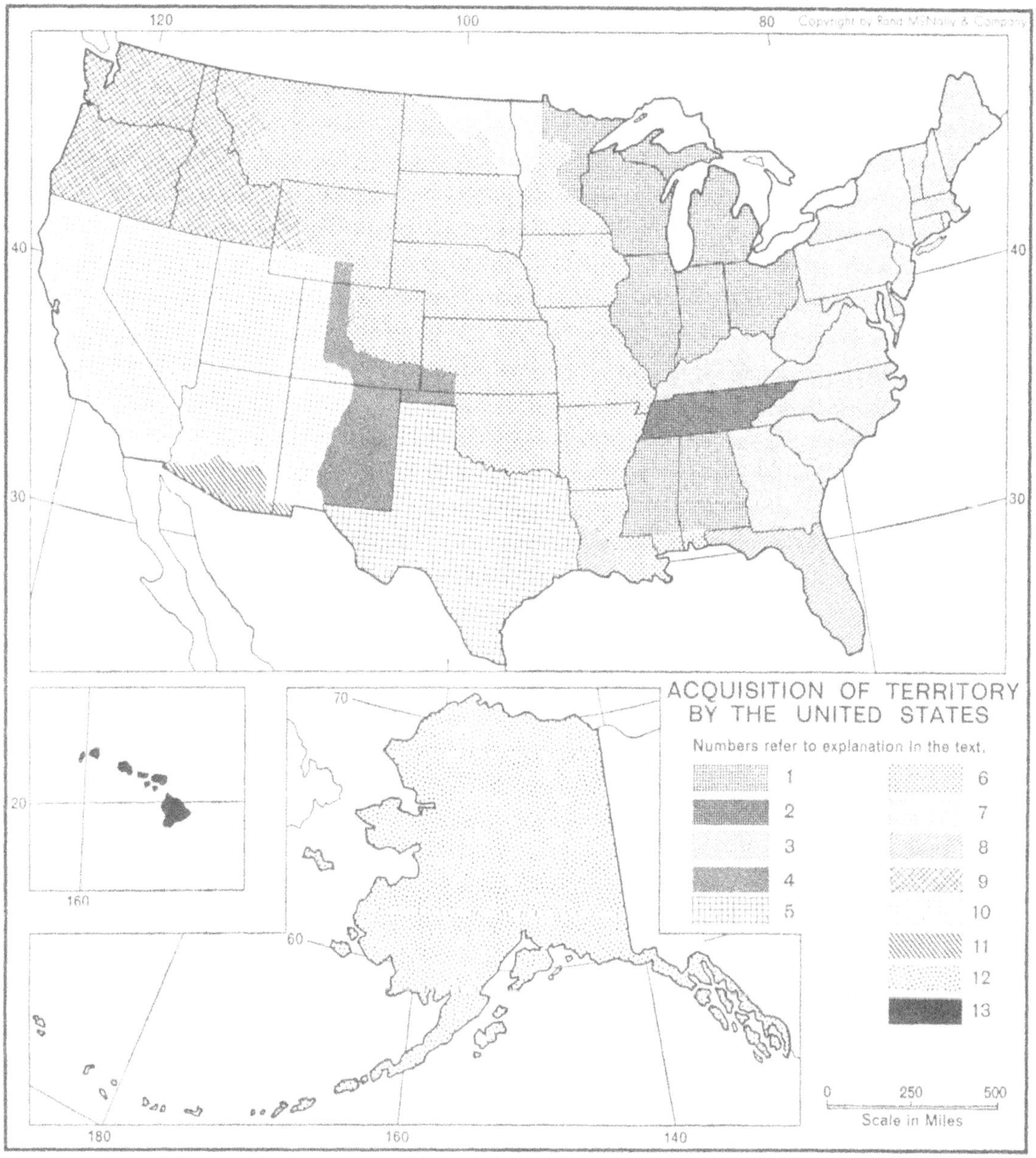

2 – 2 *Acquisition of the territory of the United States.* (Bureau of Land Management) The patterns indicate the acquisitions as follows:

Territory of the Original Thirteen States

(Treaty of 1783 with Great Britain)

1. State cessions to the United States 1781, 1784 (1788), 1785, 1786, 1787, and 1802
2. North Carolina cession to the United States, 1790; United States cessions to Tennessee, 1806 and 1846
3. The Original Thirteen States (present area) plus the District of Columbia (Maryland cession to the United States, 1788) and the new states created out of the territory not ceded to the United States: Vermont (1791), Kentucky (1792), Maine (1820), and West Virginia (1863)

Territory of the Republic of Texas

(Annexation of Texas, 1845)

4. United States purchase from Texas, 1850
5. State of Texas (present area)

Other Acquisitions by the United States

6. Louisiana Purchase from France, 1803 (delimited, 1819)
7. Basin of the Red River of the North (historians differ as to date of acquisition)
8. Spanish Treaty (Florida cession, claim adjustment), 1819
9. Oregon Compromise with Great Britain, 1846
10. Cession from Mexico, 1848
11. Gadsden Purchase from Mexico, 1853
12. Alaska Purchase from Russia, 1867
13. Hawaii annexed, 1898

2 – 3 *Cessions of Indian lands.* This generalized map shows the principal cessions with the numbers indicating the order in which they were effected. (Bureau of Land Management)

DISPOSAL POLICIES

The Federal government had no intention of being permanent landlord over the vast acquired territory. "The wholesale transfer of the public domain to private ownership was made under the (then) revolutionary political and economic idea that a nation in which the people are landlords is a nation in which the maximum of individual freedom can best be assured."[5] Short on cash but long on land, it established the policy of using land in lieu of cash to encourage the "installation of 'internal improvements' to build up the country, particularly in the newly acquired territories."[6] Fortunately, it was recognized at the same time that a more adequate system of describing and designating land would be required than those systems employed during the colonial period—a system that would permit the orderly disposition of the public domain and transfer of privately owned land from one owner to another without confusion and controversy. One of the features of the Northwest Ordinance passed in 1785 was the provision for the rectangular system of surveys which, with improvements, has been maintained to this day (Figure 2 – 4). Except for Texas, most of the western lands were subdivided according to this system (Figure 2 – 5).[7]

Settlement, development, and disposal of the public lands, however, did not follow any orderly or well-conceived plan; too little was known of the vast western country and experience in such matters was too limited. Accordingly, the depleted condition of the United States Treasury, public pressures, and trial and error guided the early policies. During the 1800's numerous laws were passed by Congress providing for the disposal of public lands to private owners and state and local governments.[8] To provide local facilities to serve settlers seeking public land, a system of district land offices was created in 1800. These were placed under jurisdiction of the General Land Office established in 1812.[9]

DISPOSAL TO INDIVIDUALS. Although never yielding a substantial proportion of Federal income, the sale of land for cash remained a major disposal policy from 1785 until the general repeal of the cash sale laws in 1891. The system was varied from time to time. In 1785 a system was established for cash sales of public domain lands,

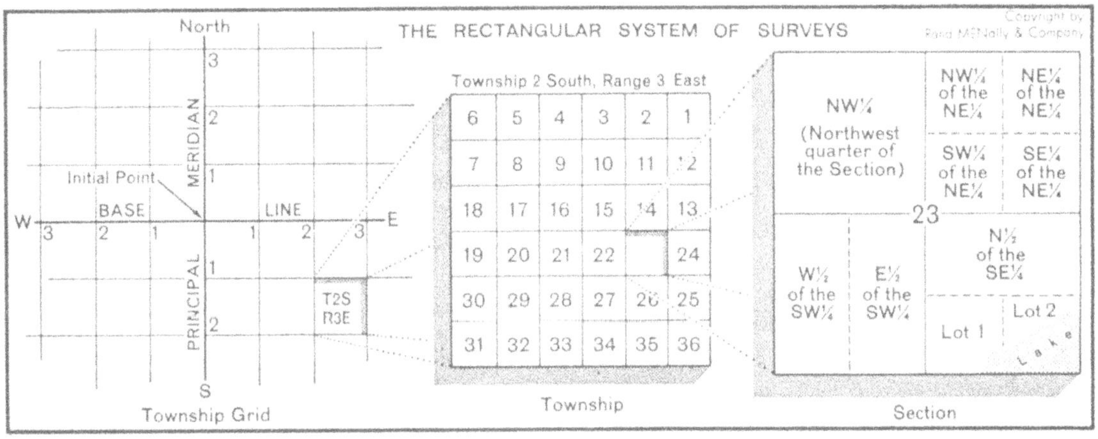

2 – 4 *The rectangular system of land surveys.* Study of these diagrams will give an understanding of this survey system which provided a great improvement over those previously developed.

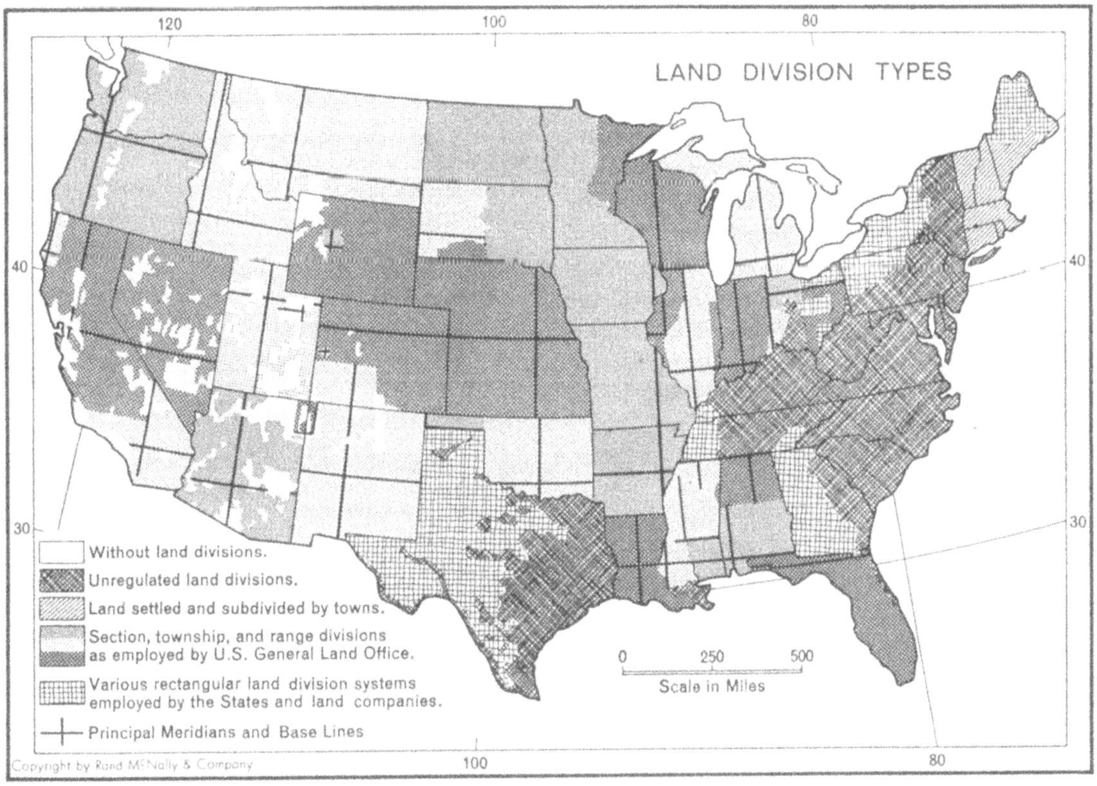

2 – 5 *Land division types.* (Agricultural Research Service)

partly in units of thirty-six square miles to land companies for resale and partly in units of one square mile. The first patent (deed) executed by the government was issued March 4, 1788.[10] The cash sales policy was modified by allowing installment payments in 1800; however, difficulties in administration resulted in its abolition in 1820. The idea of selling large blocks of land to land companies who, in turn, would sell tracts to settlers persisted for thirty-five years. Then public pressure resulted in a series of laws allowing smaller sales; in 1820 the minimum was reduced to eighty acres and in 1832 to forty acres. Most of the sales were at the minimum price of $1.25 per acre

The policies for acquisitions were further modified in 1841 with the passage of the General Pre-emption Law which permitted settlers to enter upon public lands and secure patent after complying with the required length of residence and cultivation and the payment of the minimum statutory price. Payment for land they had developed, however, did not prove to be popular with the settlers. Subsequently, they, with the aid of their champions in politics, were able to win a major battle when President Lincoln signed the Five-Year Homestead Act. This law of 1862 permitted the settler to enter upon 160 acres of public land and gain patent with only a nominal service charge after living on the land for five years, building a habitable home, and bringing part of the land under cultivation. There were several later modifications. When settlement reached the semi-arid West in 1909, the Enlarged Homestead Act was enacted allowing 320 acres to the settler. In 1912 the residency period was shortened to three years, and in 1916 the Stockraising Homestead Act permitted the settler 640 acres of range land.

The Desert Land Act of 1877 also is worthy of mention in this section. Under this act desert land capable of being irrigated could be bought at the rate of $1.25 per acre by a settler who could demonstrate that he could irrigate and cultivate it. Notable, also, were a great variety of acts which gave land as a bounty for services rendered in military engagements.

There were many other laws. Those mentioned above, however, were the most significant in fostering individual acquisition. Under the authority of the various laws permitting sales, a total of some 333 million acres of public land passed to individuals. The homestead laws accounted for transfer of 287 million acres to private ownership, and military bounties and private land claims, an additional ninety-five million acres (see Table 2-2).

TABLE 2-2

DISPOSITION OF PUBLIC LANDS, 1781-1964*

Type of disposition	Acres
Disposition by methods not elsewhere classified[1]	301,800,000
Granted or sold to homesteaders[2]	287,300,000
Granted to states for:	
Support of common schools	77,600,000
Reclamation of swamp land	64,900,000
Construction of railroads	37,100,000
Support of miscellaneous institutions[3]	21,000,000
Purposes not elsewhere classified[4]	15,200,000
Construction of canals	6,100,000
Construction of wagon roads	3,400,000
Total granted to states	225,300,000
Granted to railroad corporations	94,300,000
Granted to veterans as military bounties	61,000,000
Confirmed as private land claims[5]	34,000,000
Sold under timber and stone law[6]	13,900,000
Granted or sold under timber culture law[7]	10,900,000
Sold under desert land law[8]	10,000,000
TOTAL	1,038,600,000

* *Public Land Statistics, op. cit.*, p. 6.

[1] Chiefly public, private, and pre-emption sales, but includes mineral entries, scrip locations, sales of townsites and townlots.

[2] The homestead laws generally provide for the granting of lands to homesteaders who settle upon and improve vacant agricultural public lands. Payment for the land is sometimes permitted, or required, under certain conditions.

[3] Universities, hospitals, asylums, etc.

[4] Construction of various public improvements (individual items not specified in the granting acts), reclamation of desert lands, construction of water reservoirs.

[5] The government has confirmed title to lands claimed under valid grants made by foreign governments before the acquisition of the public domain by the United States.

[6] The timber and stone laws provided for the sale of lands valuable for timber or stone and unfit for cultivation.

[7] The timber culture laws provided for the granting of public lands to settlers on condition that they plant and cultivate trees on the lands granted. Payment for the lands was permitted under certain conditions.

[8] The desert land laws provided for the sale of arid agricultural public lands to settlers who would irrigate them and bring them under cultivation.

DISPOSALS BY LAND GRANTS. As a further aid and incentive for the development of the country, a variety of land disposal policies were passed by Congress which granted land free of charge to the states and to transportation companies. Accordingly, as each public domain state was admitted to the union it was granted land which could be used or disposed of for public purposes. These grants became increasingly liberal and included land to subsidize public works, public institutions, common schools, and higher education. Congress enacted a large body of laws to grant land for special programs. For the most part grants were made to the states, and the land was then given to private companies for the projects, or the proceeds from sales of the land were used for public purposes. Noteworthy among these grants were: those for the construction of military wagon roads (first, 1823; last, 1869), the construction of canals (first, 1827; last, 1866), the improvement of rivers (first, 1828; last, 1862), the construction of railroads (1850–1871), the reclamation of swamp lands (1849–1850 *et al.*), and the reclamation of arid lands (1894 and 1910). A General Purpose Grant in 1841 gave each of the public domain states five hundred thousand acres for use in the financing of various "internal improvements." In 1862 all states were given land to provide funds for the establishment of agricultural and mechanical colleges. In total, the various land grants to states amounted to approximately 225 million acres.

Grants also were made directly to railroad companies beginning in 1862. Before this time such grants had been made by the states, but as the transcontinental rail lines pushed into the West, grants were made directly to the private corporations organized to undertake the construction. Incentive grants to the railroads totaled 94.3 million acres (Figure 2 – 6).

MAN—RESOURCE RELATIONSHIPS

Wherever man claimed the land and brought it under use during the period of the Young Republic, the potentials of the resource base tended to decline. Certainly the idea of the long-range good of the nation entered the minds of few. These conditions, in a large measure, resulted from the general tempo and stage of development.

Agencies granting and selling land titles to settlers had no definite information concerning the productivity or capability of the land; nor were the settlers able to determine these qualities. The first settlers in new areas of the Midwest, attracted by water, transportation possibilities, and wood for construction and fuel, commonly chose sites along streams or on adjacent breaks and uplands. Later arrivals got the fertile prairies of the plains through the process of limited choice. In a like manner some of the best lands of the South were relatively late in being claimed. There was no store of experimentation basic to the development of techniques for adequate management. Each individual was on his own, with only practical experience and trial-and-error as his guide.

After 1850 rapid advances were made in the application of science and technology to manufacturing and transportation. Industrial production expanded and became diversified with machines powered by internal combustion engines. Urban growth was stimulated. These developments brought chain reactions in resource demands for fuels and a wider and wider range of raw materials. The general progress of the economy in relation to resources was such that the nation's ability to convert resources to use advanced far more rapidly than its ability or concern to maintain or improve the productivity of resources.

FAVORABLE TRENDS. It was not an entirely bleak period for conservation. A few developments

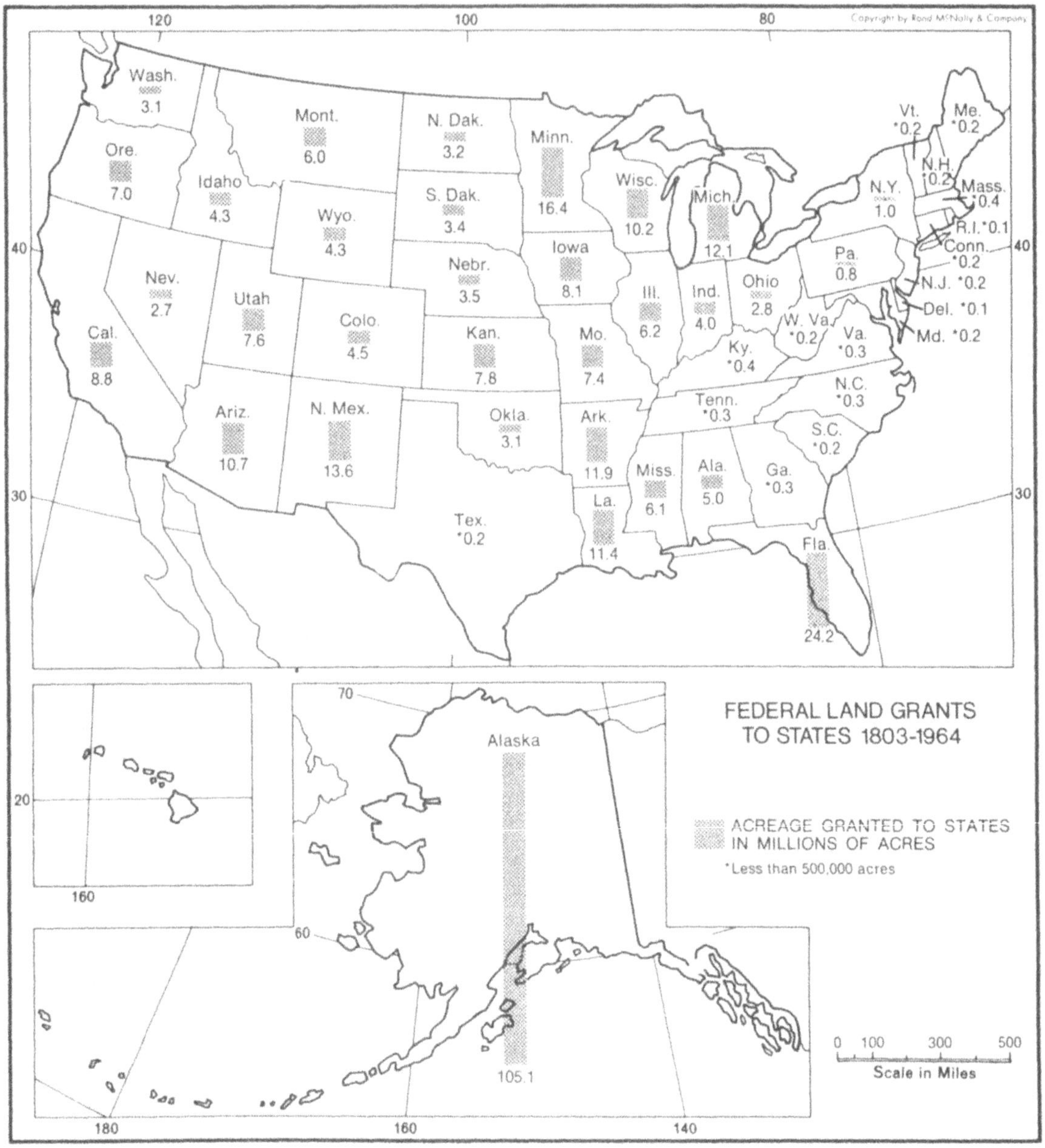

2 – 6 ***Federal land grants to states, 1803–1964.*** **(Bureau of Land Management)**

were to be the seeds for change. The Corps of Engineers, which was formed during the Revolutionary War, assumed civil functions to maintain a cadre of trained men, and also because West Point was the only institution educating engineers. During the 19th century their work was mainly in exploration, survey and mapping, the improvement of navigable streams, and the construction of canals and roads.[11] Later this agency gained much broader authority in water resource development. The Department of the Interior was created in 1849, and a number of agencies existing in other departments were transferred to it, including the especially significant Office of Indian Affairs and the General Land Office. During this period, its activities dealt primarily with administration and disposal of the public domain, but in the present century it has become the largest Federal department with multiple resource management and development concerns. The Department of Agriculture was created in 1862, but did not become an executive branch of the government until 1889. Until then its work was involved primarily with the purchase and distribution of seeds and plants. After 1884 its duties and responsibility increased rapidly. By the end of the period of the Young Republic, land grant colleges were beginning to be established, providing a higher education institution in each state which had direct concern for resources. The Congress defined the purpose of the land grant college in these words: "The leading objective shall be, without excluding other scientific and classical studies, and including military tactics, to teach such branches of learning as are related to agriculture and the mechanical arts, in order to promote the liberal and practical education of the industrial classes in the several pursuits and professions in life." Their record of achievements in the twentieth century attests to the soundness of their purpose and the value of their activity.

By the 1860's a number of individuals were beginning to consider seriously the problems of resource use and misuse. Especially noteworthy was George P. Marsh, a statesman, scholar, and farmer, who had grown up in New England, worked in the law profession, and served as a member of Congress and as a minister to Turkey and Italy. His observations in New England and the Mediterranean countries resulted in a lively interest in man's modifications of the earth's surface. Toward the end of his career in 1864, his ideas were set forth in a book, *Man and Nature or Physical Geography as Modified by Human Action.*[12] In the preface Marsh stated his objective as being "to indicate the character and, approximately, the extent of the changes produced by human action in the physical conditions of the globe we inhabit; to

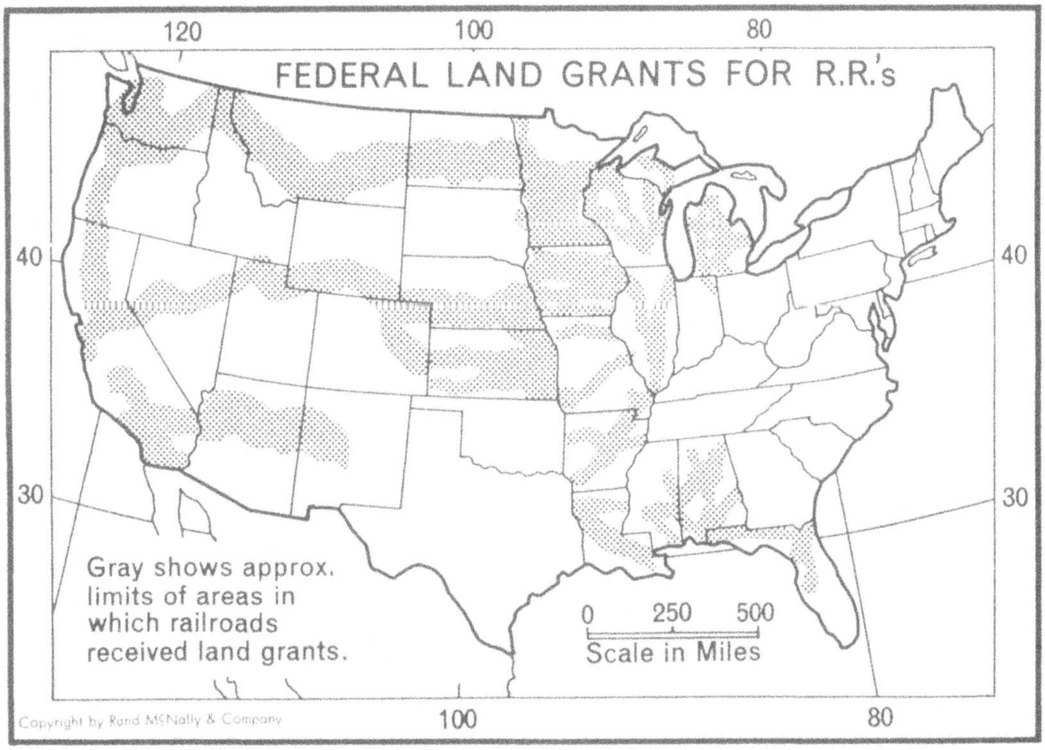

2-7 *Federal land grants for railroads.* **This generalized map shows the approximate areas in which the railroads received grants. Only a portion of the land covered by the pattern actually was granted. (Bureau of Land Management)**

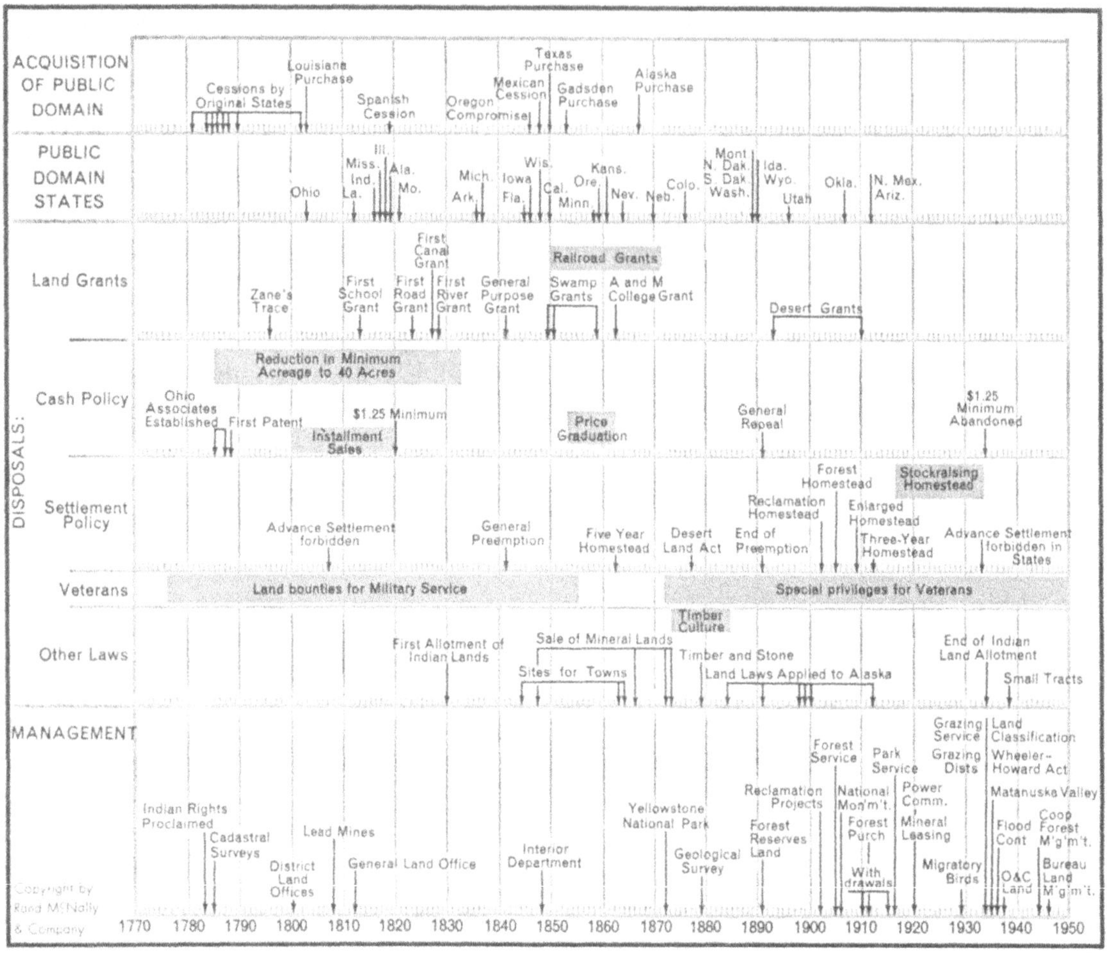

2–8 Highlights in the history of the public domain. (Land, the Yearbook of Agriculture, 1958, p. 68)

point out the dangers of imprudence and the necessity of caution in all operations which, on a large scale, interfere with the spontaneous arrangements of the organic or the inorganic world; to suggest the possibilities and importance of the restoration of disturbed harmonies and the material improvement of waste and exhausted regions. . . ." Although this book did not include all categories of resources, it was the first major synthesis on the subject and did present a mature appraisal of the major problems of the times with suggested solutions and remedies. The book was not read widely by the American public; but it did, no doubt, serve as an influence in shaping the opinions of the conservation leaders who were to multiply in number as the century drew to a close.

The Period of Awakening, 1870 – 1933

In the course of the fifty years following the Civil War, the nation gradually acquired the leadership requisite for a general public awakening to the needs for responsible resource stewardship. Through the influence of a few thoughtful and dedicated public officials and a number of scientists, several important pieces of legislation were passed; government leadership improved; resource inventory, classification, and management research were begun; and technical education programs were established in institutions of higher education.

DEVELOPMENT IN FORESTRY

Interest and public attention were drawn most strongly toward the forest resources. This is understandable because settlement had been progressing through predominantly forest land, and the results of indiscriminate cutting were widely apparent. In some sections wood and water supplies had been seriously impaired by then. The nation is indebted to the untiring efforts of a few men for drawing national attention to these problems and paving the way for improvements.

Benjamin F. Hough, a physician, became interested in forest problems as a result of work in census-taking, first in New York State and later as Superintendent of the United States Census in 1870. In 1873 he presented a paper before the American Association for the Advancement of Science on "The Duty of Governments in the Preservation of Forests." His ideas gained the support of that organization, and Hough was appointed a member of a committee which prepared a memorial to Congress calling attention to the forest problems in the United States, noting progress in Europe through government interest, and recommending some similar action by the Federal government.[13] This memorial was a major factor in the creation of a Federal commission in 1876, with Hough as the first forest agent, to study how timber growing might be encouraged and the existing timber protected. Not long after, Congress provided for organized forestry work by the Department of Agriculture, and in 1881 Congress established the Division of Forestry.

In the meantime, Carl Schurz, a German immigrant familiar with scientific forestry as then practiced in Europe, became Secretary of the Interior—the first truly forestry-minded man to hold this office. During his term, 1877-1881, the reports of the Department of the Interior were filled with forceful pronouncements and plans. Due, in a large measure, to his foreign background, most of his efforts in the interest of forest conservation were opposed. He was able, however, to establish a forest protection system with a force of special agents guarding against timber thievery and depredations on public lands.

Another German, Bernard E. Fernow, who was trained as a forester, immigrated to the United States in 1876. Upon arrival he secured employment with a mining company which had timber holdings in Pennsylvania. He became known through articles published on forestry and through his efforts in helping to organize the American Forestry Congress in 1882. Later, in 1886, he became the first formal Chief of the Division of Forestry, a position he held until 1898. During his tenure he apparently believed that forestry on the government level was still impractical in America, and that the responsibility of the Division was to preach rather than practice. He was, however, in a large measure responsible for the forest reserve system.

By an act of Congress in 1891 the President was given power to establish forest reserves from the public domain. Before the end of his term of office, President Harrison set aside forest reservations totaling thirteen million acres. No plan for management, however, was passed, and the reserves were simply closed areas. In 1897, just before President Cleveland completed his term, twenty million acres of new forest reserves were proclaimed. On July 4 of that year Congress passed the act outlining a system of organization and management for these public forests. Until 1905, however, the General Land Office of the Department of the Interior administered the forests while the Division of Forestry, giving technical advice, remained in the Department of Agriculture.

A new era in the practice of forestry began with the succession of Gifford Pinchot to the office of Chief of the Forestry Division. A dynamic individual with missionary zeal, training, and practical experience as a forester, he was the ideal leader to bring existing forces together into a program of forest conservation. Having decided to become a forester, he went to Europe upon graduation from Yale in 1889, and enrolled in the French Forest School for professional schooling (no forestry school then existed in the United States). After completing one year of training plus field observations in Germany and central Europe, he returned to the United States, the first American trained in forestry. In 1892 his practical experience began. He was hired by George W. Vanderbilt to make working plans for the management of the forests on the Biltmore estate near Asheville, North Carolina. This effort resulted in the first successful forest management work in the United States. During the following five years he served as a consulting forester and also travelled through many of the forest regions of the nation. When he took the position of Chief of the Forestry Division in 1898 he did so with an excellent background and with an understanding of the problems and a conviction that government direction was needed and possible. Under his leadership the responsibility of the Division grew rapidly. In 1901 it became the Bureau of Forestry, with authority to engage in a variety of activities, including the making of forest plans for private timber owners, tree planting, and investigation. In 1905 the forest reserves were transferred from the Department of the Interior to the Department of Agriculture, and the Bureau became the Forest Service. The reserves then numbered sixty and totaled fifty-six million acres. By the time Pinchot resigned five years later, nearly 150 million acres had become national forests, and the conservation philosophy of management and multiple use which still prevails had been established.[14] The government was now able to provide leadership in forest conservation, and continued to improve this position during the period through favorable legislation.

By the end of the nineteenth century, forestry schools had been established at Biltmore and Cornell. These were closed in a few years, but the Yale Forest School was established in 1900 and more were to follow soon. Forestry was established as a profession, and trained personnel increasingly began to work in the field. Private forest industries, however, were slow to adopt the practice of forestry because of unfavorable economic conditions and the seemingly abundant forests.

DEVELOPMENTS IN THE MANAGEMENT OF OTHER RESOURCES

During the Period of Awakening favorable developments in other resource fields gradually occurred. The activities of the Department of the Interior broadened considerably. Although the Park Service was not established until 1916, several national parks had been created by the end of the century. The first of these, Yellowstone, was set aside in 1872 "as a public park and pleasuring grounds for the benefit and enjoyment of the people."[15] Thus the desirability of establishing and protecting land for recreation was recognized. The Geological Survey was created in 1879 "for the classification of public lands and examination of the geological structure, mineral resources, and products of the national domain."[16] The work of John Welsey Powell, the first director, assures him a position among the leaders of the conservation movement. His report on the arid lands of the United States was the first realistic appraisal of the West in terms of agricultural potentials, especially through irrigation.[17] It began a long-continued agitation over the government's responsibility in irrigation development, with Powell taking a leading part. The final result was the Reclamation Act of 1902, designed to promote the conservation and utilization of one of the most precious resources of the West —water—through Federal leadership.[18]

Mining became significant early in the nineteenth century, but government concern, for the most part, was directed toward the encouragement of private production rather than conservation. Early laws, nevertheless, did exempt mineral lands from entry under usual land preemption procedures. In keeping with the idea of the revenue values of public land disposal, the Mineral Lands Ordinance of 1785 required that one-third of all gold, copper, silver, and lead mined on land obtained from the public domain be reserved for the government. The first general mining law (1866) abolished this royalty and set the precedent by which minerals are regarded as private capital belonging to the finder. In 1872 the general mining law, still in force today with minor changes, was passed. The following year the coal land law recognized the value of land containing coal deposits; accordingly, in the sale of public domain land containing coal deposits, a charge of ten dollars per acre, or twenty dollars per acre if the land were within fifteen miles of a rail line, was made. In 1906 authority was established to withdraw mineral land from entry until the value of the deposits could be established. Eighty million acres were thus withdrawn in the forty-eight states, plus nine million acres in Alaska. In 1910 the technical division of the Geological Survey was given greater status as the Bureau of Mines, providing government and industry a basic research agency.

With the establishment of the Department of Agriculture and the land grant colleges, research in all facets of agriculture began to pave the way for improvements in this aspect of the economic base. A number of states developed experimental programs after 1840, but the real impetus came with the passage of the Hatch Experiment Station Act in 1887, providing Federal aid for state agricultural experiment stations based at the land grant colleges.[19] The dissemination of information to the rural population was strengthened greatly by the passage of the Smith-Lever Act in 1914 which provided for Federal support of extension work and thereby extended the role of the land grant college from resident instruction and research to extension teaching through the county agent system.[20]

THE WHITE HOUSE CONFERENCE

Perhaps the most outstanding development of the period, in relation to the idea of focusing both public and government attention on the nation's resource problems, was the White House Conference of 1908. This conference resulted from suggestions by members of the Inland Waterways Commission to President Theodore Roosevelt. The Commission had been appointed by the President the previous year to study the public demand that rivers maintain a uniform flow for water power and navigation. In its first report, the Commission had emphasized that the problems of resources were interlocked, that water control and use had influence on soils, forests, and minerals. Later in the year, on October 3, 1907, the Commission had requested that the President call a conference—primarily a congress of governors—to consider the total problem of resource conservation. The President approved the plan, and the following year—on May 13-15, 1908—a conference was held in the White House. In attendance were the President and Vice President, members of the Cabinet and Supreme Court, members of Congress, the governors (or their representatives) of forty-one states, and many invited guests.

It was, indeed, a landmark in the conservation movement. Never before in the nation's history had there been such a gathering—political leaders, administrators, scientists, and interested citizens. The Conference was addressed by President Roosevelt and papers were delivered by scientists, governors, and eminent citizens covering most phases of the nation's resources and pointing out the problems attending wasteful use and the growing need for conservation to assure future resource reserves.[21]

The immediate results of the Conference were the appointment, by the President, of a forty-nine man National Conservation Commission, and the formation of forty-one state conservation commissions and fifty-one conservation commissions to represent national organizations. The national commission, under the chairmanship of Gifford Pinchot, prepared an inventory of the natural resources of the nation which was endorsed by the entire Commission and transmitted to the President, who in turn transmitted it with a special message to Congress. It was published as a three-volume report in 1909, the first of its kind.[22]

Theodore Roosevelt deserves the distinction of being designated the first truly conservation-minded President. His interest extended even beyond the national boundaries. In February, 1909, he convened the North American Conservation Conference, attended by representatives from the United States, Canada, Newfoundland, and Mexico, to urge the adoption and implementation of conservation principles for the entire continent. He also laid plans for bringing the powers of the world together at The Hague to consider conservation of the natural resources of the entire world, but World War I intervened.

Although Roosevelt was a champion of conservation, the Sixtieth Congress was unfriendly to most of his programs in this interest. Appropriations to support the National Conservation Committee were refused, and as a result it ceased to exist. In its short period of action, however, it had succeeded in arousing public interest in resource problems. As a result a National Conservation Association, made up of leading citizens, was formed to continue to press upon the public the need for conservation.

With the termination of Roosevelt's administration and the coming of World War I, there was some waning in the popular appeal of conservation. Nevertheless, several noteworthy implements in the interest of the movement were added before the period ended. In 1910 Charles Richard Van Hise, geologist-president of the University of Wisconsin who had a long-standing interest in conservation, published the first textbook on conservation.[23] This set the stage for introduction of conservation courses in institutions of higher education. The rudiments of conservation came to the mining industry during the late years of the period. In 1915 California led the way with the Oil and Gas Law which prohibited unreasonable waste of these products. In 1920 the Mineral Leasing Act provided for the setting aside of Federal lands having coal, sulphur, phosphate, oil, and oil shale, and for the leasing of such land to industry, with conservational practices required. The decade of the twenties saw considerable attention turned to the idea of unitized oil production, and by 1930 regulations were beginning to be imposed along this line. Congress also passed several pieces of legislation relating to forest, water, and wildlife resources which will be discussed in later chapters.

The Period of Implementation, 1933 to the Present

In the few years that have elapsed since 1933, the conservation movement has gained momentum. The need has been recognized, techniques have been developed, and an army of qualified leaders have assumed responsible positions. As a result, greater progress has been made in the establishment of conservation practices during the past three decades than during the previous three centuries. In the years immediately before World War II attention was focused upon preventing destruction and waste of natural resources—the problems of soil erosion, deforestation, and mineral waste were targets of particular concern.

1933 TO WORLD WAR II

The depression years of the 1930's were both a burden and a blessing for the nation; a burden in that severe economic strain was placed on the populace, but a blessing in that due to the economic adjustments necessary, attention was focused on internal improvements with government support of many public programs related to resource development and improvement.

Franklin D. Roosevelt was a conservation-minded President; but more important, the depression called for bold action to create work —the time was ripe for implementation. Developments along this line came rapidly after he took office. Almost immediately, the Public Works Administration was created with the Secretary of the Interior as administrator (June, 1933). It was charged by Congress with two principal objectives: "The creation of employment by large expenditures for labor (at a time when private industry was reducing the number of available jobs), and the creation of this employment by the building throughout the country of sound public works of recognized and lasting social value."[24] Thus America's first large-scale public works program was launched. Civic improvements, hospital, school, and road construction and many resource development projects by the Corps of Engineers, the Bureau of Reclamation, and other Federal agencies were included.

Inventory and planning were part of the program. The National Planning Board was created to advise and assist the Administrator of Public Works "through preparation, development, and maintenance of comprehensive plans . . . through surveys and research . . . and through analysis of projects for coordination and sequence . . . and to obtain the maximum amount of cooperation and correlation of effort among the agencies of the Federal, state, and local governments."[25] The report of this Board was submitted to the President in June, 1934.

This led to its immediate reorganization as the National Resources Board, with broader activity assigned to it. The executive order establishing the new board required that it prepare and present to the President a "program and plan of procedure dealing with the physical, social, governmental, and economic aspects of public policy for development and use of land, water, and other national resources and such related subjects as may from time to time be referred to the Board by the President."[26] Before the year had ended the Board had submitted a series of reports, the most comprehensive of which was dated December 1, 1934, and dealt with problems of land, water, minerals, the organization and timing of Public Works programs, basic data planning, and state and regional planning. In transmitting the reports of the National Resources Board to Congress, President Roosevelt expressed the desirability of having a permanent board "which would recommend yearly to the President and the Congress the priority of projects in the national plan.[27] Thus the National Resources Committee was created on June 14, 1935, to continue the work of the two previous boards which drew their authority from the National Industrial Recovery Act. This Committee continued to function as a correlating, fact-finding clearing house of vital information on natural resources under the chairmanship of the Secretary of the Interior until national attention was shifted to the wining of World War II. The values of the Committee and its predecessors to the conservation movement are inestimable. As a result of its many reports, the nation had for the first time a comprehensive inventory of the resource base and the problems relating to it. The idea of planning for the future became widely accepted. As a direct result of the Committee's work, most of the states established planning organizations, and several interstate or regional planning commissions also were created. In addition, many developmental projects were carried through as a consequence of the Committee's findings and recommendations.

Closely associated with the public works idea was the Civilian Conservation Corps which was created in 1933 and functioned until 1942. The purpose of this organization was to employ young men in endeavors of value to the nation. Initially, 1,500 camps, with a capacity of two hundred men each, were set up in various parts of the country. At their peak in 1935 there were more than 2,600 camps. They were located in state and national parks and forests and in other places where their services were useful. They developed trails, roads, picnic and camping facilities, planted trees, and did many other kinds of work directly related to the improving of resources and their use. Moreover, the program performed an important role in the education, both formal and practical, of the young men involved.

The decade of the 1930's brought Federal leadership more formally into conservation in many fields. The details of this will be discussed in later chapters, but mention of the most notable developments is needed here to complete the general picture of the movement. The Soil Erosion Service began operation in 1933, and passage of the Soil Conservation Act in 1935 established the Soil Conservation Service with the responsibility of developing and carrying out a permanent national soil and water conservation program. The Agricultural Adjustment Administration program set up in 1933 recognized national responsibility for aid to farmers on problems related to prices, surpluses, land deterioration, and land improvement. One aspect of the program, operating on the county level, provided for cost-sharing in the establishment of desirable conservation farming practices. The Taylor Grazing Act of 1934 closed a large gap in Federal land management by providing for classification of the unreserved public domain lands as to the suitability for use and for their orderly disposition, as well as by providing for the establishment of grazing districts. Several improvements were made in wildlife conservation, including the Wildlife Restoration Act of 1937 which gave Federal aid to states for acquiring and developing land for this purpose. Other noteworthy developments included the creation of the Tennessee Valley Authority, a unique experiment in the integrated development of the resources of a river basin; the establishment of a committee to study the drought area of the Great Plains and recommend solutions; and the creation of the first regional power administration (The Bonneville Power Administration) for the purpose of marketing power generated at Federal dams.

WORLD WAR II

With the beginning of World War II, the nation mustered all forces—money, manpower, and materials—and dedicated them to the singular goal of winning. Until that great conflict was over and our allies were well on their way to economic recovery, conservation had a comparatively small place in resource use programs. Great quantities of minerals, timber, and agricultural raw materials were spent in the effort. The end, of course, justified the actions and demands. Nevertheless, in any realistic appraisal of conservation gains, it must be noted that these demands for materials hastened the decline of many mineral deposits, caused marginal lands to be cropped, upset established rotation and other desirable cropping systems, retarded the adjustment of drain to growth in many forest areas, and delayed the establishment of other necessary programs in the interest of conservation.

THE PRESENT STATUS

The return of peace brought a resumption of economic growth that had been curtailed by the war. It also brought a reestablishment of government leadership in conservation with new dimensions added—cleanliness and beauty. Special presidential committees have studied water, raw material policies, and recreation.[28] There has been a general broadening of the interests of several executive departments to include extended efforts in water and atmospheric sciences, rural redevelopment, and landscape beauty; and new departments now have concern for urban and transportation problems.

FEDERAL AGENCIES AND THEIR RESOURCE FUNCTIONS. Today the role of the Federal government in conservation is more prominent than ever before. Congress has always been responsible for the budget and enabling laws, but it can probably be stated that there has been distinct improvement in this leadership relative to resources. The resource administration and action programs, however, are largely in the hands of several executive departments.[29]

The Department of the Interior is the largest of the Federal departments with resource concerns, and its responsibilities are the most diverse. Its role has changed through its more than one century of existence from that of general housekeeper for the Federal government to that of major custodian of the nation's natural resources. Its management and development programs are now pursued with a conservation outlook. Its variety of bureaus and agencies indicate its multifunctions. Included are the Bureaus of Indian Affairs, Land Management, Mines, and Reclamation; the Fish and Wildlife Service; Geological Survey; National Park Service; Office of Territories; and the Bonneville, Southeastern, and Southwestern Power Administrations.

The Department of Agriculture bears the Federal responsibility for the welfare of the nation's agriculture and the coordinated Federal agriculture program. Included under its administration are the Forest Service and Soil Conservation Service with direct conservation functions, and the Agricultural Conservation Program Service, Agricultural Research Service, Economic Research Service, Rural Community Development Service, Office of Experiment Stations, and Federal Extension Service which cooperate with the states in work which has conservation ramifications. It also has a number of other offices and services dealing with marketing, agricultural stabilization, credit, loans, and rural electrification which are of less concern to conservation, but of direct concern to farmer welfare.

The Corps of Engineers of the Army Department has civil functions in navigation and flood control developments and the construction of multiple-purpose projects from which power and other incidental benefits are derived.

The Bureau of Census, the Environmental Sciences Service Administration, and the Area Redevelopment Administration, all of the Department of Commerce, also should be noted. The first collects, analyzes, and publishes data related to all facets of the population and the economy which are basic to planning. The Environmental Sciences Service Administration was formed in 1965 by joining the long established Weather Bureau and Coast and Geodetic Survey, and several newly created science oriented agencies; it promises to contribute significantly to both basic and applied knowledge of the biological-physical subsystems of the nation. The Area Redevelopment Administration provides technical assistance to economically distressed areas to contribute to their rehabilitation.

The Federal Power Commission, an independent agency, is charged with the administration of the Federal Power Act and the Natural Gas Act. Relative to resource development, it is primarily a regulatory, fact-finding, and planning organization which has broad responsibilities in connection with the development of the hydroelectrc potentials.

In addition to those mentioned, there are also a number of other agencies, services, and authorities which have some resource-related functions. Examples may be noted in the land under the control of the defense services, the Atomic Energy Commission, and the Tennessee Valley Authority.

It will be recognized from the foregoing discussion that the Federal government now has concern for, and is providing leadership in research and management of essentially all facets of the resource base. An evolution in the position of Federal government resource responsibility may thus be noted—from that of disposer to that of steward. Although this is true, the Federal system is not without problems. Notable among them are vulnerability of programs to politics, empire-building tendencies, conflicts in interest (often between agencies of the same department), and lack of money to put programs into action.

STATE AGENCIES AND THEIR RESOURCE FUNCTIONS. The interest of the states in taking a leading role in conservation lagged considerably behind that of the Federal government. For many years such activity as did exist centered around the work of the land grant colleges.

Before 1900 several states had established forestry departments, but major interest in the total resource base came after the stimulus of the White House Conference of 1908 when forty-one states formed conservation commissions. Since that time there has been steady growth in state interest in conservation. Today every state has the power to enact zoning regulations, and several agencies which deal with resources. There is relatively little uniformity among these agencies from state to state. Some have departments of natural resources or conservation which handle several state resource functions; others have separate departments for agriculture, forestry, fish and game, water, minerals, etc. In the latter case, some have coordinating committees composed of members representing each department. Most have planning or development boards, departments, or committees.

The state functions in resource matters also vary from one state to another, depending largely on the variety and nature of their resource endowments and the ownership pattern. Fish and game law enforcement and the management of state-owned land are the province of state departments. In the forest regions, state departments enforce state forest conservation laws, and a few states carry out or sponsor management and products research. All of the western states are active in water resource management.

COOPERATION IN MANAGEMENT AMONG PUBLIC AGENCIES. Resources are complexly interrelated —mismanagement of one can adversely influence others. This, plus the fact that resources are not limited by state, national forest, or any other artificial boundary, presents a major problem because management is highly compartmentalized—scattered among a number of agencies in Federal, state, and urban departments. There are, naturally, overlaps in authority, lack of desirable synchronization due to lack of uniformity in budgets and goals and in some cases cross purposes. Yet successful programs must have concern for the regions as a whole and their total resources. They must also have concern for the increasing interaction and relations between regions. The Federal Interstate Highway System, other modern means of transportation, interregional power grids, interregional water transfers, and the like are binding the nation with even closer ties. More and more local developments have regional, interregional, and national implications. This interrelation requires the cooperation of the many agencies involved—and often the adjustments of many conflicts of interest—to achieve the delicate, harmonious balance necessary for the best possible long-range development programs.

Some gains in cooperation have been made. States increasingly are joining efforts in interstate programs; but as yet these programs are mainly concerned with individual resources. Progress on the Federal level is noted in the establishment, immediately following World War II, of regional departmental field committees to aid in the synchronization of programs. A marked degree of cooperation also has been achieved since 1946 between Federal and state agencies in several of the major river basins as the result of the formation of interagency committees. Made up of representatives of the various resource agencies and the state governors, these committees work on a voluntary cooperative basis in an effort to coordinate resource programs in the interest of balancing development, maximizing benefits, and minimizing costs. Federal-state cooperation and Federal-urban cooperation have also resulted from programs in which the Federal government pays a major share of the costs, such as the Interstate Highway System and urban renewal.

ADVANCES ON THE PRIVATE LEVEL. Since 1930 noteworthy gains have been made in conservation management of privately controlled resources. The changing complexion of farming from a way of life to a business has focused the farmer's attention on maintaining and improving his capital resource—the land. During this period farmers have been increasingly receptive to the assistance offered by the Soil Conservation Service and the Federal Cooperative Extension Service. Private forestry has improved materially. The large companies recognized the urgency as the major operations shifted to the last remaining virgin stands on the West Coast. Many now are adjusted to sustained-yield programs. The situation on smaller holdings is not so bright, but improvements have been made in the past two decades with increased profitability of tree farming, active industry promotion, and the establishment of extension farm forestry agents. Similar gains will be noted in other resource categories in the following chapters; yet there remains need for much improvement.

A principal problem requiring attention relates to individual attitude. The profit motive still is the major stimulus for individual practices in conservation management. Even the large and powerful special interest groups which have become active in the past few decades, are largely promoting programs of self-interest. Many of the urban dwellers of the nation are so far removed from the resources which support them that they have little interest in or concept of the requirements of conservation.

Therefore, although major advances have been made, complacency cannot be tolerated. The growing population and improving standard of living will require that increasing attention be given to maintenance of adequate resources. The nation has a long way to go before the desirable state of total environment conservation is reached. A major prerequisite will be the improvement of the attitude of the citizenry—a realization that man is of the earth, not simply on it.

FOOTNOTES

[1] For a well-chronicled history of conservation in the United States see Stewart L. Udall, *The Quiet Crisis* (New York: Holt, Rinehart, and Winston, 1963), 224 pp.

[2] Highlights in the History of Forest Conservation," *Agriculture Information Bulletin No. 83* (Washington: Forest Service, Department of Agriculture, 1952), pp. 1–2.

[3] Angus McDonald, "Early American Soil Conservationists," *Miscellaneous Publication No. 449* (Washington: Department of Agriculture, 1941), pp. 3–7.

[4] "Highlights in History of Forest and Related Natural Resource Conservation," *Conservation Bulletin No. 41* (Washington: Department of the Interior 1959), p. 1.

[5] "Government Land Acquisition," Oregon Edition (Washington: American Forest Products Industries, 1965), p. 6.

[6] Irving Sensel, "Brief Notes on the Public Domain" (Washington: Bureau of Land Management, Department of the Interior, 1950), p. 3.

[7] For an excellent discussion of the systems used in the eastern states see Francis J. Marschner, *Boundaries and Records, Eastern Territory of Early Settlement with Historical Notes on the Cadaster* (Washington: Agricultural Research Service, Department of Agriculture, 1960), 73 pp.

[8] For detail see T. Donaldson, *The Public Domain*, 3rd ed. (Washington: Government Printing Office, 1884), 1,343 pp. A general interpretive view is presented in Guy Stevens Collander, *The Economic History of the United States, 1765–1860* (Boston: Ginn and Company, 1909), pp. 666–92.

[9] The General Land Office, consistent with the revenue earning idea, was in the Department of the Treasury until 1849 when it was moved to the Department of the Interior. In 1812 four district offices were created to serve the Ohio Territory. As settlement moved west, the number was increased to a high of 123 in 1890.

[10] Sensel, *op cit.*, p. 6.

[11] For a good review of their activities until 1812 see Forest G. Hill, *Roads, Rails, and Waterways, the Army Engineers and Early Transportation* (Norman: Univ. of Oklahoma Press, 1957), 248 pp.

[12] George P. Parish, *Man and Nature or Physical Geography as Modified by Human Action* (New York: Charles Scribner's Sons, 1864). Later edition, *The Earth as Modified by Human Action* (1877). The John Harvard Library edition is particularly valuable. David Lowenthal, ed. (Cambridge, Mass. The Belknap Press of Harvard Univ. Press, 1965). It includes an introduction by the editor and additional footnotes summarizing the later views of the author. See also David Lowenthal, *George Perkins Marsh—Versatile Vermonter* (New York: Columbia Univ. Press, 1958), 422 pp.

[13] F. W. Putnam, Permanent Secretary, ed., "Report of the Committee on the Preservation of Forests," *Proceedings of the Association for the Advancement of Science*, vol. 24 (1874), pp. 38–40.

[14] Gifford Pinchot, *Breaking New Ground* (New York: Harcourt, Brace & Company, Inc. 1947), 522 pp. The book not only details his career as a forester, but also gives an excellent picture of the problems and general conditions attending forestry development from 1889 until 1910.

[15] "A Century of Conservation," *Conservation Bulletin No. 39* (Washington: Department of the Interior, 1950), pp. 10–11.

[16] *Ibid.*, p. 9.

[17] J. W. Powell, *Report on the Lands of the Arid Region of the United States* (Washington: Government Printing Office, 1879), 195 pp. This work is a classic, worthy of the attention of anyone interested in the details of the conservation movement.

[18] For a historical sketch of reclamation development see "Reclamation—Accomplishments and Contributions," *Committee Print No. 1* (Prepared by the Library of Congress Legislative Reference Service for Committee on Insular Affairs, House of Representatives, 86th Cong., 1st sess., 1959), 132 pp.

[19] See Alfred Charles True, "A History of Agricultural Experimentation and Research in the United States, 1607–1925 (Including A History of the United States Department of Agriculture)," *Miscellaneous Publication No. 251* (Washington: Department of Agriculture, 1937), 321 pp.

[20] See Lincoln David Kelsey and Cannon Chiles Kelsey, *Cooperative Extension Work* (Ithaca: Comstock Publishing Associates, 1955), 424 pp. This book provides an excellent analysis of the origin, development, role, and procedures of the Extension Service.

[21] *Proceedings of a Conference of Governors in the White House, Washington, D.C., May 13–15, 1908* (Washington: Government Printing Office, 1909).

[22] *Report of the National Conservation Commission*, vols. I, II, and III, 60th Cong., 2nd sess., Senate Document No. 676, vols. 10, 11, and 12 (Washington: Government Printing Office, February, 1909).

[23] Charles Richard Van Hise, *The Conservation of Natural Resources in the United States* (New York: The Macmillan Company, 1910, 1921), 482 pp.

[24] "Back of the Buffalo Seal" (Washington: Department of the Interior, 1936), p. 97.

[25] *Ibid.*, p. 108.

[26] *Ibid.*

[27] *Ibid.*, p. 109.

[28] For example see *Resources for Freedom, A Report to the President by the President's Materials Policy Commission*, 5 vols. (Washington: Government Printing Office, 1952), and *Outdoor Recreation for America*, a report to the President and the Congress by the Outdoor Recreation Review Committee (Washington: Government Printing Office, 1962). See also Task Force Reports prepared for the Commission on Organization of the Executive Branch of the Government (Hoover Commission), especially Appendix L, "Natural Resources," and Appendix M, "Agricultural Activities" (Washington: Government Printing Office, January, 1949).

[29] See for detail, *The United States Government Organization Manual*, Federal Register Division, National Archives and Records Service, General Service Administration (Washington: Government Printing Office, published annually).

3

The Character of Land Resources

"LAND IS MANY THINGS TO MANY PEOPLE."[1] To the environmental scientist, however, land is defined as the surface expression of the sum of the characteristics presented by the physical-biotic system at a given site. Because of the countless combinations possible in the associations of subsystems (climate, vegetation, soil, surface relief, water), details of site characteristics vary markedly from place to place. No evaluation classifications, taking all into account, have yet been devised. More commonly, classifications are based upon the overriding natural characteristic of interest (hence, soil, climate, topography, vegetation, and similar classifications do exist), or upon general ecological relations (like humid forested slopes, prairie plains, coastal flatwoods, sandy deserts, and so forth).

The conservationist's interest in land extends beyond appraisal of the natural characteristics of site to an interpretation of its functional utility and adaptability. Thus he is concerned with *land resources* for crop, animal, and timber production; for recreation; and for all kinds of building, residential, and service space requirements. He is, furthermore, interested in management practices necessary to maintain or improve site capability, and in the supply, the spatial distribution of the supply, and the areal, regional, and national spatial relations of the various qualities of land resources to evolving demand and established use systems.

Obviously the natural adaptability of land to the many resource uses varies significantly from site to site. To be sure, Americans have progressed far in the technology of land use, and have overcome many former natural barriers. Yet it remains true that good natural qualities in land remain a distinct asset, lowering the cost of use and providing opportunity for selectivity in uses. A given parcel of land may present either limited or varied use possibilities. It is, therefore, axiomatic that a region or nation that possesses abundant land high in the qualities of adaptability and capability is, indeed, fortunate. But variety is advantageous: oranges will not grow in the north; wheat does poorly in the optimum cotton climate; and mountain slopes attractive to ski resorts are restricting to most other resource uses. Because any region or nation has finite space, even though technology continues to improve, land conservation must be concerned with the total resource values of each parcel and the rational establishment of sustainable or improvable uses for all.

The abundant and varied land resources comprising the national domain of the United States are its most valuable natural possessions. Little of the land area is completely without utility. Most has multiple possibilities for use, and a large portion is capable of supporting several uses at the same time; this derives from an unusually favorable combination of natural attributes.

The Natural Character of the Nation's Land

There are six major groups of natural variables that influence both land use and land resource conservation. These are (1) size, shape, and accessibility, (2) surface form, (3) climate, (4) soil, (5) water, and (6) biota. In reality, these influences of nature at the site level are complexly interrelated, and adversities in any one can have a measure of veto power over the favorable qualities of the others. Thus existing combinations are the important considerations in resource values. Nevertheless, in analysis it is first essential that individual factors be understood. For this reason a brief examination of each group of variables as it applies to the United States follows.

SIZE, SHAPE, AND ACCESSIBILITY

In gross area the fifty United States encompass 2,313,772,160 acres, including 42,428,800 acres of water surface. This large area provides slightly under twelve acres per capita. Of course, not all is equally capable of sustaining human use, but size alone does increase the opportunity for variety in quantity.

Study of Figure 3 – 1 will indicate that the forty-eight conterminous states occupy a solid block, roughly rectangular in shape, extending across the North American continent essentially astride the thirty-eighth parallel. Long ocean borders on the west, east, and southeast favor easy access to the remainder of the world and provide opportunities to tap the resources

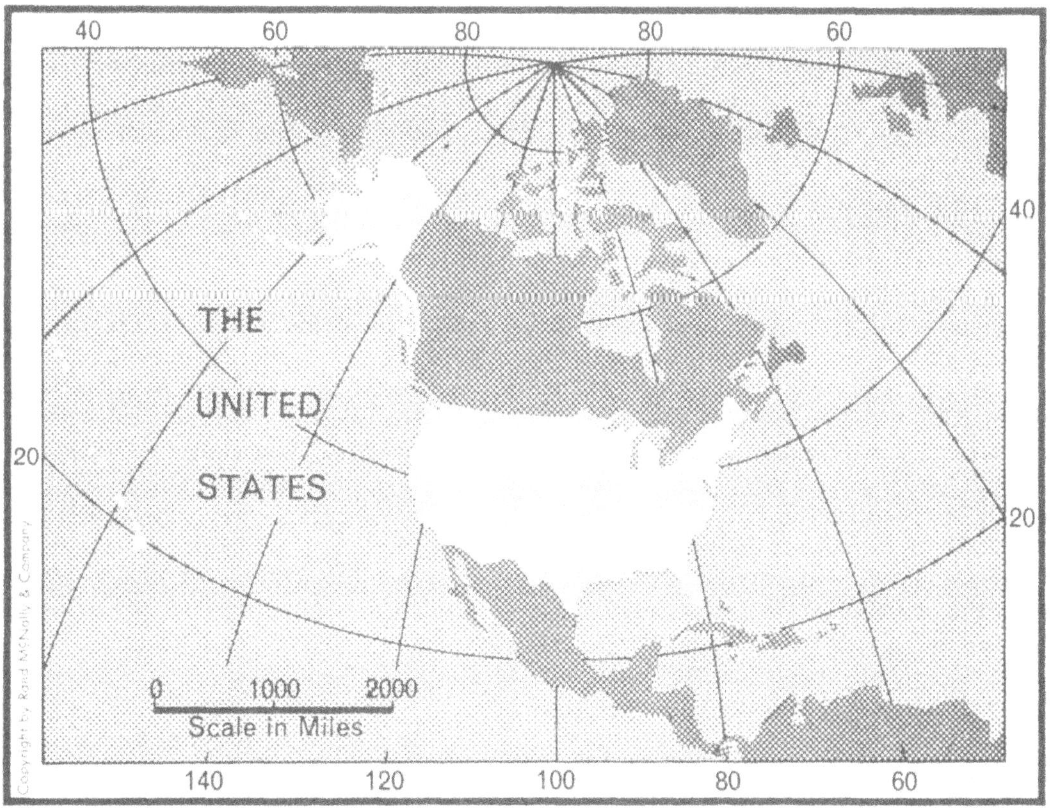

3 – 1 *The United States in relation to its neighbors.*

of the sea. The compact shape of the forty-eight conterminous states, unencumbered by insurmountable physical barriers, enhances the construction of transportation arteries and economic, political, and social organization and development. The new states, Alaska and Hawaii, are in a less favorable position in these regards; but, nevertheless, both are accessible by ocean routes, and Alaska is also accessible through a friendly country via the Alaska highway. The addition of these states to the union definitely increased the environmental variety and the resource potential of the nation.

Just as size, shape, and accessibility are important influences in the broad consideration, so are they significant in evaluation of individual parcels of land. Size influences scope of possibilities. Shape influences patterns of use layout and range of utility. Accessibility, particularly location and connectivity with respect to transportation and markets, is a major factor in economic evaluation. A small plot of physically high-quality land isolated in a mountainous area is of lower economic use-potential than a similar plot on the margin of a major urban center; a plot along an Interstate Highway midway between interchanges has lower commercial value than a plot adjacent to an interchange.

SURFACE FORM

The surface forms of land influence its utility in many ways. In the broad order forms of mountains, hills, plateaus, and plains, and their regional associations and arrangements with respect to oceans and prevailing wind and storm patterns, they influence the nature of climate and water drainage. Degree and aspect of slope locally affect temperature, run-off, and soil stability and accumulation. The detailed character of the surface forms has a marked bearing upon the kind and quality of land uses and the possibilities for sustaining them. Smooth surfaces are most favorable for machine agriculture, most types of settlements, and transportation facilities, and also enhance most other land uses. Areas with modest slopes, however, may be more favorable for certain tree crops locally providing frost protection through air drainage, and such land may be satisfactory for cropping, pastures, timber, or urban settlements. Variety in use possibilities decreases as slope increases. Nevertheless, steeper slopes, other physical factors being favorable, may be useful for timber production, grazing, watersheds, recreation, or wildlife habitat. Even areas of sierran relief may have potentials for the last three of these uses. Obviously, however, the possession of large amounts of those land types that offer the widest range for uses is most desirable. In this regard the United States is in an extremely favorable position.

The generalized pattern of the nation's terrain types is shown in Figure 3 – 2. It will be recognized that the eastern 60 per cent of the country is dominated by plains, with hills and low mountains breaking the continuity in the Superior Highlands, the Interior Highlands, the Appalachians, and the New England Highlands. The surfaces of the plains, of course, are by no means equally smooth or equally adaptable; some areas present relatively steep irregular slopes that are restrictive to crop agriculture. In the case of the hills of the Appalachians, although some of the area is restricted by adverse slopes to forest, watershed, or recreation uses, an important amount of the land is satisfactory for agriculture, particularly those phases centering around livestock. The western portion of the nation has less favorable surface forms. However, it will be noted that there is a considerable area of plains in the intermontane region and the longitudinal valleys of the Pacific Coast states. Moreover, the high mountains, especially the Rockies and the Sierra Nevada—Cascades, have served major roles in the economic development of the West, providing watersheds for irrigation development, timber and grazing land, and varied scenery and other attractions for recreation. In terms of surface form, Alaska is the least favorably endowed large area of the nation.

CLIMATE

The climatic character of land is one of the most significant influences on its utility. Temperatures, frost-free season, precipitation, sunshine, winds, and storminess in amount, regimen, and variability play individual roles. The composite of all, operating as a *climate*, is most important in the assessment of regional resource values. Every living thing thrives best under certain interrelated conditions of temperature, moisture, and light—there are extremes under which a given plant or animal will not grow, and within the range of these there is a climatic optimum for each. Climate as a major factor in soil formation has marked bearing upon the fertility, structure, and texture of soil. Thus it is a major influence on biotic potential, both in natural growth and in those plants and animals grown by man. Climate may be an enhancing,

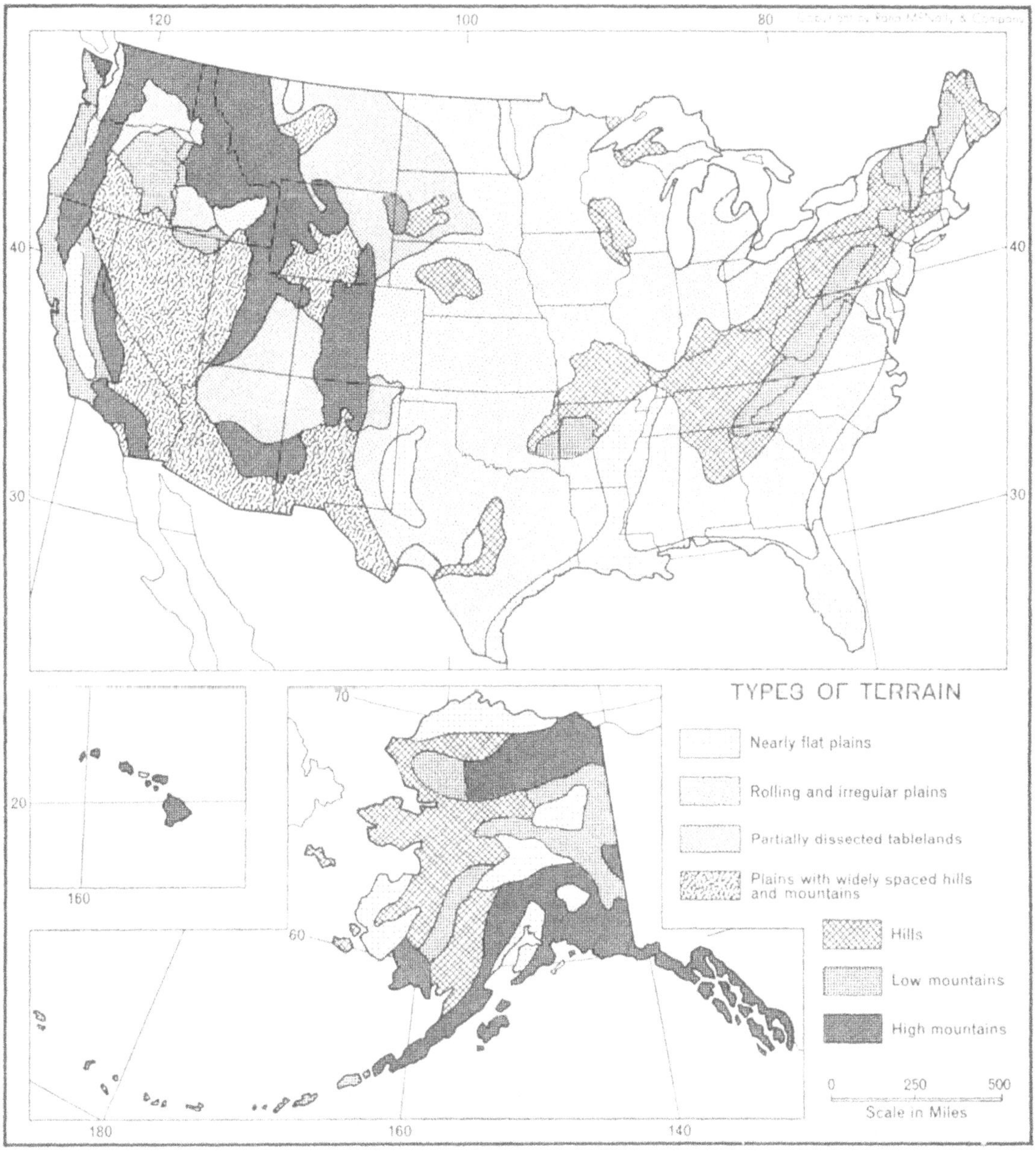

3 – 2 ***Types of terrain.*** **This map shows the nation's principal terrain types and their areal distribution. (After Edwin H. Hamond)**

detracting, or modifying factor, affecting land value, attractiveness, or input-output relationships for most other categories of land use—dwelling, factory, transportation, recreation, etc.

In the long process of populating the earth, man has found the subtropical and middle latitude climates most satisfactory for his needs. Certainly scientific man has given most attention to the use of native plants and the development of crops, animals, and other necessary resource-converting techniques and facilities for living in these climates. It is clear, therefore, that for the production of natural and cultural biota the possession of large areas within these fruitful climates is essential. Variety in climate, however, is desirable because of differences in biotic potentials and because some land uses are

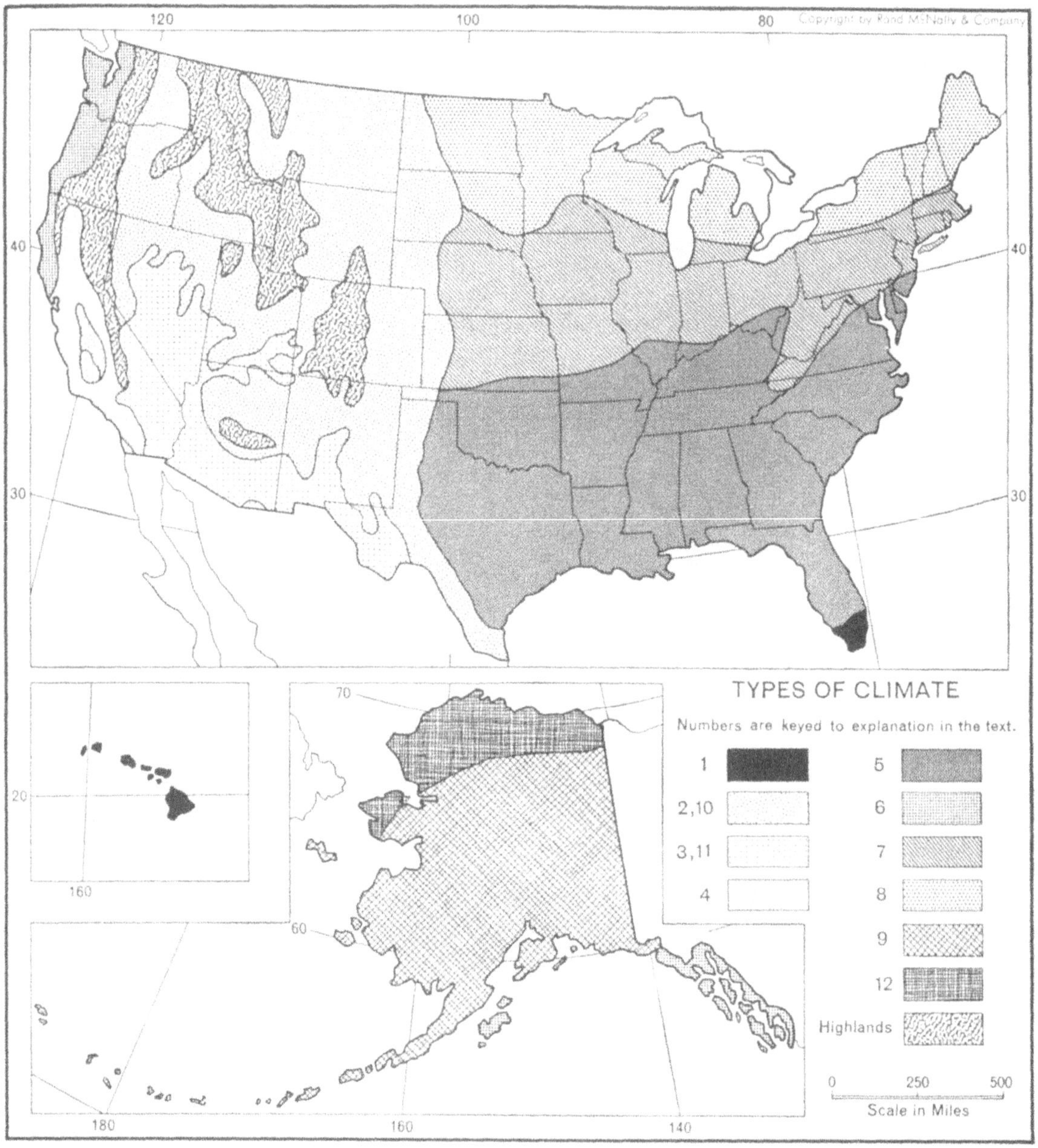

3 – 3 *Types of climate.* (After Glenn T. Trewartha)

REPRESENTATIVE AVERAGE DATA FOR CLIMATES OF THE UNITED STATES

	Warmest month	Coolest month	Frost-free days	Annual precipitation	Precipitation distribution
1. Tropical wet and dry	78-80° F	65-70° F	240-365	30-60 inches	high sun maximum
2. Tropical steppe	80-90° F	50-65° F	180-240+	10-20 inches	high sun maximum
3. Tropical desert	85-95° F	50-65° F	occasional winter frost	10 inches	erratic
4. Subtropical (dry summer)	60-80° F	45-55° F	variable — over 240	15-25 inches	winter maximum
5. Subtropical (humid)	75-80° F	40-55° F	150-240	30-60 inches	even
6. Marine	60-70° F	35-50° F	150-210	40-70 inches	wetter winter
7. Humid continental (long summer)	70-75° F	20-30° F	100-150	25-45 inches	wetter summer
8. Humid continental (short summer)	65-70° F	5-20° F	under 100	20-40 inches	wetter summer
9. Subarctic	40-50° F	—5- —20° F	occasional summer frost	10-20 inches	wetter summer
10. Mid-latitude (steppe)	65-75° F	15-30° F	quite variable	10-20 inches	wet spring — early summer
11. Mid-latitude (desert)	65-80° F	20-35° F	under 90 days	below 10 inches	erratic
12. Tundra	25-40° F	—15- —30° F	under 60 days	5-15 inches	———

based upon other than the biotic characteristics. For example, a dry, warm, sunny climate may be attractive to industries such as aircraft manufacture or urban dwelling and recreation.

The climatic base of the United States provides large areas of productive climate for biota as well as diversity and variety for other land uses. The distribution of the major climatic types is shown on Figure 3 – 3, and the accompanying legend summarizes the principal characteristics of each type of climate.[2] It will be noted that the lands of the nation include climates ranging from tropical to tundra and from wet to arid. Especially important, however, is the wide distribution of climates with high plant growth potentials in the eastern half of the country. This in combination with the other climates provides areas of large dimensions favorable to most types of land-use needs.

SOILS

Soil as a factor of land utility relates primarily to its influence on plant growth possibilities. Depth, texture, structure, plant foods, pH, water-holding capacity, and drainage are the variable characteristics which set the limitations. The higher the quality in all of these, the wider is the range of adaptability. It should be understood, however, that plants differ in their soil needs. Douglas fir thrives on soils of low organic content; tree crops, in general, need deep, well-drained soils; corn is fairly tolerant, but yields best on soils with good drainage and moisture-holding ability containing abundant plant nutrients, especially nitrates, potassium, and phosphorus; vegetable root crops, in addition to other qualities, require surface soils of medium texture and loose structure and deep mellow subsoils; many grass crops, however, do well on heavy soils. These few illustrations indicate the desirability of a nation's having soils of different characteristics.

Again it is possible to state that this nation is fortunate in the possession of both quantities and varieties. The patterns of major soils are shown on Figure 3 – 4 and brief descriptions of each are given in the accompanying legend. Attention is directed especially to the large areas of productive chernozems, prairie soils, rendzinas, alluvial soils, and gray-brown soils.

WATER

Water is a fundamental requirement for life of all kinds and for many of the activities and

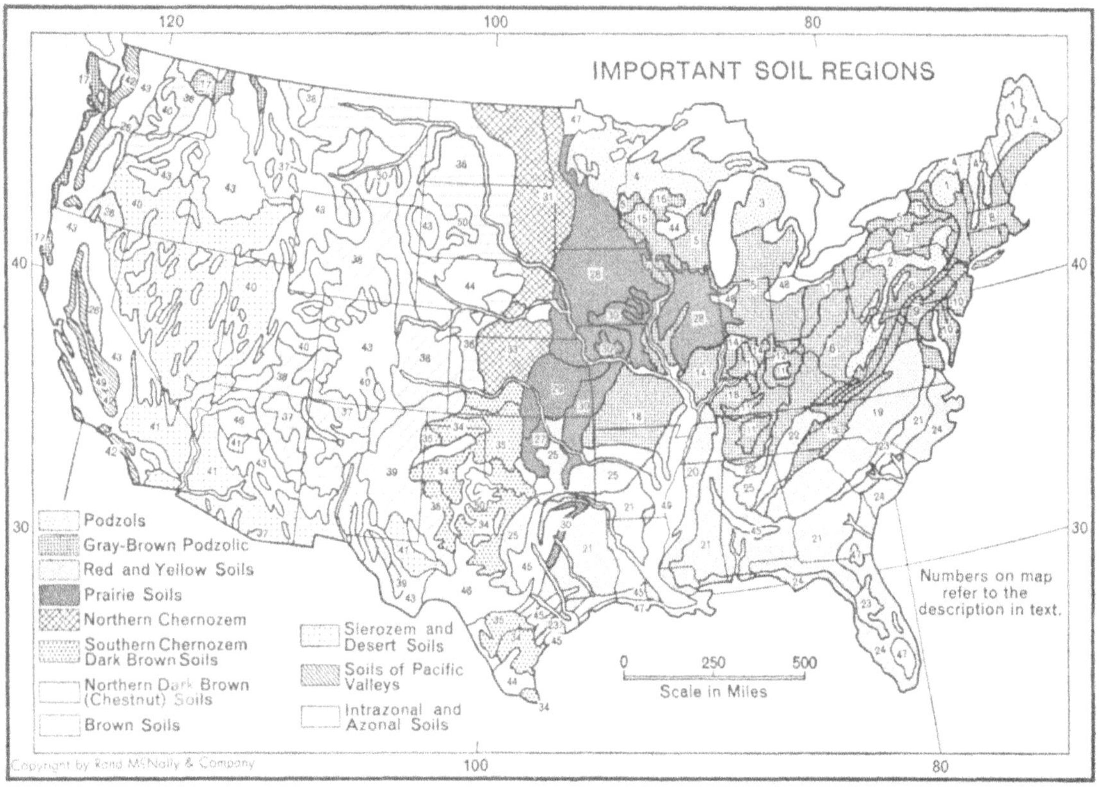

3 – 4 *Important soil regions.* (Department of Agriculture)

Legend for Map (*Agriculture Handbook 153,* Department of Agriculture).

Podzols

The profile consists of a thin organic layer above a gray leached layer which rests on a dark-brown or coffee-brown horizon. The podzol is developed usually under a coniferous forest in a cool, moist climate. Its inherent productivity for crop plants is low. The figures below refer to area numbers on the map.

1. Rough stony land, including areas of shallow podzols.
2. Chiefly loams and silt loams developed from sandstones and shales of the plateau and mountain uplands.
3. Dominantly sands and loamy sands developed on glacial drift.
4. Dominantly loams and clay loams developed on glacial drift.

Gray-brown podzolic soils

The profile has a rather thin organic layer over grayish-brown leached soil which overlies a brown horizon. The soils are generally acid, at least in the surface. These soils develop in a moist and cool-temperate climate under a deciduous forest and are inherently more productive than the podzols.

5. Dominantly loams and silt loams developed on calcareous glacial drift.
6. Brownish-yellow silty loams or stony loams with hilly relief developed on sandstones and shales.
7. Loams and silt loams developed on acid glacial drift and composed of sandstones and shale material. Some of these soils are imperfectly drained.
8. Dominantly stone and gravel loams developed on glacial drift.
9. Loams and silt loams developed mainly on the crystalline rocks of the northern Piedmont.
10. Largely sandy loams developed on the sands and clays of the northern Coastal Plain.
11. Chiefly brown silt loams developed on limestone.
12. Shallow soils developed on interbedded limestone and calcareous shales.
13. Loams and stony loams from granitic material with hilly to mountainous relief.
14. Silt loams with heavy clay subsoils, developed on Illinoian glacial till.
15. Silt loams developed largely from loess.
16. Imperfectly drained grayish silt loams with silty clay loam subsoils developed from acid glacial drift.
17. Largely loams and silt loams with yellowish subsoils developed from sandstones and shales.
18. Grayish-yellow to reddish silt loams and cherty silt loams developed from cherty limestones.

Red and yellow soils

This group of soils consists of two general types of profiles that are intimately associated. Both have thin organic layers. The profile of the red soil is a yellowish-brown leached layer over a red horizon, while the profile of the yellow soil is a grayish-yellow leached layer over a yellow horizon. Both developed under the forest in a moist warm-temperate climate. Generally, the yellow profile is more pronounced under the coniferous forest and the red under the deciduous forest.

The inherent fertility of the yellow soils is relatively low and that of the red soils is medium.

19. Dominantly brownish-red clay loams and gray sandy loams developed largely from crystalline rocks of the southern Piedmont.
20. Yellow to light brown silt loams developed on loess.
21. Dominantly gray to yellow sandy and fine sandy loams with some sands and fine sands developed from Coastal Plain materials.
22. Largely brownish-red to red silt loams and clay loams developed from limestone.
23. Grayish-yellow to light brown sands and fine sands of the Coastal Plain.
24. Grayish fine sandy loams, with some gray or black loams, developed in the flatwoods area of the Coastal Plain. Includes areas underlain by coralline limestone.
25. Grayish-yellow to reddish fine, sandy loams and silt loams, developed from sandstones and shales. A considerable portion is hilly and stony.
26. Red soils of the north Pacific slopes.

Prairie soils

The profile of the Prairie soil grades from a very dark brown or dark grayish-brown surface through brown to lighter-colored parent material at a depth of two to five feet. It is developed in a moist temperate climate under a tall-grass prairie. Inherent fertility for crop plants is high.

27. Reddish-brown soils of variable texture developed on sandstones, shales, clays, and sands.
28. Dark brown silt loams with yellowish-brown subsoils developed on glacial drift and loess.
29. Dark brown to reddish-brown silt loams and clay loams developed from limestone and calcareous shales.
30. Dark brown or grayish-brown silt loams having heavy subsoils or claypans.

Northern chernozem

The profile has a black or dark grayish-brown surface soil grading below into light-colored material that is calcareous at two to six feet. It is developed in a temperate to cool sub-humid climate under tall and mixed grasses. Inherent productivity is high.

31. Black, silt, and clay loams developed on calcareous glacial drift and associated lacustrine deposits.
32. Dark grayish-brown loams and silt loams developed from loess.
33. Dark grayish-brown silt loams with claypans developed from loess.

Southern chernozem — dark brown soils

The profiles have dark brown to reddish-brown surface soils underlain by brown or red horizons, grading below into light-colored material that is calcareous at three to six feet. These soils develop in a warm, subhumid to semiarid climate under a mixed tall- and short-grass prairie.

34. Heavy or moderately heavy dark brown soils developed from calcareous materials.
35. Predominantly red and brown sandy loams and sands developed largely from unconsolidated calcareous sands, silts, and sandy clays.

Northern dark brown (chestnut) soils

The profile grades from a dark brown surface soil into a whitish calcareous horizon at a depth of one and a half to three feet. These soils develop under mixed tall and short grasses in a temperate to cool semiarid climate.

36. Dark brown soils developed on unconsolidated, calcareous sands, silts, and clays.
37. Dark brown soils developed on heterogeneous material associated with mountainous and plateau terrain.

Brown soils

A brown surface soil grading at a depth ranging from one to two feet into a whitish, calcareous horizon. The profile is developed in a temperate to cool, semiarid climate under short grasses, bunch grasses, and shrubs.

38. Northern: chiefly brown loams developed largely on unconsolidated sands, silts, and clays.
39. Southern: chiefly light brown to gray fine sandy loams to silty clay loams of smooth relief developed largely on limestone or unconsolidated sands, silts, and clays.

Sierozem and desert soils

Grayish and reddish soils closely underlain by calcareous material. These soils develop in an arid climate under short grass and desert plants.

40. Northern: gray and grayish-brown soils of variable texture developed largely on loess and alluvial fan material.
41. Southern: gray, brown, and reddish soils of variable texture, developed largely on alluvial fans.

Soils of the Pacific valleys

42. Includes a number of variable zonal, azonal, and intrazonal soils that are too intimately associated to separate on a schematic map. These soils developed under a range of climatic and geological conditions.

Intrazonal and azonal soils

These soils may possess one of the two general types of profile: (a) The profile may express a local condition such as drainage or parent material rather than the zonal profile of the region; (b) The profile may be too immature to express a zonal type.

43. Rough and mountainous (azonal).
44. Largely azonal sands, some of which are associated with bogs.
45. Black (or brown) friable soil underlain by whitish material excessively high in calcium carbonate. These soils developed under a prairie vegetation and are known as Rendzinas (intrazonal).
46. Shallow stony soils from limestone (azonal).
47. Marsh, swamp, and bog (intrazonal).
48. Soils largely intrazonal developed on lake plains.
49. Alluvial soils (azonal).
50. Rough broken land.

---

processes developed by man. Thus water availability, in terms of form (surface, soil, or ground), quantity, and quality, is an influential factor to be measured in appraising the utility of land. Precipitation is the keystone in the water supply cycle. But temperature, soil, and bedrock character, slope, and vegetation-cover all influence availability. Nevertheless, it is almost a truism that humid lands will be well supplied with water and dry lands will be poorly supplied. Modifications may exist, however; streams may carry water through dry lands, or it may be available at reasonable depths in the water table. Moreover, the supplies of humid areas may be deteriorated by imprudent human actions, or overdrafted through concentrated settlement and use.

The character of the nation's water resources is appraised in detail in Chapter 11. Therefore, it is sufficient to make only a few generalizations at this point. The lands east of the one hundredth meridian are, for the most part, well supplied with water. Many of the problems existing today in this large area result from improper use or overuse of what is available. In much of the West, water shortage is a distinct limiting factor in land utility. Exceptions include the higher mountains and the Pacific slope of northern California, Oregon, and Washington. Moreover, some parts of the dry West are favored with streams or ground water.

Too much water on the land can be as limiting as too little. Taking the nation as a whole, a relatively small portion is restricted by poor drainage, the principal areas being on the outer margins of the southeastern coastal plain, the Mississippi floodplain, and portions of the Upper Lake states.

NATURAL BIOTA

The natural biota of an area is the expression of total environment, past as well as present. Thus it is intimately connected with the utility of the land, an indication of its capacity to produce. In this discussion attention is focused upon those forms which presently influence the

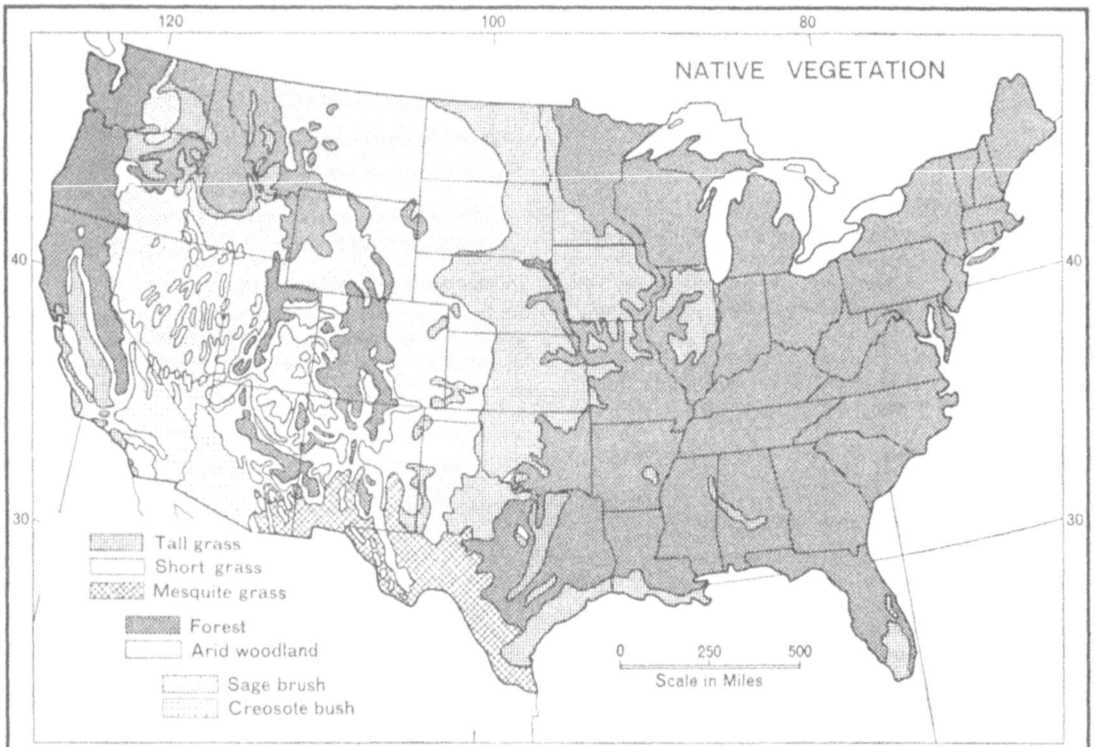

3 – 5 ***Native vegetation.*** **The vegetative cover represents an adaptation to the physical environment and consequently mirrors prevailing natural conditions. This map shows the generalized patterns of major types of vegetation before the time of settlement by Europeans. During the past three centuries major modifications have occurred as much land has been developed for agriculture and other uses. (Department of Agriculture)**

value of noncropland for timber production, grazing, watershed, wildlife, recreation, and scientific interests. Very little of the nation's land is entirely without natural biota of some utility. In most areas the forms of value are sustainable and in many cases improvable.

The varied environments present a wide range of *ecosystems* (areal associations of interrelated natural processes) which manifest interrelated biotic communities. Particularly important from the point of view of major resource values, are the large areas in forest and woodland, tall grass and short grass. The pattern of original distribution of these is shown on Figure 3 – 5.[3]

The vegetative cover of the fifty states at the beginning of colonization on the eastern seaboard is estimated to have been as follows:

TABLE 3–1

VEGETATIVE COVER, BEGINNING COLONIZATION*

Type	Million acres	Per cent of the land
Forest and woodland	1,065	47
Grassland	726	32
Desert shrub	266	12
Tundra	214	9
TOTAL	2,271	100

* "A Graphic Summary of Land Utilization," *U.S. Census of Agriculture, 1959*. Final Reports, Vol. V, Part 6, Chapter 1, Special Reports (Washington: July, 1962), p. 6.

The extent and quality of all types of vegetation have been altered by clearing for agriculture and other forms of settlement, timber cutting, overgrazing, and reseeding. Nevertheless, considerable areas with high quality native biota still remain, and there are excellent potentials for improvement in many other portions of the nation.

Resource Adaptability of the Land

Upon correlating and synthesizing the spatial relations of all the variables which influence the natural qualities of the land and the possibilities of land utility, it is recognized that they exist in systems that provide the United States with a wealth of land resources. This will be made clearer by studying the accompanying map (Figure 3 – 6) in conjunction with the following comments, both of which are derived with only minor alterations from Department of Agriculture data.[4]

The above thesis is based upon the utility of land for crop production, pasture and range, and forest products. The categories on the main map thus are based upon the utility evaluations of existing combinations of moisture, surface form, soils, and drainage characteristics. The inset showing isopleth lines of frost-free days adds a further element of refinement and should be studied in conjunction with the main map. The humid areas, covering a little more than one-half of the forty-eight states are subdivided on the basis of surface form, soils, and drainage, and the subhumid and arid areas are subdivided on the basis of these criteria plus moisture supply.

The first category includes those humid areas that possess good moisture supply; deep, moisture retentive soils; and level to gently sloping surface (Figure 3 – 7). Most of this land lies south and west of the Great Lakes and is favored with a warm to hot frost-free season of 140 to 180 days. In this area a wide range of use possibilities is allowed by nature: crop agriculture, forest production, grazing, and so on. The major portion is suited to machine farming and is especially adapted to feed and livestock production on an intensive basis. Generous and varied mineral resources and transportation plus a large population have been attractive to industrial and urban development within the area.

The second category includes the lands lying along the Gulf and southern Atlantic coastal plain which are nearly level and, for the most part, are characterized by poor drainage. Under natural conditions, the better-drained lands are covered with southern pines and the poorly drained lands with cypress and broadleaf forests; some areas, however, in the western margin and in south-central Florida are covered with tall prairie grasses. Uses based on natural productions are limited to forest products,

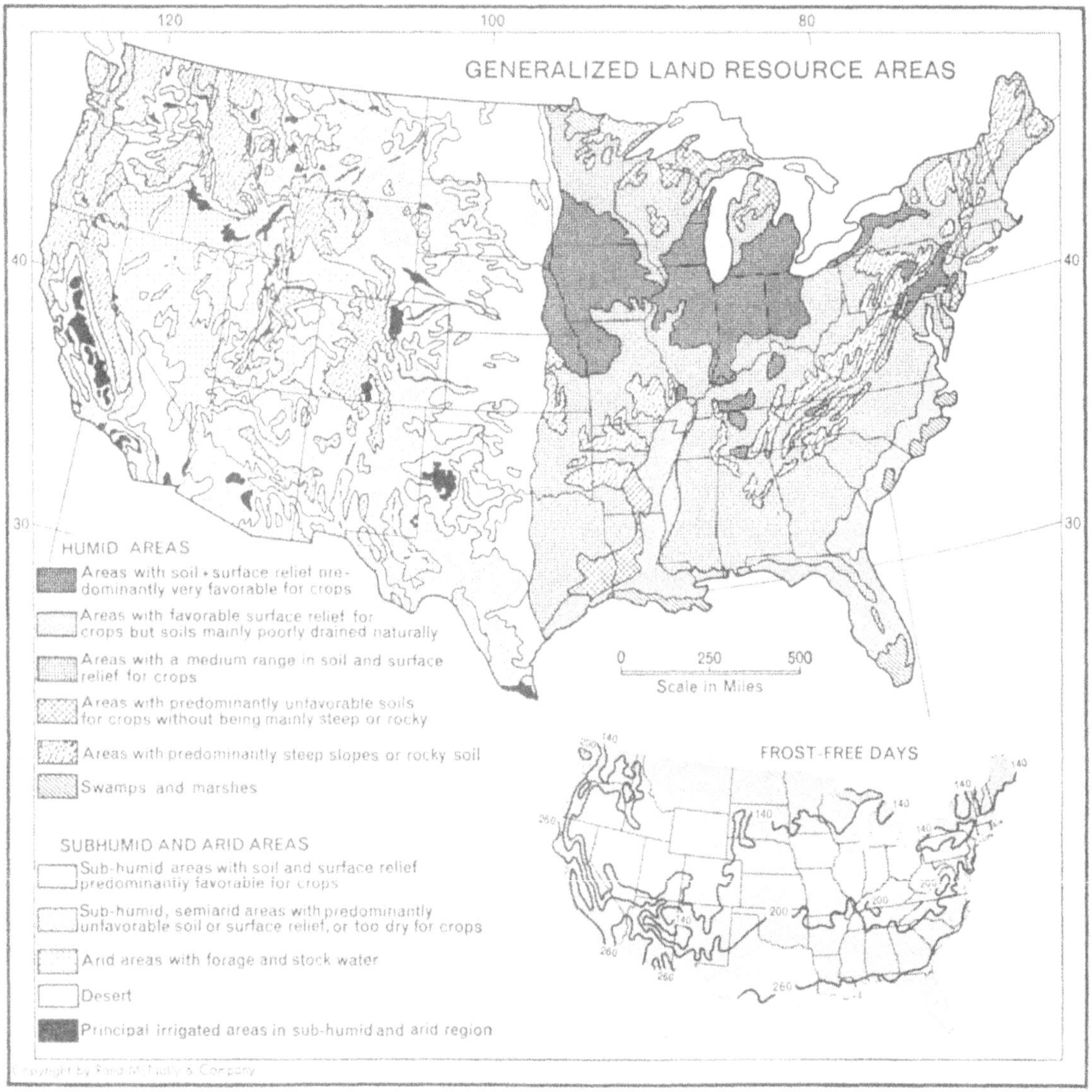

3 – 6 ***Generalized land resource areas.*** **(Land, the Yearbook of Agriculture, 1958, p. 10)**

grazing, wildlife habitat, and recreation. Much of this land has potential for upgrading to cropland or pasture use through development of artificial drainage facilities. Considerable land has been thus developed on the Mississippi flood-plain and delta and in Texas and Florida.

The lands in category three, the largest of the humid categories, are, in general, characterized by less favorable conditions of slope and soils than those in category one. Steep slopes and/or poor soil inhibit crop use in a number of places. Such areas, however, comprise some of the best forest land in the nation. There are many segments, both large and small, having smooth surfaces and relatively productive soils which are favorable to agriculture. In fact, such lands in the South provide a greater range of agricultural adaptability than those of category one, owing to longer frost-free seasons and the fact that most of the land is capable of sustaining tree production.

In category four the lands have relatively favorable surfaces but are restricted by unfavorable soils for agricultural use. They are

3 – 7 *Category one land.* This air view of Webster County, Iowa, shows a complete section of land bounded by roads on all sides. The surface, smooth to gently rolling, together with favorable soils and climate makes the land suitable for intensive machine agriculture. Corn is the chief crop, but some hay and other forage crops are grown. The photograph, made in September, shows the tone and texture pattern of crops just before harvest time. (Department of Agriculture)

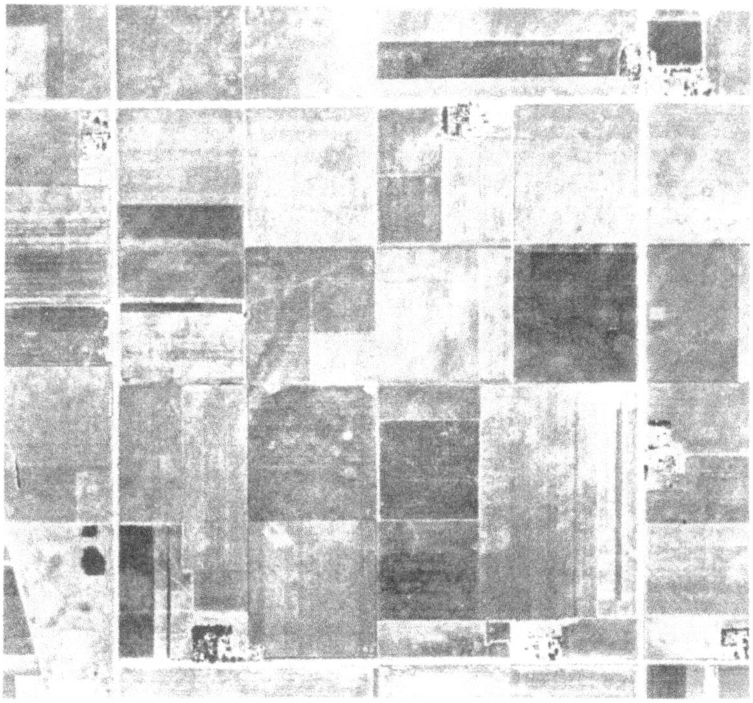

best suited for forest production and grazing. Only the major areas are indicated on the map; many others too small to show are scattered through the other categories.

Category five encompasses the humid lands with predominantly steep slopes and thin or rocky soils, principally the mountain areas of the East and the higher mountains of the West (Figure 3 – 8). These are useful for forest production, grazing in the warm season, and as watersheds. They also have important value as recreation lands and wildlife habitats.

The last of the humid categories, six, includes lands that may be characterized as swamps and marshes—considerably wetter than those of category two. Only a few are sufficiently large to show on the map, mainly the outer fringe of the Mississippi delta, the Everglades, and several areas along the seaward margin of the south Atlantic coastal plain. Under natural conditions the major use potentials are restricted to wildlife habitat, recreation, and scientific interests. It may be physically possible to drain these lands and upgrade use, but costs of reclamation and the availability of suitable land elsewhere inhibit development at present.

3 – 8 *Category five land.* Steep slopes, thin soils, rocky surface, and short frost-free season restrict the productive capacity of this area in the northern Cascades of Washington. Spectacular scenery, snowfields and glaciers, lakes, and subalpine and alpine vegetation, however, give the area important value for outdoor recreation, watershed, and wildlife habitat. (David Simons)

Lands included in categories seven through ten become increasingly restrictive for uses based upon biotic potentials. In category seven the lands are subhumid with smooth to rolling surfaces and favorable soils. Use for crop production is limited to the more moist parts or to those areas where irrigation is possible. When farming is based upon precipitation, moisture-conserving techniques are imperative, and crops are limited to those grains and forage plants which have limited water needs or which are highly efficient moisture utilizers. Much of the land is best used for grazing, and this is encouraged by high-quality native grasses and the availability of livestock water.

Category eight is composed of lands of considerable diversity ranging from subhumid to semiarid, smooth to rough surface, and good to very poor soils. These exist together in combinations which inhibit crops except on lands where irrigation is possible. Precipitation is usually sufficient to furnish a moderate amount of natural forage and water for livestock; however, precipitation variability with recurrent droughts requires caution in stocking and management of the range.

Category nine includes lands which are similar in general characteristics to, but drier than, those of category eight. Most are grazing lands with poor quality forage and low carrying capacity for either wildlife or livestock. These lands are commonly used for seasonal grazing in conjunction with categories five and eight and irrigation projects.

In category ten the lands comprise the true deserts of the nation. It will be noted that these occupy a relatively small area in the Southwest. The grazing potential is minor, but several irrigation projects have been developed which are important out of proportion to their sizes. This derives from the artificial application of water to adaptable soils under the long frostless season and high temperatures which together result in unique crop potentials.

Whereas the generalized classification just employed is a satisfactory vehicle for a broad view of the nation's land resource base, it is wholly unsatisfactory for the planning and ordering of land uses on either an area or site basis. A classification system that rates natural land values and permits local site rating on a national scale is a lack in the nation's planning tools that should be eliminated. The Soil Conservation Service's Land Capability Classification System, which is discussed later in this chapter, only partially fills the needs. Its shortcomings for the purpose suggested are illustrated by the fact that differences in climatic values are only broadly considered; hence class I land occurs in widely varying climatic environments. Obviously there are differences in resource values of land between class I in Maine and class I in Florida.

Societal Influences in Land Utility and Land Use

Whereas nature sets the broad limits of land utility under any level of technology and societal development, and these limits must be given prominent consideration in any conservational adjustment, within the framework of utility, the features inherent in the societal order largely determine how land will be used (see Chapter 1, Resource Variables). Moreover, men have found it within their technical ability to modify natural limitations. In many instances these modifications are sustainable improvements.

By drainage, wet land may be lifted to higher-use possibilities; by irrigation, dry lands may be cropped; by fertilization, yields may be improved on poor soils; by using chemicals, pests and disease may be controlled; by terracing and other practices, slope lands may be improved for agricultural or other uses; through plant and animal breeding, strains may be produced to lift environmental barriers; or a new highway may provide favorable spatial interaction, removing former restrictions of isolation. Available science and technology and favorable economic returns are the promoters in these instances. Such developments should be sound and lasting, in harmony with nature.

Too often, unfortunately, economic exigencies and a short-range view have prevailed, resulting in imprudent use and land deterioration. The blame does not rest entirely with the individual land manager and owner. A general public which has had little concern for these considerations must share the blame, and so must the government. The factor of taxation may be cited to illustrate how the government is responsible, in part, for imprudent land use. In a number of states the taxation on forest lands is unfavorable to the development of a harvest program based upon tree maturity, resulting in cutting immature timber. In most states land evaluation for tax purposes is based upon the existing highest market value, and this has worked to the detriment of farming land on the rim of expanding urban centers. In such cases farmers faced with high taxes or the possibilities of selling to urban developers at a high price often choose the latter course. Thus much quality agricultural land has been covered with streets, houses, and shopping centers while neighboring land of low agricultural utility remains open.

Recently, modern limited access highways, including the Federal Interstate System, have become a significant new force in shaping the land-use and occupance patterns of the nation. The routes, designs, and interchange locations of these rapid transit facilities are largely determined by state and federal highway engineers whose primary concerns are with the functions of the highways and economies in construction costs. Too little account is taken of the quality of land removed, and, particularly, of the impact the newly established highway will have upon the existing man-land system, especially through its impacts upon improved travel time and access, land values, taxes, field and farm parcellation, and even drainage. Too little concern, also, has been shown for the attraction of the interchange properties for commercial development. These interchanges are forming the loci for many new communities; the limited access highway and its interchanges, in this respect, compares with the river and railroad junction of the past. Unless broader participation is brought into the formulation and enforcement of land use and tax controls, these developments, with the corollary force of market economics, threaten to allow—even to encourage—commercial, industrial, and residential sprawl over significant amounts of high quality land.

LAND OWNERSHIP AND USE

Of the several societal influences, the factor of land ownership deserves special mention. Distribution by ownership classes and land uses is shown in Table 3–2 (see also 3–3). This indicates that about 72 per cent of the land estate of the forty-eight states is in private ownership. This includes more than 99 per cent of the land used for crops, 63 per cent of the land used for pasture and grazing, and nearly 70 per cent of the commercial forest land (Table 3–1).

TABLE 3–2
MAJOR CLASSES OF LAND BY USE AND OWNERSHIP, 1959*

Ownership	Million acres									
	Cropland		Grassland pasture and range		Forest land		Special use and other land		Total	
	48 states	50 states	48 states	50 states	48 states	50 states	48 states	50 states	48 states	50 states
Federal	0.8	0.8	157.1	159.1	198.5[1]	323.9[1]	50.6	281.2	407.0	765.0
State and other public[2]	1.0	2.0	40.0	40.4	33.3	34.6	43.8	44.0	119.0	121.0
Private[3]	454.3	454.8	433.0	433.5	406.7	414.3	81.8	82.7	1,375.8	1,385.3
TOTAL	457.0	457.6	630.1	633.0	638.5	772.8	176.2	407.9	1,901.8	2,271.3

* *Land and Water Resources, A Policy Guide* (Washington: Department of Agriculture, September, 1962 revision) Table 3, page 12.

[1] Includes reserved forest in parks and other special uses, and Indian forest.

[2] Excludes state grant land in process of transfer from the Federal public domain to the state of Alaska.

[3] Includes Indian cropland, pasture and range, special uses, and other land.

Under this democracy, land held in fee simple is regarded as personal property. With the exception of a few regulations affecting a small portion of the total area in private holdings—minimum forest conservation laws, rural and urban zoning laws, and land-use practices imposed by some soil and water conservation districts—selection of land uses and management practices is the privilege of the owner. Therefore, it will be understood that the existing land-use patterns in the nation result from the decisions of the many owners (see Table 3–3). Many factors in numerous different combinations influence the individual owner's choice—land capability, size of holding, comparative productive costs, outside earning capacity, capital, markets, transportation, experience, personal preference.

The authors do not wish to be misunderstood. We believe that individual ownership is desirable and that the democratic–private enterprise system is most ideal. Our plea is for more prudent judgement, based upon consideration of uses consistent with the capability of the land, total national needs, and long-range societal goals in order that the system can be enjoyed by our progeny. This may require more in the way of leadership, guidance, and perhaps actual regulation by government—the organ of the body politic. In our judgement,

TABLE 3–3

LAND UTILIZATION: UNITED STATES, 1959*

Major use	48 Conterminous states		All 50 states	
	Million acres[1]	Percentage	Million acres[1]	Percentage
Agricultural:				
Cropland	457	24	458	20
Crops and related uses[2]	(391)	(21)	(392)	(17)
Used only for pasture	(66)	(3)	(66)	(3)
Pasture (excluding cropland pasture)	[3] 630	33	633	28
Forest land[4]	614	32	746	33
Commercial	(488)	(26)	(530)	(23)
Noncommercial	(126)	(6)	(216)	(10)
Farmsteads, farm roads	10	1	10	([5])
Total agricultural land	1,711	90	1,847	81
Nonagricultural:				
Special-purpose uses	129	7	147	7
Urban and other built-up areas	[6] (53)	(3)	[6] (54)	(3)
Areas limited primarily to recreation or wildlife use	(47)	(2)	(62)	(3)
Forest land[4]	(25)	(1)	(27)	(1)
Nonforest	(22)	(1)	(35)	(2)
Public installations and facilities	(29)	(2)	(31)	(1)
Miscellaneous land	62	3	277	12
Total nonagricultural land	191	10	424	19
TOTAL LAND AREA	1,902	100	2,271	100

* *Land and Water Resources, A Policy Guide, op. cit.*, Table 1, p. 13.

[1] Acreages rounded to nearest million.

[2] Cropland harvested, crop failures, and cultivated summer fallow, soil improvement crops, and idle cropland.

[3] Open permanent pasture and range in the forty-eight contiguous states comprises 473 million acres and 157 million acres federal grassland range used for grazing.

[4] Includes forested grazing land or range, including federal forest range used by permit. The combined acreage of forest land including areas limited primarily to recreation or wildlife use (the twenty-five and twenty-seven million acres shown under special purpose uses, and embracing reserved forest land in parks, wildlife refuges, wilderness, and related areas) totals 639 and 773 million acres in the forty-eight and fifty states respectively.

[5] Less than 1 per cent.

[6] Rounding to the nearest million accounts for part of the difference between totals for the forty-eight states and fifty states.

however, the adjustments can be achieved for the most part within the framework of existing governmental authority. More desirable than regulation by law, in the interests of maintaining democracy, is that the permanent adjustment to land resources stems from the interest and acceptance of responsibility by society in total and land owners and managers in particular. In this regard, some aspects of the record of progress of the past three decades portend continuing improvement in the future.

During the 190 years in which the nation has evolved, there has developed a division of responsibility between private enterprise and government with potentially adequate checks and balances. Private enterprise controls most of the land of highest capability, and with this control goes the right of decision-making respecting use. Government can influence land-use decision through such devices as taxation practices, incentives, market controls, zoning, and by education. Government responsibility includes public ownership and direct control of use of lands totaling one-quarter of the area of forty-eight states and most of Alaska.[5] Public ownership of these lands is desirable in most cases. Most lands fit into one of the following categories: (1) lands of such low productive capacity, utility, or value that they are not attractive as private investments; (2) lands which are devoted to public purposes; (3) lands used by defense services; (4) lands used for recreation, often involving sites of unusual scenic, ecological, or historical significance, providing protection for the public from the public; and (5) national forests which include a relatively small portion of the highest producing or most accessible forest land, but which in addition to providing a stabilizing element in the forest products industry serve important roles as watersheds, grazing areas, recreation areas, and wildlife habitats.

The Conservation Classification of Land

The nation, from the points of view of available basic data and management technology, is now in a position in which the biotic uses of land can be adjusted to the land's maintainable and improvable capability. Much work has been done in the general field of land capability classification and inventory by Federal and state agencies as well as by private concerns. However, the most far-reaching work is that being carried on under the leadership of the Soil Conservation Service with the cooperation of the states and other agencies. Within the framework of fulfilling its responsibility in soil and water conservation, the Soil Conservation Service recognized the need for a national land capability inventory. This work has progressed rapidly since World War II on two levels: a general inventory for counties and states, and detailed inventories by ownership for farms and ranches participating in Soil (and Water) Conservation Districts.

The latter inventories are made in sufficient detail to provide the facts needed in selecting improvable and permanent uses, and in developing and applying soil and water conservation plans. In making the surveys, experienced technicians identify the different kinds of land, the vegetal cover or kind of use, and the climate (insofar as climate is permissive to agriculture or natural biotic production in the local area). This information is recorded in accordance with nationally established standards on aerial photographs or other base maps. The physical land factors included on the field maps are kind of soil, erosion, slope, and other locally important circumstances such as direction of slope, salinity, alkali accumulation, degree of wetness, and overflow hazard. Symbols used on the field maps have the appearance of a fraction, with facts about the soil expressed above the line and those relating to slope and erosion below the line—each symbol and position is in accordance with the national standard. The vegetal cover or land use also is mapped with symbols.[6]

In the laboratory the detailed information is carefully appraised and lands with similar characteristics, use possibilities, limitations, and management problems are grouped together into one of the eight nationally established land capability classes (Figure 3 – 9). Land suitable for cultivation is placed in classes I, II, and III. All such land must be workable, having soil sufficiently deep and free of stones for cultivation. It must be productive enough, considering climate as well as soils, to return at least moderate yields of crops common to the area under consideration. Wet lands must be drained, or drainable, and suited for cultivation after drainage. Bottom lands must be sufficiently free from overflow to make cultivation practical. Land in arid regions must be suitable for cultivation under irrigation and must have water supply. Classes I, II, and III are differentiated from each other on the basis of the factors that affect the entire set of conditions and measures necessary for safe, permanent cultivation: erodability, slope, natural drainage, permeability, liability to overflow, and in some instances natural fertility and salinity.[7]

Class I land is suitable for cultivation without special practices. It is nearly level and thus not subject to more than slight erosion regardless of treatment (Figure 3 – 10). It possesses soils that are deep and easy to work. It is free from overflows that interfere with planting, growing, or harvesting of crops, and well-enough drained, naturally or artificially, to permit at least moderate yields. If drained artificially, the system must be one that can be maintained without special practices; and if irrigated, land must be level enough to require no special practices (ordinarily less than a 1 per cent slope) and must have good permeability. In common with classes II and III, it may require replacement of nutrient elements removed by crops, cultural practices, and leach-

3 – 9 ***A composite view of the land capability classes.*** **In this landscape all eight land capability classes are shown. (Department of Agriculture)**

3 – 10 *Land capability class I land.* Smooth surface, deep, fertile soil, and good drainage suit such land for permanent cultivation and management without regard for danger of soil loss. This scene, near Sidon, Mississippi, shows soybeans on the left and cotton on the right. (Department of Agriculture)

ing, or it may require green manure crops to replenish organic matter. These requirements, however, cannot be greater than for the other two classes. Class I land is frequently, but not necessarily, the most productive class, and it is usually the most desirable because special practices are not needed.

Class II land is satisfactory for permanent cultivation with simple practices (Figure 3 – 11). The slope may be just steep enough to make water run off at a speed which will carry away soil. Some class II land is naturally wet and requires drainage; some has lower water-holding capacity than class I land. Each deficiency either places a slight limitation on use or requires some special attention year after year. Typical management practices required are erosion control, water conservation, simple drainage, simple irrigation, removal of stones or other obstacles, or correction of moderately low fertility by fertilizers or soil amendments.

Class III land is suitable for permanent cultivation with intensive conservation practices (Figure 3 – 12). Problems are similar to those existing on class II land, but are greater in degree. Safe and permanent cultivation, therefore, requires more care and usually a combination of several practices.

Class IV land is suitable for occasional or limited cultivation (Figure 3 – 13). It may be steeper than class III, more severely eroded, more susceptible to erosion, more difficult to drain or irrigate, or more open or porous and so give excessive permeability. It is not good land for row crops and is best used for permanent vegetation (hay or pasture, and orchards or vineyards if protected by cover crops). Even with care class IV land cannot be cultivated safely more frequently than once

3 – 11 *Land capability class II.* The land on this farm, near Dodge Center, Minnesota, is naturally wet, but with the installation of a tile drainage system its adaptability has been widened. Erosion danger, in this case, is no greater than on class I land. (Department of Agriculture)

3 – 12 *Land capability class III.* The rolling character of this land in Muskegon County, Michigan, requires intensive care to support permanent cultivation while maintaining the soil. The close-growing crop offers good protection. (Department of Agriculture)

every five or six years. Some of the nearly level, imperfectly drained land classified as IV is not subject to erosion, but is unsuited for intertilled crops because of the time required for the soil to dry in the spring and because of its low productivity when planted in these crops.

Class V land, nearly level, is not suitable for cultivation, but is suitable for permanent vegetation that may be used for grazing or wood products without any special restrictions (Figure 3 – 14). Stoniness and wetness are usual limitations for crop use. To fit into this class, land must not be subject to either water or wind erosion, even if the cover should be removed.

3 – 13 *Land capability class IV.* This land in Lubbock County, Texas is naturally restricted by stony, porous soil, and slope. The treatment, bench leveling and a cover of wheat and vetch, protects it from both water and wind erosion. The field is cross fenced and used for rotation grazing. (Soil Conservation Service)

Where vegetation cover has been depleted, such land may require moderate or severe restriction in use for a certain period to allow regeneration. If the cover is good, only reasonable management practices are required to maintain quality and protection.

Class VI land is suitable for permanent vegetation that can be used for grazing or for woodland with moderate restrictions (Figure 3 – 16). It is not arable. Common characteristics are moderately steep slopes, susceptibility to water or wind erosion, shallow soils, stoniness, or alkali. Definite care in management, and in some instances land improvements, are required to maintain class VI land.

Class VII is not suitable for cultivation and requires severe restrictions if used for grazing or woodland (Figure 3 – 17). Its major hazards or limitations commonly relate to very steep slopes, shallow, stony, or droughty soils, excessive erosion, or severe alkali. All class VII land has the capacity to produce useful vegetation. Use, however, must be accompanied by special care in management. In humid areas, woodland use is recommended over grazing.

Class VIII land is suitable neither for cultivation nor for the production of useful permanent vegetation that may be harvested under grazing or woodland use (Figure 3 -18). It is chiefly comprised of extremely rough, high, stony, or barren land, or swamps and marshes that cannot be drained. In any given circumstance class VIII land may be used for wildlife habitat, recreation, or protection of water supplies.

3 – 14 ***Land capability class V.*** **Wetness restricts this land on the coastal plain of Georgia from cultivation. Planted to Dallis grass and lespedeza, it makes an excellent pasture which is favored by a long grazing season. (Department of Agriculture)**

The land capability classes are further divided into land capability subclasses that indicate the dominating limitation or hazard. On maps these are shown with a small letter following the Roman numeral: e for erosion, w for wetness, s for a soil problem, and c for a serious climatic problem (Figure 3 – 18). The most specific and detailed grouping of the classification scheme is the land capability unit. Lands that can be used in the same way and require the same management practices are grouped together and designated on maps in Arabic numbers as the third digit in the description.

Accurate maps have been prepared for an important share of the private lands cooperating in Soil Conservation Districts, and guide sheets are available that detail land character and necessary conservation treatment for essentially all regions of the nation.[8] Thus it is clear that there is now a firm basis for the establishment of conservational management practices on the nation's lands.

Between 1958 and 1961, as part of a "National Inventory of Soil and Water Conservation Needs," the land capability classification was projected to include all non-Federal and nonurban lands.[9] Local committees, under the leadership of the Soil Conservation Service, produced inventories for each county by combining soil and land use data from sample areas with other sources such as soil surveys, the Timber Resources Review, and the agricultural census. The resultant data were published in state reports and in a national summary.[10]

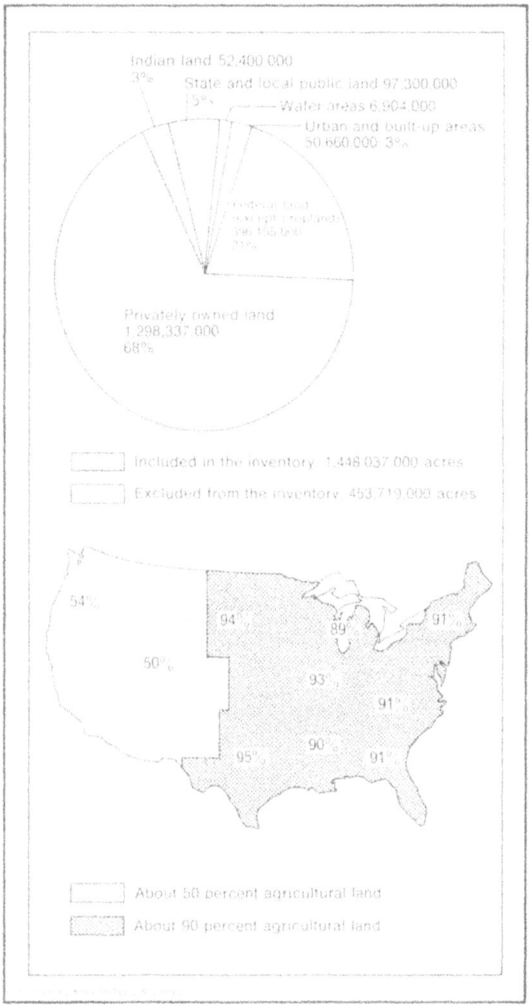

3 – 15 ***Land included in the National Inventory of Soil and Water Conservation Needs.*** **(Department of Agriculture)**

3 – 16 *Land capability class VI.* The land in this Santa Cruz County, California, scene obviously has a major limiting factor of slope. The use is for summer pasture and the cover is in excellent condition, thus the soil is well protected. (Department of Agriculture)

Table 3-4, drawn from the national summary, shows the amount of land included in the inventory, and the amount and per cent of each class together with existing land uses by class. It can be assumed that all land suited to crops under present technology was included in the inventory, because such lands are essentially all in private ownership. Because food is the basic human need, and croplands rate top position on a nation's scale of land resource needs, it is of major significance that 638,009,000 acres (classes I-III) are suited to regular crop production and another 169,181,000 acres (class IV) are capable of occasional crop production. These data, of course, are indicative only of the amount of land suited to the production of cultivated crops; they do not reveal kinds of crop adaptability.

3 – 17 *Land capability class VII.* In this typical example, located in Santa Barbara County, California, the cover is scrub oak and grass. It is highly susceptible to deterioration by erosion if not grazed with extreme care. (Department of Agriculture)

TABLE 3–4

LAND CAPABILITY CLASSES BY LAND USE FOR FIFTY STATES FOR NON-FEDERAL, NON-URBAN LAND*

Class	Thousand acres				
	Cropland	Pasture and range	Forest and woodland	Other	Total
I	27,435	3,940	3,573	1,247	36,195
II	192,923	42,851	43,426	11,279	290,479
III	152,970	66,602	77,910	13,854	311,335
I-III	373,328	113,393	124,909	26,380	638,009
IV	48,993	53,938	58,413	7,838	169,181
I-IV	422,321	167,330	183,322	34,218	807,190
V	1,773	10,525	28,920	1,832	43,051
VI	17,940	166,288	88,490	4,995	277,712
VII	5,636	138,690	144,227	7,682	296,233
VIII	66	2,523	6,518	18,136	27,242
V-VIII	25,415	318,025	268,154	32,645	644,238
TOTAL	447,736	485,356	451,476	66,863	1,451,428

* "Agricultural Land Resources, Capabilities, Uses, Conservation Needs," *Agriculture Information Bulletin No. 263* (Washington: Department of Agriculture, 1962), p. 29.

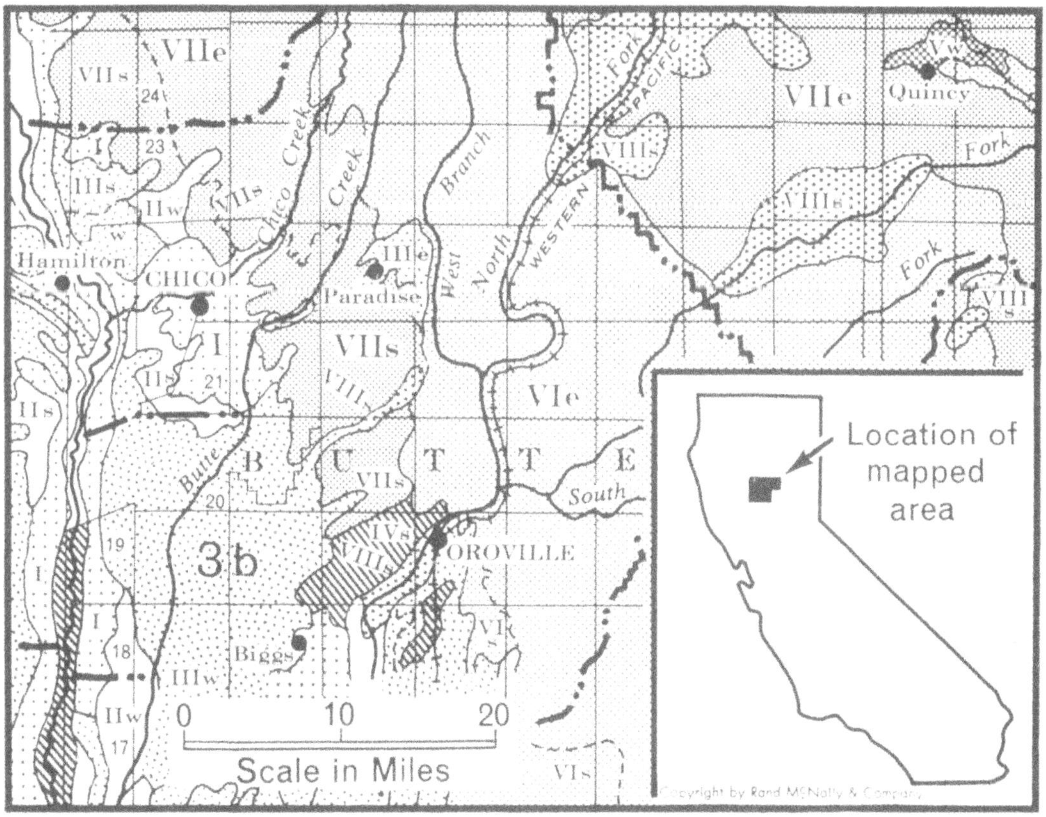

3 – 18 *A segment of the generalized land capability map of California. (Soil Conservation Service)*

FOOTNOTES

[1] See *Land, the Yearbook of Agriculture, 1958* (Washington: Department of Agriculture, 1958), pp. 2–9.

[2] Study of the characteristics and distribution of the individual elements is desirable. Maps showing these can be found in most atlases. For example, see Edward B. Espenshade, Jr., ed., *Goode's World Atlas,* 12th ed. (Chicago: Rand McNally & Company, 1966).

[3] More detailed maps are found in Chapters 6 and 7.

[4] Carleton P. Barnes and F. J. Marschner, "Our Wealth of Land Resources," *Land, the Yearbook of Agriculture, 1958.* op. cit., pp. 10–18.

[5] The policy of reserving certain public domain lands was established early by the federal government. In 1817 Congress delegated authority to the President to withdraw from entry special purpose lands—for military posts, wagon roads, and the like. In 1832 authorization was given to reserve land with extraordinary natural values, and the Creative Act of 1891 permitted the withdrawal of forest lands; other considerations then entered the picture and mineral lands, power sites, national parks, and so on, were added to the list. Moreover, before the withdrawals had ended, procurement from private owners began. The passage of the Weeks Law in 1911 authorized federal acquisition of land "for the purpose of conserving the navigability of navigable rivers" with the consent of the state concerned. Interpretation of this law has been a notable avenue to federal acquisitions. In the depression years of the 1930's, purchase authority was broadened with the National Industrial Recovery Act of 1933, the Emergency Relief Act of 1935, and the Title III of the Bankhead-Jones Farm Tenant Act. By the close of 1939, approximately 10,800,000 acres of "marginal" and "submarginal" forest and grassland and some cropland had been purchased under these three acts. Subsequently, about nine million acres of these lands were transferred to the Forest Service. During the war years (late 1930's to mid 1940's) there was considerable buying of private land for military establishments as well as government sponsored defense industries. The end of the war and the return to prosperity brought a major slowdown in federal purchases; however, a new objective, recreation land, has been the basis for some purchases in recent years. The total area acquired by the federal government through June 30, 1963, was reported by the General Services Administration as 51,787,889 acres.

[6] "Soil Conservation Survey Guide, Pacific Region" (Washington: Soil Conservation Service, Department of Agriculture, 1951), 44 pp.

[7] Many Soil Conservation Service publications cover the capability classes in greater detail than can be presented here. Where completed and published, the state bulletins are excellent. See for example, Leonard R. Wohletz and Edward F. Dolder, "Know California's Land, A Land-Capability Guide for Soil and Water Conservation in California" (Department of Natural Resources, State of California and Soil Conservation Service, United States Department of Agriculture, February, 1952), 43 pp., with map; R. D. Hockensmith and J. G. Steele, "Classifying Land for Conservation Farming," *Farmer's Bulletin No. 1853* (Washington: Department of Agriculture, 1943), 45 pp.; and J. G. Steele, "The Measure of our Land," (Washington: Soil Conservation Service, Department of Agriculture, 1951), 17 pp.

[8] These are discussed in Chapters 4 and 5.

[9] The small amounts of federal lands used for crops such as on experiment stations and game refuges were included in the inventory.

[10] "Agricultural Land Resources, Capabilities, Uses, Conservation Needs," *Agriculture Information Bulletin No. 263* (Washington: Department of Agriculture, March, 1962), 31 pp.

4

Croplands

AGRICULTURAL LANDS OF LARGE AREA, varied adaptability, and great productive capacity comprise one of the strongest elements in the resource foundations of the United States. The country's strength in this respect is unequalled by any other nation. American farmers regularly account for about 17 per cent of the world's agricultural output, compared with 13.2 per cent for China, about 10 per cent for India, and 9.7 per cent for the Soviet Union.[1]

The benefits of this nation's farm production accrue to 6.0 per cent of the world's population compared to 22.3, 13.6, and 7.2 per cent respectively for China, India, and the Soviet Union (1960 estimates).

Ability to produce the basic food crops as well as animal commodities in abundance has allowed the American people to escape the perils of constant or even periodic shortages of food which are traditional in India and China. The average family's food supply in the United States presently provides 3,160 calories per person a day, compared with recommended reference standards which range from 2,300 calories for the Far East to 2,700 per capita for Canada and the Soviet Union. Studies conducted in the late 1950's indicate that 56 per cent of the world's people (92 per cent of the Asians, 38 per cent of the Africans, and 29 per cent of the Latin Americans) have energy supply levels that are below the recommended standards. Americans not only enjoy greater quantities of food, but also benefit from better nutritional quality and balance. Starchy foods, for example, comprise but 24 per cent of the calories Americans consume, compared with 43 per cent in Western Europe, 49 per cent in Latin America, 62 per cent in Eastern Europe-U.S.S.R., and 72 per cent in Asia. Especially significant is the high American consumption of livestock products, which amount to 30 per cent of the caloric intake.[2] The livestock industry, in fact, indicates something of the luxury character of American agriculture, because much of it is supported by feed grains and forages produced on high quality land. Additionally, this nation has a higher percentage of land devoted to industrial crops than any other major nation, and it also produces commodities for export on about one out of every five acres.

Since the early beginnings of American agriculture, the wealth of land resources for farming has fostered a sense of complacency in the minds of all the citizenry—the farmers, the government, and the populace in general. This complacency has engendered poor land management practices and short-range policy solutions. Lavish land resource use has always characterized American farming. Until the 1930's, land deterioration progressed at an alarming rate. Although much attention has since been given to this waste, erosion, soil depletion, and extravagant use of water continue to lower the potentials of some areas of the nation. Yet during the past three or four decades, despite the population increase, the developments in machinery and agricultural science—in cultural practices, improvements in plants, animals, fertilizers, and feeds, and in biological controls—have allowed farm output to more than keep pace with growth in needs. Surpluses are today a knotty problem, and increasing ability on the part of the American farmer promises to extend this problem into the immediate future. In the mid-1960's, however, it was demonstrated that accumulated stocks of basic commodities can quickly be diminished with expanded foreign assistance activities.

Overproduction, for the most part, has resulted in attempts at short-range solutions. Some of these, such as acreage controls, have further complicated the problem by encouraging the ingenuity of the farmers and agricultural suppliers in upping yields. Controls also have hastened some resource deterioration, as in the instance of water mining in the High Plains of Texas to increase cotton output on restricted acreage. Perhaps one of the most serious results of the present surplus problem is the depreciation of agricultural land in the longer view of the nation's needs; the attending attitude is complacency concerning the alienation of agricultural land for other uses—broad highways, cities, industrial plants, forest production, recreational facilities, and so on. Once investments in such developments have been made, even if the land could be brought back into farming, economic considerations in many cases will be prohibitive. In consequence, an element of flexibility in the economy to meet short-term needs, as well as the greatest opportunities for rational ordering of future land uses are lost.

A long-range view is necessary if the American people are to establish and maintain a truly conservational relationship with the resource foundation. Food is basic. Its production results from the application of agricultural know-how to agricultural land, and both elements are needed. An expanded role will be required of each in the future because population increase is inevitable. This need can be demonstrated by considering probable growth within the relatively short period of the next thirty years. By the year 2000 if the population has grown to only 275 million and the cultivated land remains static, the amount per capita will have dropped from the 2.8 acres available in 1955 to 1.7 acres. If the population should double in that period, the arable land per capita would be reduced to about 1.2 acres. What will be the case a century from now? In addition to meeting its own needs, the United States may very well be called upon to supply an increased proportion of the world's food. By 2000 the world population is expected to be nearly double the 1965 total. If no new croplands are added, the one acre per capita available in 1965 will be decreased to just over one-half acre by the turn of the century.

Land wealth, in the final analysis, can be measured only in terms of the number of beneficiaries. In this light, agricultural land, a strong element in the past, present, and immediate future, could become a weaker element in the long run. A conservational adjustment will require a multifaceted program which includes (1) maintenance and improvement of the present agricultural land with diversion to other irreversible uses only on the grounds of a higher quality sustainable use or the unsuitability of the land for agricultural use; (2) continued developments in agricultural science and education to improve yields, management practice, and responsibility; and (3) a long-run program of agricultural land development to keep pace with diversions to other uses and increasing demands. Some aspects will be difficult to attain because of the very fact that this is a competitive democratic society, and economic forces are strong considerations in the way of life. Certainly it is clear that all levels of the populace must be able to make a reasonable living, including the farmer. Nevertheless, the provision for the welfare of society must be the continuing long-range goal. Thus the problems of agriculture affect not only the farmers, but all segments of the population. Conservation of agricultural land, like all resources, requires total national effort.

Extent and Character of Croplands

Sixty per cent of the land area of the forty-eight states is in farm and ranch ownership. This statistic alone indicates the major role of farmers in the management of the nation's land resources. In 1959 the 1,123,507,574 acres of land in farms and ranches in the fifty states was divided among the following major uses: cropland (including sixty-six million acres used only for pasture), 455 million acres; pasture and range land, 466 million acres; woodland and forestland, 164 (of which 92.5 million acres were pastured) million acres; and farmsteads, feed lots, farm roads, ditches, small orchards and gardens, 37 million acres. Land in farms includes essentially all of the cropland of the United States and 65 per cent of the pasture and range land which accounts for nearly 88 per cent of the feed supplied by such land. Total land use trends for the forty-eight states since the turn of the century are presented in Table 4-1.

Since World War II the total area classed as cropland has varied from a high of nearly 478 million acres in 1950 to a low of about 450 million acres in 1965. During this same period, acreage used for crops ranged from 386 million acres in 1949 to 350 million acres in 1965, and the acreage from which crops were actually harvested fluctuated between 346 million acres in 1945 to 1949 to a low of about 295 million acres in 1965. The data for 1959 may be used to illustrate the components of the total cropland. In that year 457 million acres were classed as cropland. Cropland used for crops totaled 352.5 million acres, of which 311.5 million acres produced crops that were harvested, 30.7 million acres were cultivated summer fallow, and 10.3 million acres had crop failure. Also included in the total were 16.4 million acres in soil improving crops, 13.6 million idle acres, and 65.5 million acres of cropland used only for pasture.[3]

The decline in cropland harvested in recent years has resulted primarily from the dramatic increases in per-acre yields that have attended the application of advanced technology and general efficiencies. Figure 4 – 1 illustrates the impact and relationship during the period 1950 to 1965. Domestic consumption has not increased as rapidly as per-acre yields. This factor has been coupled with another, namely a significant change in the nature of consumption. Figure 4 – 2, showing per capita food consump-

TABLE 4–1

TRENDS IN MAJOR LAND USES, CONTERMINOUS STATES, 1910-59*

Major use[1]	Million acres 1910	1920	1930	1940	1950	1959
Crops and related uses[2]	347	402	413	399	409	391
Cropland used only for pasture	84	78	67	68	69	66
Total cropland	431	480	480	467	478	457
Pasture and range[3]	693	652	652	650	631	630
Forest and woodland[4]	610	614	615	630	634	639
Other land[5]	169	157	156	158	161	176
TOTAL LAND AREA[6]	1,903	1,903	1,903	1,905	1,904	1,902

* *Land and Water Resources, A Policy Guide.* (Washington: Department of Agriculture, September, 1962, revision), Table 2, pg. 11.

[1] Acreages are for the forty-eight conterminous states. For example, excluded in 1959 are the combined totals for Alaska and Hawaii of about one-half million acres of cropland, and three million acres of grassland pasture.

[2] Cropland harvested, crop failure, fallow, and idle cropland. Cropland and pasture use relates to the preceding years, except for 1959, where they are for the current year.

[3] Grassland pasture and other nonforest range land, excluding cropland used only for pasture. Includes idle grassland which probably existed in significant acreages only before 1920.

[4] Includes forest land in parks, wildlife refuges, wilderness areas, national defense sites, etc. Includes commercial and noncommercial forest land, and forest land grazed.

[5] Includes "special land use areas."

[6] Remeasurement of the land area of the forty-eight conterminous United States in connection with the 1960 Census indicated an approximate land area of 1,902 million acres. The total land area of Alaska and Hawaii was 369 million acres. Decreases in the total acres for the forty-eight conterminous states since 1940 chiefly represent increases in large reservoirs.

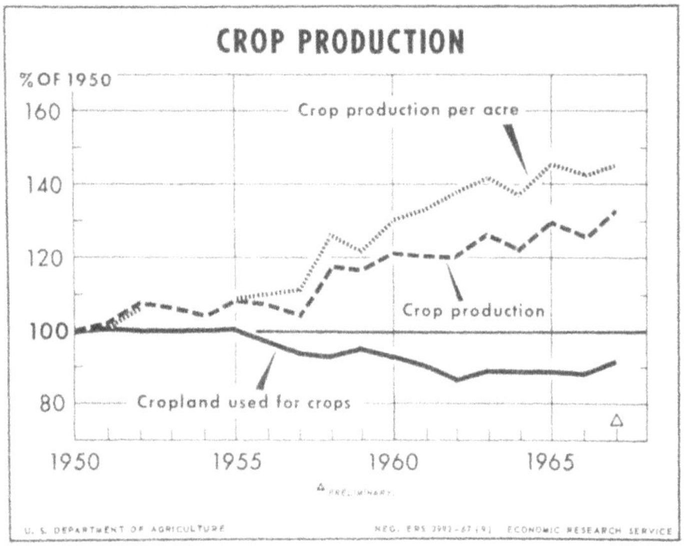

4 – 1 *Trends in crop production.* (Department of Agriculture)

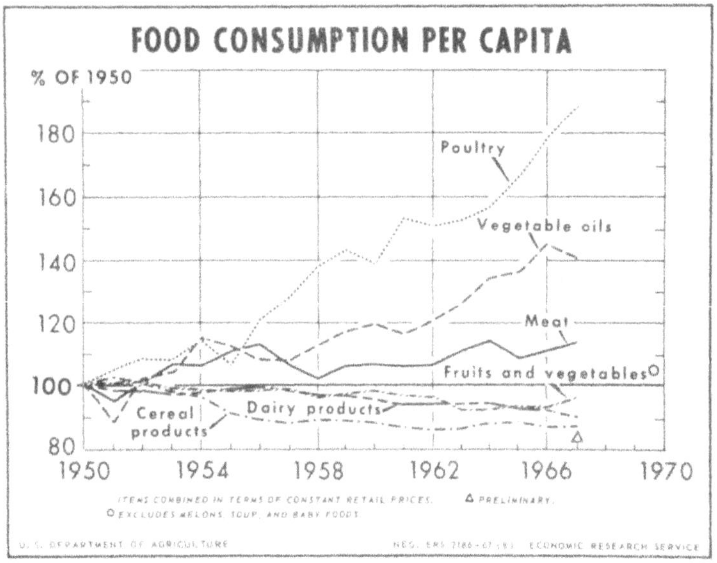

4 – 2 *Trends in food consumption.* (Department of Agriculture)

tion, indicates a noteworthy decline in cereal products; these traditionally have been large acreage users. Acreage controls and other government programs have been employed to encourage reduction in the planting of certain crops, and even to encourage the shift of some cropland to other use. Changes in types of farming, for example greater emphasis on pasture and livestock, and the decline in small and uneconomic farm units, are other contributors to cropland decline. Chronic crop failures are not a major factor in either the decline of cropland harvested or the fluctuations over short periods. The area of crop failures in recent years is typically eight to eighteen million acres. Drought is the chief cause of crop failure, and the Great Plains and mountain West normally account for about three-fourths of the total. Floods, frost, severe storms, insects, and diseases also are responsible for crop failures in some areas.

LAND QUALITY

Mention was made of the quality of land resources in the preceding chapter, but more detail on cropland quality is now necessary. It is apparent from study of Figure 4 – 3, which shows the principal cropland areas, that the adaptability of land resources for agriculture differs markedly from one portion of the nation to another. Quality, it will be recalled, results from the individual and combined influences of the following physical conditions: (1) heat-unit accumulation and regimen; (2) frost-free season; (3) annual amount, intensity, and seasonal distribution of precipitation; (4) soil characteristics; (5)surface form, including degree and direction of slope; (6) drainage conditions; (7) water supply; (8) native vegetation; (9) biological conditions such as weeds, pests, and diseases; and (10) location with reference to markets. Despite the tremendous advances made in agricultural science, the ease of and the economic limitations on the application of technological improvements remain closely tied to these physical influences of quality. The distribution of cropland is largely a reflection of the distribution of the relatively favorable combinations of these elements. Aridity, unfavorable surface form, and water supply for irrigation are the major limiters in the West. Rough surface, poor soils, and inadequate drainage place restrictions on parts of the East. The large concentration of croplands in the central plains of the nation is explained in terms of the most favorable combination of the factors noted above.

The Soil Conservation Service land capability classification system does provide a basis for evaluating land quality and adaptability when the spatial distribution of classes is shown on a map. It then becomes possible to correlate these classes with the distributions of other variables of biotic productivity. Figure 4 – 4 gives an indication of the general distribution of capability classes on non-Federal lands, as determined by the National Inventory of Soil and Water Conservation Needs. When the spatial relations of the capability class pattern and relevant climatic data are examined, it is clear that the nation is, indeed, well endowed with agricultural land resources of various qualities. There is excellent balance in land

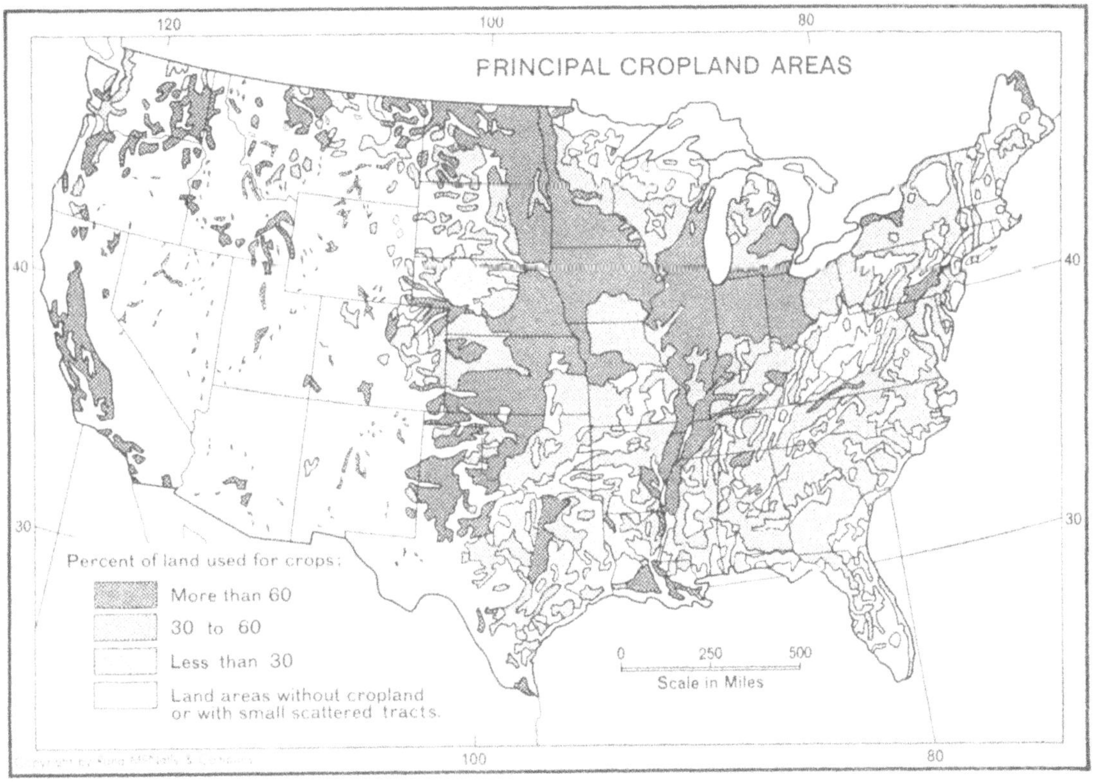

4 – 3 ***Principal cropland areas.*** **(Department of Agriculture)**

suited to warm weather, mild weather, and cool weather plants; in land suited to plants requiring long, intermediate, and short frostless seasons; and in land suited to the varying moisture requirements of crop plants. (If a shortage presently exists, it is in land suited to humid tropical plants.) Proper use and management must continue to maintain these favorable relationships.

Historical Changes in Agriculture

For the purpose of this discussion it may be said that the history of American agricultural land use can be divided into four periods.[4] The first was the Colonial period, with settlement confined largely to the Atlantic littoral. Most settlers were farmers and their primary purpose was subsistence; however, commercial agriculture had its beginning, particularly in the production of tobacco for export. Land abundance and limited technology resulted in considerable land deterioration. Period two, from the end of the Colonial period until about 1880, saw the accumulation of the land estate of the nation, its exploration, and the development of major railroads. Agricultural settlement advanced to the subhumid margin, and the range livestock industry became established in the West. The Corn Belt as a feed livestock region, and the South as a Cotton Belt appeared. Period three lasted from the 1880's until the 1920's. Within the span of forty years change occurred rapidly; cropland harvested doubled, increasing from 178 to 362 million acres. This increase was accompanied by large decreases in native grass-

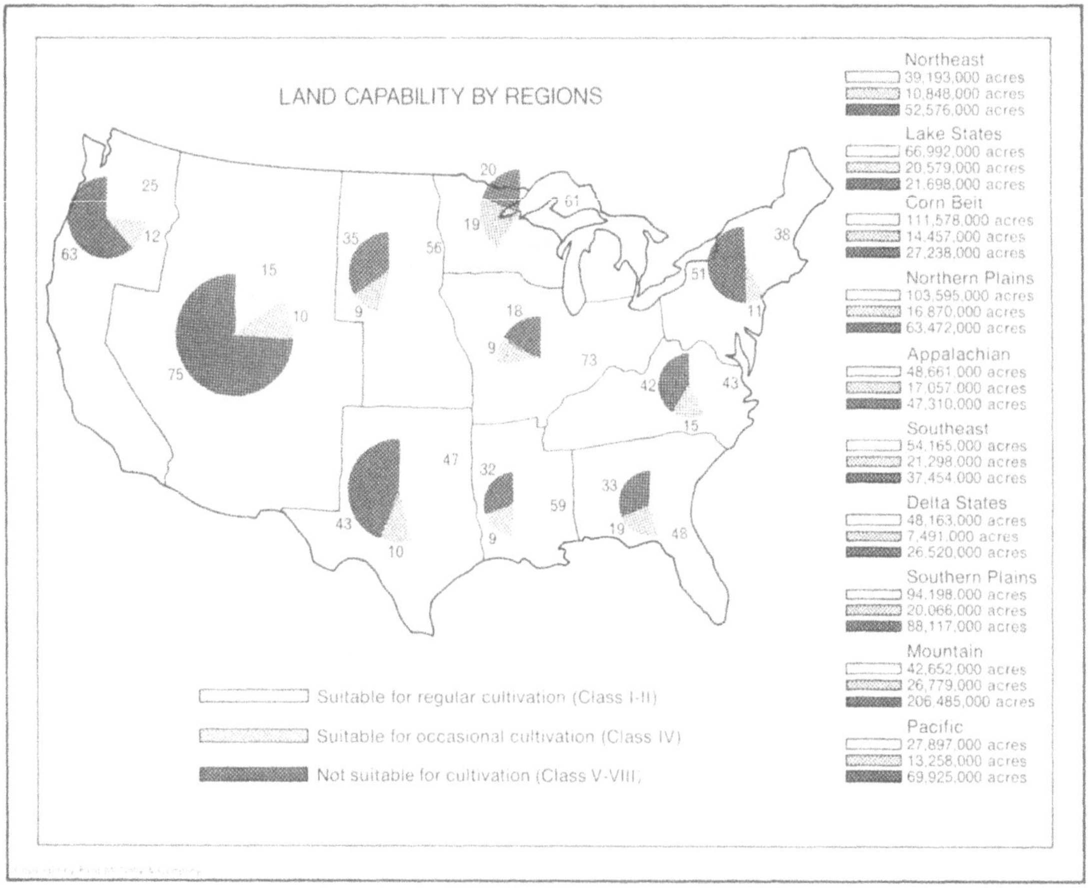

4–4 ***Land capability by regions.*** **Regional proportion of land in the several capability classes for non-Federal and non-urban land only (see Figure 3-15). (Department of Agriculture)**

lands. During the period, clearing of forest land also continued to move along rapidly. There was lively development of irrigation in a number of western valleys. In the period since 1920 there has been a greater degree of stabilization in the major categories of land utilization. There has been, however, shifting in the regional distribution of cropland. Land development has continued through drainage, irrigation, and clearing in some areas originally bypassed. In the 360 years since settlement began at Jamestown, Virginia, nearly 320 million acres of forest and woodlands and 380 million acres of grasslands have been converted to cropland, pasture, and other uses. Additional irrigated croplands have been developed and now include some thirty million acres of former semi-arid short grass and desert shrub land in the West. In the sorting processes that have accompanied the twentieth century development of agriculture, poor quality and depleted lands have been diverted to other uses and urban and service facilities have come to occupy a considerable area of former croplands.

Cropland reached a peak in the 1930's for the nation as a whole, but contrasts exist in this trend from region to region. For example, abandonment of cropland has characterized New England and the Appalachian region since 1900; the Lake States, Corn Belt, and Northern Plains have fluctuated only slightly in this respect the past thirty-five years; the Southeastern states have had a noteworthy shift of cropland to pasture and forests; land in farms and ranches has increased in the Southern Plains, but cropland harvested has declined somewhat; and the Western states had, in general, increased croplands up to 1950. In summation, then, the nation's croplands increased until the early 1930's and land in farms continued to increase until 1950. Since 1950 farmland has declined moderately with shifts to use for urban and industrial expansion and highway systems.

In addition to the land selectivity process which has characterized the recent period, a revolution has occurred in the organization of the agricultural industry. Actually, instead of one revolution, three have been occurring: a revolution in the business of agriculture, in the mechanization of agriculture, and in the science of agriculture. Farming has shifted from a way of life to a distinctly commercial activity—a business—competitive with other economic endeavors for brainpower, manpower, and capital. Farmers, like businessmen in general, must

4–5 *Machines are an essential requirement for modern commercial agriculture.* The replacement cost for the equipment being utilized in this harvest scene would exceed ten thousand dollars. (Tennessee Valley Authority)

make a profit to continue operation. The necessary capital requirements have risen with the trend of the entire economy and the growth in mechanization and agricultural science. The successful farmer today is a combination businessman, agricultural scientist, and technician. His activity calls for many elements not required by the farmer of 1920: more land, more power machinery, improved seeds, fertilizers, chemicals for pest controls, credit, and a profit to remain in operation (Figure 4 – 5). The amount of production capital per farm almost doubled between 1950 and 1958—from $17,000 to $33,500—and has continued to increase. The investment per farm worker has shown a similar trend, advancing from $9,600 in 1950 to $21,000 in 1960 (the latter figure compared with $15,900 per worker in industry).[5] These data are for the average farm. As Figure 4 – 6 shows, many commercial farms now repre-

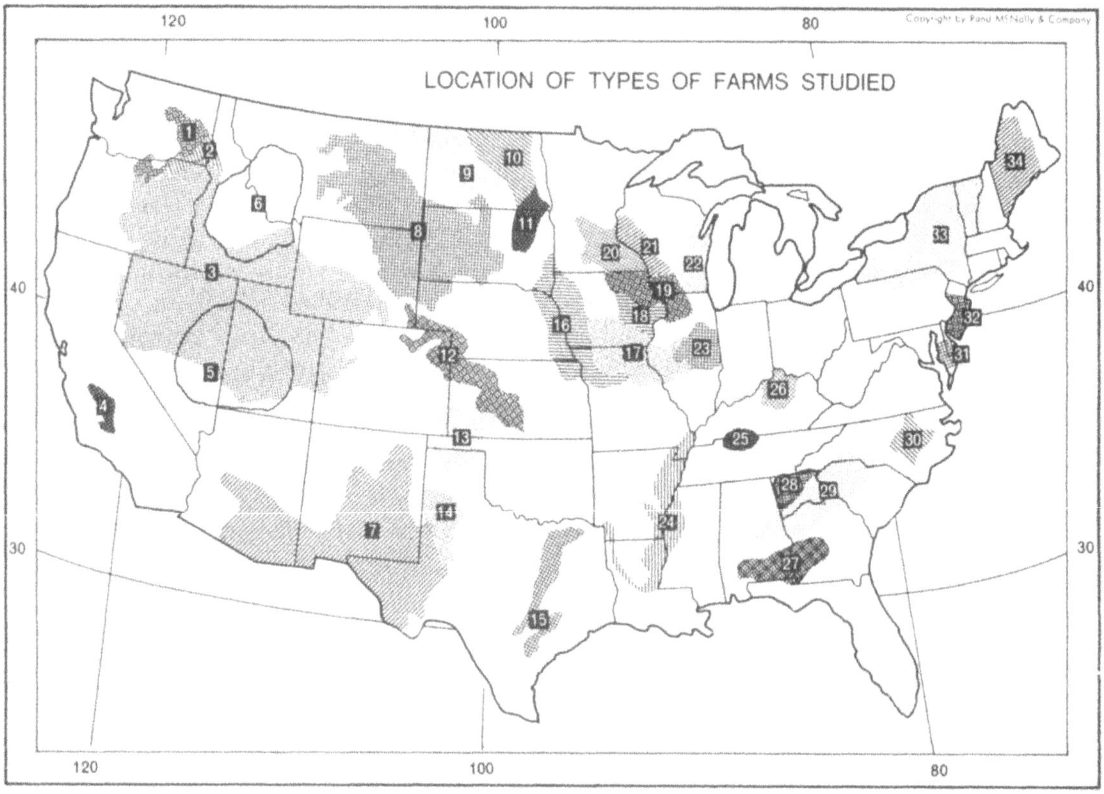

4 – 6 *Location of important types of commercial farms.* These are farms for which annual costs and returns are estimated by the Department of Agriculture's Economic Research Service. (See "Farm Costs and Returns," *Agriculture Information Bulletin No. 230*, revised annually.)

1. Wheat-fallow
2. Wheat-pea
3. Intermountain range cattle
4. San Joaquin Valley cotton
5. Utah Desert sheep
6. Northern Rocky Mountain cattle
7. Southwest sheep and cattle
8. Northern plains sheep and cattle
9. Spring wheat-fallow
10. Spring wheat-small grain livestock
11. Spring wheat-corn-livestock
12. Winter wheat
13. Winter wheat-grain sorghum
14. High plains cotton
15. Black Prairie cotton
16. Hog-beef fattening
17. Hog fattening-beef raising
18. Hog-beef fattening
19. Hog-dairy
20. Southeastern Minnesota dairy-hog
21. Western Wisconsin dairy
22. Eastern Wisconsin dairy
23. Cash grain
24. Delta cotton
25. Pennyroyal tobacco
26. Central Kentucky tobacco
27. Coastal Plains peanut-cotton
28. Georgia broilers
29. Southern Piedmont cotton
30. Coastal Plain-tobacco-cotton
31. Delmarva broilers
32. New Jersey poultry
33. Central Northeast dairy
34. Maine broilers

sent a production capital in excess of $75,000. Figure 4 – 7 shows what has been happening in terms of numbers of farms. From a high of 6,-812,350 in 1935 the number of farms declined to an estimated 3,286,000 in 1966. Average farm size, in the meantime, has increased steadily, from 174 acres in 1940 to an estimated 350 acres in 1966.[6]

It is significant to realize that in recent years about 100,000 farms account for nearly one-third of the market sales, and that slightly over 750,000 farms produce about 75 per cent of the market sales (see Table 4–2). Nearly one-third of the total accounts for less than 4 per cent of all farm sales. Thus two trends in farm size may be recognized: (1) commercial farms are being reduced in number and are growing larger; and (2) there has been considerable development of part-time and residential farms which reduce the size of the "average farm" and mask the facts of the statistical picture of farming organization.

Increasing farm size has paralleled an increase in farm mechanization and technology, and the net result has been a steady decrease in farm population. In 1960 8.7 per cent of the population were farm people, a decline from 23.2 per cent in 1940 (see Table 4–3). The farm output, however, is greater than ever (Table 4–4).

REGIONAL PATTERNS OF AGRICULTURAL LAND USE

Through the years of agricultural development and its adjustment to a highly commercial economy, broad regionalized patterns of farming types have become discernible. The Department of Agriculture has produced a map recognizing 165 types of farming areas which have been grouped into sixty-one subregions. These subregions have been summarized in a map showing eight major types of farming (Figure 4 – 8).[7] Although these major regions have remained relatively stable in recent years, there have been and will continue to be many shifts within them. For example, cotton has shifted westward, hay and feed crops have increased in many areas, the acreage of improved pasture and livestock numbers have been steadily climbing, and in New England, the Lake States, and particularly in the South there has been an increase in forest land. Since World War II grain sorghum has become a major feed crop in the subhumid zone, and soybeans a noteworthy cash crop in the Corn Belt.

SUMMARY OF FACTORS INFLUENCING AGRICULTURAL LAND USE

With the general progress of the economy, agriculture has become responsive to an increasing number of influences. The physical-biological character of the land resources determines the outside limitations of what will or will not grow. What is grown or how the land is actually used within the range of possibilities is the result of such considerations as ownership, land value and market trends, and taxes; physical, technical, and capital requirements of com-

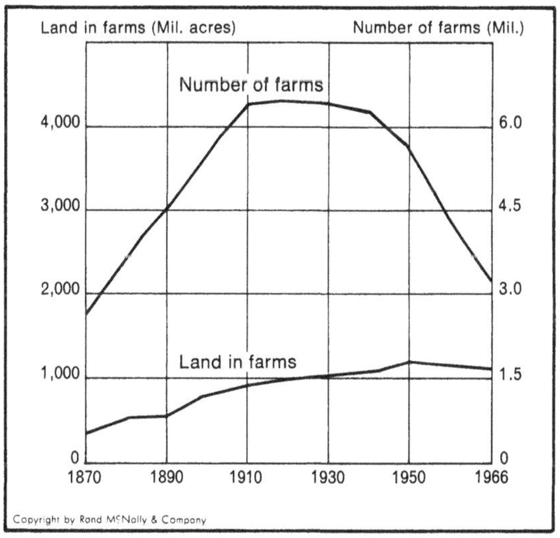

4 – 7 *Forms and Land in Farms.* **Farm numbers more than doubled between 1870 and 1910, from less than three million to more than six million. The main reason was the westward expansion of agriculture. Thereafter, the rate of increase in farm numbers was relatively stable, until the 1930's. The downtrend that became evident about 1930 sharply accelerated after 1950. Farms were down to slightly more than three million by 1966, not much above the level of 1870. The reduction in farm numbers was due largely to the consolidation of farms into bigger, more efficient units, and to some change in farm definition. Land in farms continued to expand from 1870 to 1950. Since 1950 farmland has declined moderately for use in industrial expansion, highway systems, and urbanization. (Department of Agriculture)**

modities produced; technology and machinery available; market potentials; and government policies relative to such conditions as production controls, incentive payments, and so on. Within the range of these considerations there is also the factor of farmer preference.

Despite the developments in agricultural science and machinery and in marketing requirements, it is important to remember that the land resource remains basic. Farm real estate represents three-fourths of all physical assets in agriculture (see Table 4–5).

Land is the farmer's most basic tool. Its price has been trending upward over the years. In 1965 U.S.D.A. economists estimated the average price of an acre of farm real estate (land and buildings) reached a new high of $139—slightly more than 100 per cent more than fifteen years earlier. The 1965 figure is 300 per cent more than the price of an acre sixty years ago; 750 per cent more than was paid in Civil War days.

Cropland Problems

Through the years of rapid settlement of the nation's land and then the establishment of the commercial-industrial economy, a number of cropland problems developed that if unattended will bode ill for the future of the American society. Some relate to the ability to provide agricultural products for an increasing population; others have direct bearing upon the future potentials of other resources.

LAND DETERIORATION THROUGH MISUSE

The potentials of land for crop production may be diminished or lost in several ways: through improper cultural and cropping practices, improper drainage provisions, and improper irrigation. Improper cultivation and cropping practices may cause loss of organic

TABLE 4–2

NUMBER OF FARMS AND PERCENTAGE OF ALL FARM PRODUCTS SOLD, BY VALUE OF FARM PRODUCTS SOLD, UNITED STATES, 1950, 1954, AND 1959*

Value of farm products sold	1950 Farms		1950 Percentage of all farm sales	1954 Farms		1954 Percentage of all farm sales	1959 Farms		1959 Percentage of all farm sales
	Thousands	Per cent		Thousands	Per cent		Thousands	Per cent	
Commercial farms:									
$40,000 or more							102	2.8	31.5
$25,000 or more	103	1.9	26.0	134	2.8	31.3			
$20,000–$39,999							210	5.7	18.5
$10,000–$24,999	381	7.1	24.8	449	9.4	26.9			
$10,000–$19,999							483	13.0	21.9
$5,000 –$9,999	721	13.4	22.7	707	14.8	20.5	654	17.6	15.4
$2,500 –$4,999	882	16.4	14.3	812	17.0	12.1	618	16.6	7.4
Less than $2,500	1,619	30.1	9.6	1,226	25.6	7.1	349	9.4	1.5
Commercial farms	3,706	68.9	97.4	3,328	69.6	97.9	2,416	65.1	96.2
$10,000 or more	484	9.0	50.8	583	12.2	58.2	795	21.5	71.9
Less than $10,000	3,222	59.9	46.6	2,745	57.4	39.7	1,621	43.6	24.3
Other farms[1]	1,672	31.1	2.6	1,455	30.4	2.1	1,292	34.9	3.8
ALL FARMS	5,378	100.0	100.0	4,783	100.0	100.0	3,708	100.0	100.0

* *Land and Water Resources, A Policy Guide* (Washington: Department of Agriculture, May, 1962 [slightly revised, September, 1962]), p. 26. All data based on reports for a sample of farms. Percentages computed from unrounded data.

[1] Includes part-time, part-retirement, and abnormal farms. Definition of the first two of these classes and of commercial farms with less than $2,500 farm sales differed in 1959 from previous years.

matter and soil structure, loss of plant foods, and in many cases loss of the soil itself. All are interrelated. When the first settlers carved their farms out of the virgin lands which were protected by a covering of natural vegetation, most of the potential croplands contained a legacy of natural wealth contributed by centuries of weathering of the underlying rocks, the growth and decay of biota, and alluvial and wind deposition that with proper management was maintainable. But, faced with land aplenty, limited science and technology of management, and an economic situation in which natural resource spending was encouraged, soil deterioration was a common result.[8] Some of the croplands have been cultivated for over

TABLE 4–3

FARM POPULATION AS A PER CENT OF NATIONAL AND REGIONAL TOTALS*

Area	1920			1940			1960		
	Total population	Farm population: Number of persons	Farm population: Per cent of total	Total population	Farm population: Number of persons	Farm population: Per cent of total	Total population	Farm population: Number of persons	Farm population: Per cent of total
	Thousands		Per cent	Thousands		Per cent	Thousands		Per cent
United States (including armed forces)	106,089	31,974	30.1	131,820	30,547	23.2	180,007	15,635	8.7
Regions:									
Northeast	29,669	2,537	8.6	35,977	2,411	6.7	44,678	1,119	2.5
North Central	34,197	10,158	29.7	40,143	9,349	23.3	51,619	5,836	11.3
South	33,242	17,063	51.3	41,666	16,400	39.4	54,973	7,160	13.0
West	8,981	2,216	24.7	13,883	2,387	17.2	28,053	1,520	5.4
Geographic divisions:									
New England	7,428	633	8.5	8,437	623	7.4	10,509	232	2.2
Middle Atlantic	22,241	1,904	8.6	27,540	1,788	6.5	34,168	887	2.6
East North Central	21,644	4,953	22.9	26,626	4,638	17.4	36,225	2,821	7.8
West North Central	12,553	5,205	41.5	13,517	4,711	34.8	15,394	3,015	19.6
South Atlantic	14,046	6,496	46.2	17,823	6,060	34.0	25,972	2,838	10.9
East South Central	8,901	5,257	59.1	10,778	5,283	49.0	12,050	2,494	20.7
West South Central	10,295	5,310	51.6	13,065	5,057	38.7	16,951	1,828	10.8
Mountain	3,340	1,179	35.3	4,150	1,118	26.9	6,855	678	9.9
Pacific	5,641	1,037	18.4	9,733	1,269	13.0	21,198	842	4.0

* *The Farm Index* (May, 1964), p. 15.

TABLE 4–4

FARM PRODUCTION IN RELATION TO LABOR*

	Base period or unit	1940	1945	1950	1955	1960	1964	1965[1]
Farm output	1957–59 = 100	70	81	86	96	106	112	115
Man-hours of farmwork	1957–59 = 100	192	177	142	120	92	79	75
Farm output per man-hour	1957–59 = 100	36	46	61	30	115	142	153
Total farm inputs	1957–59 = 100	97	99	101	102	101	103	103
Crop production per acre	1957–59 = 100	76	82	84	91	109	116	124
Supplied by one farmworker	Persons	10.7	14.6	15.5	19.5	25.9	33.2	37.0
Plant nutrients[2]	1,000 tons	1,178	1,787	2,772	4,507	5,643	8,093	8,474

* *The Farm Index* (August, 1966), p. 9.
[1] Preliminary.
[2] Nitrogen, available phosphorus and soluble potassium.

TABLE 4–5
VALUE OF LAND AND BUILDINGS*

Year	Per acre	Total
	Dollars	Million dollars
1850	11.14	3,272
1860	16.32	6,645
1870	18.26	7,444
1880	19.02	10,197
1890	21.31	13,279
1900	19.81	16,615
1910	39.59	34,793
1920	69.37	66,310
1930	48.52	47,873
1940	31.71	33,636
1950	64.96	75,255
1960	116.48	129,929
1965	139.00	159,400

* Data from Economic Research Service, Department of Agriculture.

three hundred years; most have been in cultivation for at least seventy-five years. Preparation of seedbeds and cultivation requires that the soil be disturbed from three to as many as ten times each crop year. With clean cultivation, removal of the crop, and stubble burning (a common practice in the wheat regions until the late 1930's) little or no organic material is added. Such pulverization of the soil and the mining of the natural organic matter results in a reduction of *tilth quality*, the physical condition which has most to do with air and water movement. Water-holding capacity is diminished, compaction and crusting increase, the soil becomes more difficult to farm and more susceptible to erosion (Figures 4 – 9 and 4 – 10). Careful cultivation practices, the return of crop residue, animal manure, and green manure can eliminate these deterioration trends.

Loss of plant foods goes with cropping, depletion of organic matter, cultivation, erosion, and failure to fertilize. Cultivation naturally tends to speed loss through leaching and oxidation. Most of the croplands of the nation are deficient in some plant foods. Nitrogen, phos-

4 – 4 Maior types of farming (Department of Agriculture).

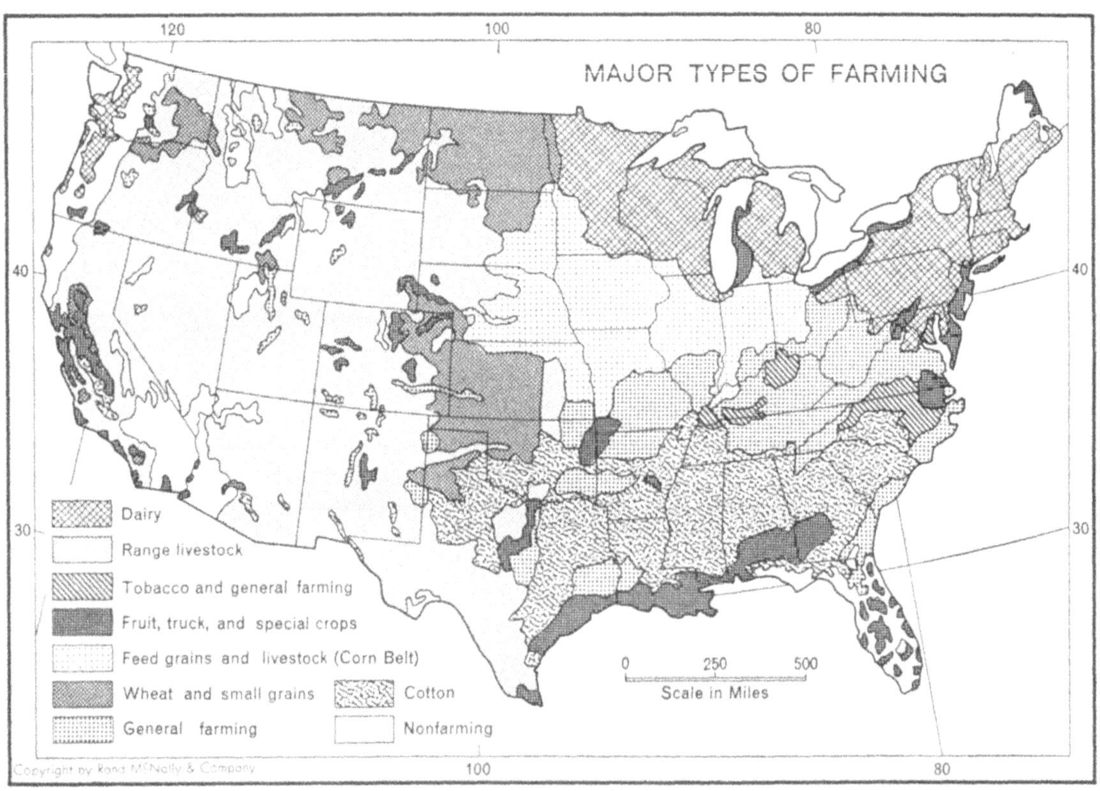

4 – 8 Major types of farming (Department of Agriculture)

4 – 9 An example of the consequence of misuse. Land use without regard for water runoff control can cause gullying such as this even on gentle slopes. (Department of Agriculture)

phorus, potassium (the major elements), sulphur, magnesium, and calcium (secondary elements) are most commonly deficient, although the absence of a number of minor elements such as boron, zinc, and copper is restrictive in some areas. Fertilization, therefore, has become necessary.

SOIL EROSION. Soil erosion is the most serious form of cropland deterioration. Unfortunately, through improper selection of land uses and improper cultural practices it has taken a significant toll. It is to be understood that erosion is a natural process, but American farmers have upset the natural balance and hastened the process. Water erosion is a leading problem, and the steeper the land and the higher the intensity and amount of yearly precipitation, the greater is the danger. Erosion can, however, progress on smoother land, sometimes almost undetected, and under low annual precipitation, especially if rainfall is of high intensity. In subhumid and semi-arid areas wind erosion becomes a major threat.

The magnitude of soil loss was not known until 1933, when the Soil Erosion Service was created. On the basis of reconnaissance surveys carried out by this agency, the Soil Conservation Service which superseded it estimated the soil erosion situation of the mid-1930's to be as presented in Table 4–6 and distributed as shown in Figure 4 – 11. In appraising these data, it seems probable that much of the fifty million acres ruined for cultivation were not originally

4 – 10 The same land shown in figure 4 – 9 after remedial work. This picture was taken three years later. The gully has been plowed, sloped 4:1, and strip-sodded to Bermuda grass. The pasture strip is regularly fertilized and the excellent vegetative cover now prevents further gully development. (Department of Agriculture)

TABLE 4–6
ESTIMATED EXTENT OF EROSION DAMAGE, MID–1930's*

Condition of erosion	Acres
Damaged cropland	
Essentially ruined for cultivation	50,000,000
Severely damaged	50,000,000
One-half or more of top soil gone	100,000,000
Slight to no erosion	100,000,000
TOTAL	300,000,000
Damaged land of all types (cropland, grazing land, forest land, etc.)	
Severely damaged to essentially ruined	282,000,000
Erosion beginning to be moderately damaging	775,000,000
TOTAL	1,057,000,000
Land not then damaged (primarily forest, swamp, marsh, etc.)	702,000,000
Damage not then defined (primarily deserts, mountains, etc.)	145,000,000
THEN TOTAL LAND AREA	1,904,000,000

*Based on Soil Conservation Service Data.

suited for this purpose. Nevertheless, the data do indicate that a major share of the lands of the nation suited to cropping are susceptible to erosion, and that only through care in management and use can their future productivity be assured.

The National Inventory of Soil and Water Conservation Needs revealed that several cropland-deteriorating problems were not being adequately treated in 1958 (see Figure 4 – 12).

DRAINAGE PROBLEMS. Too much water on the land and in the soil can place serious limitations on agricultural use, interfering with land preparation, tillage, the development of plants, and harvest operations. Drainage requirements, as they influence crop growth and response, are determined by effects of excess water on aeration, soil temperature, biological activity, structural stability, soil chemistry, and the over-all problems of land and crop management. Much of the cropland is naturally provided with adequate drainage; however, in parts of the Midwest, Lake States, Mississippi Valley, Coastal Plain, and many valleys of the West, artificial

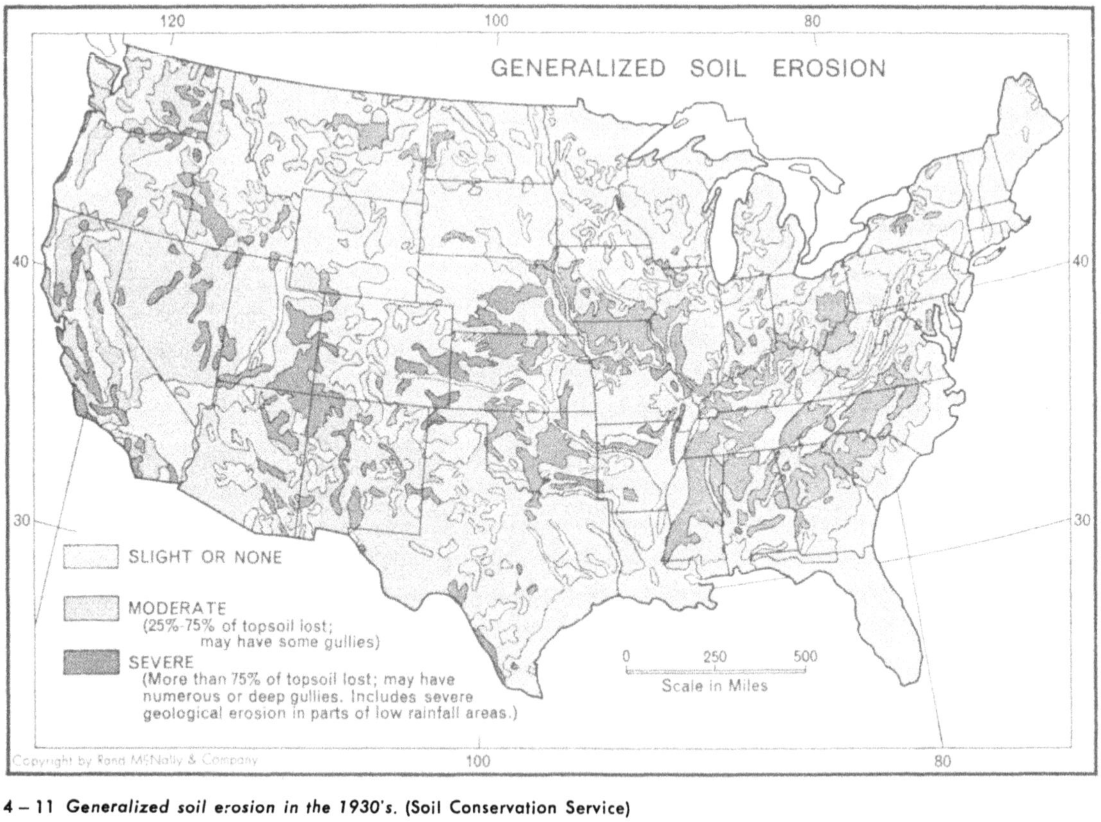

4 – 11 ***Generalized soil erosion in the 1930's.*** **(Soil Conservation Service)**

means of removing excess water by tile, ditches, and improvement of stream channels have been necessary. Many drainage projects have been developed in these areas (Figure 4 – 13). These have materially improved the cropland base. Some projects were poorly planned, and others are inadequately maintained; thus the full potential of drainage is not being realized. In a few instances drainage is not an implement of permanent agricultural land use; such is the case in the organic soils of the Everglades extending southward from Lake Okeechobee in southern Florida. Here subsidence of the land following drainage has increased the cost of development and indicated that the peaty soils can be farmed profitably for only a limited time. Drainage of undeveloped peat in some cases has permitted destruction of soil by uncontrolled fire. Moreover, because of the way drainage canals operate, at times drainage of

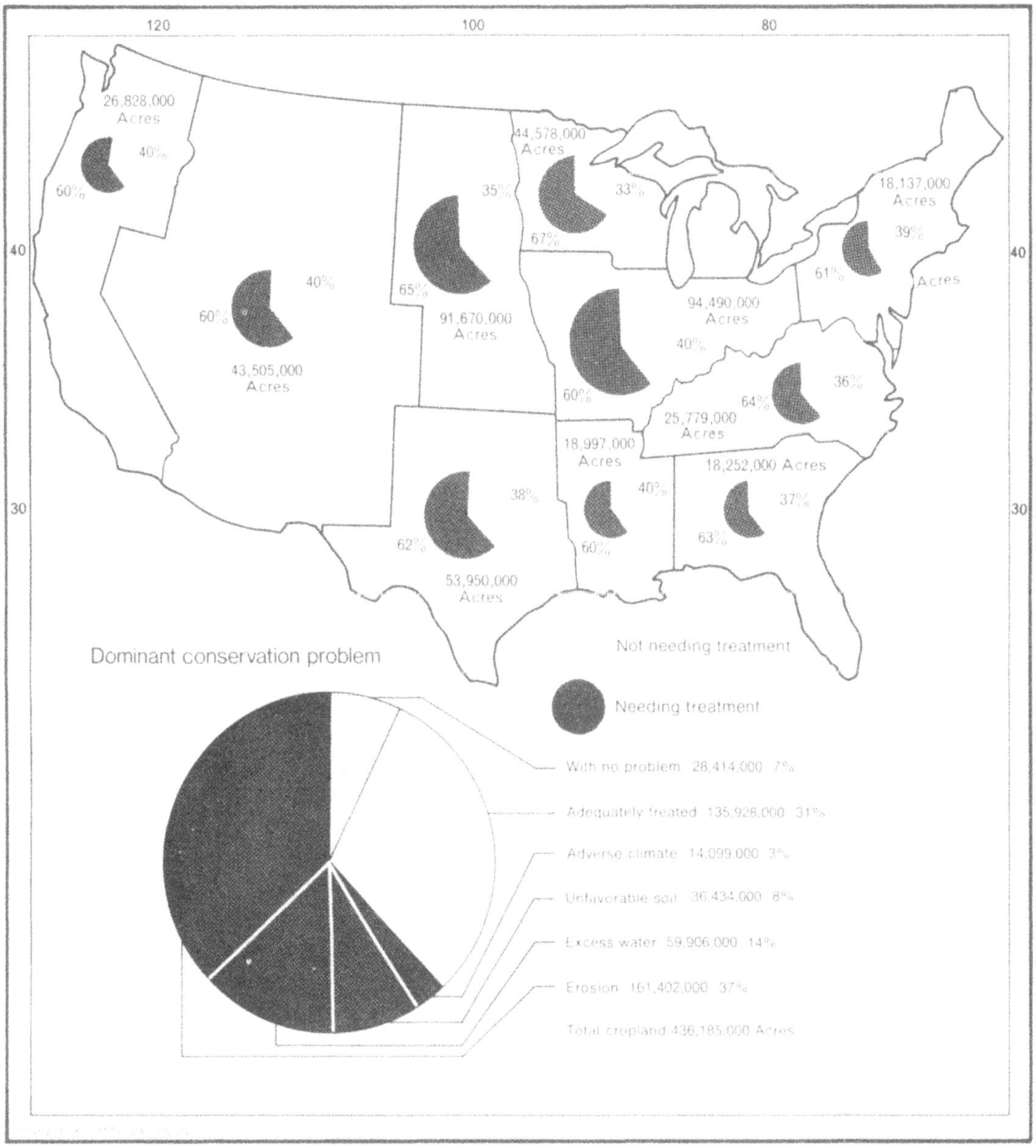

4 – 12 ***Conservation needs on croplands.*** **(Department of Agriculture)**

some farm lands involves flooding of others or drainage of some involves drought for others. Potential irrigation water is thus wasted. It also appears that extensive and indiscriminate drainage theatens injury to water supplies of municipal centers along the coast by reducing the quantity of fresh water and permitting sea water to enter city wells.[9]

IRRIGATION AND ATTENDANT PROBLEMS. Irrigation, properly planned, developed, and maintained, like drainage, is a means of improving land and upgrading use. Since World War II, the development of improved irrigation equipment (particularly portable, lightweight pipes and sprinklers), along with increased knowledge of the moisture needs of plants, greater use of fertilizers, and relatively high farm incomes (and perhaps acreage allotments) have encouraged investments in equipment and considerable development in supplemental irrigation as a means of assurance against drought periods, and have thus increased yields.

Nearly thirty-five million acres of farm land are now being irrigated. Study of Figure 4 – 14 indicates that the major distribution is in the West and in areas where total irrigation is essential. In 1959 about thirty-one million acres were being irrigated in the seventeen western states. Most of the acreage irrigated in the western states is done on the basis of projects involving dams and canals and several hundred to several thousand farmers. There are, however, many individual and small cooperative operations drawing from surface sources or pumping from wells; particularly noteworthy in this respect is the development on the High Plains of Texas which is based upon the individual farmer pumping from wells. Outside of the rice-producing areas of Louisiana and Arkansas and the horticultural areas on the sandy soils and in the Everglades of Florida, there was little development in the humid East before 1945. Crops and pasture yield at least reasonably well under normal precipitation. Individual development prevails in the supplemental irrigation practiced in the humid East as well as the humid section of the Pacific Northwest, major

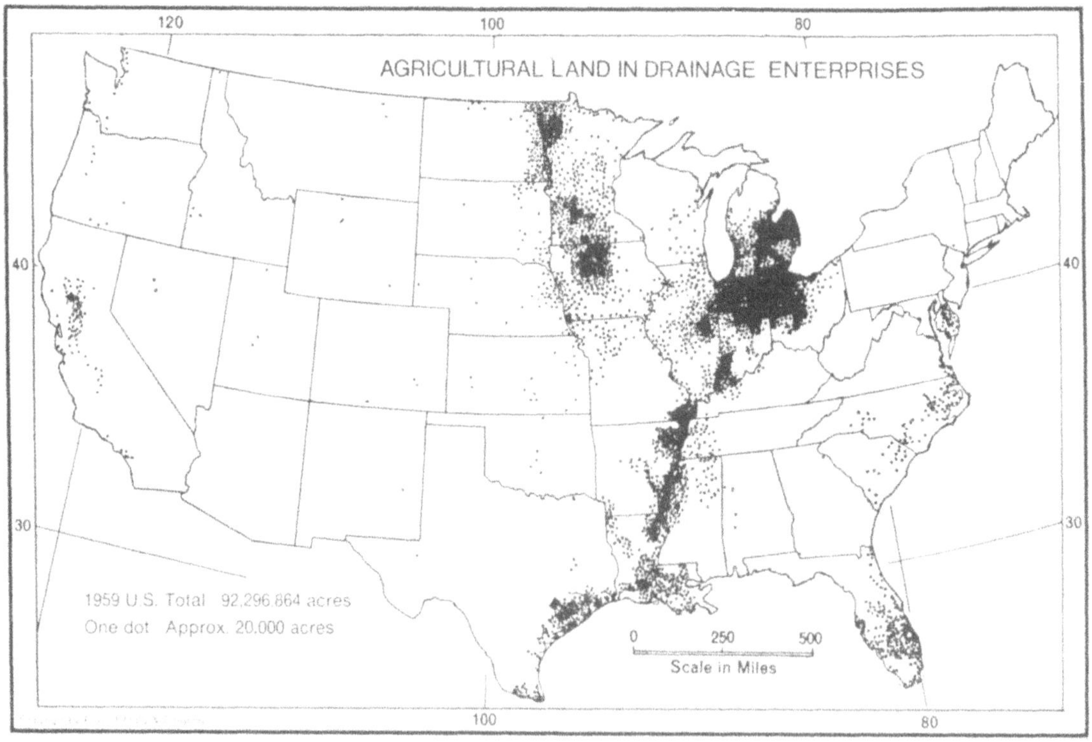

4 – 13 *Agricultural land in drainage enterprise.* (Department of Agriculture)

exceptions being in the rice areas of the South and the Everglades.

The irrigation enterprises have added materially to the agricultural land base of the United States. They are of greater significance than their area alone would indicate. In the western states the value of crops harvested from irrigated land normally amounts to one-third or more of the value of crops harvested.

Unfortunately, several types of land deterioration have been associated with irrigation development. Some result from poor planning and development, some from improper maintenance, and some from wasteful and careless application of water. The consequence, in these cases, has been soil erosion damage, saline or alkaline accumulation, water-logging, or undue loss of water. For example, about seven million acres of irrigated land in the seventeen western states are salt-affected, and thus materially reduced in productive capacity.[10]

LAND DETERIORATION THROUGH WATER DEPLETION. In a number of areas irrigation is based upon an inadequate permanent water supply. This is particularly true where the development is dependent upon ground water. In recent years widespread attention has been given to the decline of water tables. The most noteworthy example is the High Plains of Texas. The water source for irrigation in the Texas High Plains is the Ogallala formation, a combination of wind and water deposits of sand, gravel, clay, and silt laid down many thousands of years ago. Originally this formation extended westward to the mountains in New Mexico and eastward into Central Texas (Figure 4 – 15). The present area (22,400,000 acres), which has been reduced by erosion, now stands almost like an island, bounded by escarpments on the west and east and separated by the Canadian River which is entrenched in the older rocks. Hence the Ogallala formation (the "water sands") is now cut off from the replenishing water supply of the Rocky Mountains and is primarily holding water that has been in storage several thousand years. The thickness of the formation ranges from a few feet up to six hundred feet and is from sixty feet to five hundred to six hundred feet from the surface. This is the reser-

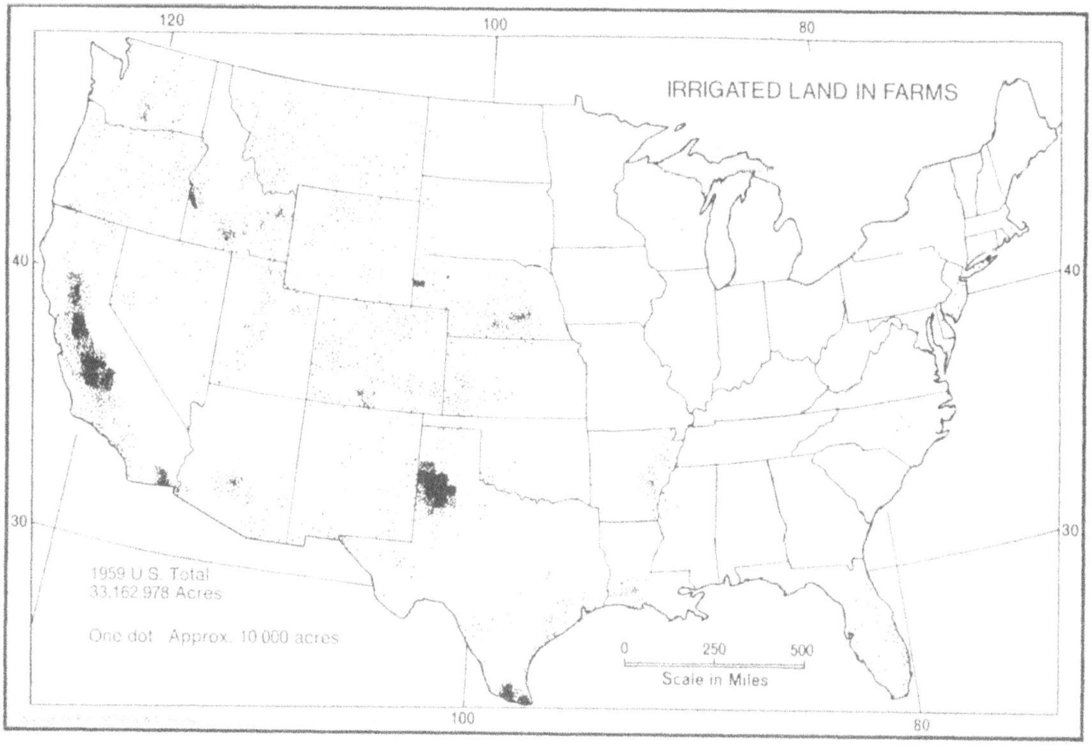

4 – 14 *Irrigated land in farms.* (Bureau of the Census)

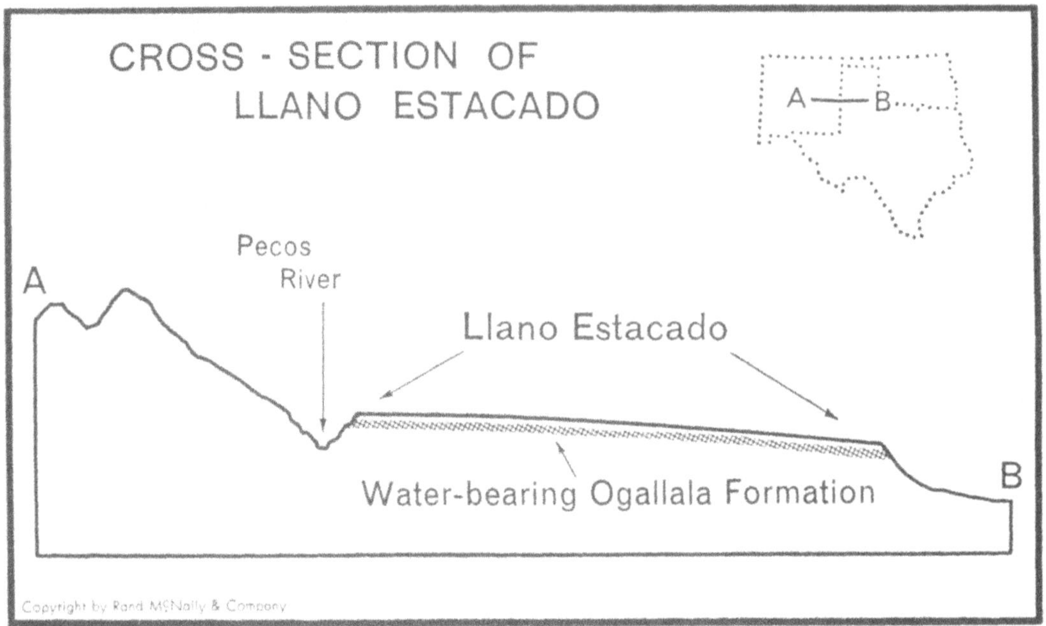

4 – 15 A schematic representation of a cross section of the Texas High Plains.

voir. Natural recharge is limited, and runoff from the area is slight. Annual precipitation amounts to sixteen to twenty inches, much of which falls in sudden, heavy showers, and the evaporation rate is up to eighty inches per year. Although the soil is receptive to water penetration, two factors are limiting: the high intensity nature of the precipitation, and the presence of an impervious clay or *caliche* subsoil. Most of the surface water, estimated to be 1.4 million acre-feet annually, collects in shallow depressions (*playa* lakes) scattered through the area (Figure 4 – 16). There are a total of about thirty-seven thousand of these lakes. A typical playa lake is about 1,100 feet in diameter, contains an average of about twenty-two acres, and collects about thirty-eight acre-feet of water annually, essentially all of which is lost to evaporation. Before the development of irrigation, the Ogallala formation was essentially in balance, with intake about equal to discharge on its exposed eastern margin.

4 – 16 *The High Plains of Texas.* This is an air view of the area following a rainy period. Water collecting in a playa is limited from infiltration by an impervious subsoil. The artificial recharge of this water into the Ogallola formation by means of wells is one of the possibilities for improving the basic supplies. Some of these depressions hold water into the summer and may thus be used for direct pumping for irrigation. (High Plains Underground Water Conservation District Number 1)

In this physical setting, agriculture has evolved from range livestock activity in the late nineteenth century to dryland farming in the early decades of this century to extensive irrigation beginning in the 1930's (80,000 acres in 1936) and developing rapidly after 1945 (1,-496,000 acres in 1949). The possibilities of increasing yields and eliminating the problem of erratic precipitation, the availability of improved irrigation equipment, rising production costs, and acreage allotments on cotton have been contributing factors. By 1965 fifty thousand wells were tapping the stored water supply, estimated in the 1950's to be more than one hundred-fifty million acre-feet, at a rate on the order of six to seven million acre-feet annually to irrigate a little more than five million acres. From the irrigated acreage came about 11 per cent of the nation's cotton crop and 35 per cent of its grain sorghum.

Upon examination of the data concerning water in storage and yearly withdrawal, it is easily understood that irrigation on the High Plains of Texas is presently a mining operation with water diminishing in quantity and increasing in cost (Figure 4 – 17). The development, however, should not be censured on the basis of being a mining operation. It should, from the conservation point of view, be appraised on the basis of the efficiency of water use and the economic future of the area. Although progress is being made, water waste is common in present practices—tail waters are still uncontrolled on many farms and closed distribution systems (primarily underground concrete tile) were used on less than 50 per cent of the irrigated land in the early 1960's. The possibilities of artificial recharge have been studied but little is being done, and the same is true of water in the playas; thus an estimated 1.4 million acre-feet, nearly one-third of the amount presently used annually, is being lost to evaporation. Another factor in water waste pertains to a general decline in the organic and fertility level of the soil on many farms. More water is required in such cases than when the systems were first developed.

If the present practices and expansion are continued, the short-range effect will be reduced well capacity and increased irrigation costs. The long-range effect will be a depleted water supply and a major reduction in the agricultural supporting capacity of the area. Present estimates are that the area will "peak out" at a little less than six million acres. Gradual reduction will then ensue until by 2020 only

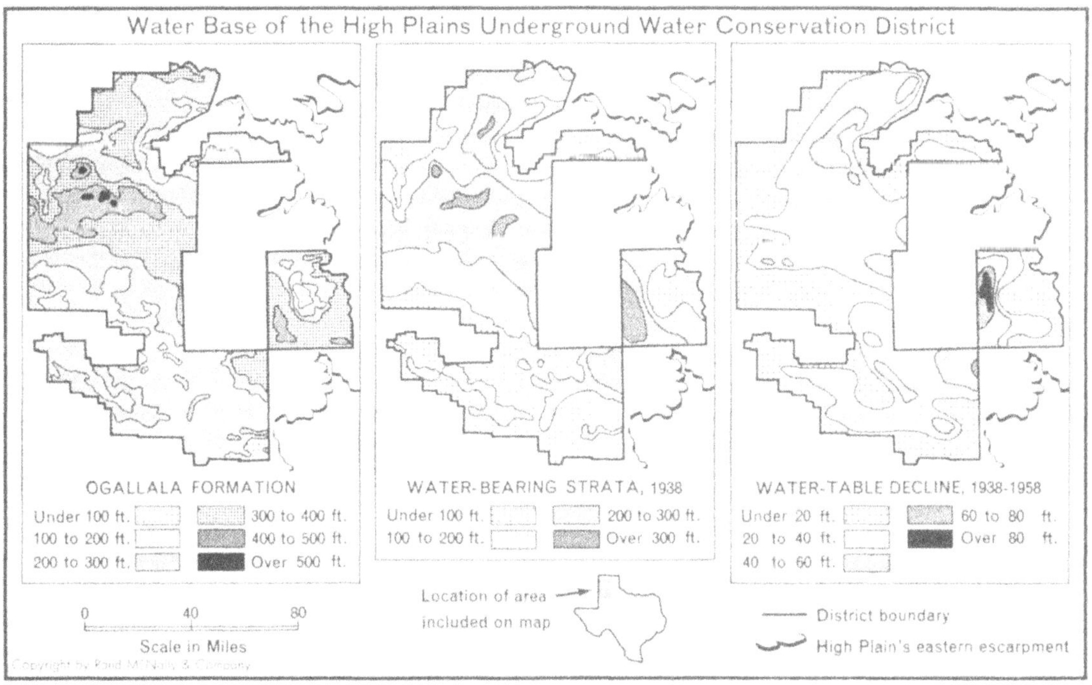

4 – 17 ***Water supply and decline on the High Plains of Texas.*** **The map on the left shows the thickness of the formation; the middle map shows the thickness of the water-bearing portion in 1938; the map on the right shows the thickness of the water-bearing portion in 1963. (After High Plains Underground Water Conservation District Number 1)**

about 2.2 million acres will be supportable with ground water. Farmers, in general, are aware of the problems. The Underground Water Conservation District, to which many belong, has done much to educate and establish group effort. A few are beginning to establish recharge facilities. State and Federal agencies also are exerting some efforts, mainly through experimentation in water-use efficiency, recharge techniques, and basic data accumulation. It is clear, however, that with all efficiencies the present level of irrigation cannot be maintained, without water importation.[11]

CROPLAND DEPLETION THROUGH ALIENATION OF USE

In the introduction to this chapter, mention was made of the fact that some croplands have been and are now being diverted to other uses. This is not necessarily undesirable. For example, in recent years, with the increase in grass farming and livestock, considerable cropland has been diverted to improved pasture, especially in the South and the Tobacco and General Farming Region. In many instances such land has some limitations which require special care for crop production: steep slopes, shallow soil, wetness, or low fertility. Because the land remains in farm ownership and is actually undergoing improvement, it can be diverted back to crop production when the need arises. In the same regions, especially on the Piedmont, there are several million acres that have been allowed to return to low-quality pasture or even brush. Such land is being protected by cover from physical deterioration; however, an important share is in ownerships too small to provide an economic unit under existing or foreseeable national agricultural conditions and therefore is in danger of being permanently alienated to industrial, residential, or forest uses.

With the upswing of the forest industries of the South, a good deal of land, including some quality agricultural land, has been planted in pine trees. A number of large timber corporations have had active land buying and planting programs in progress for several years with the goal of developing sustained yield operations. It may be difficult to return these former croplands to agriculture when the national economy requires them because corporation economics do not necessarily react quickly to national conditions, and timber production may be more profitable and equally needed. Loss of farm land to strip mining operations is also notable, although less widespread. This deterioration is discussed in Chapter 16.

The changes which have come about in this century requiring a major share of the working force to be engaged in industry, commerce, and service activities have brought major shifts in the settlement pattern of the nation. The country has been rapidly becoming a nation of urban dwellers. More and more space is being required for cities and towns, industrial and commercial sites, and transportation and recreation facilities. In view of the portion of the population involved, this is justifiable; however, much of this development is at the expense of quality cropland. The point of concern is that, in many cases, alternative lands of lower productive utility could have been diverted to these uses with prior study and planning. This topic is considered at greater length in Chapter 18.

CROPLAND IN RELATION TO OTHER RESOURCES

Despite the fact that croplands rate priority consideration in the development and use of the nation's land resources, they should be considered in terms of the total complex. Development and use of croplands should minimize despoilment or depletion of other resources. Any reduction in the capacity of other resources should be based upon perpetual, or at least long-range, values to society. These have not been, and in many cases are not now, guiding principles of the farmer or agricultural development agencies.

Water and wildlife may be cited as two major examples in which agriculture has played a role in unnecessary resource depletion. Water is perhaps the most complex of all resources, being intimately interrelated with most other resources. Water control and development require a delicate and harmonious balancing of all hydrologic, economic, and human influences within watersheds. Through cultivation and cropping practices which have had little regard for water control, American farmers have reduced the natural soil storage capacity of many watersheds. This has contributed to flood flow of streams, adding to the danger of economic developments on their floodplains, and at the same time it has reduced low season flow and thereby reduced water supply. Poor practices have contributed unduly to the sediment load of streams, and impaired water quality for power, industry, domestic, fish, recreation, and other uses. At the same time, of course, speeded and uncontrolled runoff causes soil depletion and sediment problems on the farmlands.

Damage to wildlife has come primarily through reduction in habitat (cover, feed, and water). A considerable amount of this is to be expected with agricultural use of the land. Nevertheless, some agricultural use has been poorly conceived and not even in the interest of desirable land use. Conservation farming, employing practices of grassed waterways, grassed or wooded slopes, drainage ditches, and farm ponds is compatible with the support of more wildlife on farms.

The Evolution of Programs and Organizations for Cropland Conservation

Cropland conservation, as an overt aim, received little attention in the United States until the 1930's. Abundance of land, low valuation, and the lack of inventories of supplies and of scientifically-based practices were major deterrents. It should be noted again, however, that there have always been some individuals who were interested in farming techniques which would permit a permanent agricultural adjustment.[12] Considerable private experimentation was carried on in the latter part of the eighteenth century, and before the Civil War a number of agricultural societies were formed from time to time to press for local improvement in farming practices.[13] Gains in this interest, nevertheless, remained restricted to individual farms and small areas. Cropland depletion, in general, continued.

In the early 1930's several important influences came to bear upon the problem to begin a period of rapid improvement. Noteworthy among these were the coincidental occurrence of the depression, drought in the Great Plains, and the general plight of many farmers resulting from an accumulation of agricultural problems including marketing, improving science and technology, and their impact upon the economics of farm size and land deterioration. The depression focused attention upon the need for internal improvement and brought the Federal government firmly into leadership in ameliorating the situation. Soil erosion as a menace to agriculture was formally recognized in 1929 when the Buchanan Amendment to the Agricultural Appropriation Bill for the fiscal year 1930 was adopted by Congress. This amendment provided $160,000 to be used by the Secretary of Agriculture in conducting soil erosion investigations. During the same year, regional soil erosion experiment stations were set up in carefully selected erosion and agricultural-type areas. The Agricultural Act for the fiscal year 1932 contained an appropriation ($333,000) specifically to be used in enlarging the activities of the ten established erosion and water conservation research stations. New information with respect to both the erosion process and methods of control was rapidly accumulated. Further impetus was given to erosion studies when, in June, 1933, Congress passed the National Recovery Act which provided (among many other features) for erosion control work as a means to unemployment relief. In September, 1933, the Soil Erosion Service was established as a temporary agency in the Department of the Interior to carry out the provisions of the National Recovery Act. H. H. Bennett was appointed Director of the Service, and during approximately eighteen months of operation in the Department of the Interior an extensive demonstration program was put into operation throughout the country. Forty-one soil and water conservation demonstration projects were established, and about fifty Civilian Conservation Corps camps were assigned to erosion control work under the supervision of the Soil Erosion Service. In March, 1935, the service was transferred to the Department of Agriculture and expanded to consolidate similar activities being carried on by other agricultural agencies.[14]

As a direct consequence of the work and new knowledge brought to light, two Federal agencies were established that were soon to become major forces in markedly improving management and use of the nation's croplands. These were the Soil Conservation Service and the Agricultural Conservation Program Service.

THE SOIL CONSERVATION SERVICE

As the work of the Soil Erosion Service progressed, several Congressional committees were considering legislation to establish a permanent Federal agency for soil erosion control. On April 27, 1935, the Soil Conservation Act of 1935 passed both Houses without a dissenting vote. This law specifically established within the Department of Agriculture a "soil conservation service" for the development and prosecution of a long-term program of soil and water conservation. Accordingly, the Soil Erosion Service became the Soil Conservation Service with status as a regular bureau of the Department. Through the years of its existence, its program has been broadened and realigned from time to time.

At present the Service's major responsibilities may be summarized as follows: (1) The principal duty is to assist farmers and ranchers in locally-organized, farmer-directed soil conservation districts through its planning technicians and other soil and water specialists who live and work in the districts. (2) In the carrying out of this duty, it has become responsible for leadership in the national soil survey in cooperation with the state experiment stations. (3) It is responsible for administering the flood control and watershed and river investigation activities of the Department of Agriculture. (4) It provides on-site assistance in farm drainage and irrigation for farm groups, supervises the agricultural phase of the water utilization program in the western states under the Case-Wheeler Act of 1939, and makes and coordinates snow surveys for water forecasting in the western states. (5) It has the responsibility for assisting in the national, state, and county Agricultural Conservation Program of the Department of Agriculture, contributing mainly by technical assistance to farmers participating.[15]

The over-all program is directed by the Administrator and his staff from the central office of the Department of Agriculture in Washington, D. C. State offices are maintained in each state which serve as headquarters for a state conservationist, a conservation engineer, a soil scientist, and a soil conservationist who, with their staffs, carry out administrative and technical functions to provide service to field personnel in about 300 area offices. Each of the area offices supervises several of the more than three thousand work units. Each work unit is staffed by a professional conservationist and one or more aides who work directly with the farmers and ranchers. The work units are thus the operational divisions of the Soil Conservation Service. They may or may not coincide with the districts, efficiency of technical assistance being the deciding factor.

SOIL CONSERVATION DISTRICTS. Soil Conservation Districts are legal subdivisions of the states. They are set up by farmer initiative, organized by farmer vote, and administered by farmer supervisors who serve without pay. They are operated under state law, administered by a state committee, and are without power to tax, levy bonds, or make assessments. They are empowered to request and receive help from public agencies and from private sources and are authorized to furnish assistance only to farmers who apply for it. Farmer-members have no responsibility other than to pay for and carry out the practices they decide are desirable for their farm programs. Technicians are paid by their respective agencies.

As of July 1, 1966, there were 2,977 Soil Conservation Districts in the United States (plus seventeen in Puerto Rico and one in the Virgin Islands). These contained 1,072,478,518 acres of land, which amounted to about 95.5 per cent of the total in farms and ranches, and had 2,080,231 farms and ranches cooperating. Twenty-seven states were completely covered by districts at that time.[16]

Location within a district does not necessarily mean that a farmer is a cooperator. The landowner must make a request. In assisting the member farmers and ranchers with their soil and water conservation programs, Soil Conservation Service Work Unit technicians go through four principal steps. First comes an acre-by-acre soil survey to determine capability. Second, a technician and the landowner together draw up a conservation farm plan. This provides for various alternatives in treatment and use of the land and is designed to enable the operator to proceed with the most needed measures immediately. The third step involves the applications of practices called for by the plan with the soil conservationist giving necessary technical guidance. The farmer may do the work himself; he may hire any special equipment required or obtain it through his district. The fourth step is the continuing job of maintaining the farm conservation system after it has been applied to the farm. The Soil Conservation Service provides the necessary guidance and aids with the establishment of refinements as new research provides for better crops and methods (Figures 4 – 18 and 4 – 19).

By July, 1966, a little more than 903,256,416 acres had been covered by detailed soil surveys. The scope of the total program is indicated by the summary presented in Table 4–7 for the 1965 fiscal year.[17] Because the guiding philosophy behind the Soil Conservation program is to place each parcel of land in its highest order of use commensurate with the perpetual maintenance or improvement of capability, it can be understood that this work is sound and progress had been rapid.

THE AGRICULTURAL CONSERVATION PROGRAM

The second element in the front line of the United States Department of Agriculture's conservation effort is the Agricultural Stabilization and Conservation Service. In contrast to the technical assistance program of the Soil Conservation Service, this organization aids the farmers financially. In states, counties, and communities, responsibility for program development and operation has been assigned to Agricultural Stabilization and Conservation Committees in cooperation with representatives

4 – 18 ***Farming without regard for soil and water conservation.*** **This photograph of a 320-acre Nebraska farm was taken in 1940. Gullies and evidences of sheet erosion are apparent. (Soil Conservation Service)**

of the land grant colleges, Soil Conservation Service, Forest Service, and other agricultural agencies.

Operating as a government-farmer partnership, the program shares with farmers and ranchers the cost of carrying out soil and water conservation measures intended to (1) protect farm and ranch lands from water and wind erosion; (2) improve the productivity of the nation's agricultural resources; and (3) protect and improve the source, flow, and use of water for agricultural purposes. Assistance is in the form of materials, services, and financial aid, and averages about 50 per cent of the costs.

All farmers are eligible to participate. Practices are prescribed and administered by the locally-elected farmer committees; thus they are geared to the needs of the farm and local area. A limitation of $2,500 is imposed per person establishing approved practices; however if two or more persons form a pooling agreement to serve the group, the maximum is $10,000 for each person. Usually such agreements are approved for structures required for irrigation or drainage. In recent years the national expenditure for the program has been around 200 million dollars annually.[18] Between June, 1961 and June, 1966, 2,259,577 farms participated at least once in a program; in the 1965–66 fiscal year, 1,034,496 farms participated.

It is significant to note that in this program the government has recognized some public financial responsibility in assuring the future of the agricultural resources of the nation. The basic purpose is to get more conservation practices established on farms and ranches than could be expected if operators were dependent entirely upon their own financial position.

RELATED PROGRAMS. The Agricultural Stabilization and Conservation Service administers a number of related programs. Some are aimed at production adjustment and some at the improvement of land and water management. Most noteworthy in the first instance is the commodity acreage allotment and price support program, intended to stabilize the agricultural economy by adjusting production to consumption levels in basic commodities prescribed by Congress. The conservational bene-

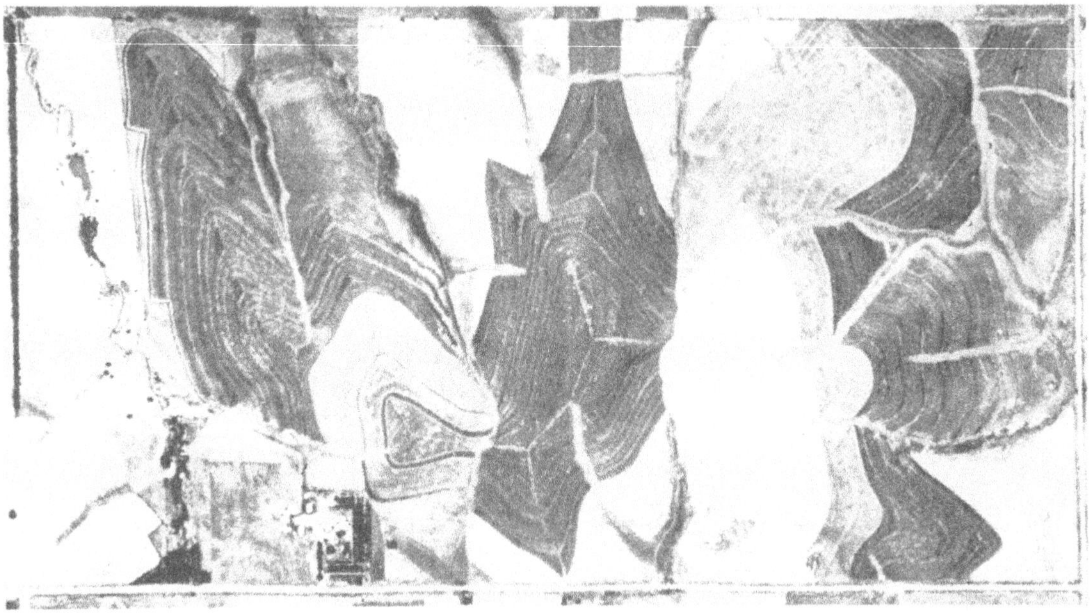

4 – 19 *Farming with regard for soil and water conservation.* This photograph shows the same farm as figure 4 – 18 as it now appears. Land use has been adjusted to land capability, and farming techniques such as contour planting and cultivating, terracing, and tie-ins to grassed waterways afford maximum protection of the soil and reduce surface water runoff. This farmer is not only protecting and improving his resource base, he has increased his land value and his income. (Soil Conservation Service)

fits are primarily indirect—some lands of low capability, for example, have been diverted into permanent cover—and, in fact, the program has resulted in adverse effects in some instances. A case in point is the rapid development of irrigation on the High Plains of Texas; in the face of allotments on cotton acreages, and the need to increase the size of their operations, the farmers have employed irrigation as a means of increasing per-acre yields. It will be recalled that irrigation in that area is based upon a water mining operation.

Between 1956 and 1960 the letting of contracts for the Soil Bank Program was admin-

TABLE 4–7

CONSERVATION PROGRESS, FISCAL YEAR 1965*

SOIL CONSERVATION DISTRICTS	Fiscal Year	To Date
Soil conservation districts (net increase)		
Number	18	2,989
Acres	14,073,002	1,753,281,000
Farms and ranches in districts		
Number	–	3,695,912
Land in farms and ranches		
Acres	–	1,065,945,137
Cooperators		
Number	113,148	2,041,182
Acres	39,220,340	664,563,511
Basic conservation plans		
Number	105,010	1,570,822
Acres	35,920,010	497,341,578
Land owners and operators assisted		
Number	1,139,487	–
Land owners and operators applying practices		
Number	726,519	–
Consultive services provided		
Number	94,022	–
Basic data for use in comprehensive plans		
Acres	3,367,607	–
Group project plans prepared		
Number	2,965	33,371
Acres	1,403,000	24,559,408
Land owners in groups		
Number	16,309	236,900
Soil surveys		
Acres	57,256,734	862,118,407
GREAT PLAINS CONSERVATION PROGRAM		
Contracts signed		
Number	3,739	19,956
Acres	5,290,996	39,227,743
Contracts terminated		
Number	1,507	4,824
Acres	2,171,484	7,426,735
Planned cropland conversion		
Acres	187,248	1,299,568
AGRICULTURAL CONSERVATION PROGRAM	Program Year	
Cost-sharing referrals received (1964)		
Number	–	359,546
Cost-sharing referrals serviced (1964)		
Number	–	349,019

TABLE 4-7—Continued
CONSERVATION PROGRESS, FISCAL YEAR 1965*

CONSERVATION PRACTICES APPLIED		Fiscal Year	On the land
Conservation cropping systems	acres	21,844,577	147,504,415
Contour farming	acres	4,525,330	41,717,487
Cover and green manure crops	acres	3,906,328	25,436,975
Crop residue use	acres	16,515,514	106,550,740
Debris basins	number	2,941	79,959
Range deferred grazing	acres	14,909,220	48,185,135
Diversions	miles	2,928	88,361
Farm ponds	number	56,104	1,420,733
Farmstead windbreaks	acres	42,468	637,243
Field windbreaks	miles	4,117	75,839
Floodwater diversions	feet	16,664	1,046,154
Floodwater retarding structures	number	661	7,423
Floodways	miles	28	296
Grade stabilization structures	number	7,292	136,852
Grasses and legumes in rotation	acres	1,568,466	17,821,663
Grassed waterways	acres	77,469	1,706,889
Irrigation water management	acres	2,659,237	10,900,269
Minimum tillage	acres	1,074,354	6,619,449
Pasture and hayland planting	acres	2,307,519	44,382,787
Range proper use	acres	64,643,787	179,723,521
Range seeding	acres	512,202	11,857,838
Recreation area stabilization	acres	1,969	16,548
Recreation area planting	acres	10,554	68,783
Stripcropping	acres	711,971	20,080,798
Terraces	miles	33,922	1,225,558
Tree planting	acres	310,317	11,837,976
Wildlife wetland development	acres	69,325	730,146
Wildlife habitat development	acres	238,148	2,794,598
LAND TREATED			
Land adequately treated	acres	22,692,192	209,309,189
LAND USE CONVERSION[1]			
Cropland to grassland	acres	1,887,901	10,108,893
Cropland to woodland	acres	123,640	1,169,392
Cropland to wildlife-recreation	acres	71,143	507,708
Cropland to "other"	acres	90,870	2,037,673
All other uses to cropland	acres	386,704	2,741,894
All except cropland to wildlife-recreation	acres	213,059	1,633,207

* "Conservation Highlights, 1965" (Washington: Department of Agriculture, Soil Conservation Service, 1966), 6 pp.

[1] Conversions on the land June 30, 1965, are estimates of conversions since July 1, 1962.

istered through the Agricultural Stabilization and Conservation Service. This program was designed to help farmers make a direct attack on crop surpluses, to retire and build up land not presently needed for crops, and to protect farm net income while these changes were being made. There were two programs: the acreage reserve and the conservation reserve. The former offered producers of "basic" crops an opportunity to reduce current production without loss of net income through payments. Farmers with acreage allotments were eligible to participate. Agreements were for one year. All farmers were eligible to participate in the conservation reserve. Under contracts running for three, five, or ten years for grassland and ten or more years for woodland, a farmer diverted cropland out of production and established these covers plus approved conservation practices, including building dams, pits, or ponds for impounding water and the improvement of wildlife conditions. Contracts allowed up to 80 per cent of the cost of establishing the practices and yearly payments.[19] Under this program about twenty-eight million acres were diverted from cropping.

Two similar programs have been authorized more recently: the cropland conversion program and the cropland adjustment program. The former was authorized by the Food and Agricultural Act of 1962; to date (1968) it has been funded only on a pilot basis in selected counties throughout the country. Its present purpose is to test "new methods for improving family farm income by converting land regularly used for the production of crops to other economic uses and thereby reducing the intensity of use of some of our nation's cropland." Agreements are made with farm and ranch owners to convert land regularly used for crops to pasture, a forest tree plantation, wildlife habitat, or an outdoor recreation facility. The program is administered by the County Agricultural Stabilization and Conservation Committees on the basis of farm conservation plans developed in cooperation with the local Soil Conservation District. For taking part in the program, cooperators receive two kinds of payments: an adjustment payment on land representing a reduction in row crops or small grains, and conservation practice cost-share payments on all land involved in the program. Under the agreements no new cropland can be developed on the participating farms. The agreement periods are five years.[20]

The cropland adjustment program, authorized by the Food and Agricultural Act of 1965, helps farmers to divert cropland to protective conservation uses under long term agreements. Under it, land may be taken out of production for periods of five to ten years. Participants receive adjustment payments calculated as a portion of the value of the crops which would otherwise be produced on the land (on price-supported crops this amounts to about 40 per cent of the county price-support loan rate; on other crops a fixed rate is paid per unit of established yield—hence acres times yield per acre times rate equals payments). Farmers also are eligible for conservation cost-share payments on diverted land (for establishment of vegetative cover and wildlife habitat, or for special practices to preserve open space and natural beauty, or for other measures to prevent air and water pollution), and may receive additional per-acre payment if they agree to permit free public access for fishing, hunting, hiking, and trapping. To avoid adverse effect upon the local economy of a county or trade area, the total land placed under agreement is limited to not over 10 per cent of the total cropland or acreage of any one allotment crop or base feed grain in any one year. Associated with this program is a provision to aid state and other government agencies to acquire cropland for non-farm uses such as the preservation of open space, natural beauty, air and water purity, and for wildlife and recreation development. Aid is also given to these agencies to establish conservation practices on land acquired; under this provision (called the Greenspan) the Federal Government may share up to 50 per cent of the cost of the land.[21]

Two regional programs are also under the administration of the Agricultural Stabilization and Conservation Service: The Great Plains Conservation Program, authorized in 1956; and the Appalachian Land Stabilization and Conservation Program, authorized in 1966. The Great Plains program applies to counties in ten states and provides a means of assistance for minimizing the hazards of farming and ranching brought on by erratic climate. It differs somewhat from other programs aimed at conserving soil and water: Operators must develop a complete long-range plan for needed changes in land use practices; they must agree to carry out the plan within ten years; cost sharing for the entire plan, ranging from 50 to 80 per cent of the cost, is earmarked when the agreement is signed, and payments are made at the completion of each step; technical assistance, by the Soil Conservation Service, is at an accelerated rate. The program covers changing unsuitable cropland to other uses, reseeding damaged or depleted range, developing livestock water, permanent fencing, control of competitive shrubs, construction of spreader ditches and dikes, planting tree windbreaks, establishment of wind and water erosion controls, and so forth. By the end of 1966, twenty thousand farmers and ranchers in four hundred counties had cooperated in the program, and the Federal Government had expended fifty-three million dollars and had a balance of twenty-seven million dollars unpaid cost-share obligations. Also, some forty million acres had been covered by complete soil and water conservation plans.[22]

The Appalachian Land Stabilization and Conservation Program applies to limited areas in designated counties in the twelve states of Appalachia (see Figure 4 – 20), and is designed to share with farm owners and operators the costs of certain soil and water conservation and development practices. Particular emphasis is given to needy farmers who are unable to bear the entire cost themselves. Contracts are made for three to ten years, and share costs are up to 80 per cent of the cost of performing the practice. Each farm is limited to cost-sharing on not more than fifty acres, and the amount usually cannot exceed $50 per acre. Every participant must furnish a conservation and development plan, in line with the general objectives of the program. Needed technical assistance is supplied by the Soil Conservation Service, except for forestry measures for which technical services are supplied by state forestry agencies.[23]

OTHER FAVORABLE DEVELOPMENTS

Since the 1930's several other influences have become significant in the advance of cropland conservation. Research agencies, both pub-

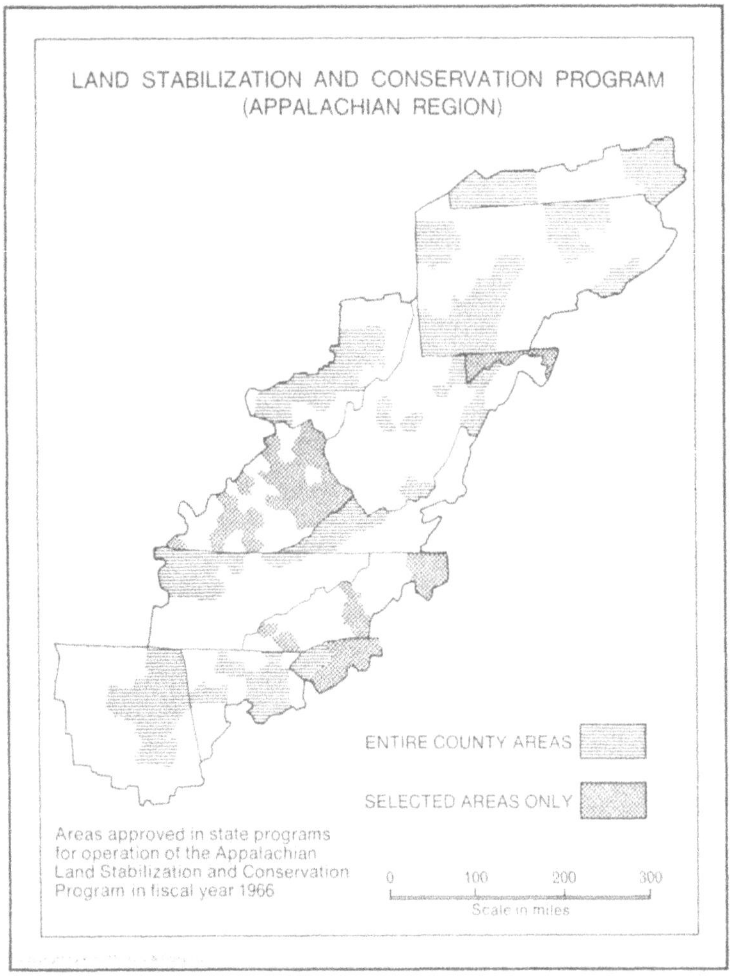

4 – 20 *Land stabilization and conservation program.* (Department of Agriculture)

lic and private, have put a great deal of effort into the improvement of crops, pasture, and range plants, and management and land treatment practices. The state agricultural experiment stations have played major roles in this work, and the Extension Service has vastly improved its farmer education program (Figure 4–21).

TVA. The Tennessee Valley Authority was created in May, 1933, as an independent corporate agency of the Federal government "to provide for the unified development of all resources of the watershed of the Tennessee River; to provide for reforestation and the proper use of marginal lands in the Tennessee Valley, to provide for the agricultural and industrial development of said Valley; and to provide for national defense."[24] It is concerned also with the economy and prospects of the area it serves with electric power (Figure 4–22).

The conservational aspects of the theory of TVA are notable in that it was the first major

4–21 Farmer education, a view of a farm demonstration field day. Such programs are conducted regularly by county agents, the Soil Conservation Service, and other public agencies and private groups throughout the nation. Seeing the actual values of desirable practices is one of the most efficient ways of inducing farmer acceptance. (Tennessee Valley Authority)

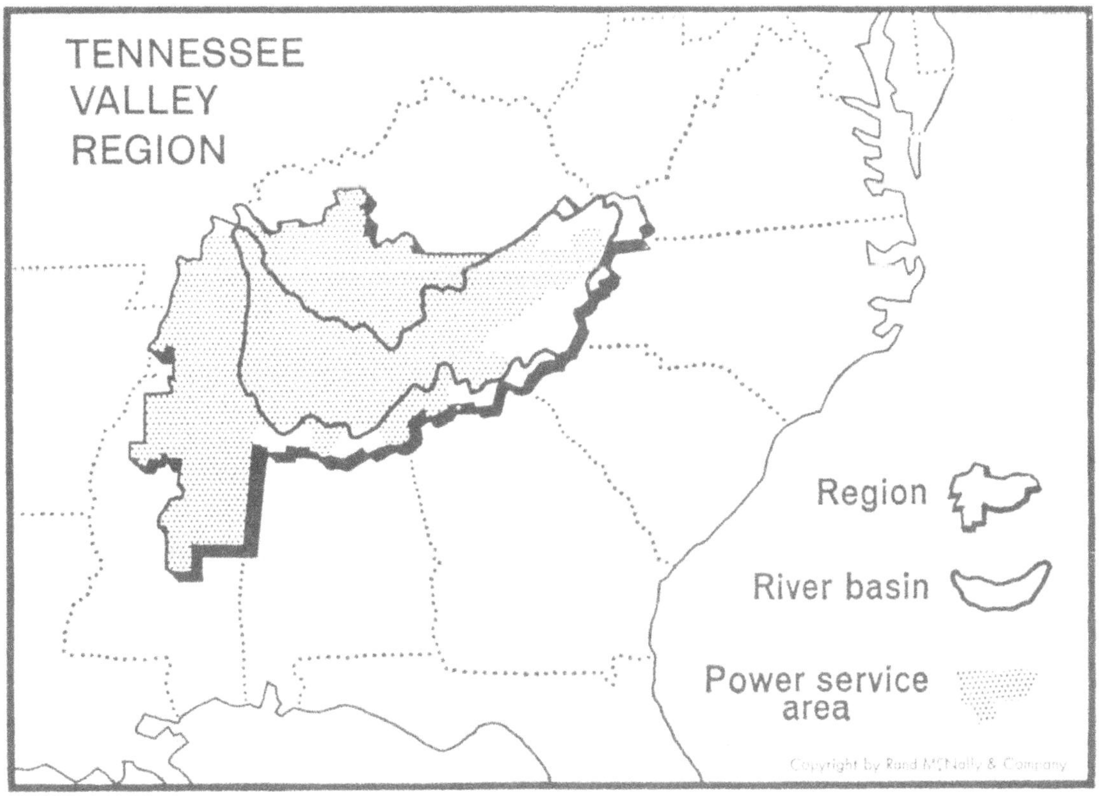

4–22 TVA, watershed and power service area. (Tennessee Valley Authority)

action program based on the recognition that a river basin is essentially a unit and should be developed as such. Despite the fact that criticisms have been and are leveled at the bureaucratic nature of its control, the conservational benefits of its integrated development program are praiseworthy.[25] Farmlands have received their share of attention, being gradually shifted toward soil and water conserving types of agriculture within a framework of an improved balance in both total land use and economy.

The work and accomplishments of TVA in fertilizer development and use deserve special mention because they have national agricultural significance. When World War I began, the United States relied on the vast natural deposits of Chile for its munition nitrates. German U-boats threatened to cut off this source of supply. Under the authority of the Defense Act of 1916, President Wilson authorized the building of two synthetic nitrate plants and a large dam at Muscle Shoals to supply the necessary electric power. Congress was aware of the fertilizer possibilities in the nitrate plants and wrote a provision in the Act for making fertilizer during peacetime. The facilities and responsibilities were turned over to TVA in 1933. Since then the fertilizer program has become an important part of the agricultural effort of TVA. Scientists now work in well-equipped laboratories at Muscle Shoals (Figure 4 – 23). Their labors are the starting point in a chain of research and testing which takes into account technical and economic production

4 – 23 *The Muscle Shoals fertilizer plant.* These properties, constructed during World War I, were assigned to TVA in 1933. The facilities have been modernized for the development of new fertilizers that will cost less and be more efficient in use. (Tennessee Valley Authority)

factors as well as the agricultural needs of every major farming area of the nation. The work is coordinated to fit in with soils and fertilizer programs of the Department of Agriculture.[26]

The fertilizer produced by TVA is used only for educational purposes among farmers and for experimentation by private industry. Once a fertilizer has been proven, TVA, working in close cooperation with the Extension Service, selects test-demonstration farms as an educational means of proving its value to farmers. Since the beginning of the program in 1935, more than seventy-five thousand farm families in forty states have participated. About thirty-three thousand of those have been "unit" farm demonstrations in which local soil associations selected one or more of their members to utilize the new fertilizers under plans worked out with their county agricultural agents. The remainder have been in "area" farm test-demonstrations in which entire rural communities have organized themselves in joint efforts of general farm and home improvement. For the most part, the latter group has been confined to the Tennessee Valley where TVA has special responsibility for watershed protection to conserve soil and water.[27] In 1966 there were 2,105 test-demonstration farms in twenty-nine states (about one-half the farms were in the Tennessee Valley states).[28]

The program thus has grass-roots ties. It has been successful in introducing new fertilizer, thereby extending fertilizer resources; in cheapening cost, thereby bringing fertilizer within economic range of more farmers; in promoting better fertilizer; and in improving products and processes.

AGRICULTURAL CREDIT. The improvement in agricultural credit has been another important development since 1930. This has enabled farmers to adopt production measures and conservation practices proven by research. Credit since that time has been generally available to all who own or can rent suitable farms and have reasonable ability to succeed in farming.

Two-thirds or more of all credit used by farmers is obtained from commercial banks, insurance companies, or private lenders. The second most important form of credit is the farmer-cooperative type of credit under the general supervision of the Farm Credit Administration. A third source is the Farmer's Home Administration, which makes loans to farmers who are unable to get satisfactory credit from other sources for purchase, enlargement, or improvement of farms. It also insures loans made by private lenders for such purposes. A fourth type of credit is available through the Rural Electrification Administration. This agency encourages the extension of adequate electric and telephone service throughout rural areas by means of long-term loans. Loans are made to local business enterprises, including both commercial and cooperative organizations, and to public bodies. Loans are repaid from the operating revenues of electric or telephone businesses.

Summary

Impressive gains have been made in cropland management during the past three decades. Government leadership and guidance have been firmly established, research has developed practices and procedures, land classification for conservation management purposes is advancing, farmer education has been improved, and credit and cost-sharing programs have been established. At the same time there has been a general improvement in the ability and outlook of the farmers with respect to conservation. The result has been the development of an important element of farmer leadership. The present practices and their adequacy will be examined in the following chapter.

FOOTNOTES

[1] "New Statistical Light on the World Farm Output," *Foreign Agriculture,* vol. 22, no. 4 (Washington: Foreign Agricultural Service, Department of Agriculture, 1958), p. 14.

[2] See Lester R. Brown, "Man, Land, and Food," *Foreign Agricultural Economic Report No. 11* (Washington: Regional Analysis Division, Economic Research Service, Department of Agriculture, November, 1963), especially Chapter IV, for elaboration of world data.

[3] Computed from various tables in the *1959 Census of Agriculture. The Farm Index* (August, 1966), p. 9 reported that crops were harvested from 302 million acres in 1965.

[4] For an excellent summary and bibliography see F. J. Marschner, "Land Use and Its Patterns in the United States," *Agriculture Handbook No. 153* (Washington: Agricultural Research Service, Department of Agriculture, April, 1959), 277 pp.

[5] Phillip F. Aylesworth, "Keeping Abreast of the Rural Community," *Agriculture Information Bulletin No. 215* (Washington: Federal Extension Service, Department of Agriculture, October, 1959), pp. 7–8; and *The Farm Index* (January, 1963) p. 6.

[6] "Number of Farms and Land in Farms" Sp. Sy. 3(1–66), (Washington: Department of Agriculture, Statistical Reporting Service, Crop Reporting Board, January 12, 1966), p. 1.

[7] For a general summary of farming in these regions see *Land, the Yearbook of Agriculture, 1958* (Washington: Department of Agriculture, 1958), pp. 109–74, or see Arthur F. Rapier and Martha J. Rapier, "Guide to Agriculture, U.S.A.," *Agriculture Information Bulletin No. 30* (Washington: Department of Agriculture, revised September, 1955), 104 pp.

[8] For both a factual and philosophical review of the situation up to the 1930's see Russell Lord, "To Hold This Soil," *Miscellaneous Publication No. 321* (Washington: Department of Agriculture, August, 1938), 123 pp.

[9] See "Soils, Geology, and Water Control in the Everglades Region," *Bulletin 442* (Washington: Agricultural Experiment Station, Univ. of Florida in cooperation with Soil Conservation Service, Department of Agriculture, March, 1948), 168 pp.

[10] "Salt Problems in Irrigated Soils," *Agriculture Information Bulletin No. 190* (Washington: Agricultural Research Service, Department of Agriculture, August, 1958), p. 3.

[11] The foregoing was originally based upon field observations and primary material obtained from Doctors Ray Billingsley and John W. Thomas, Texas Technical College and Texas Agricultural Experiment Station, in the spring of 1960. The material has been updated on the basis of various issues of *The Cross Section*, a monthly publication of the High Plains Underground Water Conservation District No. 1.

[12] See Chapter 2.

[13] For examples and a bibliography see Arthur F. Hall, "The Story of Soil Conservation in the South Carolina Piedmont 1800-1860," *Miscellaneous Publication No. 407* (Washington: Department of Agriculture, November, 1940), 36 pp.

[14] Summarized from "History of the Soil Conservation Service" (Washington: Soil Conservation Service, Department of Agriculture, mimeographed, no date), 6 pp.

[15] For greater detail see *The United States Government Organization Manual* (Washington: Federal Register Division, National Archives and Records Service, General Services Administration, published annually).

[16] Data supplied by A. B. Foster, Chief, Program Service Branch, Information Division, Soil Conservation Service, Department of Agriculture, in letter to the author dated December 7, 1966.

[17] *Loc. cit.*

[18] For details on practices and yearly accomplishments see *Agricultural Conservation Program, Statistical Summary* (given year) (Washington: Agricultural Conservation Program Service, Department of Agriculture, published annually).

[19] For details, see "The Conservation Reserve of the Soil Bank," *PA-312* (Washington: Department of Agriculture, February, 1957), 8 pp.

[20] Based upon "1964–65 Cropland Conversion Program Questions and Answers" (Washington: Agricultural Stabilization and Conservation Service, Department of Agriculture, October, 1964), 4 pp.

[21] "The 1966 Cropland Adjustment Program" (Washington: Agricultural Stabilization and Conservation Service, Department of Agriculture, November, 1965), 8 pp.

[22] From data supplied by H. D. Godfrey, Administrator, Agricultural Stabilization and Conservation Service.

[23] Based upon "The Appalachian Land Stabilization and Conservation Program" (Washington: Agricultural Stabilization and Conservation Services, Department of Agriculture, August, 1966), 3 pp.

[24] "Answers to Questions Most Frequently Asked About TVA" (Nashville, Tenn.: Department of Conservation, State Information Division, September, 1950), p. 1.

[25] See *TVA 1958, 25th Anniversary Year* and other annual reports (Knoxville, Tenn.: Tennessee Valley Authority).

[26] "Fertilizer Science and the American Farmer" (Knoxville, Tenn.: Tennessee Valley Authority, 1959), 19 pp.

[27] *TVA 1958, op. cit.*, p. 73.

[28] Data in this paragraph have been updated to 1966 on the basis of correspondence with Billy J. Bond, Assistant Chief, Western Test Demonstration Branch, Division of Agricultural Development, Tennessee Valley Authority, Knoxville, Tenn., April, 1966.

5

Cropland Conservation: Practices and Trends

THE TEST TO DETERMINE whether the nation has reached a conservational adjustment to its cropland resources must include the following questions: (1) Is land use based upon land capability? (2) Are land management practices maintaining or improving the inherent capability of croplands? (3) Does cropland use minimize the depletion of related resources, and improve them where possible and consistent with good practices? (4) Is there an active program, based upon continual projection of needs versus production capacity, aimed at providing a constant supply of agricultural commodities in required quantity, quality, and variety for a growing population? This implies programs directed toward improvement of yields as well as toward maintaining the supply of developable land.

Partial answers to these questions have been mentioned in the preceding chapters. A more specific appraisal of the current situation is provided here.

Land Use versus Land Capability

A complete understanding of the capabilities and limitations of each parcel of land is basic to the establishment of management practices which will assure permanency. Soil surveys of the agricultural lands of the nation, on a county basis, began near the turn of this century, under the leadership of the Department of Agriculture, and were completed for a major share of the croplands. These surveys were for the purpose of classifying soils according to their physical and chemical characteristics. They indicate adaptability, but not permanent capability.

LAND CAPABILITY SURVEY PROGRESS

Land capability survey work was begun by the Soil Conservation Service as a basis for establishing permanent soil and water conservation management plans on the cooperating farms within Soil Conservation Districts.[1] On the basis of work done up to 1958, the Soil and Water Conservation Needs Committee prepared an estimate of the capability of the land resources of the United States. These data, which exclude Federal land (except cropland) and urban built-up areas, are included in the two preceding chapters. It has been noted that several states have prepared generalized land capability maps, but these are on the basis of incomplete surveys. A number of the other states have prepared land resource maps as a basic step in completing the National Inventory of Soil and Water Conservation Needs (*e.g.*, Figure 5 – 1). The latter has been carried out under the auspices of the Department of Agriculture and the leadership of the Soil Conservation Service on a county by county basis. The data are available in tabular form.[2] The land capability estimates resulting from the National Inventory of Soil and Water Conservation Needs provide basic data for county, state, and national planning and programing. However, they are neither sufficiently detailed nor sufficiently accurate for applying proper management techniques to a given parcel of cropland.

The latter requires the detailed survey of land, parcel by parcel and farm by farm, now being carried on by Soil Conservation Service technicians on the cooperating holdings within Soil Conservation Districts. To date over 80 per cent of the land in private farm and ranch ownership has been covered by these detailed capability surveys. This is an impressive accomplishment, considering that the Soil Conservation Service was not established until 1935. Furthermore, as shown in the last chapter, most of the land in farms and ranches of the nation are within Soil Conservation Districts. If money to carry out the survey work continues to be available, it can be assumed that the capability of the major share of the cropland will be determined in the reasonably near future. Certainly this should be a national goal, because only when land capability is known will it be possible to make permanent land use adjustments.

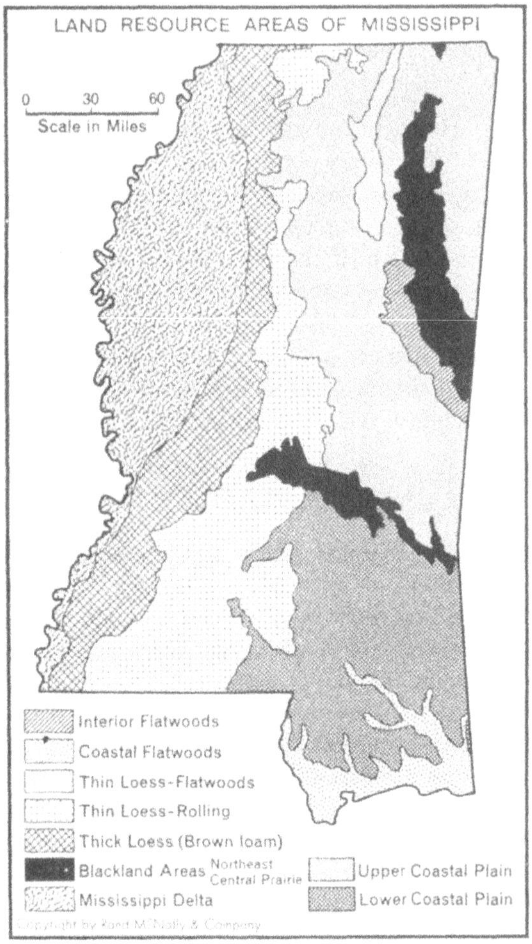

5 – 1 *Land resource areas of Mississippi*. This is an example of the type of maps a number of states have prepared. (After a map published jointly by the Mississippi Experiment Station, MSU and the Department of Agriculture)

The 1958 estimates of land use according to land capability indicated considerable disparity (see Figure 4 – 12). Some thirty-five million acres of land used for crops fell into capability classes V through VII. This is a relatively small portion of the total. The survey did suggest, however, that on nearly two-thirds of the nation's croplands improved management treatments are needed. Problems include 161,402,000 acres (37 per cent) with erosion hazards, 59,906,000 acres (14 per cent) with excess water, 36,434,000 acres (8 per cent) with unfavorable soils, and 14,099,000 acres (3 per cent) with climate problems. Even so, considering the short time in which attention has been widely given to the adjustment of land use to land capability, it is obvious that noteworthy progress has been made.

Land Management Practices for Permanent Land Use

The concerted efforts of agricultural scientists and farmers in the last three decades have established land management practices and land-use systems which, when properly applied, can go far toward maintaining or improving the productive potentials of all classes of croplands. Essentially all techniques and systems are subject to refinement as research provides new methods, machines, and crops.

CONTROL OF SOIL EROSION

A great deal of effort has been concentrated upon controlling erosion. It is impossible to control all the risks of erosion when soil is laid bare by cultivation. Even when soil is covered by vegetation, soil loss progresses, but more slowly. It is now known that a small loss can be tolerated. The establishment of an adequate control system has to be based upon understanding the erosion hazards for each field and composite of fields involved, which requires knowledge of many elements—the amount, intensity, and seasonality of annual precipitation; the duration and intensity of storms; the expected variability of precipitation over a series of years; the seasonality, velocity, and other characteristics of winds; the steepness and length of slopes; the runoff pattern on, above, and below the fields; the depth, texture, structure, and fertility of the soil; and the recent history of use. With these kinds of information it is possible to select tillage methods and crops that will best control soil loss (Figure 5 – 2). Some cases may require only contour cultivation, whereby the ridges thrown up by the plow and subsequent cultivation equipment and the rows of crops run on levels at right angles to the slopes. Each ridge and row tends to provide a small dam causing water to run along the slope, improving the opportunity for it to soak into the soil (Figure 5 – 3). Contour tillage is most effective on 2 to 8 per cent slopes not over three hundred feet long; it reduces loss to about half what it would be with tillage up and down the slope.[3] Some slopes may require that the crop selected be close-growing to minimize the exposure of bare ground. Grassed waterways, generally following natural drainages, commonly are required as an adjunct to runoff control. Cultural practices and equipment now available, when properly employed, can minimize the dangers of clean cultivation. Mulch tillage, deep tillage, rough seedbeds, and sod seedbeds are a few of the many practices that have been proven effective for given situations. In humid regions, winter cover cropping often is advisable to protect the soil and at the same time add organic matter.

Strip cropping (close-sown crops alternating with strips of other crops) slows down water flow and soil loss on slightly steeper slopes better than simple contour cropping can. Experimentation has proven that losses on slopes up to 12 per cent are reduced by about

5 – 2 Erosion control in pineapple production. **These fields are terraced and the rows of pineapples are planted on the contour. The major drainage-way remains in natural vegetation. These measures provide a high level of protection for the land. (Dole Company, A Division of Castle & Cooke, Inc.)**

three-fourths (compared with up-and-down farming), and to about one-half on steeper slopes.[4] There are three main methods of strip cropping, each suited to a particular situation: contour strip cropping, field strip cropping, and buffer strip cropping. All require that grassed waterways be used if excess water has a tendency to accumulate. (1) Contour strip cropping is adapted to well-drained soils on smooth slopes where rainfall causes erosion (Figure 5 – 4). The width of strips, the regularity of width, and tie-ins with waterways are adjusted to local conditions—generally strips are narrow where slopes are steep, and wide where slopes are gentle. A crop rotation system is normally employed in this method of cultivation. (2) Field strip cropping is used when the slopes are so irregular that strips would be difficult to till. In this method fields are laid out across the slope rather than on the true contour. Thus it is a less effective control than contour stripping and must be supplemented with adequate crop rotation and cultural practices to maximize protection. This may mean that some fields must be more or less permanently in grass. (3) Buffer strip cropping, consisting of narrow protective strips alternating with wide cultivated strips, is adapted to situations, where broad, gentle slopes are broken by short, steeper slopes or areas where soil loss has been severe. The buffer strips occupy the steep slopes or correction areas, and are usually sown to perennial grasses and legumes. This system is not so effective as contour strip cropping, and normally is applied only as a temporary measure until a more adequate system can be established.

TERRACING. Terracing, in conjunction with adequate tillage and crop selection practices, is a long established protection system that has gained in favor in recent years. This system employs a ridge, or a combination of ridge and channel, built across a slope on a controlled grade (Figure 5 – 5). The terraces intercept the water flow before it can damage the land. Graded terraces lead the water off the fields at nonerosive velocities to grassed waterways. Level terraces hold back the water, allowing it to soak into the soil. Not all soils and slopes can be successfully or economically terraced. Shallow or stony soils, highly irregular slopes, or slopes of more than 12 per cent slant are not economically susceptible to this practice.[5] Terrace construction is only the beginning of a control program. The measures must be maintained, and tillage and cropping practices geared to the needs of the particular fields.

TECHNICAL AID. It is clear from this discussion that the American farmer is required to have a

5 – 3 ***Contour planting.*** **On this sloping land, tomatoes are grown safely by planting, irrigating, and cultivating on the contour. (Department of Agriculture)**

considerable background in farming methods if he is to protect his land. He also can receive assistance from public agencies. A great deal of work has been and is being done to determine soil loss tolerance and soil loss under different systems of farming (Table 5–1). For example, by 1959 approximately eight thousand plot years of basic runoff, soil loss, and associated precipitation and related data from thirty-seven scattered Federal-state research projects in twenty-one states had been assembled by the Soil and Water Conservation Division, Agricultural Research Service. These data are the foundation for estimating soil loss under widely different cropping systems, conservation practices, and climatic conditions. Soil scientists of the Agricultural Research Service and several state experiment stations are currently perfecting a soil loss estimating equation which will have wide and easy adaptability.[6] Such information is available to the Soil Conservation Service and county agents for farmer guidance.

TABLE 5–1

RUNOFF AND SOIL LOSS ON A 2 PER CENT SLOPE FROM FOUR TREATMENTS, SPUR STATION, TEXAS, 1926-1951*

Crop	Average annual runoff in inches	Average soil loss in tons per acre
Cotton	3.65	7.2
Grain sorghum	2.76	3.8
Fallow	5.00	15.5
Grass	0.94	0.8

* "Proceedings of the Fourth Annual Conference on Water for Texas" (College Station, Texas: Water Research Information Center, the Texas A. and M. College System, September 16-17, 1958), p. 57.

The Agricultural Stabilization and Conservation Service encourages the establishment of soil and water conserving practices through cost-sharing. The gains have been heartening. Accumulative data are not readily available, but a check of one of the annual volumes of *Agricul-*

5 – 4 *Strip cropping.* On this land strips of corn are alternated with strips of oats. Notice also that the waterway has a protective cover of vegetation. (Department of Agriculture)

tural Statistics of the United States Department of Agriculture or the Agricultural Conservation Program Statistical Summary will show the impressive yearly accomplishments under the auspices of the Soil Conservation Service and the Agricultural Stabilization and Conservation Service.

EROSION CONTROL UNDER DRY FARMING

Farming in the Great Plains and the Pacific Northwest wheat region presents a number of special problems. A permanent adjustment must take into account drought, seasonally high temperatures, wind, erratic but often highly intense rainfall, and getting the utmost soil moisture from the climate as well as the soil. The farmer must also be prepared to adjust quickly to the whims of climate.

Misuse has taken a serious toll, and much of the land has been degraded in capability in these dry farming regions. The long-range need is to use each parcel of land within its capability and to treat it according to its needs for protection and improvement. Thus much land has been shifted from cropping to range use, and more should be. In the Great Plains alone about ten million acres of cropland were still not suited for cultivation in 1965.[7] Much of this land yields low returns and is subject to severe erosion even in normal years.

MANAGEMENT PRACTICES. Advocated management practices for cultivated land are reasonably effective; however, there is some difficulty in providing both wind and water erosion control practices on the same land. Stubble mulching and other tillage methods that leave crop residue on or near the surface should be used on all cultivated fields (Figure 5 – 6). In the spring-wheat region and on land to be fallowed in the winter-wheat region, it is generally best to wait until spring to till the stubble. Crop residues on the surface protect the soil from the force of high wind and improve the soil structure, permitting water to soak in more readily.

On large fields that are likely to blow, strips of sorghum (where it is tolerant of temperature and moisture conditions) or other wind resistant crops should be grown in rotation; in summer-fallow districts, strip cropping of wheat and fallow is desirable (Figure 5 – 7). On sloping lands, contour tillage or terraces are necessary, and grassed waterways are essential to dispose of runoff. Contouring, ridging, and

5 – 5 Terraces. The recently constructed terraces in the center of the picture form an outlet into established brome grass waterways. These measures will control sheet, rill, and gully erosion which were problems on this land. (Department of Agriculture)

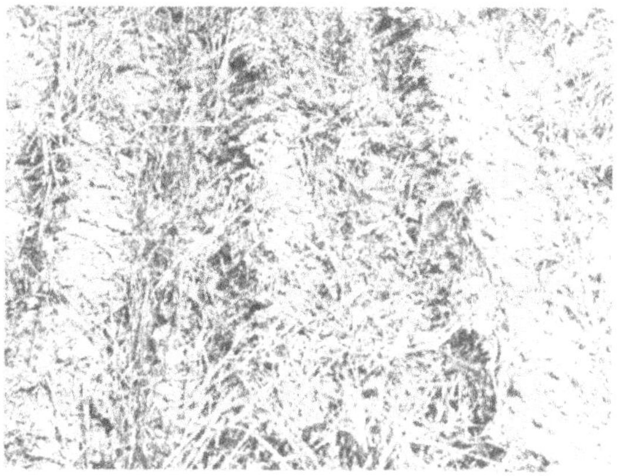

5 – 6 Stubble mulching. This is a widely used technique of soil erosion control and tilth quality maintenance in the dry farming areas. In cultivation, part of the previous crop residue is left on, and part in the soil. On this field there are 2,200 pounds of wheat residue per acre. (Soil Conservation Service)

stubble mulching also help to maintain an even snow spread, which is desirable for an even moisture supply.

When a blow begins before a crop is established, it may be necessary to apply emergency tillage (covering the fields at right angles to the wind at intervals with a lister or chisel) to bring clods or moist soil to the surface. During a drought period an emergency crop may be necessary to stop wind erosion—during such a period even weeds should be allowed to grow.[8]

USE OF WINDBREAKS. Windbreaks and shelterbelts have proved an important tool of conservation farming in the Great Plains as well as in other agricultural regions of the United States (Figure 5 – 8). They can be employed effectively, in conjunction with other practices, to protect cultivated fields against winds which might erode soil, injure seedlings, or damage a maturing crop. They also can aid in holding snow on the land to increase moisture supply. Protection of livestock, hay stacks, and homesteads, and the provision of habitat for birds and wood for fuel and fence posts are other uses.

Experience since the 1930's, when shelterbelts were first advocated for the Great Plains, has established techniques and designs which are most effective.[9] A well-designed system will protect fields and crops from prevailing winds. A typical system in the wheat regions of the Great Plains consists of a belt of three to five rows of trees on the west side extending the north-south length of the field. A series of belts one to five rows wide is located at right angles in an east-west direction at intervals of six hundred to eight hundred feet. The system is varied to suit local conditions and may, for example, be planted on the contour and tie in with drainageways. Not all areas of the Great Plains are satisfactory for shelterbelt development because of soil or moisture limitations. Moreover, there are many farmers and ranchers who do not favor their use. Typical objections are that

5 – 7 Dry-land strip cropping. On this farm strips of wheat are alternated with strips of fallow for wind and water erosion control. The strips are laid out on the contour. Notice also the use of windbreaks around the farmstead. (Soil Conservation Service)

they occupy cropland; they sap moisture from plants next to them; weeds and soil drift into them; they take time to grow and require care; and where snow drift is heavy they delay spring work. Yet many landowners who use them say that the advantages outweigh the disadvantages —that in the long run shelterbelts are money-makers.

CONSERVATION IRRIGATION

Irrigation farming in the United States, less than one hundred years old on an important scale, was begun without experience. Progressing on the basis of trial and error, farmers made many mistakes in the early development of irrigation projects and associated farming. During the past three decades, as increasing attention has been turned toward soil conservation and toward solving problems related to growth in demand for the restricted water resource of the West, attention has also been turned toward better methods of applying water to land in the interests of conservation. It is known that control of irrigation water begins at the source, and does not end until it reaches the last row in the last field, and that any water left over is cared for through adequate surface drainage. Wasted water often results in wasted soil through erosion loss, drainage troubles, leached fertilizers, and salt accumulations.

Current methodology recognizes that there are several aspects to conservational irrigation farming. These include: (1) Adequate land and soil preparation; (2) an irrigation system planned for efficiency; (3) the application of the proper amounts of water at the right times; (4) a disposal system for waste water; (5) the selection of adaptable crops and tillage systems; and (6) provisions for re-use of water (Figure 5 – 9).[10]

A farm irrigation plan must fit the land. In many cases the surface can be improved through field leveling if the land is comparatively smooth or the slope gentle and soils uniformly deep, or through bench leveling if the slopes are regular and the soils uniformly deep (Figure 5 – 10). Leveling improves water distribution, water control, and field operations. In any case the system layout must be adjusted to the lay of the land, taking into account efficient water movement and penetration and avoidance of soil loss or degradation. Thus such considerations as steepness of slope, length of runs, and contouring are significant.

Texture, structure, and depth are soil properties of special concern. By considering the relationships among these characteristics, the successful irrigation farmer can estimate available moisture storage, intake rate, and permeability—all improvable properties—through practices such as returning crop residues, use of sod crops, and avoidance of working when the soil is wet. Leveling may be able to improve spotty areas of shallow depth.

On the basis of consideration of land and soil character and increasing knowledge of the

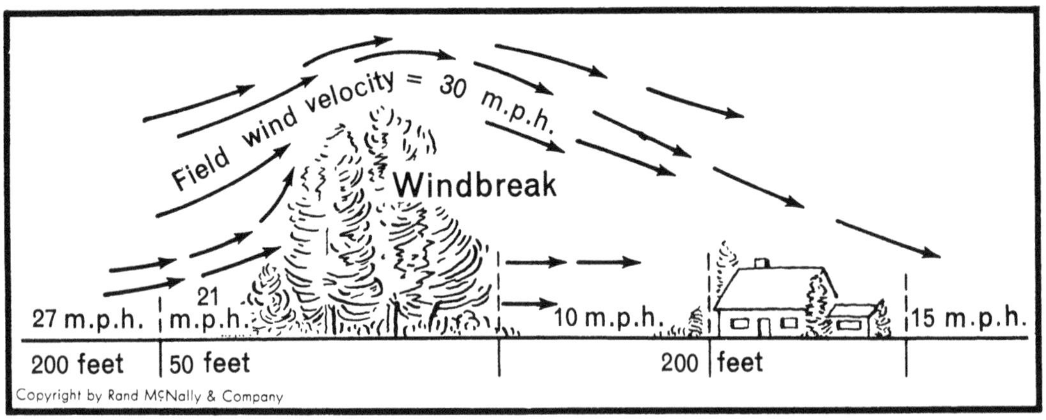

5 – 8 *A schematic representation of a windbreak.*

water needs of various plants, irrigation methods have been undergoing steady improvement. Irrigation water may be applied in four basic ways:

(1) *Flooding or border* irrigation wets the entire surface. Fields are divided into strips which are separated by low dikes and are graded in the direction of irrigation. The water advances as a sheet from one end, wetting the soil as it moves. This method is satisfactory for most soils except those of finer texture and low intake rate. It is used generally for hay, pasture, and grain crops on lands having slopes up to 3 per cent. Strips may be laid out on the contour as benches.

(2) *Furrow* irrigation, generally practiced with row crops, wets only part of the surface. Water is run in furrows between crop rows. Most commonly, the system is laid out to run water directly down a slope; however, to control erosion, contouring is sometimes applicable (Figure 5 – 11).

(3) *Sprinkler* irrigation is the application of water to the surface of the soil in the form of a spray somewhat like rain (Figure 5 – 12). This type of application is adaptable to most soils. In fact, soils that are too shallow or land that is too steep or rolling for other methods often can be effectively irrigated by sprinklers. It also is adaptable to most crops, the major exception being rice. Sprinkler irrigation, however, does have limitations—water distribution is affected by wind, power requirements are higher than for other systems, and water supply must be clean and constant.

(4) *Subirrigation* saturates the subsoil, but wets the surface very little (Figure 5 – 13). It requires complete control of the water table so that the plant root zone is kept free of excess water, but continually supplied with capillary

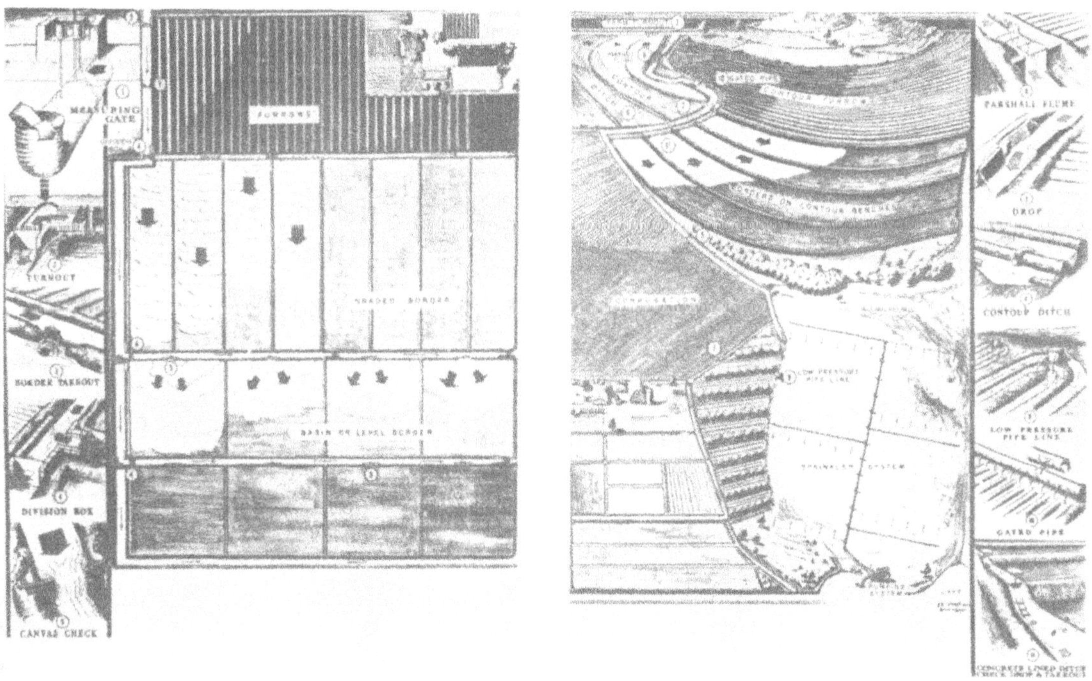

5 – 9 ***Methods and structures for irrigation.*** **These illustrations show some of the more important means of irrigating land. The numbers in the main body of the picture refer to the various types of structures on the margins used to distribute water. The variety of methods and structures now available provide the farmer with great flexibility in selecting those best suited to his farm in interest of water and land conservation, *i.e.,* those that make possible irrigation without soil erosion damage, saline, or alkaline accumulation, water logging, or undue water loss. (Bureau of Reclamation)**

moisture during the cropping season. Soils suited to this method permit rapid lateral and downward movement, yet are capable of lifting the moisture from the watertable throughout the root zone. This system is adaptable to most crops, except orchards, and has the advantage of permitting field operations without concern for the irrigation period. Very little land, however, has the special combinations of physical conditions required for subirrigation.

Whatever the system used, it should be suited to the land and water supply, and it should permit complete control. Distribution systems should minimize loss through excessive seepage and evaporation, requiring concrete lined ditches or underground pipes in some cases, and should allow fine control of subdivision and diversion. The method of application to the land should permit uniform and timely wetting of the soil. Recent research work done on water needs of plants has taken much of the guesswork out of this important phase of irrigation; guides for judging soil moisture, formulas for determining amounts of water to be applied, and mechanical devices are now available to farmers. The disposal system should control all runoff, eliminating erosion, waterlogging, and waste. It is readily understood that an adequate system requires considerable technical knowledge. This, however, is easily avail-

5 – 10 *Land leveling for irrigation.* Leveling makes possible larger fields, reduced erosion, and improved water distribution. (Bureau of Reclamation)

5 – 11 *Furrow irrigation.* This is the common method of applying water to row crops. Water is transported from the head ditch by means of siphon tubes to furrows between the crop rows. (Bureau of Reclamation)

able in most areas through the public service agencies.

It will be recognized, also, that proper system establishment is only part of conservation irrigation. Crop selection and tillage practices must be equally suited to the land being farmed. On seven million acres in the western states there also is need to improve salt-affected soils through controlled amendments and leaching.

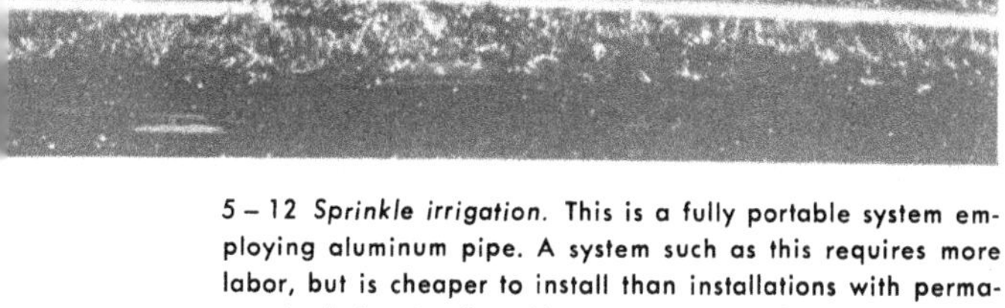

5 – 12 *Sprinkle irrigation.* This is a fully portable system employing aluminum pipe. A system such as this requires more labor, but is cheaper to install than installations with permanent buried mains. Portable systems commonly are used where rainfall supplies much of the moisture needed by crops and thus only a few irrigation periods are required. (Bureau of Reclamation)

Cropland Use versus Related Resources

There is little doubt that cropland development and use has had an adverse effect on related resources: clearing and plowing has reduced the area of forest, grass, and wildlife habitat; drainage has diminished waterfowl habitat; natural recreational attractions have been removed; irrigation has taken water from other uses and dams have obstructed fish runs; and cultivation in general has influenced water runoff and added to the sediment loads of streams and lakes. Some of this is justifiable and in the national interest because of the high priority assigned to cropland. Even though this is the case, the guiding principle for the present and future should be a true conservational approach which considers all of the elements of the resource base and minimizes the depletion of any one.

The full implementation of techniques and land-use practices now available and advocated will go far toward reducing to the smallest possible degree the despoilment of resources related to cropland. Land capability surveys of farms and ranches assign land best suited to forests and grass to those uses. With management methods now proven effective by research, these resources can be improved above natural levels in some cases (Chapters 6 and 7). The land-use decisions of farmers and ranchers can cause an abundance or scarcity of wildlife. Every year more operators are recognizing that conservation farming is compatible with improvement in wildlife and water control.

Strip cropped fields established primarily for erosion control attract about twice the number of ground-nesting birds as undivided fields because they have borders where food and cover are close together. Farm ponds, properly designed, can often serve the multiple purposes of assisting in a runoff control program, providing stock and irrigation water, providing habitat for birds, and being stocked with fish for food and pleasure. The number of farm ponds has been expanding at a rapid rate in recent years. Farmers in Soil Conservation Districts are presently stocking about fifty-five thousand ponds annually; these yield between one hundred and three hundred pounds of fish per acre each year. West of the Mississippi in South Dakota, a study showed that forty-thousand man-made ponds containing about one hundred thousand acres of water bore a harvest of 141,000 ducks. This represented a new population of ducks in the area.

Other conservational tools of cropland use which increase wildlife include windbreaks, hedge fences, grassed waterways, and rotation pastures. Most of the elements in a watershed and flood protection program are mutually beneficial to cropland, wildlife, and water. These include the techniques mentioned, as well as pasture and forest management and stream channel improvement.

All of these practices are a relatively inexpensive and necessary part of basic plans for soil and water conservation and sound land use. That progress has been notable is indicated by the fact that all of the cooperators in Soil Conservation Districts are applying at least one practice mutually beneficial to other resources.[11] Significant, too, is the recent growth of farm-based recreation, especially in the eastern districts where public lands are scarce and urban populations are nearby. Many farmers are now realizing profits from picnic areas, fishing ponds, hunting permits, horse rentals, and even the provision of the opportunity for farm vacations. As mentioned in the preceding chapter, establishment of recreation facilities comes under the cost-share programs of the Agricultural

5 – 13 *Subirrigation.* In this method of irrigation a water table is maintained at some predetermined depth with field laterals. Moisture reaches the plant roots through upward capillary movement. This method allows surface operations without concern for the irrigation period. (Bureau of Reclamation)

Stabilization and Conservation Service. (Between the time of the passage of the enabling act in 1962 and the end of 1965, 31,500 conservation district cooperators had established recreation enterprises.)

Farm Planning, Programing, and Community Cooperation

No single field nor parcel of land can be treated as an isolated element. Conservation farming requires that fields on individual farms, and the farms in a given watershed or community, be considered in their interrelationships as parts of a whole.

THE INDIVIDUAL FARM

The achievement of a conservation operation requires that consideration be given to the total farm. Land-use adjustments, field boundaries, land treatments, conservation structures, enterprise selections, rotations, and so on must be woven into a farm plan and program which will provide for protection or improvement of every parcel of land. The plan should serve as a blueprint for many years, but should be sufficiently flexible to allow shifts in response to market changes and research.

The farm plans developed with the assistance of the Soil Conservation Service recognize these needs. The illustration presented in Figure 5 - 14 is a typical example of the 2,080,231 district cooperators, as of July 1, 1966. This represents excellent progress. The goal is to have all the farms and ranches of the nation operating on the basis of similar conservation plans whether or not they are cooperators in Soil Conservation Districts.

THE COMMUNITY OR WATERSHED

It is obvious that all fields in a farm must be considered in the establishment of a conservation operation, but so must all farms in a community or watershed. Water does not stop running downhill at property lines, and wind does not stop blowing at fence lines. Soil-washing and water-control problems, for the most part, are watershed problems; their solution requires control of running water from ridge crest to stream channel. Conservation work done by one farmer in a watershed is partly or wholly undone if water and silt from an adjoining farm pour down on his land. The farmer on the Great Plains who tries to control wind erosion over it from nearby properties is faced with a similar problem, and much of his effort is wasted. Teamwork by neighboring farmers is necessary if effective and lasting soil and water conservation is to be achieved.

By working together all farmers in a community or watershed benefit by protecting soil, improving water control and moisture supply, preventing overflow or waterlogging, improving wildlife, increasing crop yields, and, in all probability, decreasing costs. Moreover, there are usually social benefits, because by working together, farmers become better neighbors. An example of an actual cooperative effort of six farmers in a small watershed is presented in Figure 5 - 15. This map clearly shows that the six farms comprising the watershed are in reality pieces which belong with each other and need to be considered together.[12]

This type of cooperation is being encouraged widely by the Soil Conservation Service, the Extension Service and public policies expressed through these agencies, the Agricultural Stabilization and Conservation Service, and other public service agencies concerned with land and water resources. In many instances farmers themselves are taking the lead in organizing for cooperative effort and mutual benefits. Certainly this is the desirable approach.

Prospects for Meeting Future Needs

An essential part of a society's conservational adjustment to its resource base is a continual assessment of needs. This assessment must project sufficiently far into the future and be sufficiently accurate to form the basis for planning and programing orderly development to minimize adjustment, unnecessary wastes, and alienations and continually to provide consumers with commodities in sufficient quantity and variety at reasonable prices.

It is impossible, of course, to anticipate future demands exactly. Agriculture has become a complex, dynamic industry with many forces shaping its future. These forces include population and income growth, technological changes both inside and outside agriculture, foreign aid programs and market developments, and shifting consumer preferences. The interacting influences of these forces are such that even short-range estimates are subject to correction. Yet land owners, farm operators, legislators, and government administrators must almost daily

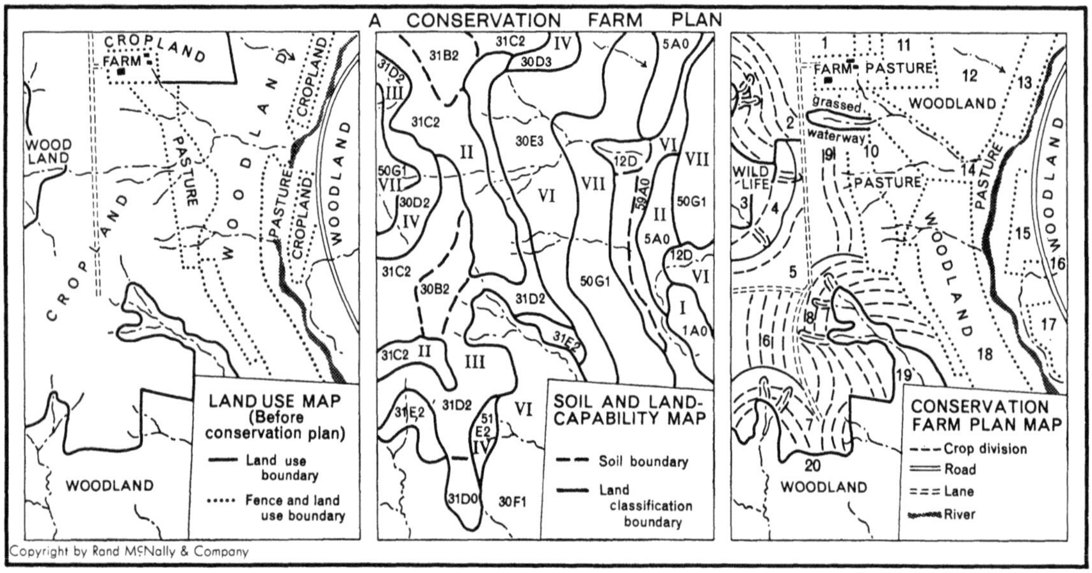

5 – 14 *A farm plan.* (Soil Conservation Service)

A CONSERVATION FARM PLAN

The farm shown in this example is a dairy farm in Wisconsin.

Soil map legend

Class I land

1AO — Arenzville silt loam, 0-2 per cent slope, not eroded.

Class II land

5AO — Rowley silt loam, 0-2 per cent slope, not eroded.

31B2 — Dubuque silt loam, deep, 2-5 per cent slope, moderately eroded.

31C2 — Dubuque silt loam, deep, 5-10 per cent slope, moderately eroded.

Class III land

30B2 — Dubuque silt loam, 2-5 per cent slope, moderately eroded.

30C2 — Dubuque silt loam, 5-10 per cent slope, moderately eroded.

31DO — Dubuque silt loam, deep, 10-15 per cent slope, not eroded.

31D2 — Dubuque silt loam, deep, 10-15 per cent slope, moderately eroded.

Class IV land

30D2 — Dubuque silt loam, 10-15 per cent slope, moderately eroded.

30D3 — Dubuque silt loam, 10-15 per cent slope, severely eroded.

31E2 — Dubuque silt loam, deep, 15-20 per cent slope, moderately eroded.

On this farm, all the fields in land capability classes I, II, III, and IV are also good for pasture, hay, trees, or wildlife food and cover. In fact, they could produce better yields of these permanent-cover crops than the fields in classes VI and VII, which are best suited to permanent cover.

Class VI land

12D — Stony colluvial land, 10-15 per cent slope, not eroded.

30E3 — Dubuque silt loam, 15-20 per cent slope, severely eroded.

30F1 — Dubuque silt loam, 20-30 per cent slope, slightly eroded.

59AO — Alluvial land, poorly drained, 0-2 per cent slope, not eroded.

Class VII land

50G1 — Rough broken land, 30-45 per cent slope, slightly eroded. This farm does not have any class V or class VIII land.

FARM PLAN MAP LEGEND: Conservation measures planned for this farm

Field	Use	Land capability class	Conservation measures
1	Crops	II	Contour tillage; 5-year crop rotation — 1 year corn, 2 years small grain, 2 years clover and timothy.
2, 5, 9	Crops	II, III	Contour strip cropping; 5-year rotation — 1 year corn, 1 year small grain, 3 years alfalfa and timothy.
4	Crops	III, IV	Contour strip cropping; 5-year rotation — center strips: 1 year corn, 1 year small grain, 3 years alfalfa and timothy. Outside strips: 1 year corn, 2 years grain, 2 years timothy and clover.
6	Crops	II, III	Contour strip cropping; 5-year rotation — 1 year corn, 2 years small grain, 2 years clover and timothy.
7, 8	Crops	III, IV, VI	Contour strip cropping; 5-year rotation — 1 year corn, 1 year small grain, 3 years alfalfa and timothy; strip near woods to be permanent hay.
13, 15	Crops	II	Contour tillage and drainage to remove excess water; 4-year rotation — 2 years corn, 1 year small grain, 1 year pasture
17	Crops	I	4-year rotation — 2 years corn, 1 year small grain, 1 year pasture.
			Conservation treatments for all cropland:
			All waterways seeded and kept in sod.
			All fields limed and fertilized as needed.
12, 18	Woods	VI, VII	To improve stocking and increase production, about 6,000 white pines to be interplanted in the openings of the two fields.
			Natural reproduction of hardwoods to be encouraged in all improvement and harvest cuttings. To improve quality of later cuts, trees cut during the first 5 years to be confined to the poorest trees, such as those with crooks, low forks, and scarred butts. Thereafter, each 5 years about 750 board feet per acre to be cut; trees to be selected so as to keep the woodland producing.
			Fields to be divided into five equal units and one unit cut each year.
			Den trees to be kept for wildlife.
16	Woods	VI, VII	Management will be the same as for fields 12 and 18 except each 5-year harvest cut to be 1,000 board-feet per acre.
19, 20	Woods	III, IV	Management the same as for field 16; about 3,000 white pines to be interplanted in the openings.
			Shrubs for wildlife to be allowed to come in along field borders.
3	Wildlife	VII	Wildlife land: open grassy and shrubby areas to be kept.
			Some hardwood trees to be cut for lumber to allow better wildlife plants to grow.
			200 conifers to be planted for winter protection for wildlife.
10, 11, 14	Pasture	III, IV, VI, VII	Conservation treatments for all pastures:
			Limed and fertilized.
			Worn-out areas seeded to grass and clover mixture.
			Newly seeded pasture to be grazed lightly the first year.
			Amount and time of grazing to be regulated.
			Pastures rotated to provide alternate grazing and resting periods of about 2 weeks each.
			Weeds and tall grasses mowed at least once each year.
			Livestock kept off legume pasture during September.
			Grazing on sweetclover regulated so that it will reseed itself.
	Farmstead		Fence to be replaced with a multiflora rose living fence.

make decisions that have impacts extending well into the future. Thus appraisals of needs and output potentials must be made, and they must be continually reviewed if they are to serve as adequate guides to rational decision making.

Estimates of Needs

In recent years the agencies of the United States Department of Agriculture have begun to perform the important service of projecting needs. The work of the Soil Conservation Service in estimating land supply and land management needs has been discussed in previous chapters. The economists and statisticians of the Economic Research Service likewise are estimating future demands, output potentials, and crop land needs resulting from anticipated technological and economic conditions and international involvements.[13] These are the present basis for the programs of the Agricultural Stabilization and Conservation Service.

Most of the projections made to date are for about fifteen years in the future. The projections made in 1959, however, did take "A 50-Year Look Ahead at U. S. Agriculture." Estimates were prepared for three levels of assumptions, all based upon 1956 as the index year: (1) A 2010 population of 300 million and an export level 52 per cent higher than in 1956; (2) a population of 370 million, a 10 per cent increase in per capita consumption, and exports at approximately the 1956 level; and (3) a population of 440 million and an export level the same as 1956. The first estimate would require a farm output increase of 1.86 over that of 1956, the second an increase of slightly more than 2.00, and the third an increase of 2.56. The second estimate, which at the time was considered to be the most realistic, would require a net increase of ninety-seven million acres of cropland as well as an increase in yields. The first projection, it was suggested, could be met by anticipated increases in yields, with cropland development necessary only to replace the land absorbed by urban and other space uses. The third projection would require as much as 230 million acres of additional cropland and the replacement of about thirty-three million acres to be absorbed by urban and other uses to meet increased needs.

A projection for 1980, published in 1962, used 1960 as the index year.[14] A 45 per cent increase in population (to 261 million), a per capita income 50 per cent above 1960, and a doubling of disposable income were assumed. Crop production per harvested acre was projected to be 56 per cent above the 1959 level.

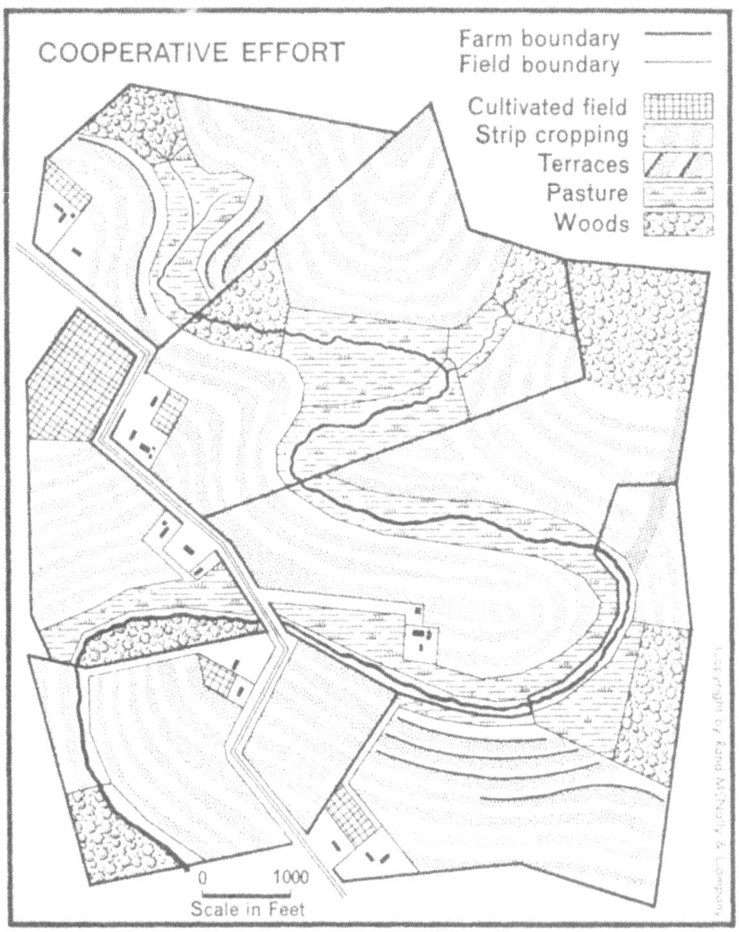

5 – 15 *A cooperative effort.* (Soil Conservation Service)

The projected cropland requirements in relation to actual uses of 1954 and 1959 are summarized in Table 5–2.

The 1966 projection, also setting 1980 as the target year, shows an even more favorable short-run relationship between resources and needs.[15] The main assumptions and aspects of this projection are indicated in Tables 5–3, 5–4, and 5–5 and the accompanying graphs (Figure 5 – 16). The significant conclusion is that 245 million people can be provided with agricultural-based needs by employing advances in technology to essentially the same amount of land that was employed during the period 1959 to 1961. It would also appear that during the next ten years American agriculture can keep pace with an export expansion equal to the increases of the 1950 to 1960 decade and still leave about 10 per cent of our developed cropland resource idle. Moreover, it is suggested that if all cropland currently diverted from production under various stabilization and conservation programs were brought under cultivation, crop output could be increased at least 60 per cent above the 1959 to 1961 average by 1980.

The Longer Future

The overriding conclusion must be that this nation is in a highly favorable position with respect to its cropland resource base. In fact, a significant economic problem of the present, overproduction capacity, undoubtedly will be carried into the immediate future. This problem may be our greatest threat to the future if we permit it to be the basis for irreversible diversion of large quantities of quality cropland to other uses.

FURTHER POSSIBILITIES IN TECHNOLOGY

Previous discussion has indicated that farm output for human use has experienced a consistently upward trend in the United States since 1870.[16] Until 1920 additions to the cropland accounted for the increase, but since that year the trend has been maintained through the applications of science and technology. After World War I the shift from using the power of horses and mules to tractors, trucks, and automobiles released large acreages from growing feed, and reduced labor requirements. Mechanical power has been accompanied by improved equipment for most farming operations. Along with these improvements have come greater use of fertilizers, better seeds and animal breeds, more efficient tillage and feeding practices, insecticides, disease controls, and various types of land im-

TABLE 5–2

USE OF TOTAL CROPLAND IN 1954 AND 1959, AND PROJECTED REQUIREMENTS FOR 1980*

	Million acres			
Land use	1954	1959	1980	Change 1959–80
Acres of crops harvested	(346)	(325)	(297)	(−28)
Acres double cropped	(7)	(8)	(6)	(−2)
Cropland harvested	339	317	291	−26
Crop failure	12	11	11	0
Cultivated summer fallow	29	31	24	−7
Total cropland used for crops	380	359	326	−33
Soil improvement and idle cropland	19	33	11	−22
Cropland used for pasture	66	66	70	+4
TOTAL	465	458	407	−51

* *Land and Water Resources, A Policy Guide* (Washington: Department of Agriculture, September 1962, revision), p. 38.

provement through irrigation, drainage, leveling, and the like. Such changes during the past two and a half decades have raised agriculture's output potential 70 to 80 per cent.[17] Although there was some change in the share of individual inputs (such as the decline in labor), the combined inputs, land, labor, and capital have changed little in the past fifteen years.

Agricultural scientists suggest that vast potentials continue through technological innovation, and recent advances would seem to support their contentions. There is little doubt

TABLE 5–3

OUTPUT, EMPLOYMENT, PRODUCTIVITY, AND INCOME*

Item	Unit	Averages 1949–51	Averages 1959–61	1964	Projected 1980
Population, July 1	Million	151.7[1]	180.5	192.1	245.3
Labor force[2]	Million	64.8	73.1	77.0	101.4
Employment[2]	Million	61.8	68.9	73.1	97.4
Nonagricultural[2]	Million	54.3	63.2	68.3	94.3
Agricultural	Million	7.5	5.7	4.8	3.1
Gross national product[3]	Billion Dollars	385.6	530.1	628.7	1,150.0
Output per man-hour[2 3]	Dollars	2.91	3.81	4.35	6.73
Disposable personal income[3]	Billion Dollars	263.1	365.9	435.8	800.0
Per capita disposable personal income[3]	Dollars	1,734	2,027	2,269	3,261

* R. F. Daly and A. C. Egbert, "A Look Ahead for Food and Agriculture" *ERS-277* statistical supplement (Reprinted from *Agriculture Economic Research*, vol. XVIII, Washington: Economic and Statistical Analysis Division, Economic Research Service, Department of Agriculture, February, 1966) p. 7.

[1] Based on forty-eight states.

[2] Includes Armed Forces.

[3] 1964 constant dollars.

TABLE 5–4

FARM OUTPUT AND RELATED DATA, SELECTED YEARS, AND PROJECTIONS FOR 1980*

Calendar year—1957–59 = 100

Item	Averages 1949–51	Averages 1959–61	1964 [1]	Projections to 1980: I[2]	Projections to 1980: II[3]
Farm output	87	105	111	151	160
Livestock production	88	104	113	147	147
Meat animals	89	105	116	160	160
Dairy products	93	101	104	119	119
Poultry products	78	107	118	152	152
Crop production	91	106	109	155	172
Feed grains	79	105	97	163	191
Hay	88	101	105	123	129
Food grains	88	106	114	162	178
Cotton	112	117	125	143	184
Oil crops	66	108	128	230	249
Cropland used for crops	107	98	94	99	106
Crop production per acre[4]	85	108	116	157	162

* R. F. Daly and A. C. Egbert, *op. cit.* p. 8.

[1] Preliminary.

[2] Production required to balance projected domestic and foreign demand.

[3] Production with all diverted cropland planted to crops, use of added production not specified.

[4] Index of crop production per acre is a ratio of total crop production to total land used for crops.

TABLE 5–5

CROP ACRES HARVESTED, YIELD PER HARVESTED ACRE, 1949–51 AND 1959–61 AVERAGES, 1964, AND PROJECTIONS FOR 1980 UNDER TWO ALTERNATIVES

Calendar year

Commodity	Average 1949–51	Average 1959–61	1964	Projections to 1980 I[1]	Projections to 1980 II[2]
Acres harvested			Million acres		
Corn for grain	73.6	67.4	57.1	64.2	78.2
Oats	37.4	26.1	20.4	13.5	11.5
Barley	10.2	13.9	10.7	11.9	10.0
Grain sorghum	8.5	14.0	11.9	16.3	21.0
Wheat, all	66.5	51.7	49.2	57.1	62.0
Rice	1.8	1.6	1.8	1.6	2.2
Soybeans	12.6	24.4	30.7	43.2	45.4
Peanuts	2.2	1.4	1.4	1.0	1.4
Cotton	24.1	15.4	14.1	11.5	15.0
Hay, all	74.3	66.9	67.9	57.3	60.3
Other crops	38.6	34.3	35.9	40.0	40.0
Total cropland harvested[3]	341.3	310.0	294.6	312.6	340.0
Yield per harvested acre			Bushels		
Corn for grain	37.8	56.2	62.1	101.1	100.1
Oats	34.5	41.1	43.2	60.0	60.0
Barley	26.1	29.9	37.8	42.0	42.0
Grain sorghum	21.4	39.4	41.1	52.7	51.3
Wheat	15.6	23.9	26.2	33.1	32.5
Soybeans	21.6	24.1	22.8	30.1	30.1
			Pounds		
Rice	2,328	3,381	4,095	5,200	5,200
Peanuts	842	1,218	1,551	2,300	2,300
Cotton	275	449	524	725	725
			Tons		
Hay, all	1.39	1.72	1.71	2.50	2.50

* R. F. Daly and A. C. Egbert, *op. cit.* p. 9.

[1] Cropland use and yields based on projected utilization.

[2] Cropland use and yields based on the assumption of full use of diverted cropland, use of added production unspecified.

[3] Does not equal total harvested acreage due to double cropping.

that tomorrow's farmer will be using improved plants, animals, and techniques and materials for environmental control. By 1980 he may be using an average of 175 pounds of fertilizer per acre; during 1960 to 1964 his average was thirty-eight pounds.[18] This increased use will be based upon more accurate formulas for specific crops and field conditions.

Despite such improvements, which will result in more efficient use and greater output from land, it is not realistic to believe that in the longer run total needs can be met completely through technology. The population will increase; there is potential demand in the low standard of living segment of the present population; and the demand for American farm products is steadily expanding abroad. Moreover, noteworthy changes have been occurring in the eating habits of the average citizen (See Figure 4 – 2). The characteristics of land qual-

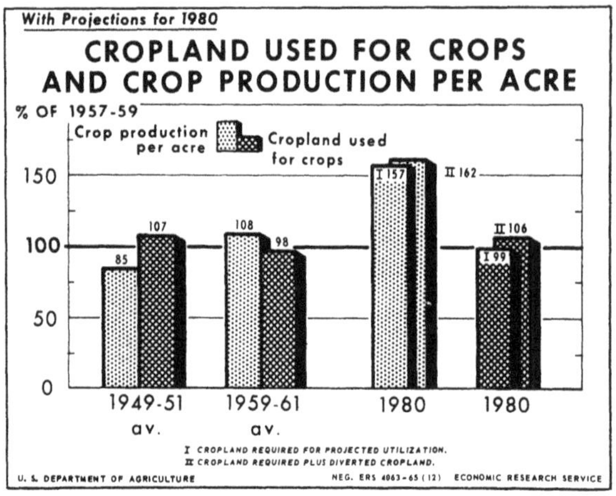
With Projections for 1980
CROPLAND USED FOR CROPS AND CROP PRODUCTION PER ACRE
% OF 1957-59
Crop production per acre
Cropland used for crops
150
100
50
0
85
107
108
98
I 157
II 162
I 99
II 106
1949-51 av.
1959-61 av.
1980
1980
I CROPLAND REQUIRED FOR PROJECTED UTILIZATION.
II CROPLAND REQUIRED PLUS DIVERTED CROPLAND.
U. S. DEPARTMENT OF AGRICULTURE
NEG. ERS 4063-65 (12)
ECONOMIC RESEARCH SERVICE

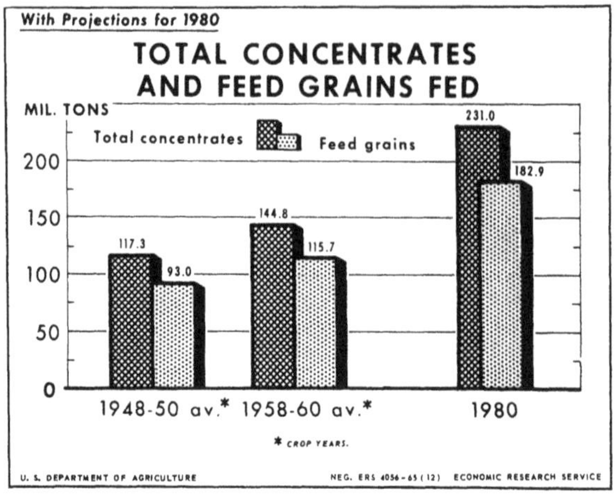
With Projections for 1980
TOTAL CONCENTRATES AND FEED GRAINS FED
MIL. TONS
Total concentrates
Feed grains
200
150
100
50
0
117.3
93.0
144.8
115.7
231.0
182.9
1948-50 av.*
1958-60 av.*
1980
* CROP YEARS.
U. S. DEPARTMENT OF AGRICULTURE
NEG. ERS 4056-65 (12)
ECONOMIC RESEARCH SERVICE

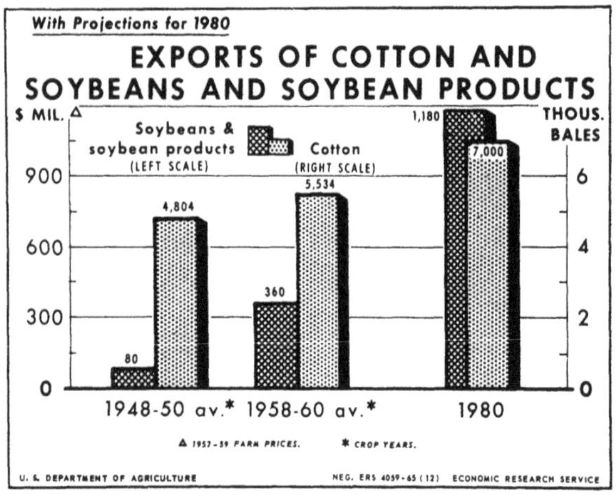
With Projections for 1980
EXPORTS OF COTTON AND SOYBEANS AND SOYBEAN PRODUCTS
$ MIL. Δ
THOUS. BALES
Soybeans & soybean products (LEFT SCALE)
Cotton (RIGHT SCALE)
900
600
300
0
6
4
2
0
80
4,804
360
5,534
1,180
7,000
1948-50 av.*
1958-60 av.*
1980
Δ 1957-59 FARM PRICES.
* CROP YEARS.
U. S. DEPARTMENT OF AGRICULTURE
NEG. ERS 4059-65 (12)
ECONOMIC RESEARCH SERVICE

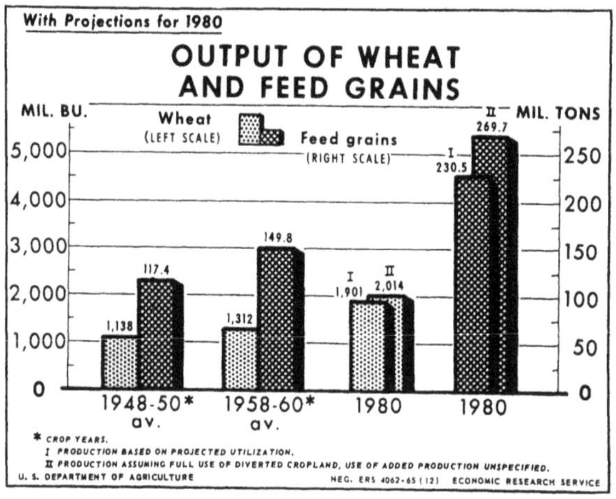
With Projections for 1980
OUTPUT OF WHEAT AND FEED GRAINS
MIL. BU.
MIL. TONS
Wheat (LEFT SCALE)
Feed grains (RIGHT SCALE)
5,000
4,000
3,000
2,000
1,000
0
250
200
150
100
50
0
1,138
117.4
1,312
149.8
I 1,901
II 2,014
I 230.5
II 269.7
1948-50* av.
1958-60* av.
1980
1980
* CROP YEARS.
I PRODUCTION BASED ON PROJECTED UTILIZATION.
II PRODUCTION ASSUMING FULL USE OF DIVERTED CROPLAND, USE OF ADDED PRODUCTION UNSPECIFIED.
U. S. DEPARTMENT OF AGRICULTURE
NEG. ERS 4062-65 (12)
ECONOMIC RESEARCH SERVICE

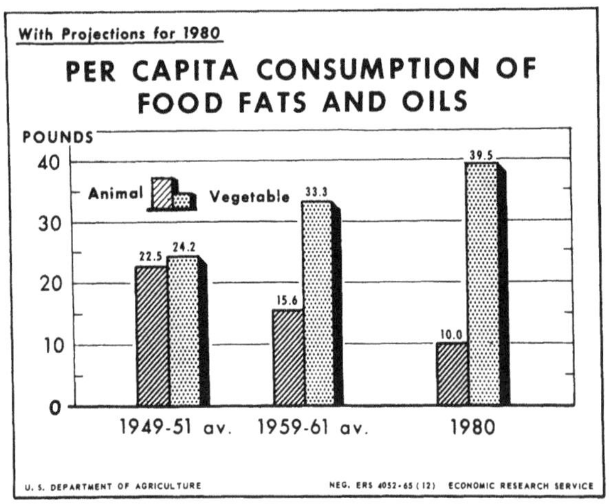
With Projections for 1980
PER CAPITA CONSUMPTION OF FOOD FATS AND OILS
POUNDS
Animal
Vegetable
40
30
20
10
0
22.5
24.2
15.6
33.3
10.0
39.5
1949-51 av.
1959-61 av.
1980
U. S. DEPARTMENT OF AGRICULTURE
NEG. ERS 4052-65 (12)
ECONOMIC RESEARCH SERVICE

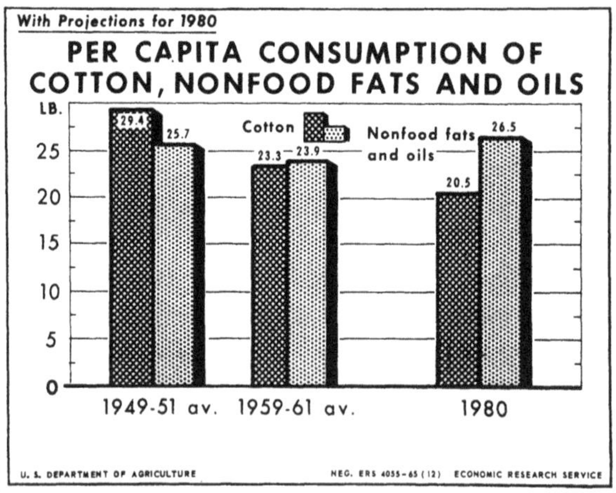
With Projections for 1980
PER CAPITA CONSUMPTION OF COTTON, NONFOOD FATS AND OILS
LB.
Cotton
Nonfood fats and oils
25
20
15
10
5
0
29.4
25.7
23.3
23.9
20.5
26.5
1949-51 av.
1959-61 av.
1980
U. S. DEPARTMENT OF AGRICULTURE
NEG. ERS 4055-65 (12)
ECONOMIC RESEARCH SERVICE

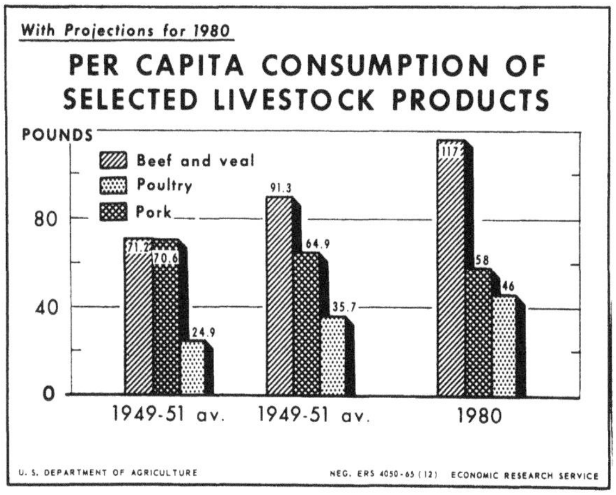

With Projections for 1980
PER CAPITA CONSUMPTION OF SELECTED LIVESTOCK PRODUCTS
POUNDS
Beef and veal
Poultry
Pork
80
40
0
71.2
70.6
24.9
91.3
64.9
35.7
117
58
46
1949-51 av.
1949-51 av.
1980
U. S. DEPARTMENT OF AGRICULTURE
NEG. ERS 4050-65 (12)
ECONOMIC RESEARCH SERVICE

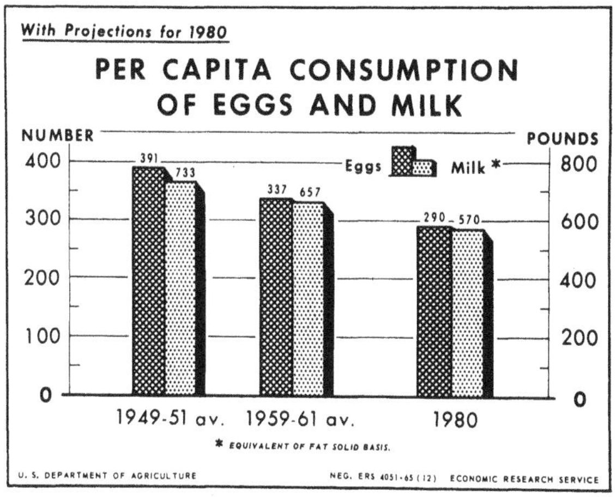

With Projections for 1980
PER CAPITA CONSUMPTION OF EGGS AND MILK
NUMBER
POUNDS
Eggs
Milk *
400
300
200
100
0
800
600
400
200
0
391
733
337
657
290
570
1949-51 av.
1959-61 av.
1980
* EQUIVALENT OF FAT SOLID BASIS.
U. S. DEPARTMENT OF AGRICULTURE
NEG. ERS 4051-65 (12)
ECONOMIC RESEARCH SERVICE

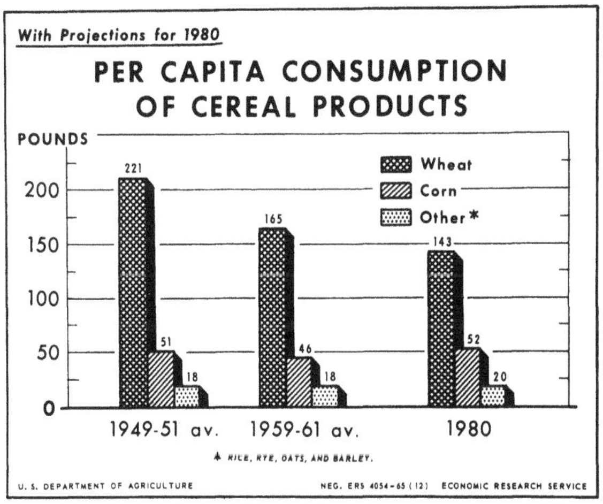

With Projections for 1980
PER CAPITA CONSUMPTION OF CEREAL PRODUCTS
POUNDS
Wheat
Corn
Other *
200
150
100
50
0
221
51
18
165
46
18
143
52
20
1949-51 av.
1959-61 av.
1980
* RICE, RYE, OATS, AND BARLEY.
U. S. DEPARTMENT OF AGRICULTURE
NEG. ERS 4054-65 (12)
ECONOMIC RESEARCH SERVICE

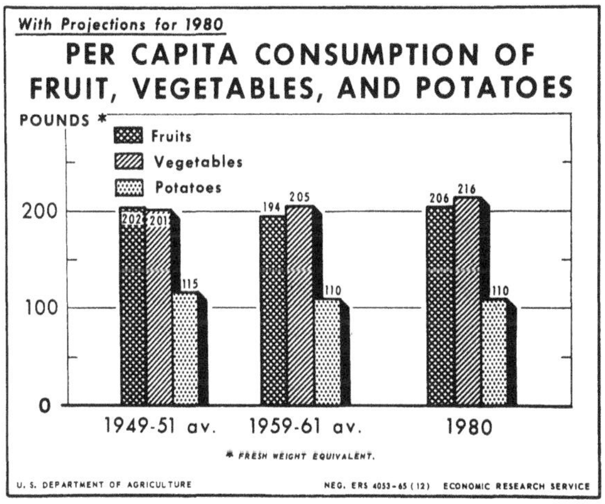

With Projections for 1980
PER CAPITA CONSUMPTION OF FRUIT, VEGETABLES, AND POTATOES
POUNDS *
Fruits
Vegetables
Potatoes
200
100
0
202
201
115
194
205
110
206
216
110
1949-51 av.
1959-61 av.
1980
* FRESH WEIGHT EQUIVALENT.
U. S. DEPARTMENT OF AGRICULTURE
NEG. ERS 4053-65 (12)
ECONOMIC RESEARCH SERVICE

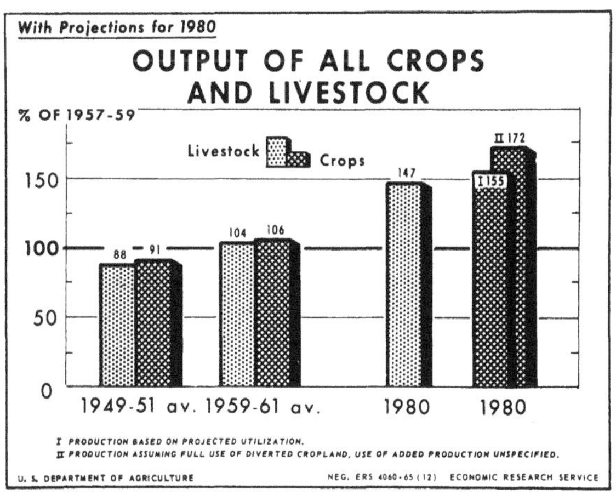

With Projections for 1980
OUTPUT OF ALL CROPS AND LIVESTOCK
% OF 1957-59
Livestock
Crops
150
100
50
0
88
91
104
106
147
I 155
II 172
1949-51 av.
1959-61 av.
1980
1980
I PRODUCTION BASED ON PROJECTED UTILIZATION.
II PRODUCTION ASSUMING FULL USE OF DIVERTED CROPLAND, USE OF ADDED PRODUCTION UNSPECIFIED.
U. S. DEPARTMENT OF AGRICULTURE
NEG. ERS 4060-65 (12)
ECONOMIC RESEARCH SERVICE

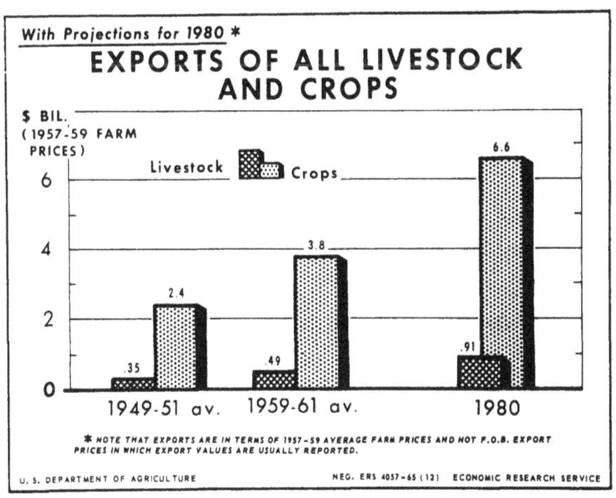

With Projections for 1980 *
EXPORTS OF ALL LIVESTOCK AND CROPS
$ BIL.
(1957-59 FARM PRICES)
Livestock
Crops
6
4
2
0
.35
2.4
.49
3.8
.91
6.6
1949-51 av.
1959-61 av.
1980
* NOTE THAT EXPORTS ARE IN TERMS OF 1957-59 AVERAGE FARM PRICES AND NOT F.O.B. EXPORT PRICES IN WHICH EXPORT VALUES ARE USUALLY REPORTED.
U. S. DEPARTMENT OF AGRICULTURE
NEG. ERS 4057-65 (12)
ECONOMIC RESEARCH SERVICE

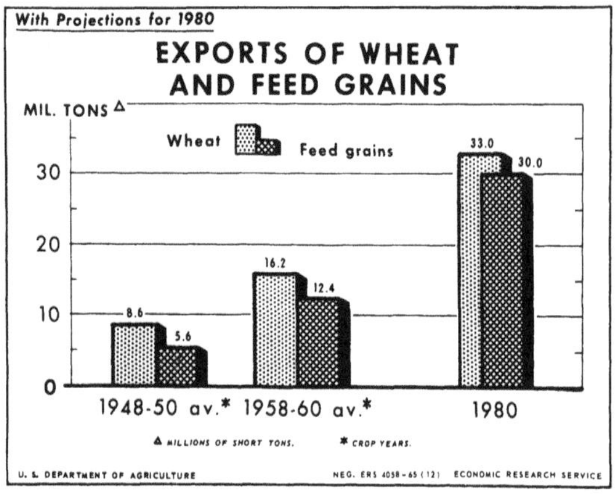

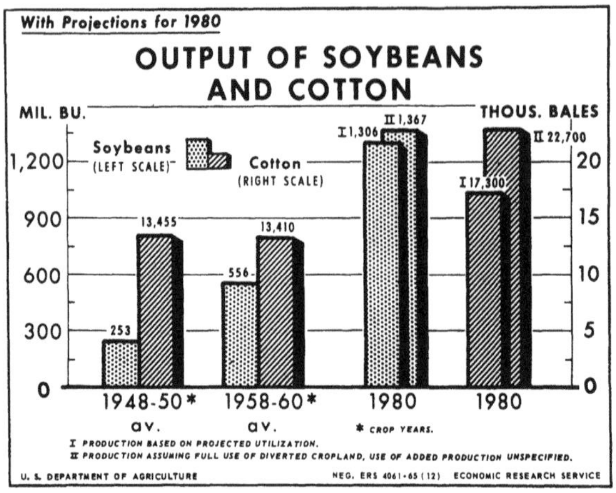

5 – 16 *A look ahead to agricultural production and consumption, 1980.* (Department of Agriculture)

ity, and capability to produce the major items in the diet, must be considered in planning the future cropland requirements. It may be noted that recent diversions of farmlands to urban uses, transportation uses, and other space uses have taken significant amounts of class I land and lands on which important shares of the early vegetables and fruits have been produced.

POTENTIALS THROUGH LAND DEVELOPMENT. Yet the nation remains in a strong position with respect to the possibilities of new croplands. As noted in an earlier chapter, the National Inventory of Soil and Water Conservation Needs revealed that 638 million acres fall within land capability classes I, II, and III, nearly 238 million acres of which are in pasture and woodland. Additionally, 169 million acres were shown to be in land capability class IV, only about one-third of which is presently cultivated. Much of the uncultivated land of these classes undoubtedly would be available if needed, but a significant amount requires clearing, draining, or other improvements to fit it for cultivation. Moreover, a considerable amount is situated in small, irregular patches which cannot be farmed efficiently with modern machinery. Furthermore, the nature of many farm units is such that they require some land to be in pasture, woodlots, or other non-cultivated uses.

Nevertheless, it would appear that the opportunity to expand cropland during the next half century is excellent. In the long run the major deterrent to further expansion may come from the competition of other uses. The demands for timber, recreation, urban development, and the like will also be increasing.

Summary

The appraisals presented in this chapter indicate that rapid strides are being made to bring the United States into a conservational relationship with croplands. Land capability survey work is in an advanced stage; land capability is being increasingly considered in land use; land management practices to maintain or improve land capability are well developed; research is advancing; farmers are coming to consider related resources; farm planning is progressing rapidly; and the Department of Agriculture is projecting needs. In the past three decades progress has been commendable, and present trends augur continued progress.

The technology and the tools for the task ahead are available, and they will improve. The probable area of shortcoming is in application and cooperation. Economic considerations too often inhibit the attainment of the desirable high level of conservation. Political considerations also are limiting factors. In some cases, human inertia and individualism still hold back implementation of technology known to be better. There also is a genuine lack of acceptance of the ethical concepts of stewardship by managers and society in general. The projections of needs for 2010 suggest that the nation can no longer tolerate any of these factors which promote land deterioration or misuse; nor can it tolerate diversion of high quality cropland to irreversible uses, particularly space uses, when alternatives exist in available lower quality land.

FOOTNOTES

[1] Hugh H. Wooten and James R. Anderson, "Agricultural Land Resources in the United States," *Agriculture Information Bulletin No. 140* (Washington: Agricultural Research Service, Department of Agriculture, June, 1955), 107 pp.

[2] These are published as state reports.

[3] *Soil, the Yearbook of Agriculture*, 1957 (Washington: Department of Agriculture, 1957), p. 293.

[4] *Ibid.*, p. 295.

[5] For details see *ibid.*, pp. 297–305, or "Farming Terraced Land," *Leaflet No. 335* (Washington: Soil Conservation Service, Department of Agriculture, November, 1952), 16 pp.

[6] The foregoing information is based on an interview with Dr. F. F. Bell, Soil Scientist, University of Tennessee. Dr. Bell is one of the leaders in this work.

[7] "The Great Plains Conservation Program, A Progress Report," PA 669 (Washington: Department of Agriculture, November, 1965), p. 2.

[8] See Tom Dale, "For Insurance Against Drought," *Farmer Bulletin No. 2002* (Washington: Department of Agriculture, March, 1950), 22 pp., and "Facts about Soil Erosion and Dust Storms on the Great Plains," *Leaflet No. 394* (Washington: Department of Agriculture, June, 1955), 8 pp.

[9] Arthur F. Ferber, "Windbreaks in Conservation Farming," *Miscellaneous Publication No. 759* (Washington: Soil Conservation Service, Department of Agriculture, August, 1958), 22 pp.

[10] See "Irrigation on Western Farms," *Agriculture Information Bulletin No. 199* (Washington: Soil Conservation Service, Department of Agriculture [also Bureau of Reclamation, Department of the Interior], July, 1959), 53 pp., and Allan W. McCulloch and Wayne D. Criddle, "Conservation Irrigation," *Agriculture Information Bulletin No. 8* (Washington: Soil Conservation Service, Department of Agriculture, May, 1950), 12 pp.

[11] See Wallace L. Anderson and Laurence V. Compton, "More Wildlife Through Soil and Water Conservation," *Agriculture Information Bulletin No. 175* (Washington: Soil Conservation Service, Department of Agriculture, March, 1958), 15 pp.

[12] The story of this cooperative effort is told in P. A. Waring, "Teamwork to Save Soil and Increase Production," *Miscellaneous Publication No. 486* (Washington: Soil Conservation Service, Department of Agriculture, December, 1949), 64 pp.

[13] See as examples, "A 50-Year Look Ahead at U. S. Agriculture" (Washington: Department of Agriculture, June, 1959), 20 pp.; *Land and Water Resources, A Policy Guide* (Washington: Department of Agriculture, May, 1962, slightly revised September, 1962), 73 pp.; and R. F. Daly and A. C. Egbert, "A Look Ahead for Food and Agriculture" *ERS-277* (Reprinted from *Agriculture Economic Research*, vol. XVIII by the Economic and Statistical Analysis Division, Economic Research Service) February, 1966), 9 pp., and "Statistical Supplement," 15 pp.

[14] *Land and Water Resources, A Policy Guide, op. cit.*, Part III.

[15] R. F. Daly and A. C. Egbert, *op. cit.*

[16] The only exception being during the 1930's, the period of drought and depression.

[17] "A Look Ahead for Food and Agriculture," *op. cit.* p. 1.

[18] *The Farm Index*, (June, 1966), p. 6.

6

Grazing Lands

NEARLY 60 PER-CENT of the land area of forty-eight of the states produces vegetation suitable as forage for grazing animals. Census data are not sufficiently detailed or complete to indicate the importance of this resource in the agricultural economy because they report only the area utilized for grazing. It has been estimated, however, that one-third of all feed consumed by domestic livestock is provided directly by pastures and ranges.[1]

In any assessment of the importance of the range and pasture lands, it must be recognized that their value is not due solely to grazing. They comprise a major portion of the watersheds of many streams, and the manner in which they are utilized and managed has direct bearing upon water supply and runoff. Moreover, they contain large areas of forest and woodland, of scenic beauty and recreational value, which also serves as wildlife habitat. Intimate relationships, therefore, exist among grazing management and management for water supplies, forest products, recreation uses, and wildlife. The values of these associated uses in any given area may be such that grazing can have an economic impact far beyond the value of livestock products alone.

In the face of increasing needs in all of these categories, it is clear that the grasslands have a highly significant role in the nation's future. Full and multiple utility can be achieved only through the establishment of management based upon the conservation philosophy. This chapter will appraise the extent and character of the grasslands and their use for grazing. Other aspects of their values will be treated specifically in succeeding chapters.

Before progressing further it will be of value to comment on several terms. In the United States, *range* and *pasture* have come to refer to different types of grazing lands without being clearly and concisely defined. Among grassland scientists the term *range* is applied to "grasslands that are a natural climax or which can be developed from a climax by natural or induced ecological succession."[2] This includes the tall-grass prairies of the Midwest, the short-grass prairies of the Great Plains, the semi-arid and arid shrub-grasslands of the intermontane West, the Pacific bunch-grasslands of the Northwest and Central Valley of California, as well as many of the natural grasslands developed by clearing and brush control in the pine region of the South. Included also are other western and southern woodland and forested areas that have sufficient grass to be profitably grazed (Figure 6 – 1). The term *pasture* has come to refer to "grassland areas that have been artificially established to domestic forage plants" (Figure 6 – 2).[3] The distinction between the terms *range* and *pasture* however, has become confused by a second mean-

6 – 1 *Range in good condition.* This scene, in the Wasatch National Forest of Utah, shows a natural vegetation complex that has been under good management. (Forest Service)

6 – 2 *Pasture on capability class II land.* This five acre field, located in Erath County, Texas, was seeded to bur clover in 1936. Twenty-eight cows graze the pasture every night the year round. In interest of weed control it is mowed three times each year. (Soil Conservation Service)

ing commonly ascribed to the latter; the word *pasture* is used also to define a fenced portion of a range that may involve meadow, a reseeded area, or simply an area of high quality forage. Further confusion is lent by the census of agriculture which, as nearly as the authors can determine, uses the term *pasture* to refer to all grasslands (including grazed woodland and forest land) in farm, ranch, and other private ownership, *grazed land* to refer to grassland (including grazed woodland and forest land) in public ownership and does not employ the term *range* at all. Thus, precise definition is not possible under present usage. For the purposes of this discussion, the authors are limited to census categories in the use of data; however, the term *grazing land* will be used, generally, to refer to all categories, and *range* will be used to mean natural grasslands that are under extensive management and use.

Extent and Character

The area used for grazing in the United States totals nearly one billion acres. This is comprised of 633 million acres in grassland pasture and range land, and 245 million acres of woodland and forest land grazed part of the year (Table 6–1). In addition, livestock annually graze about sixty-five million acres of cropland used only for pasture; and an estimated eighty to ninety million acres in planted fields are pastured for short periods each year. These include after-harvest pasturage of stubble, cornstalk, and hay fields, as well as winter cover crops and grains, grasses, and legumes planted in the fall specifically for winter grazing.

GENERAL CHARACTER OF THE GRAZING LANDS

Basic to the establishment of a conservational adjustment is a complete knowledge of the ecology of the grazing lands. Nearly every part of the nation has some grassland, and therefore many different combinations of physical conditions and plants are involved. As a result, suitable use practices in one area are not necessarily satisfacory for another.

The bulk of the grazing lands is not capable of supporting crop agriculture. The National Inventory of Soil and Water Conservation needs (see Table 3–3), however, did list 113,852,000 acres of pasture and range land of capability classes I, II, III, and an additional 53,937,000 acres of capability class IV. The remainder is restricted by limited moisture, short frost-free season, roughness or steepness of slope, poor quality soil, or poor drainage, although some could be made fit for tillage by irrigation, drainage, clearing, and careful management.

The principal native grazing lands are in the West and the lower South (Figure 6 – 3). More than three-fourths are west of the one hundredth meridian; around 225 million acres are located in the Great Plains states and occupy land too dry for crop farming; and about 500 million acres are found in the eleven western states, for the most part on land too dry or rough for cultivation. An important part of the western land is mountain woodland which is

TABLE 6–1
PASTURE AND RANGE, UNITED STATES, 1959*

Pasture and range	48 states	50 states
Grassland pasture and range:	Million acres	Million acres
Private and other non-Federal pasture[1]	473	474
Federal range[2]	157	159
Total grassland pasture and range	630	633
Forest and woodland pasture and range:		
Private and other non-Federal pasture[1]	160	161
Federal range[2]	83	84
Total forest and woodland pasture and range	243	245
GRAND TOTAL PASTURE AND RANGE[3]	873	878

* "Major Uses of Land and Water in the United States," *Agriculture Economic Report* No. 13 (Washington: Department of Agriculture July, 1963) p. 8.

[1] Non-federal pasture includes Indian, state, and local government land. From U.S. Department of Agriculture, Conservation Needs Inventory and Land Use Inventory, 1957-59 (unpublished). Excludes wild hayland harvested for hay; includes recently improved pasture.

[2] Federal range usable and available for grazing domestic livestock, compiled from records and reports of principal federal land management agencies. In addition to the acreage of federal range given here, Alaska has fifty to sixty million acres of potential range used almost solely by wildlife.

[3] Does not include cropland used only for pasture. This is included in the total cropland acreage.

moist enough for trees but generally too rough for crops or with a frost-free season too short for crops. In addition there are noteworthy (but small in total) pasture developments in essentially all of the irrigated areas. Also in the humid portion of the Pacific Northwest improved tame plant pastures afford a large share of the local grazing.

The other important region of native forage extends from central Texas along the coastal plain to North Carolina. This belt contains close to 100 million acres of uncultivated forage-producing land. It is composed of open woods, cut-over tracts, wet prairies, marshes, and abandoned fields. In recent years tame grass pastures and grazed crops are becoming increasingly significant in aiding the feed supply in some areas of the region.

The remainder of the grazing land is in the humid East, mainly in the form of tame plants in improved farm pastures. There is, however, some grazing of abandoned cultivated fields and farm woods. A more precise breakdown of the distribution of all grazing land is provided in Table 6–2.

MAJOR RANGE TYPES

The millions of acres of native forage plants can be classified in several broad range types which differ greatly in vegetative character and composition and in associated climate, topography, and soil. Such a classification is useful for basic research and the development of management practices. The range types presented here are based on a summation of material presented in *Grass, the Yearbook of Agriculture, 1948.*[4]

1. TALL GRASS. Tall-grass prairie existed throughout the Midwest and the eastern Great Plains before the coming of white man. Owing to favorable surface, soils, and climate, a preponderant share has been converted to cropland. Today about twenty million acres remain, primarily in the eastern Great Plains.

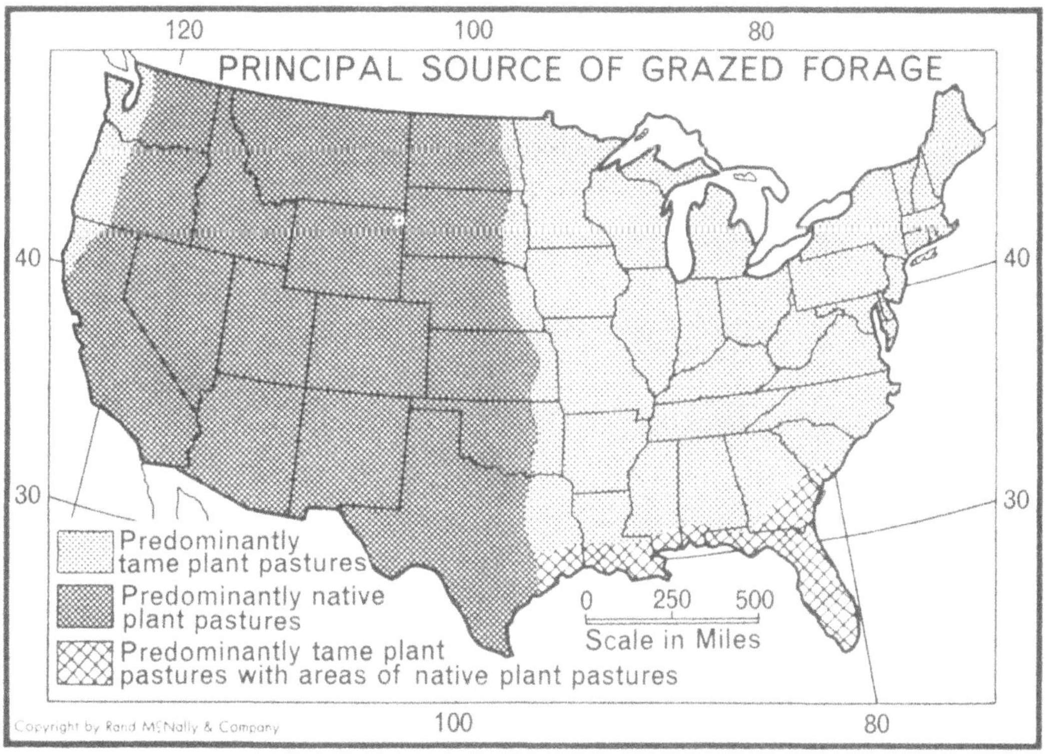

6 – 3 *Principal source of grazed forage.* **(Grass, the Yearbook of Agriculture, 1948, p. 28)**

The mixture of tall grasses and showy herbs, with an understory of short grasses and sedges, produces abundantly on fertile chernozemic soil and under annual precipitation ranging from twenty inches in North Dakota to forty inches or more in parts of Kansas and Oklahoma. As the result of severe drought in the 1930's and heavy grazing, however, low-value shrubs and weeds in many places have replaced valuable bluestems, Indiangrass, switchgrass, and some other characteristic species.

When it is in good condition its grazing capacity is high, requiring only three-fourths to one and one-half acres to supply an animal-unit-month of forage. But the tall grass does not cure well when left standing, and is therefore of little value for late fall and winter grazing. Year-long grazing of breeding cattle with supplemental winter hay is, however, a usual practice in the Sandhills of Nebraska. The tall-grass prairie has high recuperative power. Nevertheless, conservational management practices are necessary to avoid depletion.

2. SHORT GRASS. The short-grass range, stretching from the Texas Panhandle to beyond the Canadian border, and from the foothills of the Rockies into the central Dakotas with an arm extending westward from Texas across central New Mexico to northeastern Arizona, is the largest and most important grassland type in the nation. This great area naturally produces such valuable forages as the grama grasses, buffalograss, bluestem or the western wheatgrass, and needle-and-thread. Forage production coincides with the occurrence of spring and early summer precipitation, which totals about three-fourths of the annual average of thirteen inches in the Northern Plains, and somewhat less of the slightly higher annual total in the Southern Plains. The grasses have the important property of curing on the stem toward the end of the dry, hot summer and retaining much of the

TABLE 6–2

PASTURE AND RANGE BY TYPE, OWNERSHIP, AND REGION, 1959*

Region	Private, state, and Indian: Open permanent pasture and range[1]	Private, state, and Indian: Woodland pasture and range[2]	Federal range[3]	Total pasture and range[4]
	1,000 acres	1,000 acres	1,000 acres	1,000 acres
Northeast	7,991	3,464	24	11,479
Lake States	8,255	6,011	20	14,286
Corn Belt	21,772	12,629	292	34,693
Northern Plains	76,744	2,544	3,484	82,772
Appalachian	12,939	8,408	109	21,456
Southeast	13,929	18,673	1,076	33,678
Delta States	9,242	26,723	1,952	37,917
Southern Plains	108,855	29,713	1,189	139,757
Mountain	182,133	34,300	192,536	408,969
Pacific	31,130	17,830	39,718	88,678
Alaska	350	520	2,120	2,990
Hawaii	646	441	0	1,087
TOTAL	473,986	161,256	242,520	877,762

* "Major Uses of Land and Water in the United States," *op cit.*, pp. 35–36.

[1] Includes private, state, local government, and Indian land used for pasture and grazing, mostly in farms, and other non-Federal grassland range (not cropland and not woodland) not in farms.

[2] Includes forest and woodland pastured in farms and woodland and other non-Federal forest grazed or classified as usable for grazing not in farms.

[3] Includes some 159 million acres in grassland pasture and range, and over 83 million acres of woodland, and forest grazed during part of the year in the fifty states.

[4] Includes both grassland pasture and grazing land and woodland and forested land used for grazing.

nutritive quality during the winter. Grazing capacity ranges from two and one-half to four acres per animal-unit-month in the Northern Plains to five to ten acres in the Southwest. Range livestock grazing is important in the spring, summer, and fall, and in favorable winters it may continue year-long (Figure 6-4).

Drought is common throughout the areas where this type of grass is found, but especially so in the Southwest. Drought severely reduces forage production and can accelerate wind erosion. The forage species are highly resistant to both drought and grazing, but, once they have been depleted, require several years of good management and favorable weather for complete recovery. Because the area is now mostly in private ownership, management decisions are individual matters. As a result, there is considerable variety in the local quality of the range.

3. SEMI-DESERT GRASS. This range type occupies central and southwestern Texas and much of New Mexico and Arizona. Good seasonal forage production and curing qualities, along with mild winters, place it among the best year-long ranges. Variations in topography, elevation, climate, and soils, however, cause considerable differences in local forage. The grama grasses and curly-mesquite reach their maximum production in the higher elevations of the western

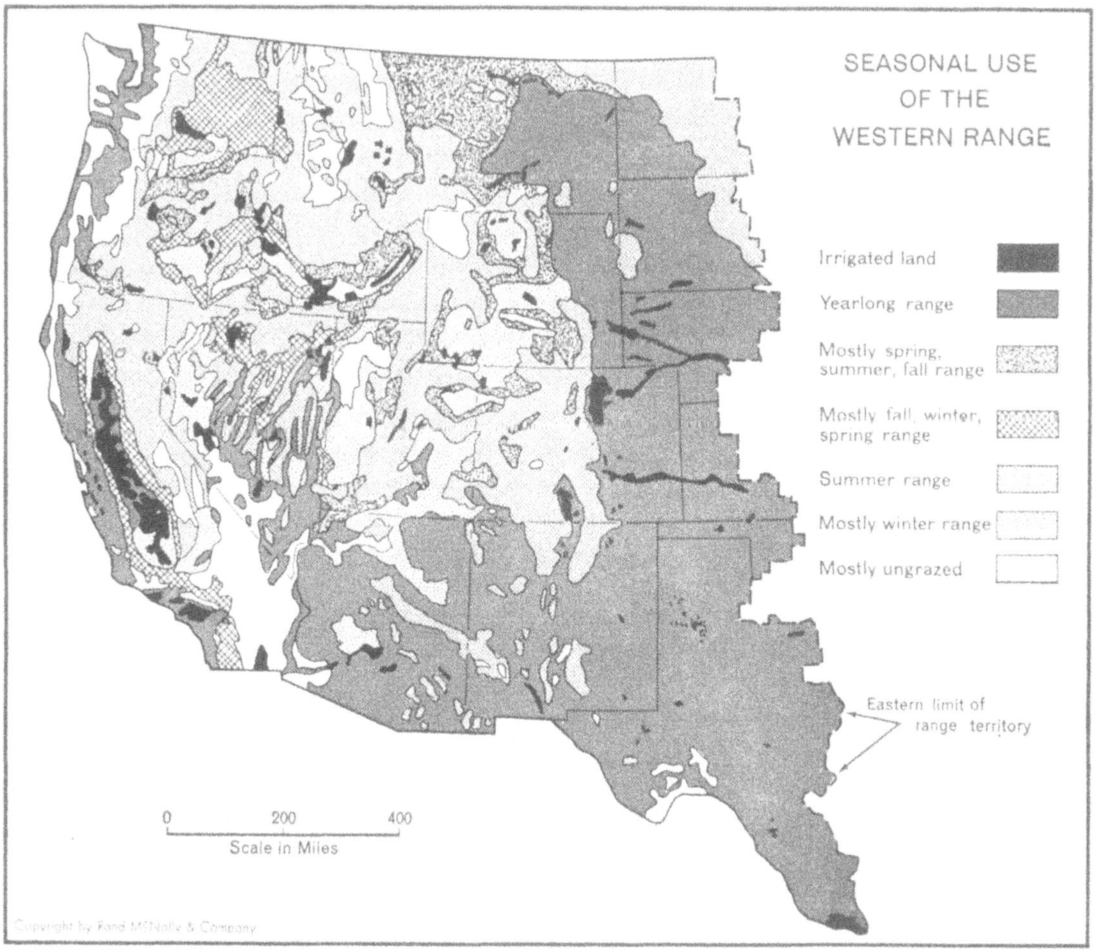

6-4 ***Seasonal use of the western range.*** **(Land, the Yearbook of Agriculture, 1958, p. 275)**

part of the area where annual precipitation reaches fifteen to eighteen inches. At lower elevations, where annual precipitation may be as low as eight inches, black grama, the three-awns, and dropseeds usually predominate, with a heavy sprinkling of shrubs. Dwarf trees, yuccas, and cacti also are common. Some shrubs, such as saltbush, mesquite, ratany, and scrub oak, are palatable, providing forage especially in the winter and spring and for year-long goat grazing. On a characteristic site this range type will support an average grazing capacity of six to seven acres per animal-unit-month.

Throughout the area where this type of grass is prominent, growing conditions are severe and drought is frequent. Overgrazing has been common, reducing the quality of the range. Lighter stocking and improvements in other aspects of management are called for over much of the area.

4. PACIFIC BUNCH GRASS. The Pacific bunch grass prairie has been changed greatly through overuse, but is still a valuable grassland. Formerly occupying much of the unforested area from western Montana through the Pacific Northwest to southern California, it has been reduced to a rather narrow belt along foothills throughout the area. The original forage plants (bluebunch wheatgrass, Idaho fescue, pine bluegrass, and California needlegrass), although well adapted to the long, dry summers and moist winters of the area, did not withstand heavy grazing well, and have been largely replaced by annuals. These annuals (alfileria, bur-clover, slender oatgrass, cheatgrass, and others) provide good forage when green. Growth begins in the fall, the plants remain green during the winter, and abundant growth occurs in the spring. As the hot weather of late spring progresses, the plants dry up and decline in palatability and nutritive quality. Thus the range is of low grazing value in summer.

Spring-fall grazing by cattle and sheep is the common use of this type of grass in the north, but in the south or where perennials remain, year-long use is feasible. Average conditions require four to six acres per animal-unit-month. It is believed that with improvements in management the productivity could be doubled.

5. SALT DESERT SHRUB. This range type, used primarily for winter grazing, occurs under low precipitation in central Nevada, Utah, southwestern Wyoming, and western Colorado (Figures 6 – 5 and 6 – 6). In the vegetation complex spiny hopsage, black sagebrush, and winterfat are the most palatable and provide considerable forage when the grass understory is unavailable because of snow. Other browse species, such as small rabbitbrush, greasewood, and horsebrush are of little forage value, and they spread rapidly when heavy grazing of the more palatable species reduces the competition for soil moisture and nutrients. In the understory, blue grama, sand dropseed, galleta, and Indian rice grass are the most important forage plants.

Although forage production is less than one-third the potential and the grazing capacity averages eighteen to twenty acres per animal-unit-month, the type is of considerable value as winter range. In many areas, however, lack of dependable stock water precludes use except when there is snow.

6. SOUTHERN DESERT SHRUB. The grazeable parts of the southern desert-shrub range are restricted to the edges of the Mohave Desert and the valleys and adjacent mesas along the Colorado, Gila, lower Rio Grande, and Pecos rivers. Most have little grazing value because of low and undependable precipitation. Growing conditions are so severe that depletion can result from only a year of too-heavy grazing. Once depleted, palatable grasses recover very slowly. Forage production and grazing occur if and when it rains (Figure 6 – 7).

The common practice is to use this range only when the annuals produce forage, and stock water is available. This saves forage on adjacent ranges, and reduces supplemental feeding requirements. During unusually favorable winters, perhaps once in five years, the warmer sections may provide winter sheep grazing.

7. PINON-JUNIPER. This open, woodland range type occupies thousands of square miles in the area extending from southwest Texas to central Oregon. It occurs irregularly just below the elevation of ponderosa pine on shallow, stony soil and rough topography, alternating with sagebrush which occupies the better soils.

Valuable forage growth is restricted mainly to the portions in Arizona, New Mexico, Colorado, and Utah. In the north, summers are dry and hot and annual precipitation is less than fifteen inches. In the southwest, precipitation is higher and comes largely in the hot summer period. Thus spring-fall grazing is common in the north, but in the south where the forage cures well, year-long use is prevalent.

The gramas, bluebunch and bluestem wheatgrasses, galleta, and shrubs such as mountain mahogany and cliffrose give the type an average grazing capacity of eight to ten acres per animal-unit-month. There is great variability, however; areas in good condition have capacities somewhat greater, but some areas in the Southwest, once productive, have been so de-

6–5 Winter range in the Big Horn Basin, Wyoming. **Large acreages of the western range look like this. Such land is restricted by climate, surface relief, and soils from crop agriculture. Under good management, however, it can be used for grazing by livestock, wildlife habitat, and watershed. (Bureau of Land Management)**

pleted by overgrazing and thickening of juniper as to have little value for livestock forage.

8. WOODLAND-CHAPARRAL. Woodland-chaparral, varying from an open forest of oak and other hardwoods with an understory of herbaceous plants and shrubs to dense brush fields, occurs throughout the foothills of the Central Valley of California. On its lower fringes the type is quite open and now supports a herbaceous annual plant cover that provides forage for fall, winter, and spring grazing. At higher elevations and in southern California, the many shrub species form thickets of little grazing value, but of high significance as watershed protection.

Introduced annuals such as alfileria, slender oatgrass, and bur-clover provide an average grazing capacity of two to three acres per animal-unit-month on the grazeable portions of the type. These plants are palatable and nutritious throughout the growing period (fall through spring), but during the summer are dry and of low value for grazing except when supplemental feed is provided.

9. OPEN FORESTS OF THE WEST. The open forests of the mountainous West constitute the second largest and most widely distributed range type. In it forests are intermingled with meadows, and open grassy areas with grasses, herbs, and browse species (Figure 6 – 8). Some spring-fall livestock grazing is possible, and big-game winter grazing is found at the lower elevations, but, because most of the type lies at high altitudes where snow comes early and stays late, summer use predominates.

Grazing capacity varies considerably throughout these open forests, but generally averages six to ten acres per animal-unit-month in summer. Meadows and alpine grasslands in good condition often have a potential capacity ten to fifteen times greater than adjacent forested range.

6 – 6 ***Typical sagebrush range in Nevada.*** **This range type has been considerably depleted through overuse. Low carrying capacity and limited stock water currently limit utility. Improvement requires the restriction of use to allow natural regeneration. In the higher precipitation areas, sagebrush control is economically practical. (Bureau of Land Management)**

10. SOUTHERN FOREST RANGES. The forested region extending from Virginia to the eastern portions of Texas and Oklahoma, although primarily important for timber growing, contains a large reservoir of native forages (Figure 6-9). High precipitation and mild winters provide a long growing season. Most of the forest range is in private ownership and, when grazed, is generally used all year long. Vegetation in this area can be grouped into four range types. The wiregrass type occurs in the longleaf slash pine flatwoods of the Southeast and is characterized by pineland three-awn, carpetgrass, and bluestems. These grasses provide good grazing from March until June or July with one-half to five acres required per animal-unit-month. About twice this acreage is required during late summer and fall, and supplemental feeding should be practiced during winter.

The broom sedge or bluestem type is characteristic of the long-leaf pine and "upland" hardwood forests of southern Mississippi, Arkansas, east Texas, eastern Oklahoma, and the Piedmont. Bluestems, panicgrasses, paspalums, and weeds provide reasonably good forage in spring and early summer. Grazing capacity varies from one acre per animal-unit-month on old fields and open forest to six acres in dense shortleaf-loblolly-hardwood forests. The grazing value of the type is very low in late fall and winter; however, carpetgrass and Bermuda grass which occur on firebreaks provide good fall grazing.

The switch cane or reed forage type is found mainly in the pond pine and river bot-

6-7 Desert shrub range in the Southwest. Low and uncertain precipitation in this area greatly limits the utility for grazing. (Bureau of Land Management)

6–8 Typical conifer and sagebrush types in the mountains of the West. **Such land plays a vital role in the western range forage supply, serving as summer range. (Bureau of Land Management)**

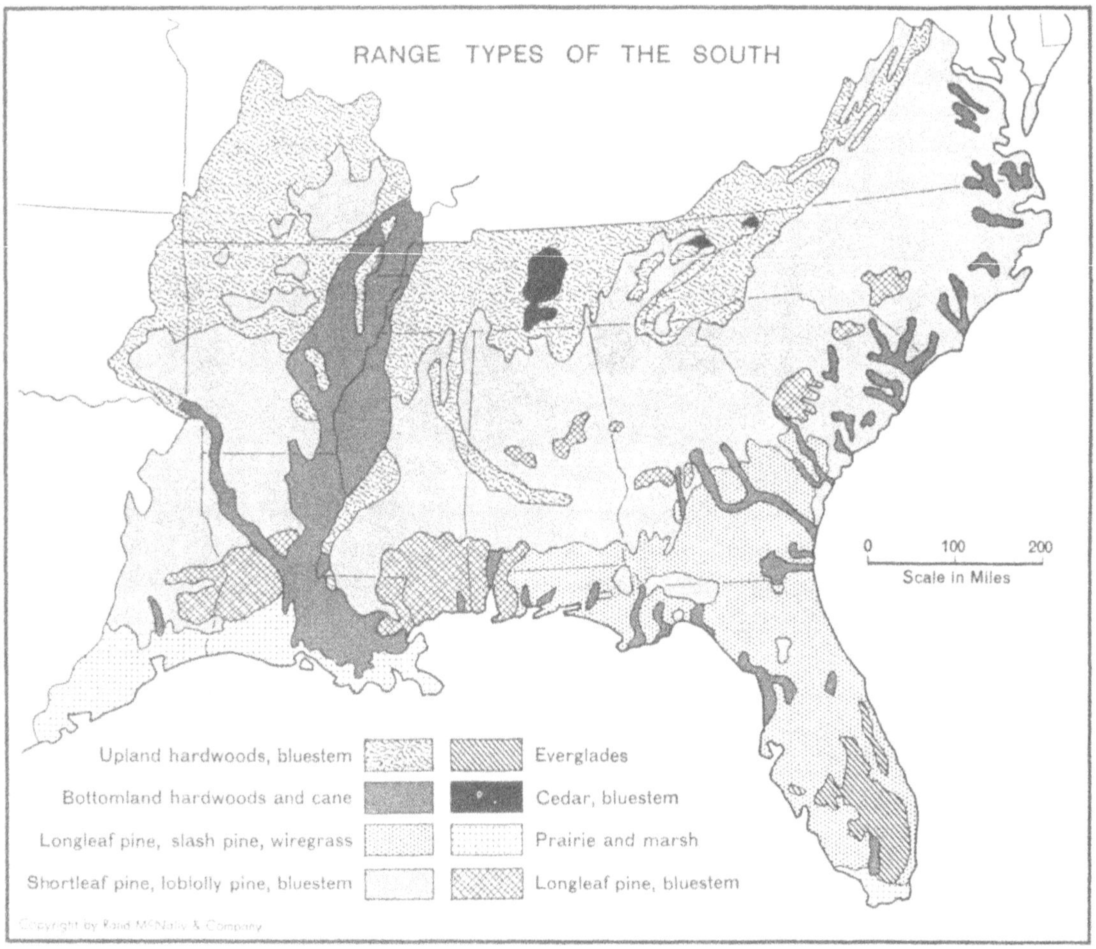

6–9 Range types of the South. **(Bulletin N.S. 9, American Society of Range Management [September, 1955])**

tom forests throughout the Coastal Plain. Grazing capacity varies from three to twelve acres per animal-unit-month from May to November. The acreage requirement during the remainder of the year approximately doubles.

The bottom-land hardwood type is typical of the Mississippi Delta and swampy areas throughout the Coastal Plain. In it shrubs and vines are valuable for winter grazing with capacity varying from eight to twelve acres per cow during the cool season. Sprouts of many of the valuable hardwoods are relished, and grazing must be carefully controlled to avoid damage.

11. COASTAL PRAIRIE. Extensive year-long grazing is provided by the coastal prairie and marshland types which occupy low, poorly drained lands along the Atlantic and Gulf Coasts and parts of the Everglades. Tall, coarse grasses, sedges, and rushes characterize these areas over which water stands at depths of several inches part of the year.

12. OTHER TYPES. Throughout the Missouri Ozarks, the Ohio and Mississippi river basins, and the Northeast are scattered small areas of native forage. These occupy mountain glades and balds, open forest areas, abandoned cultivated fields, and farm wood lots. They are comprised of varied native grasses, herbs, browse species, and hardwood sprouts, as well as introduced species. Grazing capacities are often comparatively high during the warm season.

MAJOR PASTURE TYPES

It is impossible to determine the exact area of pasture according to the strict definition, from available data. It will be recognized that some of the vegetative covers included in several of the foregoing types, such as abandoned fields, probably rate classification as pasture. Some of these are fertilized, clipped, and, in general, carefully managed. Their productive capacity is considerably greater than that of the average range types. These, however, are not reported separately by the agricultural census. Thus only inferences of their amount can be drawn from other data. Possibly as much as one-third of the total grazing lands may fit into the broader classification of pasture. Two particularly important categories of pasture are reported on by the agricultural census: Improved pasture and cropland pasture.

IMPROVED PASTURES. These pastures occupied 23,246,000 acres in 1960, distributed primarily through the humid regions of the nation. Under the various programs directed, by the Department of Agriculture toward stabilization and conservation, there has been encouragement to increase improved pasture acreage. The recent annual increase appears to be about 1.5 million acres. Improved pastures generally occupy lands that are unsuited to regular cultivation due to adverse conditions of slope, soil, or drainage. Since the late 1930's, however, some croplands of fair to good quality have been developed as permanent pastures, especially in the South (Figures 6 – 10 and 6 – 11).

In most cases, the base grasses are tame perennial or self-seeding annual species that have been planted; but it is not uncommon for native grasses to be prominent, or even for them to dominate the mixture. Over most of the South the base grasses are Bermuda grass and Dallis grass. Coastal Bermuda, however, is receiving increasing attention; carpet grass is used on moist, sandy soils; and common lespedeza and Dutch white clover are used frequently in

6 – 10 A pasture in Frio County, Texas. **This land, once in native grass, then in cultivated fields, now is a 450-acre pasture of buffelgrass. (Soil Conservation Service)**

grass-legume mixtures (Figure 6 – 12). In the North, Kentucky bluegrass is a common base grass.

The carrying capacities of permanent pastures vary so greatly that it is impossible to state averages. With adequate attention given to grass mixtures, stands, fertilization, and use, many pastures will return values comparable to local croplands; but such attention is not usually given. There are permanent pastures in the South, however, that will carry a cow per acre for a nine- or ten-month growing season and in addition allow the cutting of a ton of hay to provide feed through the nongrazing period.

CROPLAND PASTURES. Croplands used only for pasture constitute the most productive part of the pasture land area because they occupy the better soils which have been fertilized and cultivated and are comprised of high yielding plants.[5]

Most commonly, these are rotation pastures, planned to occupy the land one to five years in the interests of soil building, erosion and water control, and the support of farm livestock enterprises. Often they are planned especially for summer uses.

A variety of grasses and legumes may be planted, depending upon local adaptability and

6 – 11 *Pasture establishment.* This scene shows land recently cleared provided with a drainage system and planted to permanent pasture. (Soil Conservation Service)

6 – 12 *Planting coastal Bermuda grass.* This sprigging machine, with the crew shown, can be used to plant twelve to fifteen acres per day. (Soil Conservation Service)

the individual farm need. The pasture may be based upon perennials such as alfalfa, or annuals like crimson clover, or a mixture of grasses; or grasses and legumes may be planted.

Emergency pastures are a common category of cropland pastures. Every year some part of the nation is affected by droughts, floods, hail, insects, or plant diseases which result in crop and pasture losses. In efforts to salvage some output from land so affected, emergency pastures may be planted. Sudangrass, adapted to a wide variety of soils and climates, is a leading emergency summer forage. It is frequently mixed with soybeans in the Corn Belt and with cowpeas in the South. Sorghums also are employed for summer forage. Grains, especially oats, are good cool season emergency forage plants. Crimson clover, hairy vetch, and Austrian winter peas are also winter hardy.

GRAZING LAND ACREAGE TRENDS

The long-term trend in amount of total grazing land has been slightly downward. This is explained largely in terms of the closing of woodland and forest land to grazing as trees thicken and improved timber management progresses. Shifts of land to urban, industrial, highway development, and other nonagricultural uses are also factors in the decline.

However, the past decade has brought an increase in the portion of land occupied by pasture and open grassland, and in pasture improvement. Part of the increase has come at the expense of former cropland, some of which is land unsuited for regular cultivation. The major share has come from improving other grazing land by clearing bush and seeding, and by clearing woodland, particularly in the South.

Some further decline in amount of grazing land can be expected in the future as the result of competition from other uses. This need not reduce the forage production possibilities of the total grazing land area. Very large potentials exist through pasture and range improvement.

GRAZING LAND OWNERSHIP

Ownership is a more complex variable in grazing land management than in cropland management. Nearly one-third of the grazing land is in public ownership. Most of the better quality lands, however, are in farms and ranches. All ranches and the majority of the farms have some grazing land, thus management decisions on about 600 million acres (including cropland pasture) represent the composite thinking of several million operators.

In 1959, 243 million acres of Federal land were classified as open and usable for grazing. These lands are managed under a conservational philosophy of multiple use and are available only to permit or lease holders—about 65,000 in total. Most of this land is public domain under the control of the Bureau of Land Management and the Forest Service.

Approximately forty-three million acres of state land are available to livestock men under lease or other arrangements. State lands are primarily in the seventeen western states, and are usually of higher quality than public domain range, most having been selected from the better parts of the public domain prior to settlement.

Indian land used for grazing also totals about forty-four million acres. Much of this land is poor quality, often requiring twenty to thirty acres per animal-unit-month. Grazing preference is given to Indians, but a considerable share is available to non-Indians under permits and allotment leases.

In any given ranching unit it is common to find the grazing base made up of a mixture of private land and public land, the latter being used under various forms of permits, leases, and rentals.[6]

THE SIGNIFICANCE OF LAND IN RANCHING

Size of holding is relatively more important in a grazing operation than in any other type of farming.[7] This is especially true in ranching operations and in the West. It results from the basic fact that range land, in contrast to cropland, has comparatively low productive capacity and many acres are required for an economic unit. The fact that a rancher must be able to provide a balanced forage or feed supply for his livestock at all seasons means that he has to have several types of ranges or some cropland for feed production. This necessity may be partly alleviated by selection of a livestock specialty which has major needs coming at the time his range is best suited for grazing. Nevertheless, even a small, family-operated ranch may have or control the use of ten or twelve thousand acres. In many instances as much as 85 per cent of a ranching investment is in land.[8] Because of the variations in range quality it is, however, more common to measure ranch size by animal-unit carrying capacity than by acres of land.

Despite comparatively low productivity per acre, and competition from other livestock areas, most of the Western range land is destined to remain in grazing use. Physical restrictions on other economic utilities of the land will continue to make this necessary. As in other aspects of farming, marginal operations will continue to go out of business and be incorporated into larger and more efficient holdings. Increasing competition from other producing areas and public pressures interested in other values of the range land will encourage increasingly greater attention to improvements in management.

From Depletion to Conservation

The beginning of the human use of grass dates back to the earliest settlements. For the Indians and the early white explorers and settlers, the wide expanses of open grasslands and woodlands supported an abundance of game which was a key element in their food supply. The important use of native grasses for livestock forage was delayed for a century after the first colonial settlement.

The natural vegetation of the New England and Middle Atlantic littoral contained few plants which were suitable for pasture or hay. Some of the early ships bringing colonists brought livestock whose manure and litter contained seeds of bluegrass, white clover, and other English forage plants that were adapted to the eastern United States. These grasses which were introduced eventually spread through much of the area.[9] The agriculture of New England was patterned after the mother country; pasture and cultivation were considered separate types of land use rather than integrated farming operations. The role of forage in livestock nutrition was not understood and, therefore, the management of grasses and legumes did not become common until late in the Colonial period.

The first range cattle industry was established in the South. By the end of the seventeenth century, southern cities were receiving a meat supply from this source, and some herds were being trailed northward. By the Revolutionary War period the development of pastures and hay was lagging behind the increase in livestock; as a result, cattle were of poor quality and overgrazing was common.

Soon after the turn of the eighteenth century, settlers began to use the grasslands of the Ohio Valley—which were then considered to be unsuited for cultivation with the tools and techniques available—for range. The herds, foraging in progress, were driven to markets east of the Appalachians during the warm season. The building of the railroads ended this period and marked the beginning of the integrated raising of grains and livestock in the Corn Belt.

By the time of the Civil War, settlement was at the margin of the Great Plains, and a range livestock industry was established, particularly in the grasslands of Texas. The end of the war marked the beginning of a short period of unprecedented resource exploitation and deterioration; the victim was the western range. The wealth to be had from the native forage of that area attracted foreign capital, principally from Britain and France. Many large cattle ranches, far larger than those of today, were developed in those early years. By the 1870's the range cattle industry was "Big Business" throughout the Great Plains, and a push westward across the Rockies was under way. Enormous herds grazed on what was practically a single open "pasture" thousands of square miles in extent.[10] The cattle population soon reached, and then passed, the safe grazing capacity. This problem was climaxed by a critical drought and the severe winters of 1885 to 1886 and 1886 to 1887. Tremendous losses from which many never recovered were experienced by the large companies.

Contributing to the dissolution of the big companies was a swelling immigration of homesteaders, hastened by the western extensions of the railroads and various Federal land disposal acts (Chapter 2). Seemingly overnight the best lands were claimed and fenced with newly-developed barbed wire, and the sod was broken and planted to grains and other crops. Strong competition for the remaining range developed among the cattlemen. This condition of competition was aggravated by a tremendous increase in sheep grazing. The excessive and uncontrolled grazing that followed led to range destruction that is still apparent today.

As early as 1880 the Public Land Commission recognized the need for control of grazing, but was unable to get support for a plan to dispose of grazing lands in four-section blocks, units large enough to support small ranches. Subsequent attempts were made to secure grazing legislation from Congress, without success. Lack of unanimity among the stockmen themselves delayed remedial action. Thus depletion of the range and related resources was the general rule until the 1930's. In 1932 it was estimated that the western range had lost nearly 50 per cent of its original productivity.[11]

Even though improvements in management and some in rehabilitation have been made during the past three decades, much of the range land is well below its potential. The estimates of the conditions of privately owned range, made by the United States Department of Agriculture in the early 1960's, are shown in Table 6–3. It was suggested that ranges classified as being in good condition can be improved 25 to 50 per cent, those in fair condition can be improved 50 to 75 per cent, and those in poor condition can be improved 75 per cent or more. It was also noted that present range conditions and forage production on publicly owned rangelands are, on the average, only half or less of their potential, and that vast areas, especially in the Mountain West, have been so damaged that flood and sediment hazards and water quality and the amount of water yielded are increasing problems (as settlement intensifies). Thus there is a significant management task ahead if improvements are to be realized.

TABLE 6–3

PER CENT OF PRIVATELY OWNED RANGE IN EACH CONDITION CLASS*

Region	Excellent	Good	Fair	Poor
	Per cent	Per cent	Per cent	Per cent
West	10	20	30	40
Southern Great Plains	5	15	40	40
Northern Great Plains	10	20	40	30
Southeast	10	20	45	25

**Land and Water Resources, A Policy Guide* (Washington: Department of Agriculture, May, 1962 slightly revised September, 1962), p. 18.

THE EVOLUTION OF CONSERVATION MANAGEMENT

During the past quarter-century much progress has been made toward the establishment of grazing land conservation. Public agencies have been given distinct responsibility for conservation management of public lands and for carrying out research basic to this interest. Ranchers and farmers are recognizing the values of managing land for maximum forage production. Range and pasture management has become established as the art and science of using range and pasture lands in a manner that will maintain or improve the forage resource and simultaneously provide maximum livestock production without jeopardy to other resources or uses of the land.

PROGRESS ON THE FEDERAL LAND. The beginning of control of the use of Federal land for grazing came with the establishment of the first forest reserve in 1891. This was the turning point in conservation of public land resources. Management, however, was not instigated until the turn of the century when the Forest Service, with Gifford Pinchot as chief, was established. Roughly sixty-five million acres of national forest land are now grazed under a permit system which prescribes the grazing season, specifies livestock numbers, and compels the holder to follow prescribed practices. In 1965, 1,376,000 cattle and horses and 2,112,000 sheep and goats were permitted to graze on Forest Service land. This was the approximate level that had been allowed during the preceding decade, but it was somewhat less than half the sheep and goat numbers permitted in 1940.[12] Fees are charged in accordance with average livestock prices. Grazing in the national forests is fitted into a multiple-use management system in which forest production, watershed protection, recreation, and wildlife are considered.

The Forest Service has been active in range research since 1928 when the McNary-McSweeney Forest Research Act was passed. This act provided for the establishment of forest and range experiment stations which cooperate with other agencies.

In 1934 a major gain was made in the management of Federal grazing land. It resulted from the passage of the Taylor Grazing Act, which provided for the creation of grazing districts, and regulation of grazing on public lands by the Department of the Interior. Originally the authorization was for eighty million acres. Pursuant to Section I of the act, statewide hearings were held in the range country, and committees were designated by the stockmen to recommend areas in each state for inclusion within grazing districts. The areas recommended far exceeded eighty million acres, and indicated strong sentiment in favor of grazing control. This led to an increase in the limitation to 142 million acres in 1936.[13] In 1964, the area of Federal land within grazing districts amounted to 159,327,600 acres. Additionally, 18,053,509 acres in isolated tracts in eleven Western states and 1,988,073 acres in Alaska were leased for grazing.[14]

Following World War II the administration of these lands was placed in the hands of the newly created Bureau of Land Management. The program of this agency is "designed to facilitate the restoration, improvement, and development of associated soil, water, and forage resources to their highest practical potential; it also entails the attainment of efficient utilization of these resources, and the stabilization of the livestock industry dependent upon them. This requires the protection of the public domain from deterioration by improper grazing practices, fire, and other destructive forces, and the institution and maintenance of orderly use. Activities other than grazing that must be continued and improved under the program include wildlife management, timber and woodland management, and recreation. Strong support and cooperation must be given also to sound management practices on intermingled or adjacent land used with the federal range."[15]

Local experience is drawn into management policies through locally elected advisory boards of representative range users. Aid and advice is solicited from local sportsmen's organizations, state management agencies, and the Fish and Wildlife Service on wildlife management.

The grazing land is divided into two categories: (1) Lands within the grazing districts; and (2) lands outside these districts. The more concentrated portions of the public lands have been included within fifty-eight grazing districts (Figure 6-13). These districts, for the most part, are from three to nine million acres in size. In 1963 18,237 operators held 19,423 licences or permits to graze 7,400,917 head of livestock in the districts to the extent of 12,051,772 animal-unit-months. Big game animals, totaling 1,429,191, grazed an additional 1,663,329 animal-unit-months on district lands. On unreserved public domain lands—designated as Section 15 land—which occur as scattered tracts, grazing privileges are allowed through leases. In 1963, 9,080 leases were issued to 8,277 operators for the grazing of livestock on these tracts. The size of leases varies from a few acres to as many as four hundred thousand acres. Seventy-eight leases were issued for the grazing of 5,078 cattle and horses and 17,491 sheep and goats, a total of 94,438 animal-unit-months, on the lands administered by the Bureau of Land Management in Alaska.[16]

The fees on grazing district land are on an animal-unit-month basis and are adjusted to livestock prices. Recently this fee has been 33 cents per head of cattle. No charge is made for livestock under six months of age. Grazing rentals for Section 15 lands are on an acreage basis varying from 0.2 to 88 cents per acre—roughly comparable to the fees charged within grazing districts.

The Forest Service and the Bureau of Land Management are now in a position to provide the administration and remedial work necessary to bring the Federal range lands back to their potential utilities. The task, however, is far from complete, especially on the lands under Bureau of Land Management control. The range administered by the latter agency includes a major share of the poorest land in the West, and deterioration was far advanced when management began. In 1951 it was estimated that 50 per cent of the range was in a severe state of erosion, 32 per cent was eroding moderately, and only 18 per cent was in a condition of slight or no erosion.[17] Time, continued concerted effort, and money will be required to bring these lands into a satisfactory balance.

PROGRESS ON PRIVATE RANGE LANDS. The conservation management of private range land has accomplished considerably less than the management of croplands. Nevertheless, through the efforts of the research agencies, the Extension Service, the Soil Conservation Service, and enlightened ranchers, the problem has been recognized, remedial and improvement techniques developed, and progress is definitely being made.

The Soil Conservation Service and the Agricultural Stabilization and Conservation Service operate in the range country as they do in the farming areas (Figure 6 – 14). These agencies have been leading forces in the improvements accomplished. Study of the annual reports of the United States Department of Agriculture indicates noteworthy yearly gains. In fact it can be noted that in a few instances forage production has been raised above the natural potentials. But for the range as a whole the remaining task is great.

PROGRESS IN PASTURE IMPROVEMENT. The accomplishments in pasture development and improvement on private land are a bright spot in

6 – 13 *Grazing districts under the administration of the Bureau of Land Management.* (Bureau of Land Management)

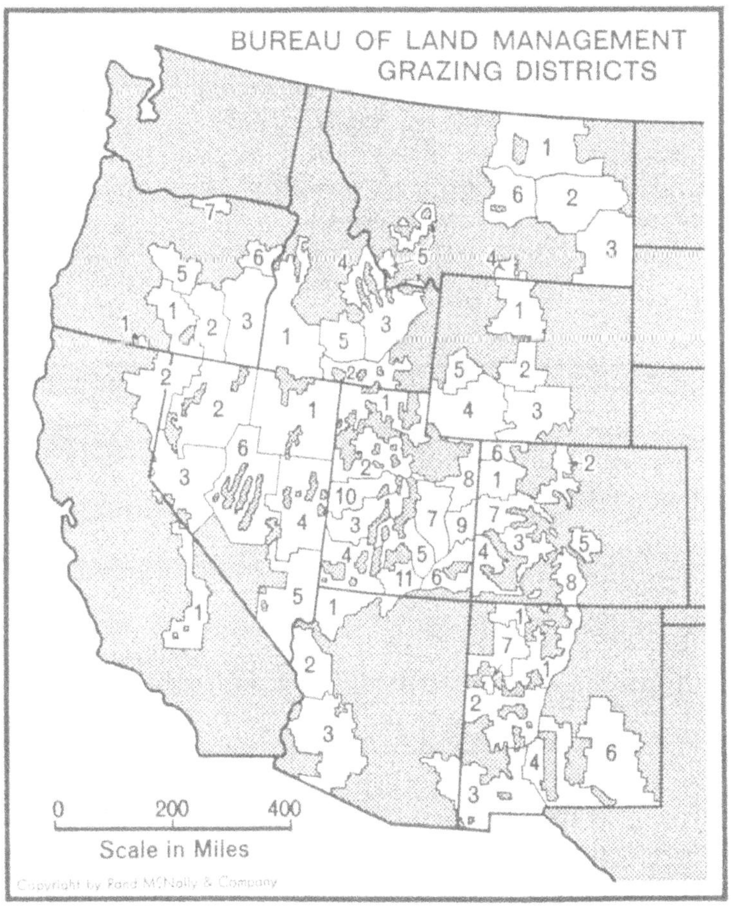

the grazing land picture. Grassland farming is now recognized as a profitable and desirable adjunct to cropland farming. As noted earlier, recently an average of 1.5 million acres of improved pasture have been added yearly. It was noted in Chapter 4 that a number of Federal programs have been encouraging this development.

Much attention has been given to development and planting of grasses, grass mixtures, and grass-legume mixtures with high production and nutritive qualities. Fertilization of pastures also is becoming common. As a result, the production capability of many pastures is equal to that of cropland.

Grazing Land Conservation Techniques

Attention in this section is directed toward conservation techniques on range lands. It will be noted that improved pastures are, in themselves, a conservation technique, protecting land from erosion while building the soil. The object of good pasture management is to get the greatest amount of forage possible at the right time and of the quality desired.

Conservation of range land, in large measure, is applied plant ecology, manipulating the response of soil moisture and natural vegetation to grazing and other influences. Range management requires a thorough knowledge of what plants need to develop and maintain themselves, the kinds of plants that fit together to make a good range, the kinds that will hold the soil and water and produce a high yield of forage, at what season each plant will be eaten by animals, and how much grazing can be tolerated. Each plant species has different requirements and must compete with others. The difference in time of growth, type of root system, and amount and type of growth are characteristics that help plant species compete with one another.

Knowledge of all these features allows the rancher to adjust grazing so that plants will produce the maximum forage. It also makes possible the planning of use in a manner that will enable the plants to remain vigorous and productive and reseed themselves.

RANGE MANAGEMENT

A ranch management program in action normally will require several distinct practices to achieve a conservation relationship with its range. These commonly include proper stocking, rotation-deferred grazing, fire protection, the keeping of feed on hand to achieve sustained grass production, fencing, care in salt

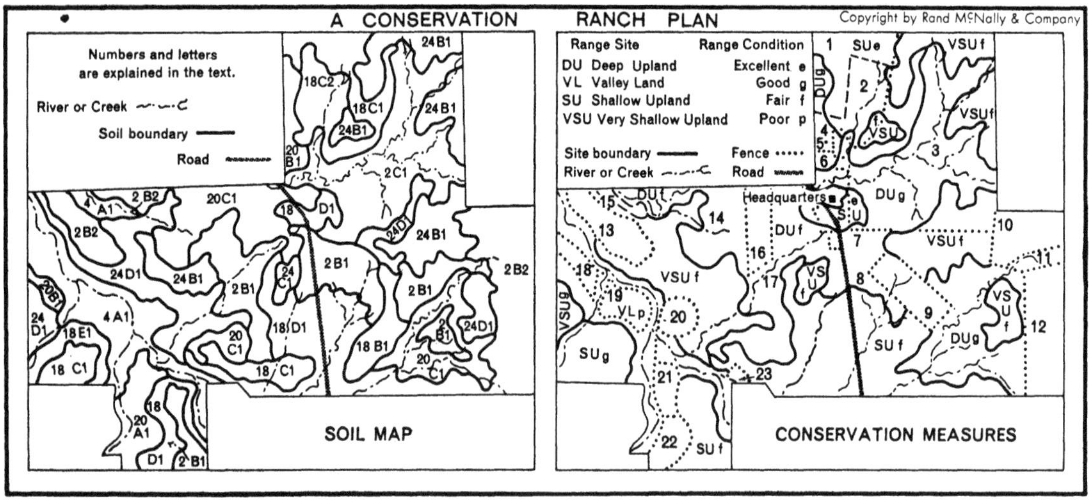

6 – 14 *A ranch plan.* (Soil Conservation Service)

A CONSERVATION RANCH PLAN

The ranch shown in the example here is a 2,235-acre ranch in the western United States.

Range site	Map symbol	Kind of soil	Land capability class
Valley land	4A1	Catalpa clay, nearly level	I
	20A1	Denton clay loam, nearly level	
Deep upland	2B1	San Scba clay, glently sloping, slightly eroded	II
	2B2	San Saba clay, gently sloping, moderately eroded	
	20B1	Denton clay loam, gently sloping, slightly eroded	
	2C1	San Saba clay, sloping, slightly eroded	III
	20C1	Denton clay loam, sloping, slightly eroded	
Shallow upland	18B1	Denton clay, shallow, gently sloping, slightly eroded	
	18C1	Denton clay, shallow, sloping, slightly eroded	IV
	18C2	Denton clay, shallow, sloping, moderately eroded	
	18D1	Denton clay, shallow, strongly sloping, slightly eroded	VI
	18E1	Denton clay, shallow, moderately steep, slightly eroded	
Very shallow upland	24B1	Brackett gravelly clay, gently sloping, slightly eroded	VII
	24C1	Brackett gravelly clay, sloping, slightly eroded	
	24D1	Brackett gravelly clay, strongly sloping, slightly eroded	

SOIL MAP LEGEND

Slope

A — Nearly level, 0 to 1 per cent
B — Gently sloping, 1 to 3 per cent
C — Sloping, 3 to 5 per cent
D — Strongly sloping, 5 to 8 per cent
E — Moderately steep, 8 to 12 per cent

Erosion

1. Slight erosion
2. Moderate erosion

LEGEND FOR CONSERVATION MEASURES: Cultivated Fields

General Practices

Weeds in tame pastures will be mowed high to avoid cutting grass needed for grazing.

Fire lanes will be maintained along public roads and other hazard areas.

Valley range site:

1. Big bluestem
2. Indian grass
3. Switch grass
4. Little bluestem

Deep upland range site:

1. Little bluestem
2. Indian grass
3. Big bluestem
4. Side oats grama

One hundred to two hundred pounds of phosphate per acre will be applied each year to sweet clover fields.

Terraced fields will be cultivated on the contour.

Grain fields will not be grazed after March 15.

Special Practices for Each Field:

The field numbers are the same as the numbers in circles on the map above.

Field 1, 65 acres
Native hay meadow will be protected from fire and grazing. Stubble of 4 to 6 inches will be left when mowed.

Field 2, 58 acres
Hay meadow will be seeded to native bluestem mixture.

Field 4, 8 acres   *Field 7*, 4 acres
Headquarters (black on the map).

Field 5, 12 acres
Hubam sweet clover for temporary pasture. Will be terraced, excess water to empty on northeast pasture.

Field 6, 30 acres
Horse trap.

Field 9, 30 acres   *Field 16*, 40 acres
Three-year crop rotation, Hubam sweet clover for seed, oats—Hubam sweet clover and cane for hay.

Will be terraced, excess water from *Field 9* to empty on southeast pasture and that from *Field 16* on North Creek pasture.

Field 10, 50 acres   *Field 19*, 26 acres
Field 12, 80 acres   *Field 20*, 15 acres
Field 13, 50 acres   *Field 21*, 20 acres
Field 18, 18 acres   *Field 22*, 27 acres

These 286 acres will be used in a 4-year crop rotation of oats, wheat, and corn. Oats and wheat will be overseeded with Hubam sweet clover.

Fields 10 and *12* will be terraced after pasture outlet strip (*Field 11*) is established. Outlet strip (170 rods) will be fenced.

Field 23, 6 acres

Wildlife area: Eroding gully will be fenced, planted to KR bluestem, and kept as a wildlife area.

PASTURES

General Practices

The goal of range management will be to improve all range to good or excellent condition.

The goal on each range site will be to increase the key forage plants as follows:

Shallow upland range site:	Very shallow upland site:
1. Little bluestem	1. Little bluestem
2. Side oats grama	2. Side oats grama
3. Hairy grama	

About 1,500 pounds of hay will be fed per animal-unit during cold weather and during early spring calving.

Salt will be fed all year. During February and March a salt and cottonseed mixture will be fed on the range in portable boxes or feed bunks, moved to undergrazed areas to help distribute the grazing.

Fall-planted small grain will be lightly grazed in winter along with native range. Grain fields will not be grazed after March 15.

Livestock will be moved between pastures to improve the grass.

The proportion of the year's growth left and the condition of the range, whether it is going up or down, will be used to determine the number of animals grazed. Animals will be moved before half of the grass is taken.

Special Practices for Each Pasture:

The numbers after the pasture names are the same as the circled numbers on the map above.

Northeast pasture (3), 445 acres

Southeast pasture (8), 442 acres

Dams and spillways of the two stockwater ponds will be repaired, sodded, and fenced.

Tame pasture (11), 7 acres

Terrace outlet will be constructed. Field will be planted to Bermuda grass and fenced.

North Creek pasture (14), 453 acres

Sixty-five acres of bottom land will be cleared of trees and undesirable brush and seeded to Bermuda grass, bur clover, and rescue grass.

Dam and pond spillway will be seeded and fenced and the pond stocked with fish.

Two hundred rods of fence will be built between pastures 14 and 17, 110 rods around *Field 19*, 100 rods on the side of *Field 13*, and 50 rods around stock pond. Pasture will be grazed only in winter until new seedings in the bottom land and in *Field 15* are established.

Field 15, 8 acres.

Area will be seeded to native bluestem mixture for grazing.

South Creek pasture (17), 420 acres

Eroding gully (*Field 23*) will be fenced to protect from grazing and then planted to KR bluestem for use by wildlife.

placement, and water development to get more effective grass management.

Ranchers have come to realize that proper stocking is a key consideration in a range conservation program, because grass is their product and livestock their manner of harvesting and marketing it. Since the beginning of the range cattle industry, the rancher's philosophy respecting stocking has evolved from numbers of animals, through pounds of meat, to net return per acre (Figure 6 – 15). This indicates the current regard for maintaining productive potentials of the forage. Much experimental work has been carried out by public agencies; thus the rancher has available proven research on which to base his stocking decision. Proper stocking requires that numbers be flexible to adjust to vagaries of precipitation and other climatic conditions.

Through rotation-deferred grazing, pressure can be taken off part of the land each year and uneven grazing can be reduced. A common system is to divide a given holding into three units of equal forage production. Grazing begins in the same unit once every third year, and in any case, not until the ground is firm and key grasses have had a chance to recover and store up food reserves in their roots. In this system one-third of the range is deferred until after seed maturity of perennial grasses.

The need for fire protection is obvious. A burn destroys forage, and more important, destroys the seed source. The lands intermingled with Federal range come under a guard organization; stockmen groups in many other areas also support lookout systems for early detection and warning. For suppression of fires, the Federal agencies and wise ranchers have fire-fighting tools in strategically located caches. Discing or plowing fire breaks along transportation arteries and range divisions are common preventatives.

The keeping of one month's feed in addition to winter needs is a sound practice. This allowance can carry the herd through a late spring which delays the use of the range until the grasses are ready.

By fencing, the ranges can be divided into equal forage production units and livestock distribution controlled. Location of salt away from water, and frequent changing of its position can lead livestock into areas that are not being grazed. Frequent riding to break up cattle concentration and move animals to desired areas is usually a necessity. Water development is another way of attracting animals to lightly grazed or ungrazed areas (Figure 6–16).

RANGE IMPROVEMENT TECHNIQUES

The practices noted above will go far toward maintaining a range in good condition. Remedial work and improvement require additional techniques. These vary with the range condition and the locale.

BRUSH CONTROL. The development of new techniques, machinery, and chemicals have made brush control practical in many areas. Owing to cost, brush control is applicable primarily on lands of high productive potential where brush competition and overgrazing have greatly reduced forage growth. Methods employed depend upon size and density (Figure 6–17). Rootcutter bulldozer attachments are useful where the brush stand is comparatively thin. Cables dragged between two large tractors are effective on stands of one-stem plants, larger trees, and on sandy soils. On small brush, large discs may be effective. Some ranchers have improvised equipment, including various chopping, beating, and mowing machines. Hormone sprays, oil sprays, and arsenicals also are proven controls in some cases. Airplane spraying is used on dense stands, ground equipment on open or thin stands and hand pouring on small areas of thin stands or single- or few-stemmed trees. Caution is required in chemical control because of the possible damage to forage plants and the loss of livestock and desirable wildlife.

Recently a good deal of interest has been shown in the control of *phreatophytes* (useless trees and shrubs whose roots reach the water table or the capillary fringe immediately above) to eliminate water waste. These occupy about fifteen million acres in the western states, often on land capable of producing good forage.[18]

CONTROL OF NOXIOUS AND POISONOUS PLANTS. Poisonous plants always have grown on the

6–15 Rehabilitation through controlled livestock use. The area on the left has been abused. (Bureau of Land Management)

6–16 Water development on the range. This hillside pond is constructed in adobe material. A diversion terrace allows a controlled amount of drainage to be brought into the pond without excessive use of the spillway or increased silting. (Soil Conservation Service)

range. Overgrazing, however, increases their number and forces livestock to eat them. Measures for control are closely related to better range management practices. The better the stand of forage plants the less likely the concentration of poisonous plants or of livestock eating them. Hand grubbing, mowing, and chemicals have given fair control of some of the species.

MOISTURE IMPROVEMENT. Because much of the range is in areas of limited and erratic precipitation, efficient use of available moisture is a major facet of range conservation. The quality and character of the vegetation cover has much to do with the receptiveness of the soil. If the range is grazed too closely or the land is bare, the upper layers tend to become compact and most water runs off. Moreover, under a cover of vegetation, soil temperatures are reduced and evaporation lessened. A good forage cover, with few woody plants, is the best assurance of efficient water use.

Where machinery can operate on land with poor stands, contour furrowing can be employed to increase soil receptiveness to runoff and stimulate forage growth. In many cases range pitting is effective for these purposes. For pitting, heavy equipment scoops out shallow discontinuous pits about sixteen inches apart.

Water spreading is another technique applicable where drainage ways are present. Dams are built in water courses with gradient ditches or terraces leading water out to gentle slopes where it is released through a system of dikes. This method is usually fairly costly, and, therefore, must be based upon sound economic planning. Increases in forage production of five to ten times in five years are not uncommon.

RESEEDING. Full restoration of considerable range land will require more than good management and the techniques described. Approximately seventy-five million acres have been so badly depleted that they will have to be re-

6 – 17 ***Examples of brush control technique.*** **Scene 1 shows a bulldozer at work. Scene 2 shows an area in shin oak which has been poisoned by spraying. Scene 3 is an illustration of the hand method of killing larger trees. (Soil Conservation Service)**

seeded artificially if they are to recover in the next twenty-five or thirty years. This need is recognized, and a good deal of research work has been carried out. There is extensive rancher interest, as evidenced by the 2,276,189 acres on which the cover was improved under the Agricultural Conservation Program of 1964 (the same year range-land improvement through deferred-grazing was carried out, under this program, on 1,244,892 acres) (Figure 6 – 18).[19]

Present practices employ extensive techniques. Except on abandoned cultivated fields or cleared plots, seedbeds are not generally prepared. Specially adapted grain drills are used commonly, but on large tracts airplane seeding is employed, and for small, rough areas the broadcast method is most practical. Native plants seeds are considered most effective.

RANGE MANAGEMENT VERSUS RELATED RESOURCES

Management of range land for maximum forage production fits within the desirable conservation goal of the nation. It provides watershed and soil protection and improvement, enhances wildlife habitat, and improves scenic value. The major questionable aspect relates to range use of land that is best suited for timber production. A guiding rule in these instances could well be: If the range use upsets regeneration and significantly restricts timber growth, the land in question should be grazed with extreme care.

Outlook

The period of indiscriminate and excessive use of range resources in the United States probably depleted the ranges more seriously than the waste of any other category of land resources. Excellent progress has been made during the past three decades, but the remaining task is large. Range scientists, however, believe that with the application of proven practices range forage can be increased one-third by 1975.[19] In view of the increasing demand for animal products this increase will be required.

6 – 18 *Range improvement through reseeding.* This land was plowed to kill sage types and seeded with crested wheatgrass. (Bureau of Land Management)

6 – 19 *Improved pasture as a conservation technique.* A few years ago two fields in Madison County, North Carolina, were equally gullied. The owner on the left entered a TVA test demonstration program and by the use of phosphate and lime produced a cover that halts erosion and permits an increase in the livestock carrying capacity. (Tennessee Valley Authority)

Increasing forage production from planted pastures probably offers the greatest relative possibilities (Figure 6 – 19). The use of pasture in conservation and balanced farming enterprises has been increasing at a rapid pace. But major increases in acreage will be at the expense of crop and forest lands; in fact, these uses can be expected to compete strongly for present pasture acreage within a few decades. There are, however, large potentials inherent in pasture improvement through better plants, more use of fertilizers, more supplemental irrigation, and better management in general. Present indications are that the implementation of these techniques, coupled with improvements in other feeds, can meet the nation's needs for at least the next half-century.

FOOTNOTES

[1] "Major Uses of Land and Water in the United States," *Agricultural Economic Report No. 13* (Washington: Department of Agriculture, July, 1962), p. 8. *Land and Water Resources, A Policy Guide,* (Washington: Department of Agriculture, May, 1962, revised September, 1962), p. 17 reported that (in 1960) "the portion of annual feed units supplied by range and pasture was: Dairy cattle, one-third; beef cattle, three-fifths; and sheep and goats, over five-sixths. About 10 million head of deer, elk, and antelope also obtain forage on ranges and pastures."

[2] Jack R. Harlan, *Theory and Dynamics of Grassland Agriculture* (Princeton, N. J.: D. Van Nostrand Company, Inc., 1956), p. 216.

[3] *Ibid.*

[4] E. J. Woolfolk, D. F. Costello, and B. W. Allred, "The Major Range Types" (Washington: Department of Agriculture, 1948), pp. 205–11.

[5] "Major Uses of Land and Water in the United States," *loc. cit.,* states "An average acre of cropland used for pasture produces five to six times as much as an acre of permanent grassland pasture (private and other non-federal pasture in farms), and twenty-five to thirty times the feed from an acre of unimproved woodland pasture and Federal range (not in farm)."

[6] See M. L. Upchurch, "The Role of Land in Western Ranching," Land, *The Yearbook of Agriculture, 1958* (Washington: Department of Agriculture, 1958), pp. 167–74.

[7] See recent issues of "Farm Costs and Returns (with Comparisons), Commercial Family-Operated Farms by Type and Location," *Agriculture Information Bulletin* (Washington: Agricultural Research Service, Department of Agriculture, published annually).

[8] Roy L. Donahue, Everett F. Evans, and L. I. Jones, *The Range and Pasture Book* (Englewood Cliffs, N. J.: Prentice-Hall, Inc., 1956), 406 pp. Contains a short review of grasslands in American history, pp. 5–18.

[9] For a review of this development as well as an appraisal of the underlying problems of the Great Plains see "The Future of the Great Plains," *Report of the Great Plains Committee* (Washington: Government Printing Office, 1936), 194 pp.

[10] *Managing the Federal Range* (Washington: Bureau of Land Management, Department of the Interior, July, 1954), pp. 1–2.

[11] *Statistical Abstract of the United States,* (Washington: United States Department of Commerce, 1966), p. 678.

[12] *The Taylor Grazing Act of June 28, 1934,* with Amendments to September 1, 1955 (Washington: Bureau of Land Management, Department of the Interior, 1956), 14 pp.

[13] "Public Land Statistics, 1964" (Washington: Bureau of Land Management, Department of the Interior, 1965), p. 148.

[14] "Managing the Federal Range," *op cit.,* p. 3.

[15] Various tables in "Public Land Statistics, 1965" *op. cit.*

[16] *Rebuilding the Federal Range, A Resource Conservation and Development Program* (Washington: Bureau of Land Management, Department of the Interior, 1951), p. 8.

[17] Herbert C. Fletcher and Harold B. Elmendorf, "Phreatophytes–A Serious Problem in the West," *Water, the Yearbook of Agriculture, 1955* (Washington: Department of Agriculture, 1955), pp. 423–29.

[18] *Agricultural Conservation Program, Statistical Summary 1964* (Washington: United States Department of Agriculture, Agricultural Stabilization and Conservation Service, March, 1966), p. 3.

[19] Interview with Dr. D. F. Hervey, Professor of Range Management, Colorado State Univ., March, 1960.

7

Forest Resources

One-third of the United States' area is classified as forest land, and it supports a variety of forest vegetation types. The forests constitute a major segment of the nation's physical environment. Moreover, the forests are a prime example of the interrelatedness between the physical and cultural environments. The type, quality, and productivity of forest vegetation are dynamic expressions of at least four groups of complexly interrelated conditions: (1) Site conditions, especially the amount and seasonality of moisture and solar heat energy, fertility and physical character of soil, slope and aspect of the surface; (2) biotic competition; (3) influences of insect and disease epidemics, fire, and catastrophic weather phenomena; (4) the actions of man.

There are few areas of the forest ecosystem where these interrelated influences remain in equilibrium for long periods and the forest association remains unchanging; instead, the ecosystem is typically characterized by change in one or more of the conditions which results in successions of vegetation associations. For example, Southern pine forest types tend to become hardwood associations after the pine is harvested unless the hardwood reproduction is suppressed by fire or other controls. Changes in the forest ecosystem may result in significant responses expressed by changes in quantity and quality of water, erosion, wildlife, natural beauty, or values for outdoor recreation. The introduction of the Dutch elm disease in combination with the European elm bark beetle, for example, threatens to kill the elm trees of the nation's forests. Harvest of a timber stand under some conditions may result in increased erosion. Destruction of a forest by fire can wipe out the economic base of a whole community.

It is clear that intensified utilization of the nation's forest lands involves many complex ecological problems and social considerations; nevertheless, the forests are without doubt maintainable and improvable, and in addition they have multiple-use values ranging from aesthetic considerations and major values for outdoor recreation and watershed, to commercial timber production. Perceived thus, as *resource*, the forests are a challenge to man's capability in management to accomplish rational, continuously viable rapport with the forest bioenvironment, and with the societal environment simultaneously.

Forest lands were early recognized as a major factor in the national economy. To the "red man" the forest provided food, shelter, and raw material for clothing. For the first settlers from Europe the forest provided the very foundations of the economy: Logs for building, game and other foods, fuel for warmth and cooking, and commercial products such as masts and barrel staves for export. The forests then must have seemed endless, for they stretched from the rockbound New England coast and Virginia shores inland to the Appalachian Mountains and beyond toward the unknown lands to the west. Few minds in the colonial days envisioned a time when the forest lands available for exploitation would be clearly limited, or when pressure on forest land capability would increase so greatly that thoughtful leaders would be asking if enough forest land remains to meet the country's needs.

Today the forest land is more important than ever. Timber provides a long list of commodities including lumber, plywood, hardboard, pulp for most of the paper, cellophane and rayon, naval stores, and firewood. From trees come chemicals which are basic for many commodities including plastics and drugs. Forest outdoor recreation has grown to major importance. Watershed values of forest lands are increasingly recognized.

As the nation's land has become more fully occupied and the forests more important, there has been a major change in attitude toward forests. The colonist usually saw the forest as something standing in his way which had to be cleared to permit agriculture. Later, commercial logging was characterized by harvest, and little or no concern for regeneration. The forests seemed endless, and there was abundance resulting in an economic condition that did not foster conservation. Timber was "mined" and the operators moved on to new lands. In the nineteenth century exploitative logging swept across the north from the forests of Maine to the white pine stands of the Upper Lakes. Paul Bunyan stalked the woods and was symbolic of the bigness and the romance of early logging.[1] Early in the twentieth century the pine forests of the South began to bear the brunt of logging, and by the time of World War I the center of exploitation had migrated to the Pacific Coast forests. Until the last decade of the nineteenth

7 – 1 *Timber awaiting conversion to products at a Virginia mill.* Forest trees form not only a desirable aesthetic aspect of the nation's living environment, but in addition provide timber and pulpwood harvests which are the basis for a great industry. Direct employment in the woods is comparatively small, involving about 300,000 persons; but the wood-using industries supported form the third largest census industrial group, employing 1.57 million persons (exceeded only by the food products group's 1.64 million, and transportation equipment's 1.62 million in 1963). Commodity output of the wood-using industries has steadily increased to meet rising market demands. A notable trend is the increasing use of wood in products that require more processing and permit use of lower quality timber than lumber, for example for veneer, laminated and particle boards, and woodpulp leading to many products such as paper and cellophane. (Courtesy, The Chesapeake Corporation of Virginia, photograph by Thomas L. Williams)

century, the few citizens who realized the impending need for conservation had little influence.

In the twentieth century, major advances have been made in public attitudes, in science, and in technology, toward conservational management of the nation's forests. The era of exploitative "mining" of timber without regard for regeneration is yielding to the more mature management concept of growing timber as a crop. The practice of "skimming the cream" by harvesting only high grade trees is being replaced by efforts toward improvement of forest quality and more complete utilization of timber cut. Responsible managers and citizens now realize that it is not the having of vast "virgin" forests that is important; rather the important goal is the maintenance of productive use of the nation's forest land for timber production and for other aspects of multiple-use such as watershed and outdoor recreation. There is value in retaining areas of "virgin" forest for scientific reasons. There is now a well developed forestry profession with an increasing number of professional forest managers, both in government and in industry. Especially significant is the employment of foresters by private industry, which has increased from less than a thousand in the 1940's to over 8,000 in 1965. (Foresters employed by public agencies—state and Federal—total about 10,000.) Finally, there is evidence of a broadening concern for forest resource problems among legislators, bankers, and scholars. Although problems and challenges remain, these accomplishments portend a good future for the forests of the nation.

Here the discussion of the forest resource and conservation need is concerned only with timber production. It is organized in four parts: (1)The review of the present situation; (2) an appraisal of the probable demands on forest resource; (3) a summary of activities of private and public agencies in the interest of forest conservation; and (4) a review of conservational practices.

The Forest Lands Today

When the white man first arrived in North America, about half of the area of the conterminous forty-eight states was forested. Increasing population, urbanization, and opening of agricultural lands have reduced this original 950 million acres of forest land by about a third, to 638 million acres. The addition of Alaskan (118 million) and Hawaiian (two million) forest areas brings the present national total to 758.8 million acres of forest land, of which 508.8 million are classified as commercial, sixteen million are reserved mostly in parks, and 234 million are unproductive for timber.[2]

Much has been written about the waste of forest by early pioneers, and there was waste. It should be noted however, that much of the forest land reduction has been to provide necessary farmlands and space for reservoirs, powerlines, highways, and urban developments. Also, many areas which were once ruthlessly cut, such as the white pine forest of the Lake States, are being rehabilitated.

THE FOREST MACRO-REGIONS

The United States is fortunate in having a wide variety of forest types. Forest vegetation generally develops where precipitation is at least twenty inches annually, and where there is at least a three to four months' growing season. The specific vegetative association, however, is the result of complex interrelationships, not only with climate, but with soil, slope, light, catastrophies of weather, insect epidemic, fire, and influences of man. Pure stands of one species are uncommon; rather the forests are varying mixtures of species. Forests also differ in age groupings; some types tend to be composed of trees of more or less the same age, while in others individual trees are of different ages. Forest types are identified according to the combinations of species, and nearly two hundred types are recognized by foresters. These many types are here grouped into six macro-regions (Figure 7 – 2). Each region has its characteristic environments and tree species requiring different management for conservation, and each region has its composite of advantages and disadvantages.[3]

PACIFIC FORESTS. The Pacific forests, including coastal Alaska, account for about one-eighth of the nation's commercial land, but support nearly two-thirds of the standing softwood sawtimber resource and produce nearly half of the softwood harvest. This is the only region in which harvest is still significant in old-growth timber stands, and as a result it accounts for only about one-fifth of the nation's annual growth. Over half of the forest land is in Federal ownerships, and farm woodlots are less important than in any other region.

The Douglas-fir subregion of Oregon and Washington west of the Cascade Mountains is one of the finest timber growing areas of the world (Figure 7 – 3). Annual precipitation ranges from 40 to 150 inches concentrated in a mild winter with a dry summer. The frost-free season is one hundred to two hundred days, and soils are generally favorable. Trees grow rapidly and to great size, commonly two hundred feet tall and six to ten feet in diameter. Annual growth of one thousand board feet per acre

7 – 2 The original macro-forest regions of the United States.

occurs on the best lands, and stands of mature timber commonly contain one hundred thousand board feet per acre. Douglas-fir dominates large areas, with western hemlock and western red cedar common associates. Douglas-fir grows best under full sunlight after the early seedling stage, and does not reproduce well as an understory. Logging, therefore, is carried out by clear cutting in patches. Regeneration is achieved by natural seeding from marginal or seed-trees, or by artificial seeding or planting. The forest, therefore, tends to be in even-aged stands. A Sitka spruce and hemlock type occurs on wet north coast sites.

Douglas-fir is the leading lumber species of the nation, accounting for 25 to 28 per cent of the lumber. Distance to markets and rugged terrain are handicaps in logging. Low humidity (30 per cent or less) in summer forces occasional shutdown of logging and is a serious fire danger. Winters are mild, with little snow on lower slopes permitting year-round operation.

The pine subregion occurs east of the Cascades in Oregon, Washington, and California, where annual precipitation is only fifteen to thirty inches. Ponderosa pine grows in near-pure stands that are open, almost park-like, with trees of varying ages. On higher slopes Douglas-fir and other conifers grow. Ponderosa pine is able to reproduce in the moderate shade of this open forest. Thus, selection harvest is generally practiced, forming a notable contrast to the adjacent Douglas-fir subregion. This harvesting difference is an instructive example of the concept that silviculture practices must be adapted to the optimum for the forest type (Figure 7 – 4).

The California forests are notable for the magnificent redwoods which grow along the north coast within a coastal fog area. Inland, the Sierra forests are continuations of Douglas-fir and ponderosa types, and support the principal timber harvest.

Coastal Alaska, while accounting for only 10 per cent of the area of the state, contains the only significant commercial forest land (about six million acres). The forest occupies a narrow coastal fringe rising sharply from the water and extending only a few miles inland (Figure 7 – 5). Growing conditions are excellent, with annual precipitation 70 to 150 inches, and a growing season of 100 to 150 days with long hours of summer daylight. The forest is dominated by western hemlock and Sitka spruce. Most of the better stands are in the Tongass and the Chugach national forests. Intensification of management has been slow because of the small local market and remoteness of the area, but

7 – 3 *The Douglas-fir forest.* A typical view in the Willamette National Forest on the west slope of the Cascades in Oregon. (Forest Service)

interest has recently increased and several pulp mills have been completed. Rugged terrain is a limiting factor, even though there is obvious access by water. In the interest of Alaska, all timber is required to be processed in the state.

SOUTHERN FORESTS. The southern forest, extending in an arc from Virginia into eastern Texas, comprise the second major tree growing area of the nation. In the early twentieth century the southern pine stands were heavily exploited, and by 1935 they were almost completely cut over. Since then forestry has experienced a rebirth which has been aided by the advent of sulphate pulping. About 60 per cent of the area is commercial forest land, accounting for about two-fifths of the nation's total, half of the annual growth, and two-fifths of the timber harvest. It is a region of rapid tree growth, favored with thirty-five to seventy inches of precipitation well distributed through the year, high humidity, a frost-free season of from 200 to 350 days, plus ample heat.

The southern pines (longleaf, shortleaf, loblolly, and slash) occupy only 40 per cent of the area. The longleaf and slash, famed as the naval stores producers, are co-dominants along the eastern Coastal Plain on sandy, droughty areas. Loblolly occupies the moister portions and shortleaf, being most tolerant of cold, ranges farther inland. Moist bottom lands develop hardwood types with gums, oaks, hickories, and the cypress. Inland, the pine forests generally give way to the oak-pine type since better soils favor the hardwoods over the pines. Throughout much of the area, control of hardwoods is a major problem of silviculture.

The rapid growth potential and emphasis on pulpwood permits harvesting cycles as short as twenty-five to thirty years. Trees are small compared to the Pacific forest, and equipment can be smaller and less expensive. The gentle terrain ranging from level-to moderate-slope lands is favorable to tree growing, and harvesting costs are low (Figure 7 – 6). Proximity to eastern markets favors the South. Farm ownerships account for 40 per cent of the total area, and many farmers have come to emphasize tree growing, which is commonly under contract with large pulp companies. The Tree Farm movement of the forest industry has become a major development in this area.

NORTHERN FORESTS. The northern forests extend from Maine through Minnesota with a southern extension along the upper slopes of

7 – 4 *The ponderosa pine forest.* **Note the contrast with the Douglas-fir forest. The stand is characterized by varied-aged trees as shown in this picture taken near Sisters, Oregon. The stumps, clearly visible, reveal the selective harvest that has recently been completed. (Forest Service)**

the Appalachian Mountains. The climate is severe, but well distributed annual precipitation of thirty to forty inches and the frost-free season of at least one hundred days are ample. Tree growth is slow by comparison with the South. Moderate terrain favors logging, and much harvesting is done in winter to utilize the snow cover. Proximity to markets, especially to the pulp paper centers, is a major asset.

As a result of considerable range in environmental conditions, the forest types differ considerably within this region. Most of Maine and the high mountain areas are dominated by spruce and fir which are an extension of the boreal forests of Canada (Figure 7 – 7). White and red pine are co-dominants in the upper Great Lakes portion from Minnesota to Michigan, but are mixed with aspen forest and hardwood types. The white pine forest lands, exploitatively logged before 1900 and burned over, are being rehabilitated. The southern portion of the northern forest region is a transition between the conifers to the north and the central hardwood to the south, often known as the birch-beech-maple-hemlock forest.

CENTRAL HARDWOOD FOREST. The central hardwood forest is intermediate between the southern pine and the northern forests. For the most part it is dominated by oaks and hickories. The combination of smooth terrain, fertile soil, thirty to sixty inches of annual precipitation (well distributed but with summer maximum), and ample warmth during a frost-free season of

7 – 5 *Coastal Alaska Forest.* **The rugged character of coastal Alaska which contains most of the state's commercial timber is shown by this view in Tongass National Forest. The lake is at the headwater of the creek entering the head of Port Houghton. (Forest Service)**

one hundred to two hundred days has favored the deciduous hardwoods over the conifers. The region is large, but much of the original forest has been cleared for agriculture and the remaining forest is second growth occupying only about 20 per cent of the land area. Major

7 – 6 Southern pine forest. This stand of longleaf pine was photographed near Kirkland, Georgia. Note the smooth terrain and the open park-like character of the forest. The tree being measured is twenty-two inches in diameter. (Forest Service)

7 – 7 The northern forest. This is a 70-year-old white pine stand in New York. These forest lands were among the first exploited by commercial logging in the nation. Today they are increasingly being managed especially for pulpwood production. (Forest Service)

stands are on the Ozarks in Arkansas and the Appalachian Mountain areas. Many fragmented small ownerships and farm woodlots produce important hardwood timber harvests (Figure 7 – 8).

ROCKY MOUNTAIN FOREST. The high elevation of the Rockies with lower temperatures and increased moisture give rise to forests chiefly of conifers dominated by ponderosa and lodgepole pine, Engleman spruce, and Douglas-fir important in the northern Rockies. The entire region is one of low growth potential due to short growing season and low annual precipitation. Forest vegetation tends to be open and park-like, and stands generally contain only modest volume. Terrain is rugged and presents difficulties for harvesting as well as management (Figure 7 – 9). Remoteness from market results in limited development in most areas. Consequently, although the region makes up 13 per cent of the nation's commercial land, it accounts for only 6 per cent of the harvest.

TROPICAL FORESTS. The tropical forests of the nation occur in southern Florida and Hawaii. In Florida the varieties are chiefly palms and mangrove and have little commercial value, though there are related scenic, scientific, and erosion control values. The state of Hawaii contains about a million acres of commercial forest land (about 25 per cent of the state area). About half is in state ownership. Typical of humid, tropical forests, there are many species of trees, but only a few have commercial value; the two most important are Koa, or Hawaiian mahogany, and the ohia lehua. In recent years a number of exotics have been introduced, chiefly eucalyptus. Overcutting and lack of management have left the forest land seriously depleted. Fortunately, the State Division of Forestry is initiating a program of conservation which should result in much improvement.

QUANTITY OF FOREST LAND

Approximately one-third of the national territory is forest land: about 759 million acres out of the total of 2,271 million acres of the fifty states and territories. One-third of this total

7 – 8 The central hardwood forest. A typical stand of white oak in Missouri. Hardwood forests such as this are certain to be more intensively used in decades ahead. (Forest Service)

forest land, however, was reported by the 1962 inventory to be noncommercial. These lands include areas incapable of growing timber in commercial volume, such as high mountain slopes and reserved lands such as military reservations, national parks, and monuments, sixteen million acres of which are capable of commercial tree growing.

The noncommercial forest lands play a major role in the total resource balance. Their role, however, is concerned with nontimber facets of forest influences and values: watershed management, provision of wildlife habitat, aesthetic values such as scientific museums in the wilderness concept, and recreation in national parks. The discussion in this chapter will be concerned only with the commercial forest lands and their conservation needs for production of timber commodities.

COMMERCIAL FOREST LANDS. The commercial forest lands totaled 509 million acres, or about 2.7 acres per capita in 1966. But this figure fails to reveal the high quality of some of the forest land in terms of tree growing capability. The United States is actually generously endowed with forest resources. Study of Table 7–1, which gives data by sections of the nation, reveals the variation in importance of forest land in each section as well as the pattern of national distribution. New England and the Douglas-fir subregion have three-fourths of their total area classified as commercial forest land, while the South, excluding Texas, has close to 60 per cent. In contrast, two vast regions have only a small percentage: the Rocky Mountain, and interior Alaska.

The eastern half of the conterminous forty-eight states accounts for 74 per cent of the com-

7–9 *Rocky Mountain forest.* **Difficult terrain as well as distance to major markets are limiting factors as illustrated by this Colorado forest scene. (Forest Service)**

TABLE 7-1
TOTAL FOREST AREA, COMMERCIAL AND NONCOMMERCIAL*

Region and sections	Total land	Total forest	Commercial forest
		million acres	
NORTH	629.2	178.4	171.8
New England	40.4	31.8	31.4
Middle Atlantic	87.3	46.6	43.9
Lake States	209.1	55.2	52.4
Central	292.3	44.7	44.1
SOUTH	512.7	220.3	201.1
South Atlantic	76.3	49.0	47.6
East Gulf	72.0	46.3	44.8
Central Gulf	89.6	53.7	53.4
West Gulf	274.8	71.4	55.3
ROCKY MOUNTAIN	555.3	143.3	65.6
Northern Rockies	215.4	55.0	39.3
Southern Rockies	339.9	88.3	26.3
PACIFIC COAST	237.0	109.8	69.3
Oregon–Washington Douglas-fir[1] subregion	35.1	28.9	26.0
Pine subregion[1]	69.2	24.9	20.1
Coastal Alaska	32.5	13.5	5.8
California	100.2	42.5	17.4
INTERIOR ALASKA	333.0	105.0	?
HAWAII	4.1	2.0	1.1
ALL UNITED STATES	2,271.3	758.9	508.8

* *Timber Trends in the United States* (Washington: Forest Service, Department of Agriculture, 1965), p. 79 and p. 140.

[1] *Timber Resource Statistics for the Pacific Northwest* (Portland: Forest Service, Department of Agriculture, 1965), p. 1.

mercial forest land area of the nation. The Pacific Coast, including coastal Alaska, accounts for only 13 per cent. The implication for management if it is to meet increasing future demands for timber products is clear. It is the East that must receive intensified forestry practices.

Commercial forest land is a subjective resource concept, depending not only on the tree growing capacity of the bioenvironment, but also on a complex of interrelated economic and political influences such as market price, cost of operations, changing technology, accessibility, and government actions. Nevertheless, the amount of commercial forest land as defined by the Forest Service is the basic quantitative measure from which appraisal of the forest land resource adequacy must begin.[4]

In the future, the area of commercial forest land available for forestry is not likely to increase; rather there is considerable evidence foreshadowing a decrease. Among the competing demands for land use that are likely to reduce the area of commercial forest land one should note the following: The needs for superhighways, power lines, air fields, and water storage reservoirs; the pressures for additional reservation of land for recreational use as in wilderness areas and national parks; expansion of farm lands in some areas; and finally the continued urban expansion which in some places is at the expense of forest lands. To be sure, some low-capability farm lands continue to revert to forest use, and some reserved lands return to commercial forest status.[5] Intensified commercial tree growing is not likely to be much extended to poorer lands now classified as noncommercial; the evidence suggests that there will be a continuing trend to increase the intensity of management applied to the better tree growing lands—lands capable of producing commodity returns commensurate with additional investment. In balance, then, there is likely to be a modest reduction in the area of commercial forest land.[6] Assuming that the nation's population reaches the projected 325 million by the year 2000 (the population level used by the 1962 forest inventory for its analysis), per capita commercial forest land will have decreased from 2.7 to about 1.5 acres. The important question for the nation is: How can these commercial forest lands be so managed that they will produce sufficient forest commodities in the decades and centuries ahead?

COMMERCIAL LAND BY FOREST TYPE. The wealth in variety of the forest resource is illustrated by the inventory of forest lands according to dominant and co-dominant species. The significance of each type is shown in Table 7-2. An important condition revealed is that the forest land area dominated by hardwood species (269 million acres) is greater than the area of softwood forests (240 million acres). The western regions support most of the softwood types (125 million acres) followed by the South (81 million acres). The hardwoods are almost entirely in the North (138 million acres) and the South (120 million acres). The South, generally thought of in terms of the southern pine species, contains more hardwood than softwood forest land. At present the timber harvest of the nation is about 30 per cent hardwoods, 70 per cent softwoods. Here is a prospect for conservation: the nation needs to give greater attention to utilization of hardwoods which occupy over half of the commercial forest land or to their replacement by softwood species.

QUALITY OF FOREST LAND

The commercial forest lands also vary greatly in their inherent capacity to support tree growth and thus produce wood. *Site quality* is the term used by forest scientists and managers to indicate this relative productive capacity of a forest area. It is, of course, an expression of the complex combination of conditions of the environment including amount, variability, and combinations of such elements as moisture availability, temperature, wind, soils, slope, and aspect. Precise expression of this complex of environmental conditions in terms of capacity to support wood growth is difficult, and research is continuing in the effort to improve the techniques of measuring and expressing site quality. In the United States *site index* is the commonly used measure of site quality. The site index number expresses growth capacity determined by the height of dominant and co-dominant trees at age one hundred years for western species and usually at age fifty years for eastern species. Tree height is used as the indicator, because research has shown that the ultimate height of trees is little affected by density of stand. Moreover, height can be readily and objectively measured. The site index is commonly given in even tens of feet, but is not limited to them. Site quality is sometimes also indicated in *site quality classes*, indicated by Roman numerals I through VI, with each class representing a range of index values (see Tables 7–3 and 7–4). Utilizing the standard tables of heights by age for each site index, the

TABLE 7–2
COMMERCIAL FOREST LAND BY MAJOR FOREST TYPE*

Forest type	Million acres	Sections of major occurrence
Oak–Hickory	116.0	Central, Atlantic, and South
Loblolly–Shortleaf pine	58.0	All South except Florida
Oak–Gum–Cypress	37.8	Mainly South
Douglas-fir	37.4	Half in Douglas fir subregion
Ponderosa pine	36.0	Rocky Mountain and Pine subregion
Maple–Beech–Birch	33.3	Chiefly in North
Oak–Pine	26.9	South; 25–50 per cent pine
Longleaf–Slash pine	26.0	Chiefly Florida and Georgia
Aspen–Birch	23.7	Chiefly Lake States
Elm–Ash–Cottonwood	20.4	Lake States and Central
Spruce–Fir (East)	19.6	New England and Lake States
Lodgepole pine	15.8	Northern Rocky Mountain, Colorado, East Oregon
Fir–Spruce (West)	15.6	Rocky Mountain, California, Pacific Northwest
White, red, and jack pine	11.1	North, chiefly Lake States
Western hardwoods	11.1	Rocky Mountain, Pacific Northwest, Hawaii
Hemlock–Sitka spruce	10.0	Coastal Alaska and Douglas fir subregion
Western white pine	5.0	California and northern Rocky Mountains
Larch	3.5	Northern Rocky Mountains
Redwood	1.6	North California coast

* *Timber Trends in the United States, op. cit.*, p. 146–48.

TABLE 7–3

AVERAGE TOTAL HEIGHT OF DOMINANT AND CO-DOMINANT TREES, DOUGLAS-FIR, BY SITE CLASSES*

Age (years)	Site class V		Site class IV			Site class III			Site class II			Site class I		
	Site index		Site index			Site index			Site index			Site index		
	80	90	100	110	120	130	140	150	160	170	180	190	200	210
	feet		feet			feet			feet			feet		
20	21	24	26	29	31	34	37	39	42	44	47	49	52	54
30	37	41	46	50	55	60	64	69	74	78	83	88	92	96
40	48	54	60	66	72	78	84	90	96	102	108	114	120	126
50	56	63	70	77	84	91	98	103	112	119	125	132	139	146
60	63	70	78	86	93	101	109	117	124	132	140	148	156	163
70	68	77	85	94	102	110	119	127	135	144	152	161	170	178
80	73	82	91	100	109	118	127	136	145	154	163	172	181	190
90	77	86	96	105	115	125	134	144	153	163	172	182	192	201
100	80	90	100	110	120	130	140	150	160	170	180	190	200	210
110	83	93	104	114	124	135	145	155	166	176	187	197	207	218
120	85	96	106	117	128	138	149	160	170	181	192	202	213	224
130	87	98	109	119	131	141	152	163	174	185	196	207	218	228
140	88	99	110	121	133	144	154	166	177	188	199	210	221	232
150	89	101	112	123	134	145	156	168	179	190	201	213	224	235
160	90	102	113	124	136	147	158	170	181	192	203	215	226	237

* Richard E. McArdle, Walter H. Meyer, and Donald Bruce, "The Yield of Douglas-Fir in the Pacific Northwest," *Agriculture Technical Bulletin No. 201* (revised) (Washington: Department of Agriculture, 1949), p. 12.

forester is able to evaluate site quality for any given forest area by determining the age of the growing trees and their height.

The reason for considering site quality in forest conservation studies is that it provides a quantitative basis for evaluating forest land for wood production. (Compare this site quality classification for forest land areas with the Soil Conservation Service Land Capability Classification discussed in Chapter 3.) Evaluation of land by site index and tables of tree heights by years also permits calculations with which to construct other tables, such as volume of wood per acre. The very large differential in productive capacity between the site quality classes is graphically illustrated in Figure 7 – 10.

TABLE 7–4

AVERAGE TOTAL HEIGHT OF DOMINANT AND CO-DOMINANT TREES, LOBLOLLY PINE, BY SITE INDEX*

Age (Years)	Site index 60	70	80	90	100	110
	feet					
20	32	38	43	48	54	59
30	45	52	59	67	74	81
40	54	63	72	81	90	99
50	60	70	80	90	100	110
60	64	75	85	96	107	118
70	67	78	89	100	112	122
80	69	80	92	103	115	126

* "Volume, Yield, and Stand Tables for Second-Growth Southern Pine," *Miscellaneous Publication No. 50* (Washington: Department of Agriculture, 1929) p. 53.

As yet no national scale map of site quality by classes has been compiled; however the Forest Service did make a preliminary estimate of growth capacity by classes as part of the 1962 survey (Table 7–5 and Figure 7 – 11). The estimate of acreage for each growth capability class is preliminary, and additional research is needed to refine the criteria and field work to determine the exact areas of each class; nevertheless, the estimates are significant.

In terms of national totals, it is significant that only a third of the forest lands are considered capable of producing eighty-five cubic feet or more per acre. The data clearly reveal the importance of the South and Pacific Coast forest regions in high quality forest lands. In regional terms only the Pacific region (56 per cent) and the South (39 per cent) have a major portion of their land in high productivity classes. The fact is that not all forest lands are equally attractive to intensification of forestry through investment of capital and effort in a private enterprise economy. In the future, industrial forestry is likely to concentrate intensified management programs on the more productive lands. Public management, with its lesser emphasis on maximizing profits, probably will continue to dominate on the least productive areas.

THE RESOURCE IN GROWING STOCK

Forest land is the fundamental resource; nevertheless, an appraisal is also needed of the growing stock from which the annual crop is harvested. Two different inventories commonly are made: *sawtimber volume*, which is reported in board feet for the usable saw log portion of live standing trees of sawtimber size only; and *growing stock volume*, which is reported in cubic feet of sound wood in all live trees including sawtimber and pole timber.[7]

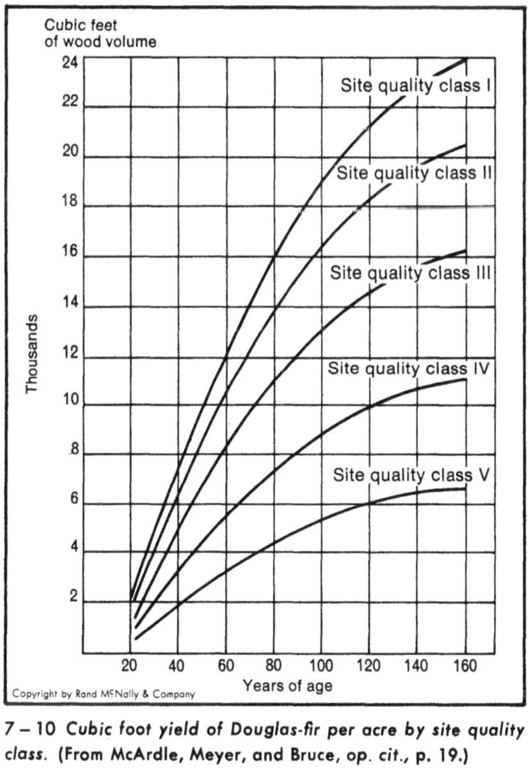

7 – 10 Cubic foot yield of Douglas-fir per acre by site quality class. (From McArdle, Meyer, and Bruce, op. cit., p. 19.)

THE SAWTIMBER RESOURCE. The growing resource in sawtimber-size trees is the most significant resource because this class is sustaining the annual harvest. As of 1962, the Forest Service reported that about 80 per cent of the total timber cut came from sawtimber-size trees.

The resource in sawtimber according to the 1962 survey was 2,537 billion board feet (Table 7–6). The dominance of the Pacific Coast forests in sawtimber is notable, for with only 13 per cent of the commercial forest land, they have 66 per cent of the softwood sawtimber and 55 per cent of all sawtimber. The Douglas fir subregion alone, on 5 per cent of the commercial land, has 33 per cent of the softwood sawtimber. In the Rocky Mountain areas, 13 per cent of the land has 16 per cent of the sawtimber. Almost all hardwood was in the eastern half of the nation; the North, with 34 per cent of the land, has 51 per cent of the hardwood but only 3 per cent of the softwood. The South, with 40 per cent of the commercial land, has only 11 per cent of the softwood sawtimber but 39 per cent of the hardwood (Figure 7 – 12).

GROWING STOCK. The growing stock resource was estimated to be 628 billion cubic feet in 1962. The distribution is similar to that for sawtimber, although the eastern areas are shown to be in a slightly better position. For example, the South has 21 per cent of the nation's growing stock, but only 16 per cent of the sawtimber stand. The North has 22 per cent of all growing stock, but only 12 per cent of the sawtimber. In contrast, the Pacific forests, because of their larger inventory of sawtimber trees, have only 41 per cent of the growing stock volume (Table 7–7).

GROWTH TO HARVEST COMPARISON

The nation has a large "capital" resource in the growing trees on its commercial forest lands, but there is need to evaluate this growing stock in terms of its annual growth compared to annual cut. Moreover, in the context of timber commodity production it is essential to realize that the trees are a crop rather than a

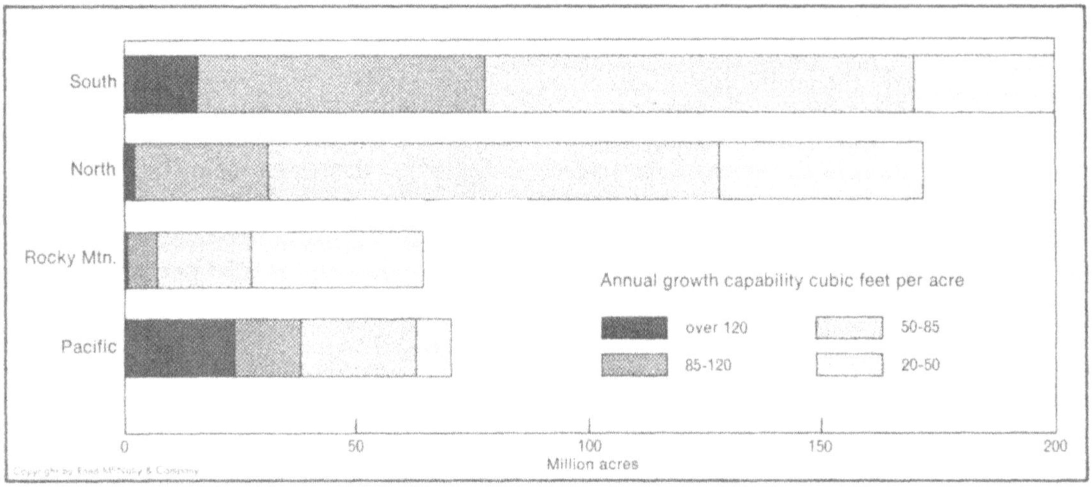

7 – 11 *Growth capability of commercial forest land.*

TABLE 7–5
AREA OF COMMERCIAL FOREST LAND BY GROWTH CAPABILITY CLASSES*

Annual growth class (in cubic feet per acre)	U.S.	North	South	Rocky Mountain	Pacific
	Million acres				
over 120	43	2	16	1	24
85–120	117	30	63	9	15
50–85	232	95	93	20	24
20–50	117	45	29	36	7
TOTAL	509	172	201	66	70

* *Timber Trends in the United States op. cit.*, p. 80.

"capital" or fixed resource to be "mined." The nation's wood needs could not long be met by harvests of old growth or "virgin forests." It is therefore a mistake to decry the cutting of mature timber stands when their orderly harvest is actually necessary to begin growing new crops for the future. The important requirement is that commercial forest lands be kept growing quality trees to supply the annual harvests.

Estimates of volume of net annual growth are made by the Forest Service for sawtimber stands in board feet and for growing stock in cubic feet. The growth estimates are based on detailed appraisals of a large number of field plots which, however, total only a fraction of one per cent of the commercial forest land. Although the surveys are designed to limit sampling errors to a maximum of ten per cent, it should be recognized that estimating net growth on the nation's 509 million acres of commercial forest lands is a complex and difficult problem. Because of changing criteria and data revisions, specific comparison between estimates published from the 1945[8], 1952[9], and 1962 national inventories is of questionable validity; consequently, this discussion of growth to harvest relationship is based only on the 1962 inventory.

Comparisons of net annual growth and cut of sawtimber are shown in Table 7–8. The important national relationship is that net softwood growth is estimated as 97 per cent of the annual cut, indicating significant improvement over 1952 when the growth to cut ratio was estimated at only 77 per cent. Hardwood saw-

TABLE 7–6
NET VOLUME OF SAWTIMBER BY REGIONS AND SECTIONS*

Section and region	Total	Billion board feet Softwood	Hardwood
New England	54.4	30.4	24.0
Middle Atlantic	103.7	13.7	90.0
Lake States	59.7	18.2	41.5
Central	91.9	4.3	87.6
TOTAL NORTH	309.8	66.6	243.1
South Atlantic	117.6	54.7	62.9
East Gulf	69.9	44.7	25.2
Central Gulf	93.6	49.7	43.9
West Gulf	131.0	75.6	55.4
TOTAL SOUTH	412.1	224.8	187.3
Northern Rocky Mountain	271.2	269.6	1.5
Southern Rocky Mountain	147.2	139.2	8.0
TOTAL ROCKY MOUNTAIN	418.4	408.8	9.5
Oregon–Washington			
Douglas-fir[1]	702.2	667.9	34.3
Pine[1]	207.2	206.7	0.5
California	303.9	302.3	1.6
Coastal Alaska	182.6	180.9	1.6
TOTAL PACIFIC COAST	1,395.9	1,357.6	38.0
Hawaii	0.7	–	0.7
TOTAL UNITED STATES	2,536.8	2,058.0	478.8

* *Timber Trends in the United States op. cit.*, p. 155–56.

[1] *Timber Resource Statistics for the Pacific Northwest, op. cit.*, p. 7.

timber growth exceeds cut. The low ratio of growth to cut in all the Pacific forests leaves room for concern. Yet it is only fair to note that the low ratios are not from lack of conservational concern in these areas; they reflect the existence of large reserves of old growth timber that produce little net growth. In the Pacific region forest managers are still in the process of converting from harvesting of old growth stands to planned crop rotation.

There is, however, considerable reason to believe that a portion of these remaining old growth reserves ought to be retained as a "living timber reserve" to provide a long-range balancing influence in the nation's forest economy. Evidence from field studies suggests that mature stands can be held in reserve without a net loss, although salvage of diseased and dead trees is desirable. A ten-year sample study on a twenty-five acre, 250-year-old Douglas-fir stand in western Oregon revealed gross annual increment of 1,582 board feet offset by annual mortality of 1,156 board feet for a net annual gain of 426 board feet.[10] The Pacific forests in the 1962 inventory accounted for only 35 per cent of the annual softwood sawtimber growth, but provided 62 per cent of the harvest. The South accounted for 47 per cent of the growth, but contributed only 23 per cent of the harvest. In hardwood, the southern and the northeastern forests dominate.

Comparisons of net annual growth and cut from growing stock are shown in Table 7–9. The pattern is similar to that of the sawtimber portion, but the overall forest situation appears more favorable. On a national basis, the growth in softwood is over one-fourth greater than the cut, and hardwood growth is double the cut.

Only the Douglas-fir subregion, California, and Coastal Alaska are cutting more than the annual growth. The South is revealed as the nation's leading wood commodity supplier, providing 42 per cent of the total cut compared to 36 per cent from the Pacific forests. The conclusions one may draw from these estimates of growth are perhaps two: (1) There is need to assure quality regeneration of Pacific forests following harvest of old growth timber areas; (2) increased utilization of hardwoods ought to be fostered.

QUALITY OF STOCKING

An inventory measure of critical importance for future timber supply is the condition of tree stocking on commercial forest land. The 1962 inventory reported conditions according to

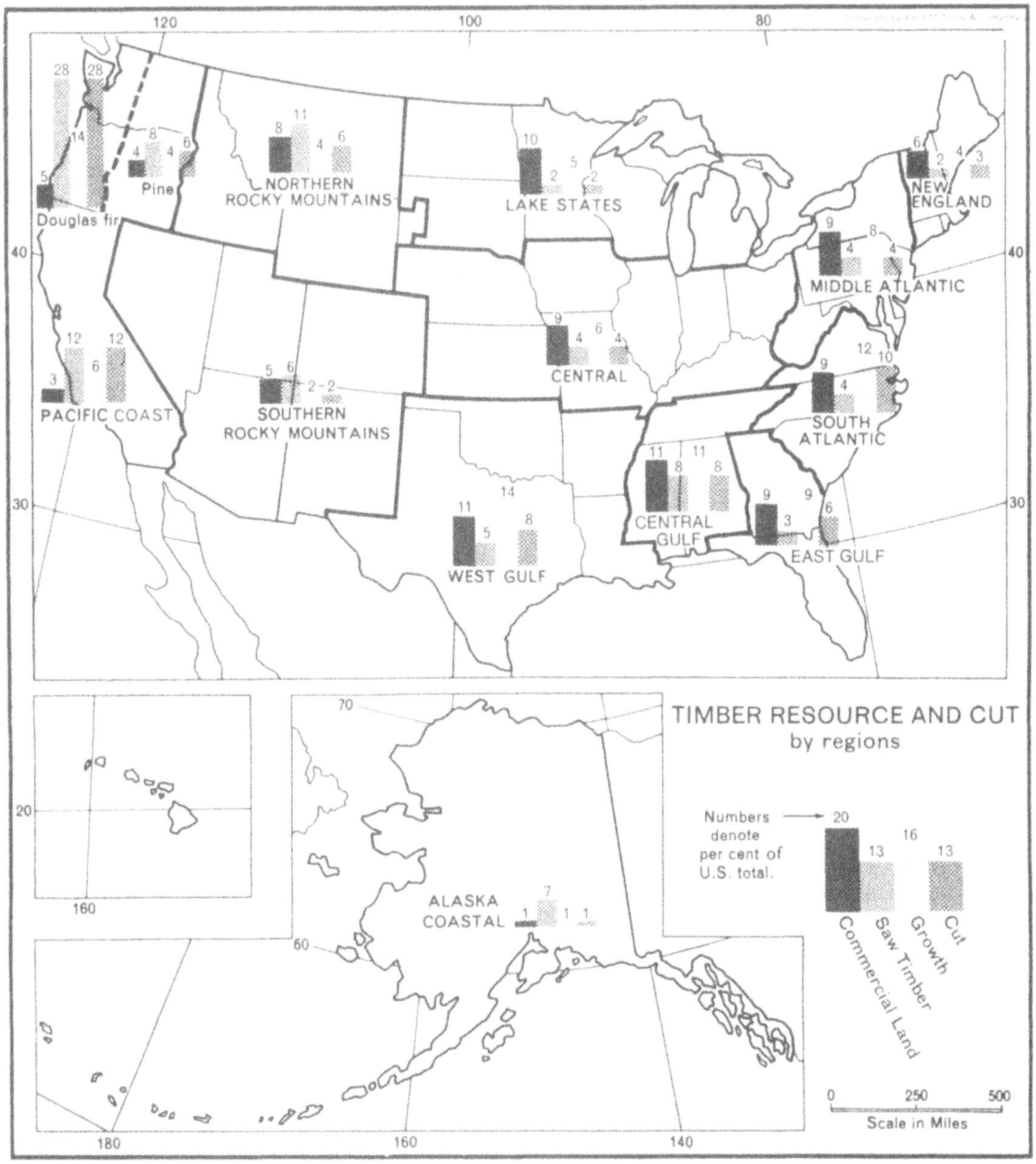

7 – 12 Commercial land, sawtimber volume, sawtimber annual growth and cut by sections.

TABLE 7–7
NET VOLUME OF GROWING STOCK*

	Total	Billion cubic feet Softwood	Hardwood
New England	30.7	16.0	14.7
Middle Atlantic	46.6	5.7	40.9
Lake States	31.9	8.4	23.5
Central	27.3	1.3	26.0
TOTAL NORTH	136.5	31.3	105.2
South Atlantic	40.5	17.0	23.5
East Gulf	24.7	14.9	9.8
Central Gulf	30.4	13.4	17.0
West Gulf	38.5	17.4	21.1
TOTAL SOUTH	134.1	62.7	71.4
Northern Rocky Mountain	61.7	60.9	0.9
Southern Rocky Mountain	37.0	32.5	4.5
TOTAL ROCKY MOUNTAIN	98.7	93.4	5.4
Oregon–Washington			
Douglas-fir[1]	123.0	112.4	10.6
Pine[1]	45.0	44.8	0.2
California	35.1	34.7	0.3
Coastal Alaska	55.3	54.9	0.4
TOTAL PACIFIC COAST	285.4	246.8	11.5
TOTAL UNITED STATES	627.9	434.1	193.8

* *Timber Trends in the United States, op. cit.*, p. 154.

[1] *Timber Resource Statistics for the Pacific Northwest, op. cit.*, p. 7.

stand size classes. About 40 per cent of the commercial lands supported sawtimber stands, a third supported poletimber, 27 per cent supported young regeneration growth, and 7 per cent was nonstocked land.[11]

In addition to the stand size class, the volume of growing sawtimber is a significant index to the economic character of the land. Table 7–10 reveals the area and distribution by three classes of timber volume. The significant indication is that over half the nation's commercial forest land is occupied by stands with less than 1,500 board feet volume per acre, which therefore are probably uneconomical to harvest at this time.

Regional differences in volume of stocking are clearly indicated. The Pacific forests have two-thirds of the land in the highest class, in contrast to the South which has only 12 per cent in this class but has nearly two-thirds in the less than 1,500 board feet class. The average volume of sawtimber in Pacific forests is estimated to be 19,850 board feet, eight times the average volume estimate for the South and four times the national average. These differences, of course, reflect the historical record of cutting as well as the inherent capability of the land. Especially in the Pacific region, large areas supporting mature stands of timber were reserved in public ownership before the first period of major harvest.

The most serious problem condition is in the low ratio of actual to optimum stocking on half of the nation's forest lands. About 112 million acres of commercial forest were estimated to be less than 40 per cent occupied with growing stock trees, including thirty-five million acres of nonstocked lands. The problem of inadequate stocking is most serious in the Lake States, California, Rocky Mountains, and East Gulf

TABLE 7–8
NET ANNUAL GROWTH AND CUT OF SAWTIMBER, 1962*

Section and region	Softwood			Hardwood		
	Growth (Million board feet)	Harvest (Million board feet)	Growth to cut ratio	Growth (Million board feet)	Harvest (Million board feet)	Growth to cut ratio
New England	1,159	922	1.26	808	426	1.90
Middle Atlantic	463	430	1.08	3,853	1,329	2.90
Lake States	909	453	2.00	1,695	847	2.00
Central	267	76	3.51	3,321	1,642	2.02
TOTAL NORTH	2,799	1,881	1.49	9,676	4,245	2.28
South Atlantic	3,916	2,578	1.52	2,822	2,212	1.28
East Gulf	3,773	1,929	1.95	1,140	895	1.27
Central Gulf	4,253	1,665	2.55	1,913	2,146	0.89
West Gulf	5,006	2,234	2.24	2,507	1,715	1.46
TOTAL SOUTH	16,948	8,406	2.02	8,382	6,968	1.20
Northern Rocky Mountain	2,428	2,869	0.85	1	9	0.11
Southern Rocky Mountain	1,034	953	1.08	107	8	13.38
TOTAL ROCKY MOUNTAIN	3,462	3,822	0.90	108	17	6.35
Oregon-Washington						
Douglas-fir[1]	6,829	13,409	0.51	791	379	2.09
Pine[1]	1,945	2,733	0.71	7		
California	3,502	5,879	0.59	24	41	0.59
Coastal Alaska	379	617	0.61			
TOTAL PACIFIC	12,655	22,638	0.56	822	420	1.96
TOTAL UNITED STATES	35,866	36,748	0.97	18,988	11,654	1.63

* *Timber Trends in the United States, op. cit.*, p. 175–76.

[1] *Timber Resource Statistics for the Pacific Northwest, op. cit.*, p. 19.

sections (Figure 7 – 13). A large share of these lands present problems such as brush or other ecological conditions that inhibit regeneration. Nevertheless, it is clear that a fifth of the nation's commercial forest land is essentially idle from the point of view of timber production, and another quarter is only 40 to 70 per cent utilized by desirable growing trees.

OWNERSHIP OF THE FOREST RESOURCE

One of the most important considerations in evaluation of forest land resource (and a major problem) is the spatial and societal reality of ownership. There are nearly five million owners of forest land, and a large majority of these ownerships are small tracts which are not well managed for timber production. Unlike the ownership pattern of most forests of the world outside of the United States, the largest portion of commercial forest land is held in private ownerships: 72 per cent as compared with 28 per cent public ownership.

PUBLIC OWNERSHIP. The pattern of ownership of the nation's commercial forest land is shown in the map (Figure 7 – 14). Public ownership is dominated by Federal holdings; 22 per cent of the nation's total commercial land, or 113 million acres. The national forest system managed by the Forest Service accounts for most of the Federally owned land—ninety-six million acres. The rest of the land owned by the Federal government is divided as follows: Indian Service, 6.5 million; Bureau of Land Management, 5.4 million; and other Federal lands, 4.5 million. State lands represent 4 per cent of the public commercial forest lands, while county and municipal lands represent nearly 2 per cent. The Federally-owned lands, however, contain about half of the sawtimber and 40 per cent of the growing stock.

TABLE 7–9
NET ANNUAL GROWTH AND CUT OF GROWING STOCK, 1962*

Section and region	Softwood Growth (Million board feet)	Softwood Harvest (Million board feet)	Softwood Growth to cut ratio	Hardwood Growth (Million board feet)	Hardwood Harvest (Million board feet)	Hardwood Growth to cut ratio
New England	538	266	2.02	463	133	3.48
Middle Atlantic	150	119	1.26	1,536	367	4.18
Lake States	298	155	1.92	854	315	2,71
Central	58	21	2.76	938	322	2.91
TOTAL NORTH	1,044	560	1.86	3,792	1,136	3.34
South Atlantic	1,031	749	1.38	982	448	2.19
East Gulf	1,097	642	1.71	422	194	2.18
Central Gulf	1,079	499	2.16	707	623	1.13
West Gulf	1,155	603	1.91	996	479	2.08
TOTAL SOUTH	4,362	2,492	1.75	3,107	1,744	1.78
Northern Rocky Mountain	581	485	1.20	5	2	2.51
Southern Rocky Mountain	285	161	1.77	59	3	19.67
TOTAL ROCKY MOUNTAIN	866	646	1.34	64	5	12.80
Oregon-Washington						
Douglas-fir[1]	1,468	2,001	0.73	264	57	4.65
Pine[1]	579	461	1.26	4		
California	644	932	0.69	7	15	0.47
Coastal Alaska	64	97	0.66			
TOTAL PACIFIC	2,755	3,491	0.79	275	72	3.88
TOTAL UNITED STATES	9,027	7,191	1.26	7,238	2,957	2.45

* *Timber Trends in the United States op. cit.,* p. 172–73.

[1] *Timber Resource Statistics for the Pacific Northwest, op. cit.,* p. 19.

TABLE 7–10
COMMERCIAL FOREST LAND BY CLASS OF SAWTIMBER VOLUME*

Class in volume (Board feet per acre)	U.S. (Million acres)	North	South	Rocky Mountains	Pacific
1. over 5,000	105	17	23	20	45
2. 1,500–5,000	118	35	51	21	11
3. less than 1,500	286	120	127	25	14
	509	172	201	66	70

* *Timber Trends in the United States, op. cit.,* p. 87.

Federal ownership is concentrated in the West, where public domain forests were reserved before homesteading or sale. State forests and community forests (about four thousand of them) are important chiefly in the Northeast, especially in areas where tax delinquent lands have reverted to county and state.

PRIVATE OWNERSHIP. Private ownerships include forest industry holdings, farm woodlots, and a large number of other individual ownerships. In 1962, forest industry, with sixty-seven million acres in about 25,000 ownerships, managed 13 per cent of the nation's commercial land. Over one-half of the ownership is in the

South, 15 per cent in the Pacific Northwest, 12 per cent in New England, and 5 per cent in California (Figure 7 – 15). For the most part, these ownerships are large in size, averaging 2,200 acres, but with two-thirds of the area in ownerships of over fifty thousand acres.

The farm category of ownership is notable, since it includes 151 million acres of forest, accounting for 30 per cent of all commercial forest land. There are over 3 million ownerships accounting for 48.6 acres each. In all eastern regions, most farm forest acreage is held in under one hundred acre tracts, while in the West most of the area is in five hundred to five thousand acre tracts. These farm forest ownerships support only about 10 per cent of the softwood sawtimber but about 40 per cent of the hardwood sawtimber, reflecting the dominance of the South (52 per cent) and North (37 per cent) in this type of ownership.

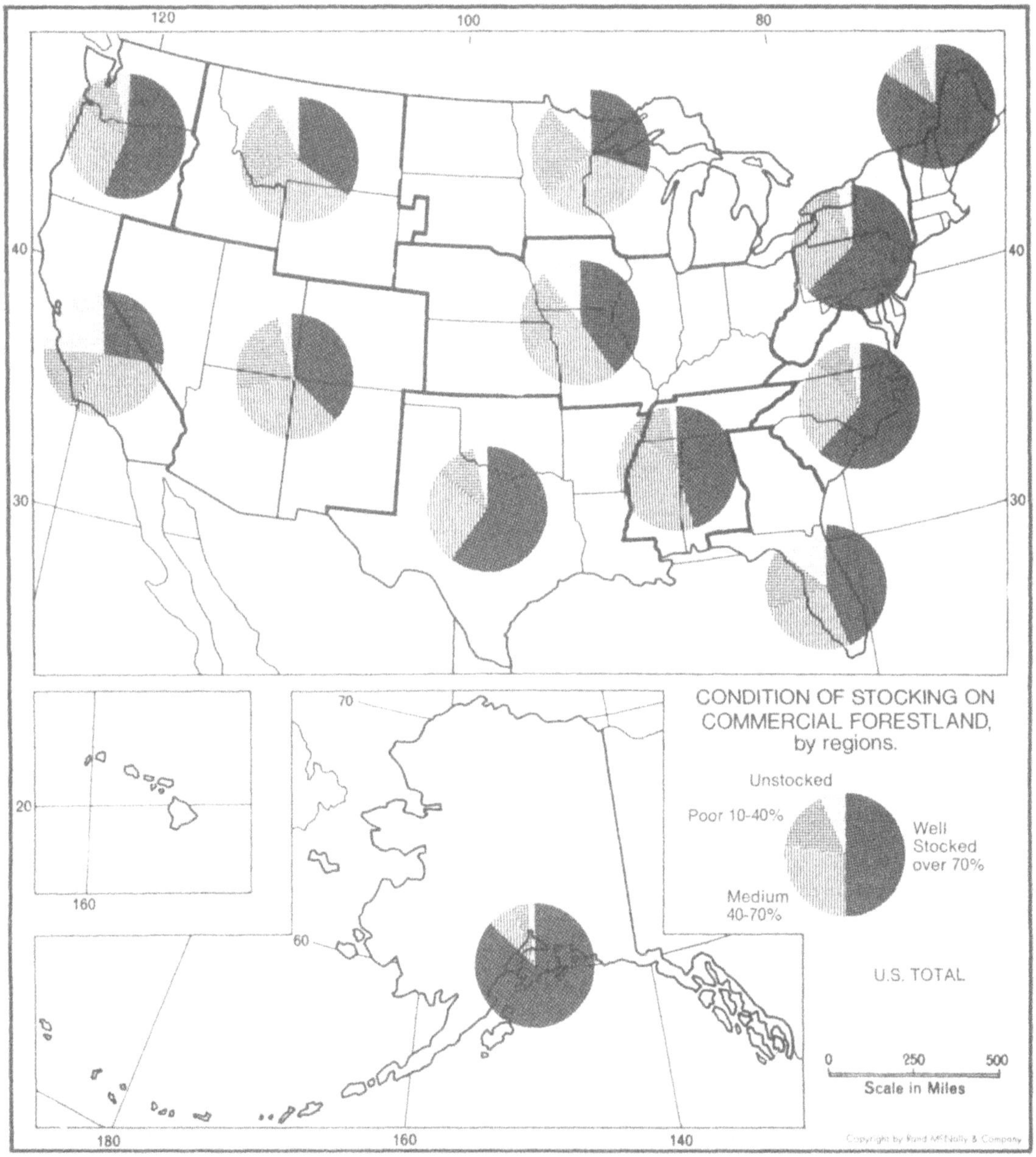

7 – 13 Per cent occupance by growing stock trees on commercial forest land, 1962.

Other private holdings (149 million acres total or 29 per cent of all commercial lands) include tracts of many sizes held by over a million owners from all walks of life. It is notable that this category is larger than the public holdings, and twice as large as industrial holdings. These small private holdings and the farm ownerships tend to rank low in intensity of management. Because of their small size, fragmented and often intermingled location, these private and farm holdings are a major problem needing particular attention in the national program of conservational management.

Future Requirements

The question of how critical is the need for forest conservation requires appraisal of the condition and capability of the resource today

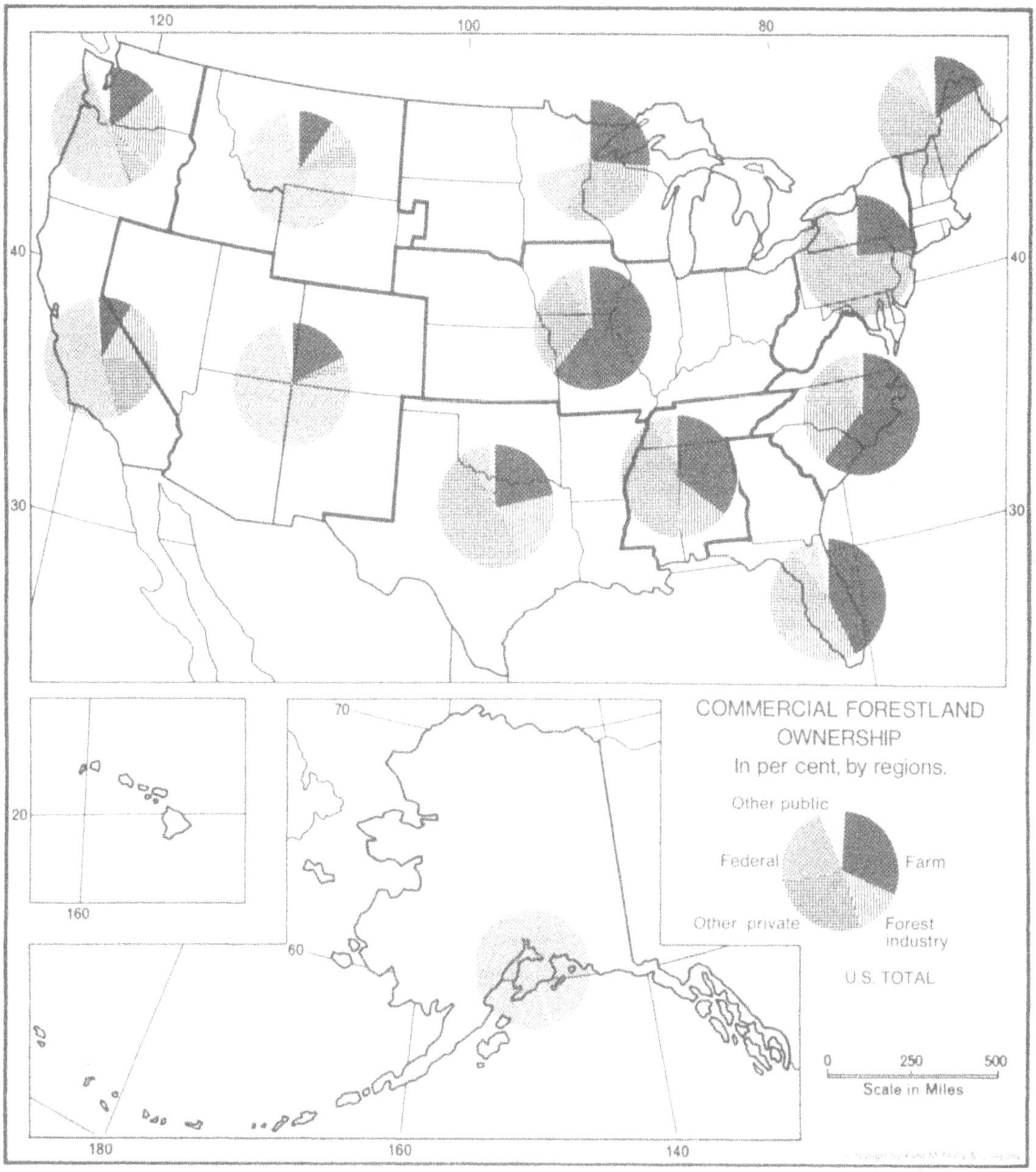

7 – 14 **Ownership of commercial forest lands, 1962.**

in terms of probable needs of the citizens of the future. Various projections and estimates have been made on timber commodity needs for future national requirements.[12] The United States Forest Service, as part of the 1962 forest resource inventory, carried out studies and made projections of demand for the year 2000 for several levels of assumed population and economic indices. The resultant demand for forest harvest at the median level with population of 325 million is shown in Table 7–11 and is compared with 1952 and 1962 consumptions (Figures 7–16 and 7–17). Use of wood for fuel is projected to decline to about half of the present level. The greatest increases are projected for pulp and for veneer; both are expected to increase at least fourfold. The projected fivefold increase in hardwood utilization for pulpwood is notable.

7 – 15 *A southern pine plantation.* This twelve-year-old shortleaf pine plantation near Mineola, Texas, is typical of development in the South. Trees were hand planted on class III land, spaced six feet apart and have been pruned to improve log quality. (Soil Conservation Service)

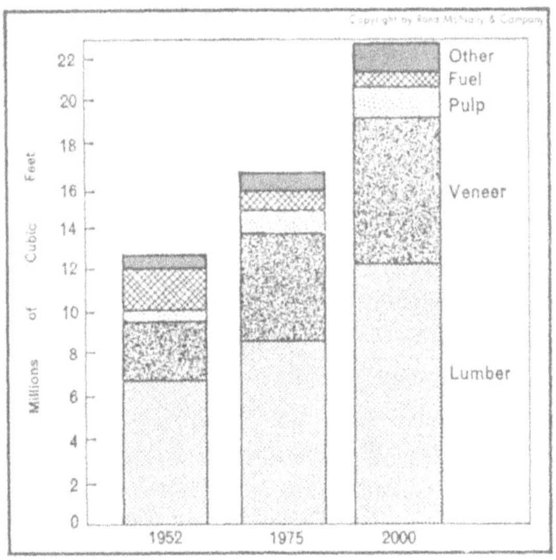

7 – 16 Timber harvest, 1952, 1962, and projected requirements for 2000.

TABLE 7–11

HARVEST FROM GROWING STOCK
1952, 1962 AND PROJECTED NEED
BY USES*

Uses	1952	1962	2000
	Millions of cubic feet		
Sawlogs	5,801	4,936	6,870
Softwoods	4,602	3,947	5,490
Hardwoods	1,199	989	1,380
Veneerlogs	392	725	1,980
Softwoods	219	579	1,630
Hardwoods	173	146	350
Pulpwood	1,655	2,353	7,390
Softwoods	1,407	1,725	4,420
Hardwoods	248	628	2,970
Miscellaneous industrial wood	579	380	370
Softwoods	278	203	200
Hardwoods	301	177	170
Fuelwood	966	517	250
Softwoods	232	93	40
Hardwoods	734	424	210
Logging residue	1,364	1,237	1,960
Softwoods	749	644	890
Hardwoods	615	593	1,070
Total timber cut	10,757	10,148	21,600[1]
Softwoods	7,487	7,191	14,400
Hardwoods	3,270	2,957	7,200

* *Timber Trends in the United States op. cit.*, p. 70.
[1] Total adjusted upward to compensate for expected smaller size of trees available in year 2000.

No one can foretell the future with complete assurance, but increasing population and a rising standard of living suggest greater demands for wood. There are, of course, several unknowns. For example, competition from other raw materials such as plastics, glass, and metals might reduce demand for wood. Yet the use of wood pulp for paper and as the base for new materials seems certain to increase. The role of imports in meeting national demands for timber and pulp is also uncertain; tropical forest areas undoubtedly will increase in significance, and may become important sources of imported timber and pulpwood. Nevertheless, the conclusion that requirements for forest harvest will be about double the present rate seems reasonable.

The implication of the projected need is obvious and deserves the concern of all citizens: the present level of forest productivity will be inadequate in the future. Certainly forest land management is much improved over fifty years ago, and annual growth has been shown to be nearly equal to the current level of harvest. The needs projected for the future, however, will demand a harvest of no less than double the present level, and on no more land—probably on somewhat less land. Is such a large increase in productivity possible? If so, how can these lands be made to grow timber at double the current rate? The need for a dynamic conservational management of forest lands is evident.

POTENTIAL GROWTH POSSIBILITIES

Forest scientists of the United States Forest Service believe that it is technically possible to produce the projected wood needs on the available land. Their studies conclude that if the best practices extended to all lands, the annual growth capacity for the nation's forest land would total some fifty billion cubic feet, including at least two hundred billion board feet in sawtimber growth. One might well term this the *theoretical maximum.* A lower level termed the *realizable growth* is shown in Table 7–12, and is more reasonable and probable. This estimate of capability is about twice the present level of annual growth, but is slightly lower than desired to assure that the forests can sustain the projected demand, especially in the sawtimber class.

There appears to be no doubt that the forest lands can be made to sustain annual harvests of at least twice the present volume. The future clearly depends on how well the forest lands are utilized as a result of the foresight and decisions of the owners, the knowledge and skill of the foresters, and the wisdom of an informed public opinion expressed through government. A relatively small number of foresters and landowners actually will be doing the job, but the entire nation has a vital stake in how well it is done. The prospects and avenues toward conservation are reviewed and evaluted in the next chapter.

7 – 17 Sawtimber growth needed by species groups.

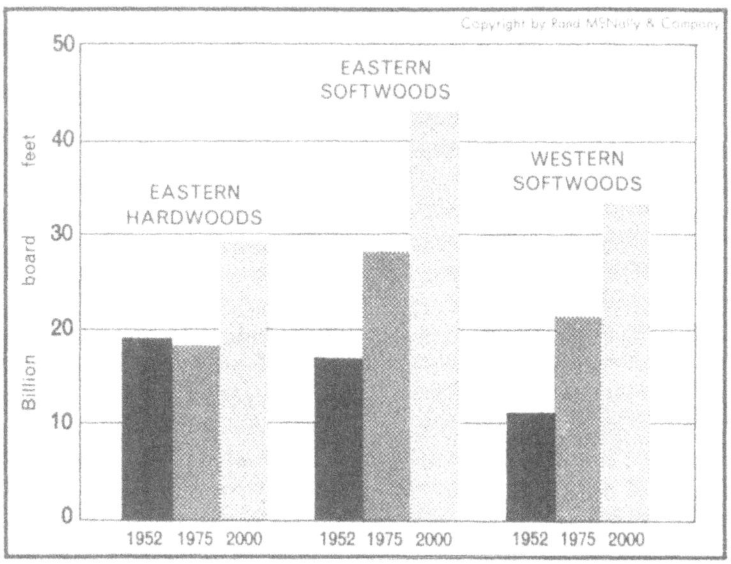

TABLE 7–12

REALIZABLE ANNUAL GROWTH OF WOOD*

	Sawtimber class			Growing stock		
	Realizable	Needed for year 2000 Billion board feet	Growth in 1962	Realizable	Needed for year 2000 Billion cubic feet	Growth in 1962
Eastern hardwoods	30.5	29.1	18.1	10.2	7.3	6.9
Eastern softwoods	39.6	43.1	19.7	9.7	8.3	5.4
Western softwoods	30.1	33.2	16.1	7.3	6.4	3.6
Western hardwoods	0.5		0.9	.3		0.3
TOTAL.	100.7	105.4	54.8	27.5	22	16.3

* *Timber Resources for America's Future* (Washington: Forest Service, Department of Agriculture, 1958), pp. 477 and 482. 1962 growth from *Timber Trends in the United States op. cit.*, p. 98, 175.

FOOTNOTES

[1] For the legends of Paul Bunyan see James Stevens, *The Saginaw Paul Bunyan*, (New York: Alfred A. Knopf, Inc., 1932), 261 pp. and Harold W. Fenton, ed., *Legends of Paul Bunyan*, (New York: Alfred A. Knopf, Inc., 1947), 418 pp.

[2] *Timber Trends in the United States.* (Washington: Forest Service, Department of Agriculture, 1965), 235 pp. This document is the source for 1962 data.

[3] For greater detail on the silvicultural aspects of United States forest regions, see John W. Barrett, ed., *Regional Silviculture of the United States* (New York: The Ronald Press, 1962), 610 pp.; for the characteristics of individual species see H. A. Fowells, compiler, "Silvics of Forest Trees of the United States" (Washington: Department of Agriculture, *Agricultural Handbook 271*, 1965), 762 pp.

[4] The 1962 national forest inventory defined commercial forest land as land "suitable for management to grow crops of industrial wood generally capable of producing in excess of twenty-five cubic feet per acre of annual growth."

[5] Note that the Agricultural Act of 1956 provided for tree planting on farm lands as a soil bank conservation reserve. Nearly three million acres have been planted under this program, especially in the Southeast.

[6] See Chapter 3, where a complete analysis of the overall land use trend of the nation is presented.

[7] A *sawtimber tree* is a minimum of eleven inches in breast-high diameter outside the bark (nine inches in the East for softwood). The tree also must include at least one merchantable saw log, eight feet long in hardwood and twelve feet long in Western conifers. A *poletimber tree* is at least five inches in diameter, but less than sawtimber size. *Growing stock* is defined as all live trees including saplings and seedlings. *Growing stock volume* is computed for sawtimber and poletimber trees from the stump to a minimum four-inch top outside the bark.

[8] "Forests and National Prosperity," *Miscellaneous Publication No. 668* (Washington: Department of Agriculture, 1948), 99 pp.

[9] "Timber Resources for America's Future," *Forest Resource Report No. 14* (Washington: Department of Agriculture, 1958), 713 pp.

[10] Carl M. Berntsen, "Productivity of a Mature Douglas Fir Stand," *Research Notes, No. 188* (Portland, Oregon: Pacific Northwest Forest and Range Experiment Station, June, 1960). See also Robert W. Steel and Norman P. Worthington. "Increment and Mortality in a Virgin Douglas Fir Forest," *Research Notes, No. 110* (Portland, Oregon: Pacific Northwest Forest and Range Experiment Station, April, 1955). This study on a 350-year-old Douglas-fir stand reveals actual net loss.

[11] The definitions adopted for the inventory were these: *sawtimber stands* have at least 10 per cent stocking with live trees of which over half are of poletimber size or larger, and of which over half are sawtimber size. *Poletimber stands* also have at least 10 per cent stocking with live trees of poletimber size or larger, but the poletimber size trees exceed the sawtimber stocking. *Nonstocked* areas have less than 10 per cent stocking with live trees of all size classes. The percentage stocking is in relationship to stocking considered necessary for full utilization of the site. The field examination involves standardized procedures and educated judgement, but the measure is either the number of trees or the basal area of desirable trees per acre. In number of trees, full stocking generally requires 800 to 1,000 for young reproduction, 600 to 750 at four-inch diameter breast-high size, 150 to 210 at ten-inch size and 30 to 75 stems at twenty-inch size, depending upon site quality and forest type.

[12] For example, see "America's Demand for Wood 1929–1975," Stanford Research Institute Report for Weyerhaeuser Company (Tacoma, Washington: Weyerhaeuser, 1954), 95 pp. See also *Timber Trends in the United States, op. cit.*, part 1, pp. 1–74.

8

Forest Resources Conservation

A GREAT DEAL HAS BEEN ACCOMPLISHED by private and public effort toward achieving a mature, forward-looking stewardship of forest lands. Nevertheless, the need for dynamic conservation is evident from the preceding review of the forest resource status and capabilities in comparison with the probable needs of the future. This chapter, then, deals with conservation of the forests. It is presented in two parts: first, a review of the activities and responsibilities of various agencies and organizations, both public and private; second, a review of specific conservation actions which are practiced and others which show promise.

Federal Government in Forest Conservation

The Federal government has long played a vital role in forest conservation. Although there were leaders who foresaw the need for forest conservation as early as the first years of the colonial period, little of significance in conservation regulation was accomplished until near the end of the nineteenth century.[1] In 1871 a Federal act provided funds for protection of timberland reserved for the navy. In 1873 the Timber Culture Act granted 160 acres on the Great Plains to homesteaders who planted 40 acres to trees. In 1876 Congress, under pressure from conservation leaders, especially in the American Association for the Advancement of Science, authorized the appointment of a special agent to study the country's forest conditions. Dr. Franklin B. Hough was appointed, and in 1881 a Division of Forestry was created under the Department of Agriculture. In 1886 Dr. Bernhard E. Fernow, one of the notable figures of early forestry, became chief of the Division.[2] The first reservation of public domain lands for forest was made in 1891 by proclamation of President Harrison in establishing the Yellowstone Timberland Reserve (1,239,040 acres now part of the Teton and Shoshone national forests). Before the end of his term, President Harrison had set aside thirteen million acres, and in 1897 President Cleveland added some twenty million acres. Management of these forests was authorized by Congress in the Forest Management Act of 1897. Although this Act is the basis for the present National Forest Service, because it first provided for management rather than preservation, it is interesting that the lands were under the Interior Department in the General Land Office, while the Forestry Division in the Agriculture Department had no forests. In 1898 a leader of early American forestry, Gifford Pinchot, became chief of the Forestry Division. Under his leadership, agitation for joining the forest reserves with the Forestry Division finally led to the establishment of the United States Forest Service, which today carries the major burden for management of the public forest lands.

THE NATIONAL FOREST SERVICE

The United States Forest Service was created in 1905 by acts of Congress which transferred the public forest lands to the Department of Agriculture and changed the name of the Division to United States Forest Service. Gifford Pinchot thus became the first chief of the Forest Service. Pinchot had tremendous drive and a missionary zeal for pioneering the kind of forestry that he believed to be in the public interest. Fortunately, a friend and political force for conservation, Theodore Roosevelt, was then in the White House, and these two men forged the foundation of the National Forest System and much of its management policy. Roosevelt rendered a great conservation service in reserving by proclamation some 148 million acres of forest land recommended by Pinchot and his foresters.[3] Pinchot made his contribution through his zeal for the public interest and by taking forestry into the woods to show what could be accomplished by management.[4]

The national forest lands, managed under multiple purpose-sustained yield practices for the long-range benefit of the nation, total about 182 million acres (including noncommercial forest land) located in 154 national forests in thirty-nine states and Puerto Rico. Nineteen national grasslands established in 1960 (3.8 million acres) also are part of the National Forest System (1965). Most of the lands included have always been in the public domain, having been reserved by proclamations. Additional lands, especially most national forest lands in the eastern half of the nation, were acquired under authority of the Weeks Law of 1911 which provided for purchase of forest land in the interest of watershed protection on navigable streams. The law was amended in 1924 to allow purchases for timber production and protection. In recent years provisions have been made for exchange of lands with private enterprise as well as other government agencies in the interest of better management.

IMPORTANCE OF NATIONAL FOREST LANDS. The responsibility and significance of the National Forest Service in forest conservation is clearly revealed by the following data: It manages about 19 per cent of all commercial forest land which accounts for 45 per cent of sawtimber and 37 per cent of growing stock, and provides about 15 per cent of the national annual harvest of timber. The harvests are made by private enterprise with the timber sold as stumpage through public auction. In 1965, about twenty-six thousand sales harvested eleven billion board feet. In addition, the national forests are managed for other benefits such as watershed, wildlife, range land, minerals, and recreation. It has been estimated that national forest lands provide half the stream flow in the West, are the major water source for 1,800 towns and over 600 hydro plants, provide habitat for one-third of the nation's big game, and provide range for one-fifth of the sheep and one-eighth of the cattle in the country.[5] Mineral exploration and development is significant. Most of the national forest lands are open for recreational use, and estimates of visits in recent years are about 150–200 million annually. (See chapter 9.)

The Forest Service lands have been revenue-producing lands. In November, 1958 fifty-three years after the creation of the service, the billionth dollar was received from sales and services. In recent years total receipts have been 125–150 million dollars per year, about 90 per cent from timber revenue. Political units in which the forest lands are located receive proportional benefits. (It is projected that these revenues will double by the 1970's.) In lieu of taxes, a quarter of the revenue is distributed to the counties in which the forest areas are located for support of schools and roads. These payments to the counties, plus the local expenditures for roads (10 per cent of receipts in addition to the 25 per cent) and fire control generally exceed the taxes the lands would pay if subject to ordinary land levies. The National Forest Service lands are administered through staff offices in Washington, D. C. and nine regional offices.

Federal research began with the Forestry Division, but significant support came later.

The first forest experiment station was established in 1908 in Arizona. Now there are eight forest and range experiment stations plus an Institute of Northern Forestry and an Institute of Tropical Forestry all of which carry on major programs of research in silvics, management practices, watershed study, and forest economics. In 1910 the Forest Products Laboratory was established at Madison, Wisconsin, and today this station is an outstanding center for research of all kinds of wood problems, utilization, and products.

INTENSIFYING FOREST SERVICE MANAGEMENT. The Forest Service plans to raise the level of management on the public forests with the goal of achieving a sustained yield level of about 21.1 billion board feet annually in the year 2000, which would be about 20 per cent of the projected national harvest needed (Figure 8 – 1). In addition, more emphasis is to be given to nontimber production aspects through the development of recreational facilities, watershed, wildlife, and range management programs. The following points indicate the objectives for timber production:[6]

1. Increase harvest to the full sustained yield level.
2. Accelerate cutting of stagnant growth stands to release new growth and increase salvage harvest.
3. Raise the level of regeneration. Plant three-quarters of the 4.4 million acres now understocked but considered to be replantable.
4. Double the work in insect and disease control.
5. Expand adequate protection against fire from the present twenty-five million acres to cover 125 million acres.
6. Expand rodent control treatment as far as is economical.
7. Improve the salvage program and erosion control efforts.
8. Improve the stand condition on one-third of the total acreage by thinning, pruning, etc.

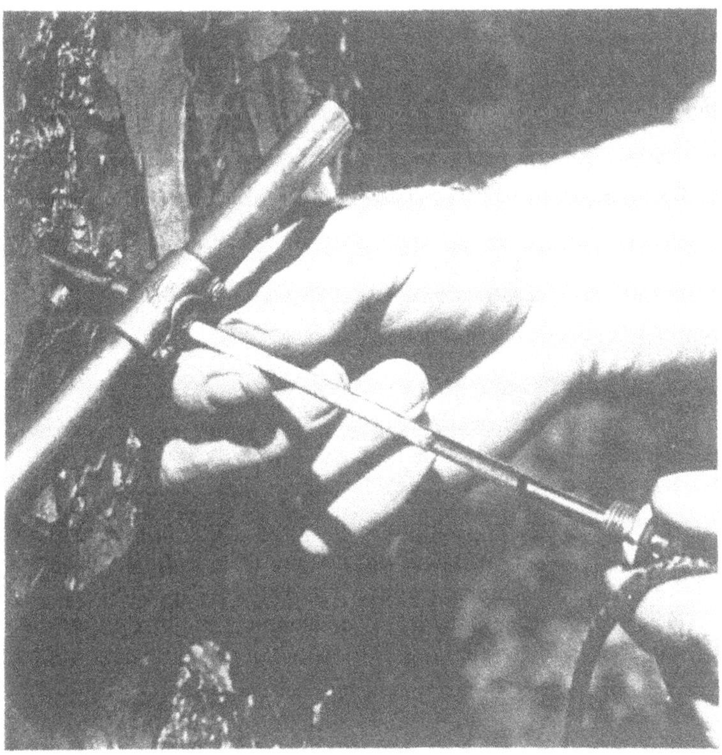

8 – 1 *Measuring tree growth with increment borer.* Intensification of management requires specific knowledge of tree growth. This close-up shows a core study of annual rings. (American Forest Products Industries)

9. Develop more adequate inventory and management plans for all areas, including coordination of sales administration with multiple use values.
10. Develop needed road systems, increase forest roads from 149,000 miles to 542,000 miles. Trails would be reduced from 112,000 to 80,000 miles.

This is a forward-looking plan for intensified management that needs and deserves support of the public and specifically of the Congress, through provision of adequate funds to do the job.

THE SOCIAL SIGNIFICANCE OF THE NATIONAL FOREST SYSTEM. The National Forest System was established by farsighted leaders with a mandate for management to provide the long-range greatest good for the public interest through practices of sustained yield and multiple use. Thus, on the national forest lands cutting cycles are usually longer than those of corresponding private enterprise lands, because maximizing the long-range timber production per acre can be considered above the short-range dollar return. In recent years pressure from private enterprise for increased levels of harvest has resulted in intensification of management on national forest lands, but the Forest Service has wisely resisted overexpansion of harvest rate.

The National Forest System and its sound policies ought to be guarded in the interest of conservation and because too-rapid liquidation of forest stands can have far-reaching social costs of local, regional, and even national significance. There is, however, continuing need for improvement and intensification of management. The area of commercial forest land in national forests is probably adequate, and no major program for additional national forest land is now under consideration.[7] Minor adjustments, however (exchanges of land, purchases, etc.), for purposes of improving management are needed. The Forest Service lands not only provide assurance for the future, but also provide competition between private and public forestry which may stimulate research and management improvements. Monopoly by government, just as by private enterprise, is generally undesirable.

OTHER FEDERAL AGENCIES

A number of other Federal agencies have direct responsibilities in forest management, especially the Bureau of Land Management and Bureau of Indian Affairs.

BUREAU OF LAND MANAGEMENT. The Bureau of Land Management of the Department of the Interior was created in 1946 by combining the General Land Office and the Grazing Service. Thus forest lands in the remaining undesignated public lands came under the management of the Bureau of Land Management in its Division of Forestry. In total, the Bureau is responsible for 5.4 million acres—only about 1 per cent of the commercial forest land, but supporting nearly 4 per cent of the sawtimber stand.

Nearly all of the lands are in the West. The most notable portion is some two million acres known as the O and C Revested Lands located in western Oregon. These lands, given in 1866–1869 as railroad grants to the Oregon and California Railroad, were repossessed in 1916–1919 by the Federal government because the provision of the grant had not been met. Also included in this transfer were reconveyed lands of the one-time Coos Bay Wagon Road Company of Southwestern Oregon. In 1937 Congress provided that these lands be managed forever on a sustained yield basis. The distribution of Bureau of Land Management commercial forest lands is: 2.5 million acres in Oregon, chiefly in the Douglas fir subregion; 2 million acres in the Rockies; and about a quarter million acres each in Coastal Alaska, California and Washington.

BUREAU OF INDIAN AFFAIRS. The commercial forest lands in the Indian Reservations total 6.5 million acres, are about 1.3 per cent of the nation's total land, and support about 2 per cent of the standing sawtimber. Most of the lands are in the West, chiefly in the Rocky Mountain region (three million acres) and the ponderosa pine subregion of Washington and Oregon (two million acres). The major eastern area is in the lake region (one million acres). Congressional acts of 1909 and 1910 provided for orderly harvest and sale of Indian forest resources under sustained yield principles and for the benefit of the Indians. In recent years some Indian reservations have been terminated and forest lands purchased by the Federal government.

MAJOR FEDERAL ACTS FOSTERING FOREST CONSERVATION

Many laws and actions at the Federal level have been instrumental in fostering the advance of forestry to its present status both on public and private lands. A few of the more important actions since creation of the Forest Service in 1905 are indicated chronologically in the following list:

1911. THE WEEKS LAW. Authorized purchase of forest land deemed necessary for protection of watershed of navigable streams. Also established the basis for Federal-state cooperation in fire protection.

1924. THE CLARKE-McNARY LAW. Extended purchase authority of the Weeks Law to include land needed for timber production. Provided for cooperation of Forest Service with state and private ownership for fire protection. The Federal government provided not more than half the cost. The law also provided for farm forestry extension and cooperation in providing forest nurseries and planting trees on farms.

1928. THE McSWEENEY-McNARY ACT. Established basis for most of the research work in the Forest Service. It was especially aimed at developing full use of forest lands including rehabilitation of nonfarm lands.

1937. THE NORRIS-DOXEY COOPERATIVE FARM FORESTRY ACT. Provided basis for the extension forestry program of aid to farm owners of forest and woodlot. Replaced in 1950 by the Cooperative Forest Management Act.

1937. O AND C SUSTAINED YIELD ACT. Provided for management of the O and C lands as timber growing lands.

1944. PUBLIC LAW 273. Authorized agreement for operation of sustained yield units combining public and private forest ownerships.[8]

1947. FOREST PEST LAW. The initial step in establishing the Federal program of pest control in forests.

1950. COOPERATIVE FOREST MANAGEMENT ACT. Authorized Federal cooperation with states in aiding and providing technical services to private land owners and industry.

1955. MULTIPLE SURFACE USE ACT. Provided for improved management of Federal lands by recognizing the right of government to retain use of surface on mining claims. In effect, separated surface rights from mineral rights.

1960. MULTIPLE USE ACT. Reaffirmed basic mandate of the Forest Service and defined the multiple use concept.

1962. McINTIRE-STENNIS COOPERATIVE FORESTRY RESEARCH ACT. Recognized need for additional effort in forestry research and provided matching funds for research in Land Grant Universities and other qualified forestry schools.

1964. ROADS AND TRAILS SYSTEM FOR THE NATIONAL FORESTS ACT. Authorized provision and maintenance of forest road systems needed for management of the national forests.

Federal responsibility in forestry has always centered around two functions: (1) The management of public forest reserves in the long-range interest of the nation; and (2) the provision of forward-looking research, aid to state and private ownership, and cooperation in protection and education for conservational forestry. Although many have wanted more Federal regulation (Gifford Pinchot among them), Congress has left the regulation of private forest lands to the wisdom of the states.

The States in Forest Conservation

The role of the states in forest conservation stems from their ownership and management of state lands, responsibility for regulation of private land-use practices and timber harvesting, and their taxing power. Many states have developed active programs of forestry—all now have departments of forestry; many have significant research programs. All states cooperate with the United States Forest Service on forest fire prevention programs. Most states have enacted laws requiring minimum forestry practices to assure regeneration as part of harvesting permits. Ownership of land by states is, however, of modest significance, because states own and manage only 4 per cent of the nation's commercial forest land. A more significant aspect of the states in forest conservation is in programs of Federal-state cooperation.

FEDERAL COOPERATIVE PROGRAMS

One of the notable programs of aid to small forest owners has been provided by the development of the Extension Forester Program as part of the Federal-state supported Cooperative Extension Service. In 1960 there were ninety full-time extension foresters employed in forty-six states providing information, demonstration, and technical advice to private forest and farm forest owners. Authority for Federal-state cooperation with small forest owners provided in the Norris-Doxey Act, and later by the Cooperative Forest Management Program (1950), has led to a major program of aid through the state forester's office in forty-nine states (Figure 8 – 2).

Related to these programs are many other Federal acts, i.e., the Agricultural Conservation Program grants aid to landowners for planting and improving forests on a cost-sharing basis.

STATE CONSERVATION REGULATION

Most of the states now have regulations providing for minimum standards of forestry on commercial lands. Oregon was one of the first states to require conservation practices, and because most state regulations are patterned after the Oregon law, the following highlights are given as indicative of state regulation—the details, of course, vary.

8 – 2 *Forester aiding private owner in the forest.* The increasing availability and recognition being given to professional management is a favorable sign for the future. Here a forester provided under the Federal-State Cooperative Forest Management Act of 1950 is offering advice to the owner of a small forest concerning animal damage. (Oregon State Board of Forestry)

The Oregon Forest Conservation Act (enacted in 1941 and subsequently strengthened), administered by the state forester, applies to all non-Federal land supporting merchantable timber. In the ponderosa pine region, where the selection system is the general practice, no tree with a breast-high diameter under sixteen inches may be cut, and a minimum of four seed trees twelve inches or more in diameter must be left per acre. In the Douglas-fir subregion, where clear cutting is practiced, 5 per cent of the area of each quarter section or at least two seed trees per acre with eighteen inches minimum diameter must be left as seed source. Owners may submit other regeneration plans for approval; for example, a company may harvest by clear-cutting and immediately plant. A $25-per-acre performance bond may be imposed to assure compliance, and these funds may be used by the state for reforestation if acceptable reproduction is not achieved in reasonable time. No commercial harvesting may be undertaken until a permit is issued by the state forester.

TAXATION BY STATES AND CONSERVATION

Taxation of forest resources is the major facet of state influence. How forests are taxed, chiefly by state and local government, is critical to conservation and constitutes a major problem. Nonowners, legislators, and the voting public thus have important direct influence on forest management practices through tax legislation.

Forest land should bear its fair share of the tax burden, but it seems clear that the public interest is best served through tax laws and interpretations which favor long-range, intensified management over short-range practices such as harvesting immature trees or low-level management. The problem is how to accomplish the desired objective. In all states, forests pay one or more assessments as property, yield, or severance taxes, and in addition, income taxes apply to gains that accrue from tree growing.

PROPERTY TAX. The property tax is the chief income source for local government and the principal tax on forest value in land and trees. The problem for the owner is that property tax is a cash outlay each year, while the forest, especially in early years of the tree growing period, is producing potential income for a distant future. Thus, the property tax becomes an annual investment on which interest charges accumulate over a long period. The effect of the property tax is to encourage early harvest of timber and to shorten the tree growing cycle. While this effect may be desirable in some cases, the danger for conservation lies in the incentive given to harvest immature timber to avoid the property tax. In most areas tax problems center around two factors. First, where trees are taxed in addition to the land, it is argued that the tax is repeated annually on the same crop in contrast to the farm crop that is taxed only once. Second, there is concern over inequalities of evaluation and taxation especially between adjacent counties and states and in comparison with other types of land use. Much study is being given these problems, and it is generally thought that, in spite of the objections, the property tax will continue to be the main source of local government income.

If the property tax is to serve the interest of conservation, the assessment ought to be based on the relative capability of land, with consideration being given for terrain and accessibility values as well as tree growing capability. Trees would not be taxed until harvested, and the best forest lands would be taxed more than poor lands, thus fostering higher level management.

YIELD TAX. The principal departure from the property tax is the yield tax, which is generally a percentage of gross stumpage value of timber harvested. The yield tax has the advantage to the tree grower of deferring payment until money is available. For conservation it has the advantage of avoiding pressure for quick harvest, but it has the undesirable feature of favoring easy-to-harvest timber while handicapping less favored timber. For the government, the yield tax has the unpopular feature of not providing a stable tax base.

SEVERANCE TAX. A third type of tax assessed against forest harvest is the severance tax which is a revenue measure, generally not a substitute for property tax. A good example is the State of Oregon severance tax of five cents (1961) per thousand feet of timber harvested, levied for support of forestry research. Severance taxes, when small, have no unfavorable effect on conservation practices.

Forest Industry in Conservation

Private enterprise holds the most important responsibility in the nation's forest conservation picture because 72 per cent of the commercial land is held in private ownerships. Just as the government was slow in achieving its present program, so intensification of management by private enterprise evolved slowly. The American Forestry Association was organized in 1875 and the Society of American Foresters was founded in 1900. In 1882 the first American Forestry Congress was held in Cincinnati. The Western Forestry and Conservation Association was founded in 1909. All attest to early interest in sound forestry.

Even though the early record of private enterprise is associated chiefly with harvesting and little concern for regeneration, there were industry leaders who recognized that the era of *growing* trees as a crop was to come. After World War I many factors contributed to the evolution of a new outlook by the forest industry. Among the more significant influences one might note are the realization that available stumpage at a low price was about at an end, passage of the forest industry code (adopted in 1933 as part of the National Recovery Act) providing for reproduction after logging, advances in the science of forestry, and changes in the economics of forestry. In the 1920's industry started hiring foresters to supervise the growing of trees.

Now nearly all forest industry companies have professional foresters managing their forests for continuous production. The level of management varies with company policy, but most are carrying on practices far above requirements of state forest laws.

The American Forest Products Industries Incorporated was organized in 1941 and has become the national forestry educational organization representing all segments of the industry. Among its significant activities are the Keep America Green Program, carried on through state organizations, a forest fire prevention education program, and the Tree Farm Program.

THE TREE FARM CONCEPT

The Tree Farm concept is significant evidence of the change in outlook by private enterprise. The idea—at least the specific name—originated with the Weyerhaeuser Company in 1941 on lands they were reforesting in western Washington. For conservation, the movement has importance.[9]

Table 8–1 shows the great expansion of

TABLE 8–1

CERTIFIED TREE FARMS BY FOREST SERVICE SECTION AND STATE*
JULY 1, 1966

Section and state	Number	Total area (thousand acre)	Per cent of private commercial forest land in Tree Farms
New England			
Connecticut	106	64	3.5
Maine	705	2,160	12.7
Massachusetts	329	107	3.7
New Hampshire	519	523	12.4
Rhode Island	63	10	2.5
Vermont	183	285	8.4
Middle Atlantic			
Delaware	8	7	1.8
Maryland	107	39	1.5
New Jersey	41	19	1.0
New York	300	322	2.9
Pennsylvania	938	430	3.7
West Virginia	124	479	4.6
Lake States			
Michigan	1,075	1,433	11.3
Minnesota	1,645	754	10.0
Wisconsin	939	1,537	14.9
North Dakota	203	7	2.4
South Dakota	263	43	8.9
Central States			
Illinois	303	29	0.8
Indiana	496	79	2.2

TABLE 8–1 (Continued)

CERTIFIED TREE FARMS BY FOREST SERVICE SECTION AND STATE*
JULY 1, 1966

Section and State	Number	Total area (thousand acre)	Per cent of private commercial forest land in Tree Farms
Central States (Continued)			
Iowa	161	11	0.4
Kansas	20	1	0.1
Kentucky	218	124	1.2
Missouri	152	237	1.8
Nebraska	97	8	0.8
Ohio	645	163	3.4
South Atlantic			
North Carolina	1,342	1,746	9.4
South Carolina	645	2,155	20.4
Virginia	626	1,159	8.0
East Gulf			
Florida	911	6,631	40.8
Georgia	2,370	7,350	30.0
Central Gulf			
Alabama	2,546	6,915	33.3
Mississippi	3,526	3,278	20.2
Tennessee	886	1,364	11.0
West Gulf			
Arkansas	1,521	4,023	21.5
Louisiana	1,847	4,308	27.6
Oklahoma	157	995	20.7
Texas	1,785	3,881	34.5
Northern Rocky Mountain			
Idaho	589	1,050	34.2
Montana	63	1,024	21.1
Wyoming	2	41	4.8
Southern Rocky Mountain			
Arizona	4	70	51.9
Colorado	9	79	2.5
Nevada	3	16	20.8
New Mexico	13	210	11.7
Utah	3	10	1.5
Pacific			
California	369	2,913	36.2
Oregon	494	5,064	49.1
Washington	779	4,757	52.0
TOTAL	30,130	67,907	19.0

* Based on data from American Forest Products Industries, Inc., July, 1966. Percentage computed from total of farm, miscellaneous private, and Forest Industry commercial forest acreage as reported January 1963 in *Timber Trends in the United States. op. cit.*, pp. 141–142.

Tree Farm area and the pattern of development in the nation. The relationship of Tree Farm land area to that of all private commercial forest land is given as a gross indication of its importance with reference to the total possible. In considering the percentage it should be noted that some well managed industrial forest lands have not been designated as tree farms, and that about 40 per cent of the private lands are in farm ownerships in some cases not fostering designation as Tree Farms.

Forest lands which bear the "Tree Farm" symbol have been certified by a regional industry association or a state committee of the American Forest Products Industries. To qualify, the owner must have developed and demonstrated the ability to carry out a plan of forest management for timber production, the lands having been inspected by a qualified forester and approved by an industry association forest practice committee or a state Tree Farm Committee. There are no tax advantages, but the public relations benefits are of major importance.

Perhaps the single most important value of the Tree Farm concept is to be found in its influence in causing many small private owners to believe in and act on the principle of growing trees as a crop. From the public interest standpoint the Tree Farm sign is certified evidence that the forest land is being managed with at least a minimum of professional guidance.

MULTIPLE USE OF INDUSTRY FOREST LANDS

Industry in recent years has increasingly given attention to the multiple-use characteristics of forest land. Watershed considerations are becoming integral parts of forest management planning. Recreational possibilities are being given special attention, and it is estimated that a large portion of industry-owned commercial forest land is open at least part of the time to the public for recreational use. In response to the mounting pressure by urban populations for outdoor recreation, many large industry owners are beginning to provide picnic and camping facilities on their forest lands.

In recent years six to ten million visits were recorded on industry lands. Some large owners are giving attention to the needs of wildlife as well as to the hunter and fisherman.[10]

Small Ownerships

A large share of the nation's commercial forest land is in small private ownerships, many of which, for reasons ranging from lack of capital to lack of interest, are not being managed to maximize wood production. Much is being done, however, through Extension Forestry, the Tree Farm concept, and Federal-state aid programs to advance management on small tracts. In some areas, as in the South, where the silvicultural cycle is short, there has been major development of small tract management under contract to the large pulp mills. There remains, however, a major problem and potential in establishing management on most small forestland ownerships.[11]

Conservation by Reducing Losses in Growing Timber

Annual mortality in growing trees caused by insects, disease, fire, and destructive weather phenomena is very high (Figure 8 – 3). The 1952 Timber Resource Inventory estimated annual mortality to be 3.5 billion cubic feet in growing stock, including 12.5 billion board feet in sawtimber. Ten years later the 1962 inventory estimated annual mortality at 5.6 billion cubic feet including 19.7 billion board feet in sawtimber. Only a small portion, 5–10 per cent, of this mortality volume is now being salvaged. In addition, there are major growth losses, especially through damage to seedlings and saplings, and losses in potential growth of growing stock through reduction in vigor which are not included in the mortality estimate. The 1952 survey estimated this growth loss to be 7.7 billion cubic feet including 31 billion board feet in sawtimber. Growth loss estimates were not made as part of the 1962 inventory.

Annual timber loss due to mortality, not including estimates for growth loss, thus is at least equivalent to a third of the annual net growth, or to about half the annual national timber harvest. Efforts directed to reducing mortality in growing stock trees therefore may prove to be one of the most effective approaches to forest conservation.

INSECT-CAUSED LOSS

The timber loss due to insect ravages is enormous. Estimated annual mortality due to insects is over five billion board feet of sawtimber (about a tenth the annual harvest for commodities), and additional loss in potential growth may be as much more. In addition, epidemic outbreaks have killed trees totaling fifty-two billion board feet.[12]

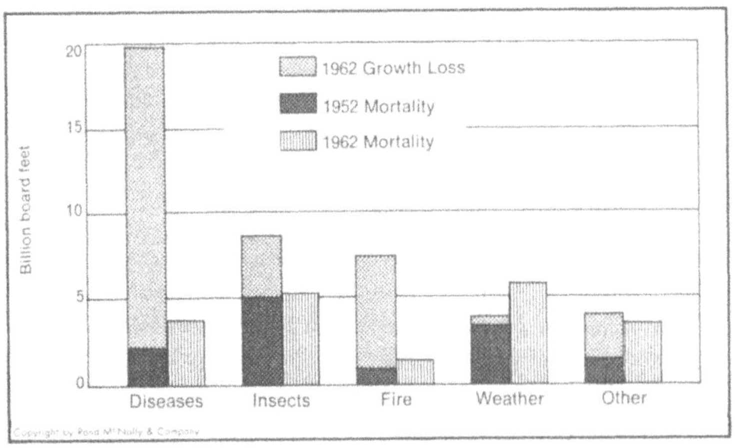

8 – 3 Estimate of sawtimber mortality loss by various causes.

Bark beetles alone kill about four to five billion board feet of sawtimber annually—90 per cent of the mortality due to insects (Figure 8 – 4). Insects such as the spruce budworm, sawfly, and the gypsy moth cause widespread damage through loss of foliage. In recent years, spruce budworm activity has reached epidemic proportions in many areas of the Pacific West and Rocky Mountain forests.

Intensified control programs began with the 1947 Pest Control Act which provided funds for Federal surveys and for control measures, and since then research and action have been steadily increased to reduce this major loss. The control measures fall into three groups: First, efforts to prevent epidemic development through control of the environment by such means as sanitation logging to remove host trees; second, biological control measures usually instigated by introducing natural enemies of pests—for example, the colasoma beetle to control the gypsy moth; third, chemical control by application of insecticides—aerial spraying with DDT has been found effective in preventing epidemic development, and about 1.5 million acres are sprayed annually at a cost of about one dollar per acre (Figure 8 – 5). Potential damage to forest wildlife and fisheries is a serious problem and insecticides require care in application.[13] Direct chemical treatment of host trees and stumps with penetrating chemical killers has recently been developed, but is expensive (2 to 2.5 dollars per tree). Recently scientists have developed new attractants to lure beetles so that they may be brought into smaller area concentrations and then killed. If these experiments are successful in perfecting economical techniques, and if lures can be perfected for other insects, the present practice of wholesale spraying of forest areas may not be required, thus solving one of the major inter-resource conflicts of our time.

8 – 4 ***How bark beetle kills trees.*** **This view, after the bark has been pried off, reveals how the bark beetles bore numerous channels in the cambium layer, killing the tree. The beetles are active in the warm season and multiply rapidly; one beetle can have hundreds of descendants in a single summer. The tree needles first become brown and the bark loosens; finally the tree dies. (Forest Service)**

Insect control will always be a major problem, because it is unrealistic to expect eradication of insects, but through research leading to better methods and integrated, intensified control measures, plus the maintenance of healthy and vigorous forests, outbreaks of epidemic proportions can be minimized and annual loss much reduced.

DISEASE-CAUSED LOSS

Annual loss of growing timber by disease in recent years has been three to four billion board feet. This is one tenth the size of the nation's annual harvest. The number of diseases is large and increasing as new ones continue to develop. Some of the disease epidemics have been disastrous; for example, the chestnut blight, thought to have been introduced from Asia, has essentially wiped out the chestnut from the eastern forests. Heart rots, especially in older trees, account for 75 per cent of loss by disease and is one of the reasons for harvesting mature stands. Blister rust, dwarfmistletoe, birch dieback, fusiform rust, and Douglas-fir root rot are but a few of the common diseases (Figure 8 – 6). Control measures are being developed by intensified forest pathology research, but indications are that the major conservational results will come from attention to harvest of mature trees before disease sets in, from salvage of dead trees, and from removal of weak trees by early thinning.[14]

FIRE-CAUSED LOSS

Although not the greatest problem, the most spectacular losses are from fire. Fire not only destroys sawtimber, but also destroys seedlings and causes widespread areas to be idled by lack of regeneration. It kills wildlife, and it reduces recreational use and watershed values. Many tragic fires were recorded in the early history of the country, such as the Miramichi fire in New Brunswick, Canada–Maine in 1825 that burned three million acres. In comparatively recent years the disastrous Tillamook fire of northwest Oregon burned a quarter of a million acres and twelve billion board feet of fine timber.[15]

Protection has reduced loss by fire, but still several million acres are burned in an average of one hundred thousand fires each year. The direct loss of sawtimber to fire totals over a billion board feet annually. In spite of increased

8 – 5 *Aerial spraying for insect control.* **Widespread spraying with DDT in concentration of about one pound per gallon of solvent has been found to be as much as 95 per cent effective in control of certain epidemic insect outbreaks. The cost is generally one dollar per acre. The control of side effects of such spraying on wildlife, fish, and humans is a difficult problem. (American Forest Products Industries)**

protection, about half of the fires occur in the South on private lands, and about half of these are caused by incendiaries and smokers, and a quarter from debris burning (Figure 8 – 7). It should be noted that controlled use of fire may be an important silvicultural technique to improve seed-bed conditions and reduce fire hazard through slash burning, and in some areas to suppress undesirable brush and hardwoods that interfere with reproduction of desirable species.

The tragedy of fire is heightened because most of the fires are preventable (Table 8–2). Only 9 per cent are caused by lightning. Man is the firebug. Because eight-tenths of the fires are man-caused, the solution lies in somehow causing all people to be careful, especially with increasing numbers of the population using the woods. The "firebug" incendiary constitutes a difficult problem in which behavioral scientists may be able to help foresters.

TABLE 8–2
CAUSES OF FIRES, 1917-1966*

Incendiary	26%	Campers	6%
Debris burning	18%	Machine use	8%
Smokers	19%		
Lightning	9%	Unknown	14%

* *1965 Forest Fire Statistics*, (Washington: Forest Service, Department of Agriculture, 1966), p. 48.

Since 1945 Smoky the Bear has been "campaigning" and has been supported by ten million dollars in donated advertising. The "Keep Green" programs now nationally sponsored by the American Forest Products Industries were started in Washington state. Under the Clarke-McNary Act, Federal-state governments cooperate in protecting over 450 million acres of state and private land at annual cost of seventy-five million dollars, of which the Federal government pays about twelve million. An additional 350 million acres are protected in national forests (Figure 8 – 8).

New techniques for fire suppression are being developed, including airplane dropping of wetting agents, especially chemicals such as sodium calcium borate. Forest closures in time of fire danger and suppression teams at strategic points are essential to reduce fire loss.

8 – 6. *Dwarfmistletoe, a parasite that attaches itself to living trees.* By "robbing" the tree of nutrients, it reduces vigor causing significant growth loss over a period of years, and may lead to death of the tree. The picture shows a female plant well established on a ponderosa pine tree. (This is not the decorative mistletoe.) Mistletoe is a major problem on western conifers especially hemlock, ponderosa pine and larch, and in the Lake states on spruce and jack pine. (Forest Service)

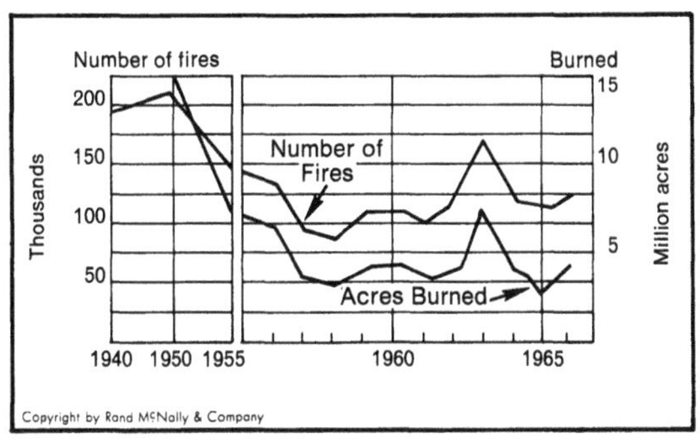

8 – 7 Fire-caused loss in timber has been better controlled since 1940. The danger, however, continues to be great, requiring unrelenting care in use of fire and continual protection.

Additional protection facilities and organization are needed on some of the land for better detection and more efficient suppression. Public support for needed fire protection funds is essential to conservation.

Conservation by Improving Growth

The second approach to meeting future wood requirements is through growing more and better trees. This can be done by extending intensified management to many more acres of forest land, especially in the small ownerships, and by raising the level of existing management intensity on most forests. It is evident that the level of management can vary from a minimum of merely protecting the forest against fire to high-level intensity near the limits of forest science and technology. Each owner, public or private, makes a decision on his specific level of management intensity based on his needs and the legal and economic realities of his area. Conflicts in interest between short-range and long-range gains and between personal and the public interest are inherent in the problem. Intensification of management is perhaps inevitable when it becomes economically advantageous; nevertheless, it needs encouragement by all citizens, especially through wisely conceived forest taxes. There also is need for increased support for research and field service by public and private agencies.

Immediate application of the most advanced practices of forest management to all forest lands cannot reasonably be expected, but there can be appropriate societal regulations that make it economical for owners to use conservational practices; the public in turn may properly expect that all owners practice at least a minimum of management and, if need be, require it by law in the public interest.

8 – 8 ***Fire control observation tower.*** **The lookout tower placed at a strategic observation point and manned at all times in the fire season is a vital part of every forest conservation program. (Forest Service)**

THE SUSTAINED YIELD CONCEPT

Basic to conservation of forests is the concept of sustained yield. In short, it means that the harvest does not exceed the net growth. The possible volume of annual harvest varies with intensity of management from a minimum supported by unmanaged growth to a maximum limited by the site capability when managed under the best of forestry science and technology in tree growing. Such a maximum will remain uneconomical for a long time. In between these extremes there are economical levels of sustained yield. Certainly forests under medium- to high-level economical management can sustain harvest volumes that are considerably greater than those of the unmanaged forest.

Establishment of optimum level sustained yield management on a forest area is the work of the professional foresters and requires careful planning, which usually includes the following: (1) an inventory of lands and growing stock to determine quantity and distribution of resource; an estimate of annual growth is essential; (2) a long-range plan for forest management, establishing the rate of harvest, an orderly sequence of timing of areas to be cut, reforestation programs, and protection; (3) a road system adequate to serve the harvesting operations and also provide for protection for all of the lands (Figure 8 – 9).

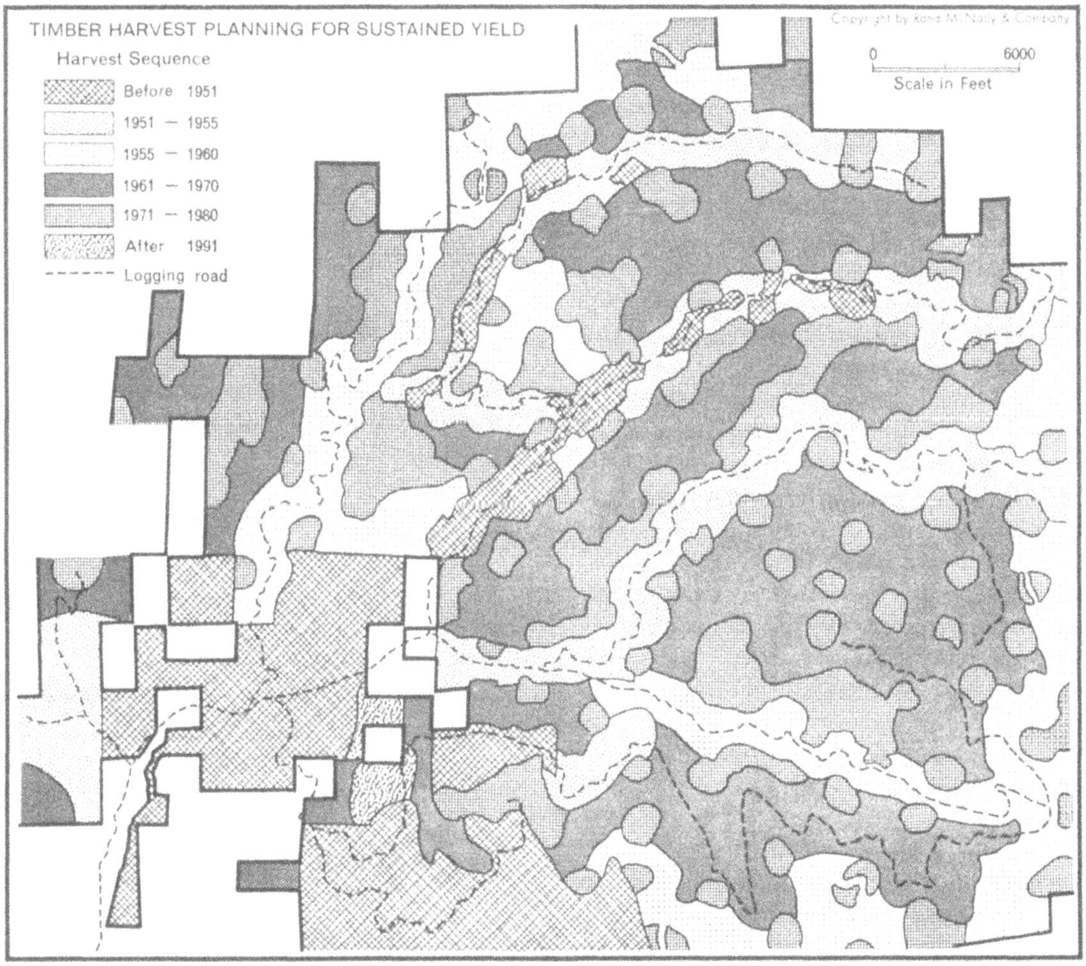

8 – 9 ***A sustained-yield plan.*** **This map shows what is essential to implement a sustained plan of forestry. Note the road system for protection and the programing of harvest as indicated by the dates in various sections. (Weyerhaeuser Company)**

The term "allowable cut" is often used to refer to the maximum volume of timber that may be harvested under a sustained yield program in any given period. This is not to imply that a specific volume must be harvested each year or from any single acre, but rather that over a period of a few years there should be a balance. Nor does it imply that each acre be harvested in any special period. Usually the sustained yield concept is applied to a large area, sometimes called a "working circle," or a unit, within which the periodic allowable cut for the entire unit may come from a small part in any one year.

The desired goal for a sustained yield program is to so manage the total forest land that there will be a sequence of different age "tree classes" in the growing stock, theoretically one for each year of the rotation. The needed harvest then can be cut from the "class" that becomes mature each year. Cutting cycles (rotation) vary in length depending upon the tree species, site quality, speed of regeneration, size and quality of timber desired, and management need for economic efficiency. Thus, southern pine, which grows rapidly and is harvested for pulpwood, is often managed on twenty-five to thirty-five year cycles, while Douglas-fir, grown for sawtimber, is usually managed on sixty to one hundred year cycles. It is worthy of special note that conservational management is not achieved by just maintaining old-age forests. Actually, mature stands are susceptible to disease and insect mortality, and a common first step in the management of an old-growth area is to harvest mature trees to start growing a new crop for the future and begin establishing the desired sequence of age classes. Yet, the nation *does* need to retain a sufficient reserve of old-growth—a responsibility chiefly held by the national forests in which need for economic efficiency is secondary to national welfare.

Sustained yield is a sound concept, but increasing requirements for timber in the future cannot be met by a program that provides only a balance between harvest and present-day growth. If the nation's forest lands are to meet the projected needs, a larger share, if not all of the commercial forest lands, will have to be utilized on a much higher level of intensity.

PRACTICES FOR IMPROVEMENT OF STAND

Conservation as applied to the forest for timber production implies management by man to improve the forest and increase the "crop" over what nature would produce alone. Only a few important practices are noted.

THINNING. Thinning is a significant practice, for it provides income from sale of small trees to be used as posts, pulpwood, or fuel, and at the same time improves the remaining stand. The primary objective generally is to reduce density, releasing the best trees to grow more rapidly and in better quality (Figure 8 – 10). Table 8–3 gives a specific example in southern pine. At age twenty years it is desirable to have 250 trees per acre, but at age forty years only 137 and at age eighty years only 89 trees per acre are desirable. Periodic thinning reduces the number of growing trees to the desired level. Not only is the final harvest increased and improved in quality, but intermediate harvests of wood have been made that otherwise would have been partly lost. The timing of thinning, the optimum number of trees per acre for each species, and site quality are still subjects of research study.[16] In any case, practical wisdom gained from experience is essential in the selection of trees to cut to accomplish desired stand improvement.

Two special types of improvement cutting that have conservational aspects are sanitation cutting and salvage cutting. *Sanitation cutting*

TABLE 8–3
DESIRED NUMBER OF TREES PER ACRE*

South Carolina site index 80				
Age	Trees per acre	Crown diameter	Diameter breast high	Cords per acre
20	250	13.2 feet	9.4 inches	19.7
30	170	16.0	11.7	29.0
40	137	17.8	13.3	38.2
50	118	19.2	14.5	43.0
60	106	20.3	15.4	46.5
70	97	21.2	16.2	50.0
80	89	22.0	16.9	51.1

* J. J. Wiley, Jr., "Control Techniques for Managed Even-Aged Stands," *Journal of Forestry*, vol. 57, no. 5 (May, 1959), p. 346.

is aimed at removal of diseased or insect infected trees to reduce danger to healthy trees. *Salvage cutting* refers to the removal of trees which have been damaged or killed by fire, lightning, insects, or windthrow when there is still a market value.

PRUNING. Pruning lower limbs to reduce knots is a practice that has potential as a means of upgrading the quality of logs. High initial cost for which return is long deferred limits use of this improvement practice in the United States (Figure 8 – 11).

HARVEST PRACTICES AND CONSERVATION. The way timber is harvested has important conservational implications, not only for the ecological succession, but also for other forest values such as outdoor recreation and watershed management. The objective of the harvest is to provide society with needed timber commodities and profit for the industry, but the owners also have an obligation to society to provide for adequate and prompt regeneration. In addition, foresters recognize that there is need in strategic places for modification in silvicultural systems to minimize disturbance to aesthetic and other outdoor recreation values, to minimize acceleration of erosion and to maximize watershed protection. Such modifications will likely result in economic cost to the operator, but nevertheless should be carried out within reason and in critical locations in the interest of the public welfare. A good example is the leaving of buffer zones of timber along scenic highways which, for example, might require modification of a clear-cutting harvest system. Forest soils and slopes especially vulnerable to accelerated erosion need to be recognized and, if necessary, logging methods should be modified. Foresters and logging operators are giving attention to the control of erosion through planning of logging roads so that they are not in stream beds, providing adequate drainage, endeavoring to keep bulldozers off steep slopes, and otherwise planning logging to minimize soil disturbances. Watershed protection and improvement is another area in which conservational forest harvest practices can be important. An excellent illustration of what might be done to improve

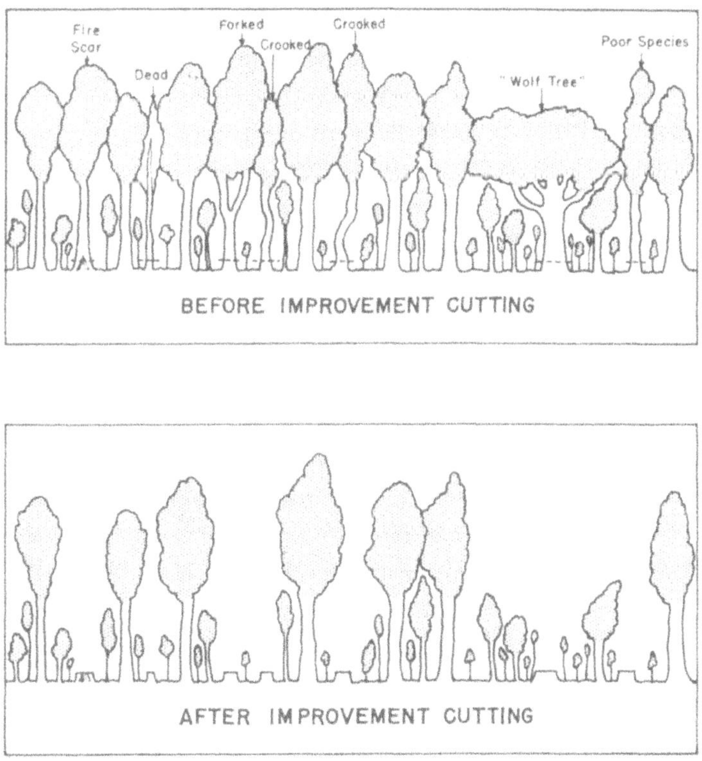

8 – 10 ***A diagrammatic illustration of how growing stock can be improved by selective thinning to remove inferior trees and at the same time to provide growing room for remaining trees.*** **(After "Managing the Small Forest,"** ***Farmers Bulletin*** **No. 1989 [Washington: Department of Agriculture, Sept. 1957], p. 7.)**

watershed resource values of a forest by modification of the silvicultural system is shown in Figure 8 – 12.

A number of silvicultural systems have been developed which in general are directed toward aiding processes of natural reproduction through the method of the main harvest, or reproduction cutting. How adequately the system selected suits the silvics of the tree species, the site conditions, and how well it is carried out are of direct importance to the efficiency and quality of reproduction, and thus to conservation (Figure 8 – 13). Out of the continuum of variations in silvicultural systems, foresters generally identify the following:

Clear-cutting is a system applicable to species capable of reproduction in full or near full sunlight, such as Douglas-fir. It is designed to produce or maintain even aged timber stands. In such stands thinnings are usually made beginning after twenty or thirty years, but the major harvest is by clear-cutting in which all trees are cut. The stands are usually cut in patches (Figure 8 – 14). The size of a clear-cutting area depends upon the tree species and site conditions, but is determined chiefly by the estimate of the maximum size area that can regenerate naturally; minimizing dangers from windthrow and fire is also considered. In Douglas-fir twenty to sixty acres is a normal clear-cutting area. Natural regeneration is provided by the surrounding trees.

8 – 11 *Pruning trees.* Pruning of the lower limbs to as much as fifteen feet above the ground results in improved quality of log through reduction of knots. The photograph also shows how the forest has been thinned to release growth of remaining trees. (American Forest Products Industries)

A variation of clear-cutting is commonly known as the *seedtree method*, and differs only in that within the harvest area a small number of seedtrees are left, usually on strategic high points. A second variation, that of clear-cutting followed immediately by artificial reforestation by seeding or planting is becoming more and more common.

The *shelterwood* system is also designed to produce or maintain even aged stands of timber, but reproduction is established before the main harvest is completed. In practice the mature stand is harvested in two or more major cuts made a few years apart. The first cut removes a significant portion of the stand, opening it to create favorable conditions for reproduction, and leaving high-quality vigorous trees to supply seed for the reproduction stand. After the reproduction has been established, the remaining mature trees are harvested. Strip-shelterwood cutting is a variation in which the total stand area is divided into strips for successive harvest by this method.

The *selection* method is a continuous harvest system designed to maintain uneven aged stands. It is applicable for the most part to species that are capable of reproduction in moderate shade. In practice the largest trees, either singly or in groups, are selected for harvest, which in turn releases remaining trees and favors reproduction. The selection system is especially profitable where terrain is gentle so that road building and maintenance costs are low. Care in limiting damage to young trees is necessary.

ARTIFICIAL REFORESTATION

Harvesting methods usually provide for natural regeneration; but, unfortunately, nature does not always handle the job adequately. There is wide variation in quantity and quality of annual seed crop, for reasons not yet understood. In Douglas-fir, for example, it is generally conceded that only one year in six or ten is a good seed crop year. Moreover, when there

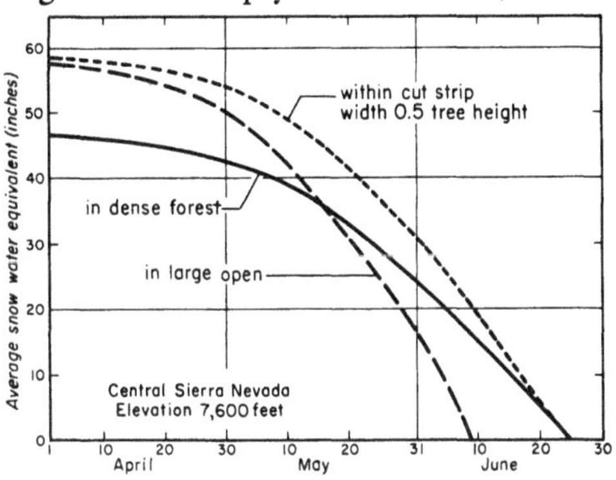

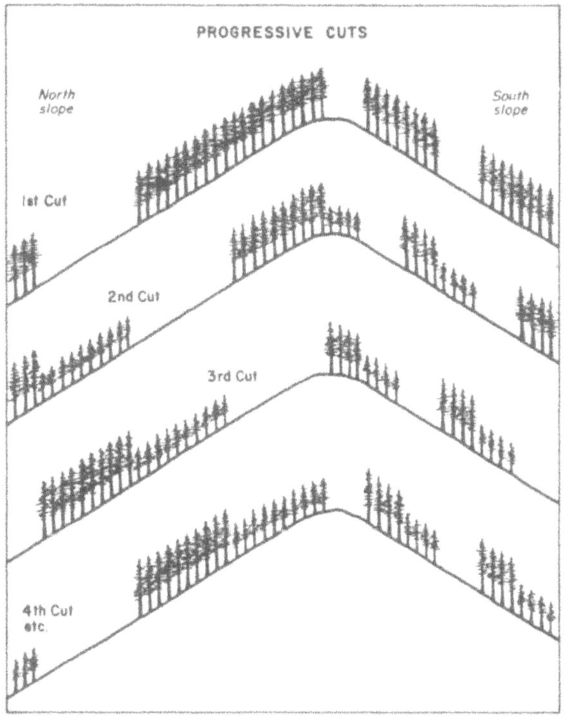

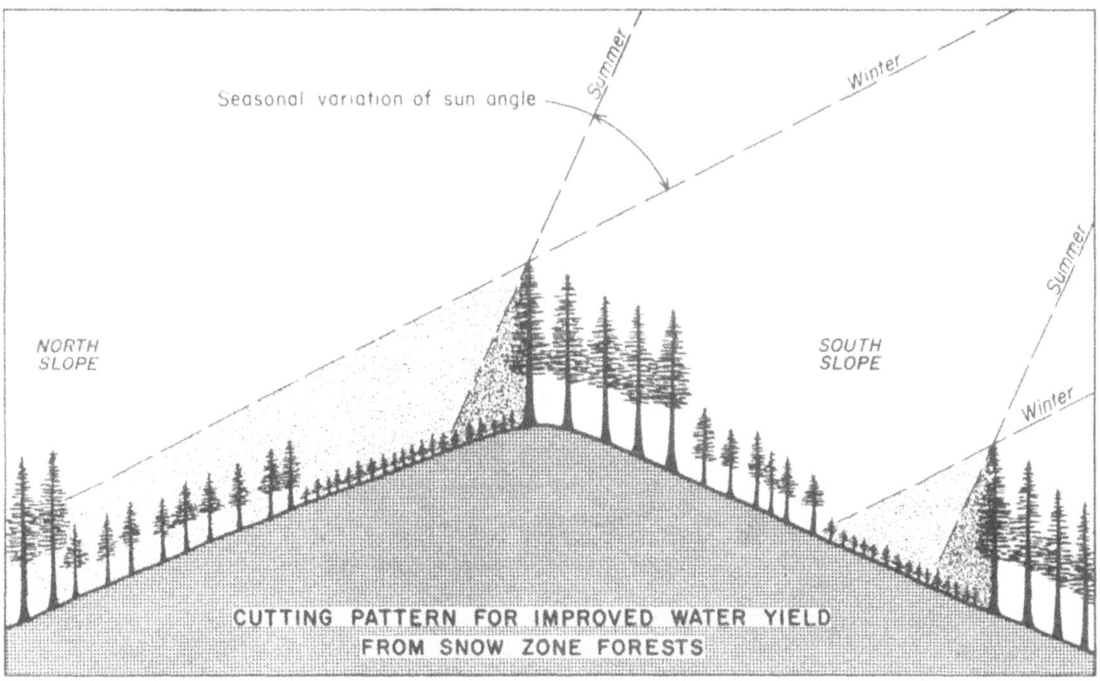

8 – 12 ***Idealized diagrams illustrating modification of the silvicultural system to maximize snow accumulation and to delay snowpack melt for increased water yield.*** **This study in northern California indicates best results from successive cuttings in east-west strips about half the tree heights in width. The narrow openings maximize shade and the sequence of cuttings is planned to minimize back radiation. The curves of snowpack water-content indicate that such narrow strip cuttings would result in 12 inches of additional snowpack-water by April 1, and 20 inches additional on June 10 after all snow has melted from the large clear areas. (Henry W. Anderson, "Managing California's Snow Zone Lands for Water,"** ***Forest Service Research Paper PSW 6,*** **Berkeley, California, 1963), p. 28.**

is a good seed crop, rodents may consume the seeds or seasonal weather conditions may inhibit establishment of seedlings. Thus industry in its effort to shorten the growing cycle for economic reasons, is turning strongly to planting and to artificial reforestation through seeding by air. In addition, it has been shown that the nation has a large area of commercial forest land that is poorly stocked for satisfactory crop production (about one-fourth of the total land). The obvious answer is that artificial reforestation is needed. The job is great, but its accomplishment is important.

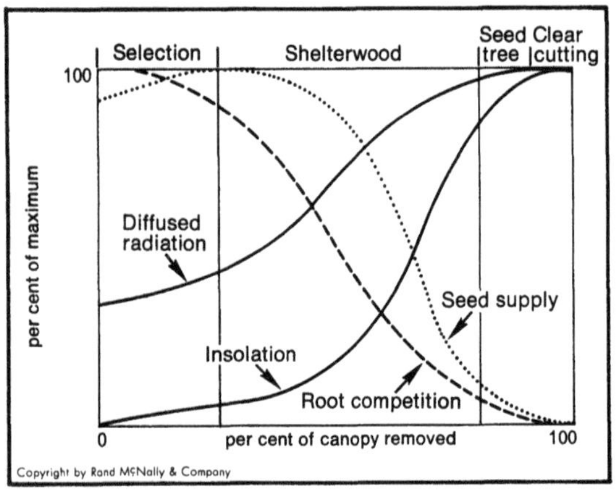

8 – 13 ***Diagrammatic comparison of the effect of the principal reproduction harvest methods on certain conditions for natural reproduction.*** **Obviously the clear-cutting method is the extreme in reducing root competition and in opening the "seedbed" to insolation. At the other extreme, the selection method leaves maximum root competition and provides maximum seed supply. (After David Martin Smith in *The Practice of Silviculture*, 7th ed. [New York: John Wiley & Sons, Inc., 1962] p. 363.)**

PLANTABLE AREA. The 1952 inventory by the United States Forest Service estimated the areas that needed planting and on which planting could reasonably be expected to succeed. In total, 51.9 million acres were so identified as plantable; about 84 per cent were in the East, especially in the Central, Lake, and Southeast regions and mainly on private lands, and 16 per cent were in the West chiefly in California and the Pacific Northwest, about half on Federal lands. The 1962 inventory reported 7 per cent of commercial forest lands to be nonstocked and another 15 per cent to be only 10–40 per cent stocked, making a total of over 100 mil-

8 – 14 ***The block, or staggered setting, system of harvesting in Douglas-fir.*** **In this panorama of the Vance Creek Area of Olympic National Forest, the first cutting has been made. The clear-cut areas will be regenerated by seed from surrounding stand. Note the pattern of roads established for the harvest which then provide for protection and later harvest of remaining timber. (Forest Service)**

lion acres of poorly stocked commercial forest land.

Artificial reforestation, of course, is costly, but the expense is justified by the need to reestablish growing stock on forest lands which nature is not likely to reforest. In the forest industry the expense of seeding and planting is economically justified by the efficiency in establishing reproduction, thus avoiding years of essentially idle forest land and shortening the silvicultural cycle from reproduction to harvest. In addition, artificial planting is helpful in establishing reproduction of desirable trees before shrubby growth or undesirable trees take over. Cost varies with terrain and seed bed conditions, but in general hand planting costs 25–75 dollars per acre; seeding costs 15–25 dollars per acre. Seeding by helicopter costs 10–15 dollars per acre and although it is less expensive, it presents many problems: rodents and birds eat the seeds, heavy rain may wash them away, or they may fall on places unsuitable for seed bed (Figure 8 – 15). Hand planting of nursery-grown stock is more certain; nevertheless, losses do occur, chiefly due to adverse weather or animal or insect damage. It is most advantageous to plant the trees with optimum spacing, usually eight feet apart or about 600–1000 trees per acre (Figure 8 – 16).

Industry and federal nurseries growing seedlings for transplanting produce billions of trees annually. In recent years about two billion trees have been planted annually to establish new plantations or reproduction on about two million acres. Large areas are being planted, especially in the South where level to rolling terrain favors machine techniques (Figure 8 – 17). In addition, industrial forest land owners commonly resort to hand planting of seedlings to fill in partially-stocked problem areas.

ECONOMICS OF TREE GROWING AND CONSERVATION

An inhibition to full implementation of forestry science and technology, and thus to intensification of management on forest lands, is the economic factor. In management, decisions to invest in improvement practices necessarily are based on conclusions derived from analysis of the complex of economic factors including cost of land, labor costs, capital invested in roads and equipment, interest rates, taxes, and reasonable profit needed by industry. It is also

8 – 15 *Aerial seeding.* Aerial seeding is less costly than planting, but success is often low. Seeds are now treated with toxic chemicals such as endrin using a latex adhesive and sometimes are aluminized to ward off birds. (Oregon State Board of Forestry)

8 – 16 *A tree nursery.* To supply the billions of trees needed for hand planting both industry and government maintain nurseries such as this one for growing the young trees which are then transplanted to the forest. (Weyerhaeuser Company)

necessary to recognize that, in spite of major advances in science and technology, tree growing is still vulnerable to major physical losses. The possibility of fire, catastrophic weather phenomena, epidemic disease, insect or rodent ravages finally translate into major economic risks. Moreover, over the long period of time common to growing trees as a crop, uncertainties of future markets, the physical risks and competing opportunities for capital investment become important considerations. A farmer invests, for example, one thousand dollars in planting a crop and gets a return on his money the same year. A forest owner, however, investing one thousand dollars in tree planting at the same time may receive no return for twenty years, and the chief return may be delayed as much as eighty years. In twenty years the one thousand dollars compounded at a modest annual interest of 6 per cent is worth 3,200 dollars; in eighty years the value is 105,800 dollars which the harvest must return. The large, well-capitalized forest industries with stockholders spanning several generations are most able to sustain conservational investment programs involving such long-time cycles. Banks still are not as willing to lend money for planting forests as they are for an agricultural crop.[17] All citizens, not just the owners and managers, need to appreciate the fundamental economics of tree growing, for only an enlightened public interest can assure regulations and aids that will favor the intensified management needed for the future.[18]

PROMISING LINES OF RESEARCH ON TREE GROWING

Forest scientists today are engaged in greatly intensified programs of research along

8 – 17 *Tree planting in the South.* The near level terrain favors application of semi-mechanized planting as illustrated. With the machine, about six acres are planted per day with about a thousand trees per acre. (Forest Service)

many lines looking forward to the demands of the future. Forest genetics studies are concentrating on tree improvements, and prospects include possibilities for developing strains that will resist major diseases, insects, and unfavorable environments or grow better wood faster.[19] Forest soils are being studied, and research will be helpful in improving success with plantings as well as management in general. The use of fertilizers, especially nitrogen, is under study.[20] In time, forest trees are likely to be fertilized just as farmers now fertilize wheat or corn.

Conservation by Better Utilization

A third conservation avenue is through more complete utilization of the timber cut. About a quarter of all timber harvested is not utilized in any way (Figure 8 – 18). About half is left in the woods, and half lost in milling, mostly in production of lumber. In total, 2–2.5 million cubic feet of volume cut from growing stock is unused annually. Some loss is inevitable; nevertheless, here is a significant possibility for increased utilization. At present the two major approaches to realizing fuller utilization are through the integrated milling center and the use of chippers at small mills and in the woods.

THE INTEGRATED MILLING CENTER. A number of large companies have made vigorous attacks on waste in utilization by establishing integrated milling centers. The goal is to utilize every log to its fullest benefit (Figure 8 – 19). At such a center, incoming logs are sorted by use quality. Sawlogs go to the sawmill; peeler logs to the plywood plant; pulplogs, waste slabs, and trimmings from the saws go to the chippers and to the pulpmill; sawdust from the saw kerf goes to the prest-o-log plant or pulpmill; a small amount of bark and waste wood goes to the boilers as fuel.

Improved utilization has extended to small mills too; many now chip wood once burned and ship the chips by special railcars and trucks to pulpmills. The increasing use of sulphate processing (*kraft*) permitting the use of resinous wood has been a major stimulant to this important conservational practice. As late as 1950 consumption of chipped residue was unimportant, but by 1965 it had multiplied more than tenfold to about eleven million cords (850 million cubic feet), and accounted for a fifth of the national total pulpwood consumption.

Sorting of logs in the woods is another means of achieving better utilization which has become significant in the industry. At sorting centers, logs are appraised in terms of their highest commodity value in keeping with mill needs and capabilities. Thus logs of low quality that would be largely wasted at sawmills go directly to pulp mills, sawlogs are sent to sawmills, and peeler logs go to plywood mills. As a result, a high level in quality of use is being achieved. In addition, product research is going forward on many fronts through the efforts of government and industry to find new uses and new ways to develop better utilization.

UTILIZING FOREST RESIDUE. Utilization of wood now being left in the forest is largely a question of economics of handling, and is more difficult to accomplish than the utilization of milling waste. The material is usually small and hard to handle economically. Chippers in the woods appear to be the most hopeful prospect (Figure 8 – 20). One should also note that if this material is to be utilized there must be not only an ade-

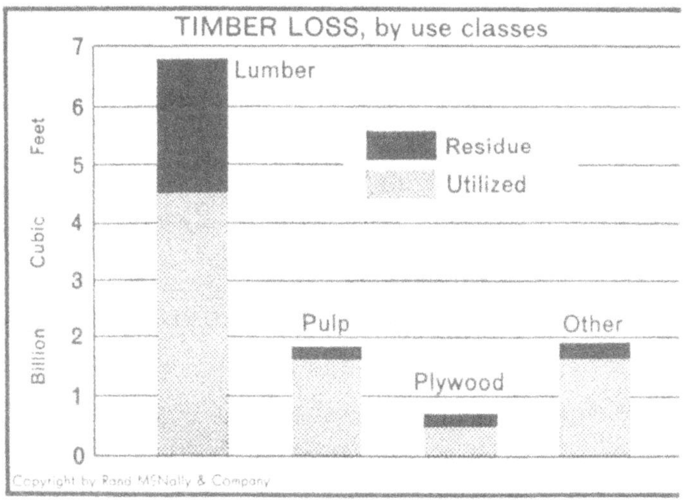

8 – 18 Wood loss in utilization, 1952.

quate supply, but also a market within economical distance. The failure to use limbs, tops, inferior logs, etc., reflects the richness of the resource foundation as well as the high material living standard which inhibits utilization of this "waste" by making it uneconomical. In time the need may justify more complete removal of wood from the forest.

Conclusion

The nation's forest land resource is still generous in quantity and in quality, but it is clear from the evidence that continuation of present levels of utilization cannot meet the projected wood needs of the future. Intensification or expansion of management within both the public and the private forest industry, therefore, appears to be essential if future wood needs are to be met by domestic production.

The practices that appear to offer most promise for increasing wood harvest are here summarized in five concepts. First, there is the possibility of extending more intensive management to a larger portion of the millions of small private ownerships of commercial forest land, three-fourths of which are in the eastern half of the nation. Because low level management is generally considered to be correlated with small size and inadequate capital, it may well be that a solution needs to be sought in developing some form of cooperative forest management to maximize wood production. Second, there are potentials for increase in applying more intensive management to a larger share of the public commercial forest lands through investment in adequate access roads, stand improvement practices, and increased harvests in old

8 – 19 ***An integrated milling center.*** **Complete use of all the wood is fostered by fully integrated plants such as this one at Springfield, Oregon. Logs come in mainly by truck from company Tree Farms on the nearby Cascade Mountain slopes and are sorted in the mill pond. The major units noted in the picture are: (a) sawmill, (b) green lumber sorting sheds, (c) dry kilns, (d) planing mill, (e) lumber storage and shipping, (f) plywood plant, (g) kraft pulp and container board plant, (h) power plant, (i) chipper plant, (j) presto-log plant. (Weyerhaeuser Company)**

growth stands. Third, there is the prospect for increase through further reducing wood loss in harvest residue and in milling, as, for example, through the use of chippers. Fourth, there is considerable evidence from forestry research indicating potential for increased wood supply through application of the best knowledge of forestry science and technology on lands that are now being managed, both public and private. Fifth, there is the potential that could be realized by artificial reforestation of the large acreage of poorly stocked land not now contributing to annual net wood growth commensurate with site quality. Finally, it is important to recognize that increasing wood harvest to meet projected needs is a problem that involves not only forestry science and technology, but also the whole of society through the need to create and maintain societal and economic conditions that encourage land owners and managers to invest in intensified management practices. Forestry in the context of commercial wood production, for the most part, has become an enterprise of an investment character. It follows that the degree of expansion or intensification of management, at least in the private sector, will be a function of expectations of profit.

8 – 20 *Chipper operating in the woods.* **A major development in increasing the amount of raw material supply for paper making is the use of chippers in the logging area. The chipper handles small diameter (16-inch maximum) and low quality logs formerly left as logging residue. The machine picks up the log and loads it on a conveyor chain; the bark is removed, the log is chipped to ¾-inch-long chips, and the chips are blown into a specially built highway van for transport to the pulpmill. Portable chippers are especially valuable in thinning operations where the volume harvested at any one site is small. (Crown Zellerbach Company)**

It is also evident that wood production is no longer the only forest resource concept that merits major consideration in conservation of the forest lands of the nation. The magnitude of population and projected increases inevitably leads to the necessity of evaluating forest lands in a broader context. Even though the major concern of management on most commercial forest land, both public and private, no doubt will continue to be production of wood, the growing needs of society foreshadow need for more integrated management to sustain forest land values for water supply, wildlife, and recreational usage, as well as for wood supply.

All possible forest usages, of course, cannot be equally maximized in a given area, and indeed some are not compatible. Two established forest uses, in fact, separate the designated land from utilization as commercial forest land. Wilderness usage excludes commercial forestry in maximizing the scientific and recreational values of forest land. National parks exclude commercial timber harvest and hunting, but are managed for recreational use and protection. Management of commercial forest lands for wood production, however, need not be incompatible with most other uses. The need is for forest land owners to become land managers integrating considerations for all values.

The ultimate conservation objective for commercial forest land is to assure a continuously adequate supply of wood to meet present and future needs of the nation, and at the same time to sustain the quality of forest land environments appropriate to the needs of society in wildlife, water supply, and outdoor recreation. In seeking definitive answers for decision making to implement such multiple use consistent with the public welfare as well as the private interests, both public and private commercial forest land owners and managers are facing a major challenge.

FOOTNOTES

[1] For readily available briefs on the subject see "Highlights in the History of Forest Conservation," *Agricultural Information Bulletin No. 83* (Washington: Forest Service, Department of Agriculture, 1952), and W. N. Sparhawk, "The History of Forestry in America," *Trees, the Yearbook of Agriculture, 1949* (Washington: Department of Agriculture, 1949), pp. 702–14. See also Chapter 2.

[2] See Andrew Denny Rodgers III, *Bernhard Edward Fernow—A Study of North American Forestry* (Princeton, N. J.: Princeton Univ. Press, 1951), 623 pp.

[3] Such presidential proclamations were prohibited by a Congressional act in 1907, mainly as a reaction of western Congressmen who opposed reserving so much public land. Now new Federal forests must have approval of Congress.

[4] For a stimulating account by a forester who began his service under Pinchot, see William B. Greeley, *Forests and Men* (Garden City, N. Y.: Doubleday & Company, Inc., 1951), 255 pp. See also Gifford Pinchot, *Breaking New Ground* (New York: Harcourt, Brace & Company, 1947), 522 pp., and John Ise, *The United States Forest Policy* (New Haven: Yale Univ. Press, 1920), 395 pp. *Journal of Forestry*, vol. 63, no. 8, August, 1965, dedicated to Gifford Pinchot, contains interesting accounts of his career.

[5] "Development Program for the National Forests," *Forest Service Miscellaneous Publication No. 896* (Washington: Forest Service, Department of Agriculture, November, 1961), pp. 4, 5.

[6] *Ibid.*

[7] For a notable report in the depression years which recommended vast increases in public forests see "The Copeland Report," *Senate Resolution 175* (Washington: Government Printing Office, 1932). See also "Forest Lands in the United States," The Bankhead Report of 1941, *Senate Document No. 32*, 77th Cong. (Washington: Government Printing Office, 1941). This report also recommended increases in public ownership, but less sweeping ones.

[8] For an example of action under this program see David T. Mason and Karl D. Henze, "The Shelton Cooperative Sustained Yield Unit," *Journal of Forestry*, vol. 57, no. 3 (March, 1959), pp. 163–68.

[9] For example see J. Granville Jensen, "Tree Farming in the Douglas-Fir Region: An Evaluation," *Yearbook of the Association of Pacific Coast Geographers*, Vol. 17 (Cheney, Washington: the Association, 1955), pp. 21–26.

[10] The estimate of industrial lands open for public use was made by the American Forest Products Indus-

tries. The Crown Zellerbach Corporation operating tree farms in the Douglas-fir region is an example. In 1965 the company reported an estimate of 38,350 man-days of hunting and a bag of 3,339 big game on its thirteen tree farms.

[11] For a review of activity see W. S. Swingler, "Improving Small Woodlands in the United States," *Unasylva*, vol. 13, no. 2 (1959), pp. 67–76. For a notable guide to possibilities for the southern farmer see Solon L. Barraclough and Alfred Pleasonton, "*Data for Planning Woodland Opportunities on West Tennessee Farms*," *Bulletin 276* (Knoxville: Univ. of Tennessee Agricultural Experiment Station, November, 1957), p. 64. For a detailed analysis of the problem see Charles H. Stoddárd, *The Small Private Forest in the United States*. (Washington: Resources for the Future, Inc., 1961), p. 171.

[12] Warren V. Benedict, "Every Forester has a Stake in Forest Insect Spraying," *Journal of Forestry*, Vol. 57, No. 4 (April, 1959), pp. 245–49.

[13] See John L. George, "Effects on Fish and Wildlife of Chemical Treatments of Large Areas," *Ibid.*, pp. 250–53. For a thorough analysis of the problem see R. L. Rudd, *Pesticides and the Living Landscape* (Madison: Univ. of Wisconsin Press, 1964), 320 pp.

[14] See George M. Harvey, "Heart Rots of Douglas-fir," *Forest Pest Leaflet No. 73* (Washington: Forest Service, Department of Agriculture, September, 1962), 8 pp.

[15] For a listing of major fires and discussion of the fire problem see "Protecting the Forest from Fire," *Agricultural Information Bulletin No. 130* (Washington: Department of Agriculture, 1954, revised, 1965), 29 pp.

[16] For example see Norman P. Worthington and George R. Staebler "Commercial Thinning of Douglas-fir in the Pacific Northwest," *Technical Bulletin No. 1230* (Washington: Department of Agriculture, 1961), 124 pp.

[17] With the tree-growing concept established, credit sources are developing. A 1953 amendment to the Federal Reserve Act permits national banks to make loans on managed forest tracts up to 40 per cent of the timber value, but only for two years. Life insurance companies have gone further into this field of credit than banks and have outstanding loans of tens of millions of dollars which are mostly in the South.

[18] For a nontechnical treatment of the economics of forestry see Albert C. Whorrell, *Economics of American Forestry* (New York: John Wiley & Sons, Inc., 1959), 441 pp. For a provocative analysis of management see Ernest M. Gould Jr., "Fifty Years of Management at the Harvard Forest" *Bulletin No. 29* (Petersham, Massachusetts, Harvard Forest, 1960), 29 pp. Also see Donald F. Flora, "Uncertainty in Forest Investment Decisions," *Journal of Forestry*, vol. 62, no. 6 (June, 1964), pp. 376-80.

[19] François Mergen, "Forest Tree Breeding Research," *Unasylva*, vol. 13, no. 3 (1959), pp. 129–37.

[20] T. E. Maki, "Forest Fertilization Possibilities in the United States," *Better Crops*, vol. 42 (October, 1958), pp. 16–24.

9

Wildlife Resources

Wildlife forms an important segment of the total resource inventory and provides a variety of aesthetic, recreational, and economic benefits. In the early decades of the American nation wildlife was important for food and clothing. Today this facet of wildlife value is much diminished, but some direct economic values still remain. In rural areas, wildlife still forms part of the diet. There is a trapping industry engaged in harvest of pelts for fur, and fishing has a commercial as well as recreational aspect. Birds have a modest role in control of insects.

In the twentieth century wildlife plays only a small direct role in the material economy, but indirectly it is increasingly significant in the economy and in the enrichment of living. Most of the population lives in cultural environments characterized by edifices of concrete, fuel-driven machines, noise, and a fast pace in which they are harassed by many pressures and tensions of the economic world. It is not surprising, therefore, that so many have turned to the out-of-doors for relaxation and recreation. Here wildlife forms an essential element. The recreational value of wildlife is well known to millions of avid hunters and fishermen. The economic benefits to local and national economies derived directly and indirectly from recreational hunting and fishing have grown to major significance. In addition, millions of citizens derive pleasure from the aesthetic values of wildlife which are essentially outside of economic evaluation. Who can measure the pleasure derived from watching and studying wildlife in the natural habitat—a flight of geese, or a deer in the forest?

Before men became so numerous, wildlife populations were regulated through natural checks and balances. Now man is a major unbalancing influence as expanding urbanization, highways, and intensification of farm land use have increasingly reduced wildlife habitat and altered ecosystems. Also important has been the growing efficiency of the hunter with the shotgun, telescopic sight, and high-powered repeating rifle. The vast herds of buffalo that once grazed the plains were slaughtered and now exist only in a few protected herds. Beaver trapping once supported a major industry. Some species of birds were nearly exterminated for harvest of feathers, and the passenger pigeon exists only in history.

There is, therefore, evidence suggesting a need for wildlife conservation, and also suggesting that values of wildlife (economic, recreational, and aesthetic) are of ample importance to justify conservational management to assure the continued viability of the nation's wildlife for future generations. The status of wildlife and conservational programs for management are here reviewed in four sections: First, an overview of the present status of the wildlife resource in quantity and variety; second, a review of the values of wildlife in society; third, the evolution of wildlife conservation, including a review of agencies and organizations; and fourth, an overview of practices of wildlife conservation.

The Quantity and Variety of Wildlife

The wildlife resource is abundant, varied, and unusually rich in species as a result of the large size of the nation combined with its latitudinal spread from the arctic slopes of Alaska to the tropical realms of Florida and Hawaii, variety of altitudinal zonations, and environments ranging from low-lying marshes to alpine pastures. For this brief review, the many species of wildlife resources are grouped on a utilitarian basis under the headings of big game, small game, upland birds, waterfowl, fish, and others which include songbirds and other nongame wildlife.

BIG GAME

Big game include deer, antelope, elk, moose, bear, mountain sheep, peccary, and the mountain lion. These animals, for the most part, inhabit the mountain and forest range. Table 9–1 lists the more important species, with environment and estimates of the population.

Deer are by far the most numerous of the big game animals, with a population approximating ten million. The whitetailed variety is considered by some wildlife specialists to be the most important and most popular of all big game animals in North America. It has a wide range and is thought to be more common now than before the coming of the white man. Deer are the most widely hunted big game animals and about two million are killed annually (Table 9–2). Yet in some areas the deer are so abundant that starvation may at times take a greater toll than the hunter. By comparison other big game are not numerous (Figure 9–1).

The antelope, at one time comparable to the buffalo in population, is found in the open

TABLE 9–1

BIG GAME SPECIES, ENVIRONMENT AND POPULATION*

Species	Range	Environment	Population magnitude
Black bear	Throughout North America	Chiefly forest	75,000
Grizzly bear	Western North America	Chiefly forest	1,000
Brown bear	Northwest North America	Chiefly forest	
Polar bear	Arctic	Arctic slopes	
Elk	Western North America	Forest and associated open land	100,000
Mule and Blacktailed deer	Western United States	Forest	2,000,000
Whitetailed deer	North America	Forest	5,000,000
Moose	Northern United States–Canada	Forest	10,000
Caribou	Alaska	Tundra	1,000
Antelope (Pronghorn)	Western United States	Semi-arid range	100,000
Cougar	Throughout Western United States	Forest	
Jaguar	Southwest United States	Woodlands	
Peccary	Southwest United States		150,000
Bighorn sheep			10,000
Mountain goat			10,000
Bison			2,000

* "Big Game Inventory 1964," *Wildlife Leaflet No. 470* (Washington: Fish and Wildlife Service, 1965). 4 pp.

TABLE 9–2
KILL OF GAME SPECIES*

Species	1961	1962	1963	1964
Whitetailed deer	867,764	954,766	1,129,716	1,095,570
Mule deer	814,977	749,894	620,515	685,322
Blacktailed deer	158,253	154,809	152,927	152,835
Wild turkey	75,398	65,343	82,306	91,168
Elk	77,905	64,221	74,752	84,309
Pronghorn antelope	75,645	79,014	77,847	75,534
Barren ground caribou	30,000	20,000	21,000	25,000
Black bear	20,430	25,817	21,392	24,187
Moose	13,363	10,150	8,933	10,136
Peccary	5,643	5,676	7,268	6,622
Mountain goat	990	1,500	1,568	1,545
Bighorn sheep	320	366	348	1,242
Dall sheep	1,000	665	905	938
Grizzly bear	515	505	560	673
Polar bear	156	196	167	251

* "Big Game Inventory 1963," *Wildlife Leaflet No. 461*, and " Big Game Inventory 1964," *Wildlife Leaflet No. 470* (Washington: Fish and Wildlife Service, 1964 and 1965), 4 pp.

9 – 1 ***Rocky mountain mule deer.*** **Part of a herd of deer on the Malheur Refuge of eastern Oregon. Deer frequently move to lowlands and fringes of settlement in search of forage during winter. (Fish and Wildlife Service)**

ranges of the West. The antelope is one of the fleetest of the game animals, and speed is its protection against predators, but not against the bullet. About seventy-five thousand are killed each year, mainly in Wyoming and Montana. The elk which once ranged from Atlantic to Pacific, was near extinction about the turn of the twentieth century, and only a few herds remained in the West. Now protection has resulted in increased numbers, and about 75,000 are killed annually, chiefly in Idaho, Montana, and Wyoming. Caribou and reindeer inhabit the tundra and are hunted by the natives more for food than for sport. The reindeer was imported to Alaska near the end of the nineteenth century and today is found from western Alaska to near the mouth of the Mac Kenzie River. The wild sheep and goats are among the more elusive species; they are difficult to bag and highly prized by hunters as trophies. The bighorn sheep bands are now small. About 1800 the bighorn sheep population was estimated at about two million; in 1960 Buechner estimated the population at about eighteen thousand.[1] The mountain goat inhabits remote mountain areas, and difficulty of access for hunters is part of his protection (Figure 9 – 2). The moose, among the larger big game animals, inhabits the coniferous forests, especially in Alaska, Montana, and Wyoming where much of his food comes from vegetation associated with ponds, marshes, and lakes.

The black bear is found in almost all of the states, and the brown bear mainly in the West. The grizzly, depleted in population, has a major stronghold in Yellowstone National Park. Although bears are most often killed while hunters are in pursuit of other game, some twenty thousand bears are bagged annually, chiefly in Washington, Idaho, and Michigan. The na-

9 – 2 *Mountain sheep in the Colorado Mountains.* **The sure-footed mountain sheep at home in the high rock-ribbed slopes are seldom seen by the automobile tourist. (Fish and Wildlife Service)**

tional parks are the best places to observe bears; there they are protected and have become a major tourist attraction and occasionally a nuisance.

Among the larger mammals classed as predators rather than big game are the mountain lion, wolf, and coyote. Most of the remaining mountain lions inhabit the southwestern mountains. The wolf has been hunted as a predator except in the confines of national park boundaries where he is protected. The coyote has been the subject of unrestricted hunting; however, up to the present he appears to be holding his own against man's exterminating efforts. The peccary, or wild pig, which ranges through the southwest is a significant game animal, especially in Arizona.

SMALL GAME

Many species of small game form part of the wildlife resource. Rabbits, ranging from the small cottontail to the jack rabbit, are found in a great variety of environments, and millions are killed each year. In the semi-arid lands of the West, the jack rabbit is generally classed as a pest and thousands are slaughtered in rabbit drives or by poison. Squirrels are second to rabbits in hunter popularity; furthermore, they supply pelts. Several million squirrel pelts are harvested each year; New York and Wisconsin each account for over one million skins. Beaver, marten, and otter were once the principal species of fur bearers, but today the muskrat is by far the most important. There has been a shift in the area of trapping activities from the historic trapping grounds of the West to the marshlands of the Southeast. Louisiana is the leading pelt producer today.[2]

The beaver, largest rodent of the United States, was a major objective of early trappers in the pioneer West. Unrestricted trapping, logging, and clearing of forests for agriculture once threatened the beaver. Fortunately, with timely protection and the start of new colonies, numbers have increased to the point where trapping is allowed, and about 200,000 are now taken each year, chiefly in Alaska, Idaho, Minnesota, Montana, Wisconsin, and Oregon.[3]

The muskrat, an inhabitant of marshlands, is the most important fur bearer, with trappers marketing five to six million pelts annually. The mink, associated with waterbodies, and usually with its favorite food, the muskrat, possesses one of the most valuable pelts in the fur market today. About 300,000 pelts from wild mink are marketed annually. Over one million nutria are harvested annually, chiefly in Louisiana. The oppossum, prized for sport and food and usually considered a southern mammal, is becoming common in the United States. Other fur bearers of importance are weasel, skunk, raccoon, fox, fisher, fur seal, and sea otter (Table 9–3).

The sea otter has been protected since 1911, and is a good example of the result of overhunting. Before 1911 the pelt was especially prized and brought prices of about two hundred dollars each as raw fur.[4] At one time, sea otters were found along the entire length of the Pacific littoral, now the remainder are chiefly in the Aleutian Islands area, and a few are found near Monterey, California. The fur seal is notable among mammals taken for their pelts, and

TABLE 9–3
WILD FUR ANIMAL CATCH, 1963–64 SEASON*

Muskrat	4,994,253
Raccoon	1,366,814
Nutria	1,309,216
Mink	317,048
Beaver	191,245
Oppossum	167,353
Fox	157,039
Skunk	45,985
Bassarisk	34,289
Weasel	24,625
Otter	17,419
Coyote	16,081
Marten	7,054
Lynx	5,399
Badger	4,137
Fisher	2,350
Timber wolf	976
Wolverine	287

* "Big Game Inventory 1964," *Wildlife Leaflet No. 471* (Washington: Fish and Wildlife Service, 1965), 4 pp.

is an outstanding example of the value of conservational management. At one time there were over three million in the north Pacific area, but they were slaughtered without regulation and were threatened with extinction. (The herd at one time reached a low of 150,000.) The Pribilof Islands are their major remaining habitat in the United States. In 1911 a treaty which afforded protection to and regulated the kill of fur seals in the north Pacific was ratified by the United States, the U.S.S.R., Canada, and Japan. A new treaty was ratified in 1957, and with the amendments of 1963, regulates management and harvest of the herds. Today, the seal herd has reached a population of 1.5 million. The harvest supervised by the United States government has been 75,000–100,000 pelts with a value of about two million dollars (Figure 9 – 3).

UPLAND GAME BIRDS

Upland game birds are found throughout the nation. Habitats range from fence rows of farms, grain fields, woodland and brush areas to forests (Table 9–4). Two of the commonly hunted birds are exotic species:The ring-necked pheasant introduced from the Orient, and the Hungarian partridge introduced from Europe.

The Hungarian partridge is an open field bird, roosting on the ground, thriving on grain, and finding cover in hedgerows and the like. The pheasant also thrives in farm fields and associated cover areas. Unlike the monogamous partridge, the pheasant is polygamous; thus there is a surplus of cocks for the hunter. Popularity of these birds has motivated game commissions to devote much research to their

9 – 3 ***Fur seal harem, St. Paul Island, Pribilof Islands.*** **The fur seal is polygamous, and each bull seal has a harem of as many as sixty cows. With protection now afforded, the herds have increased and there is an annual harvest of surplus males. (Fish and Wildlife Service)**

TABLE 9–4
GAME BIRDS

Species	Range	Environment
Bobwhite	Eastern United States	Brushy areas
Ring-necked pheasant	Exotic	Open areas
Hungarian partridge	Exotic	Open areas
Valley quail	Pacific Southwest	Brushy areas
Mountain quail	Pacific coast	Brush and timber
Turkey	Eastern United States	Forest and brush
Ruffed grouse	Throughout United States	Forest
Blue grouse	Eastern United States	Forest
Greater prairie chicken	Central and Eastern United States	Open field and meadow
Sage hen	Western United States	Sage brush
Ptarmigan (willow and rock)	Alaska	Tundra

9–4 *Mongolian pheasant.* Recreational hunting of pheasant and other upland game birds is a popular sport, especially in the northern portions of the nation. There is no native pheasant, but several have been successfully introduced, notably the Mongolian and Chinese ring-necked pheasant. The male bird is strikingly colored with tones of green and orange-red, and has a tail some fifteen inches long; the female is a dull earthy color, and the tail is half as long. The Mongolian pheasant shown here is a bit darker and larger than the Chinese pheasant. (Oregon Game Commission)

management and propagation on game farms for release (Figure 9 – 4).

Other game birds of significance to the hunter are the grouse, quail, sage hen, prairie chicken, and wild turkey. Valley quail are popular with hunters in the West. The birds prefer considerable cover, roost in trees or shrubs, and are seldom in open areas. Mountain quail occupy higher elevations, especially around cutover areas of forest. The bobwhite (quail) range in the eastern and southern United States in woodland and various brush cover areas. Grouse of one species or another range the entire United States, almost always in forest habitat, living on coniferous needles and buds (supplemented by berries in the summer) through most of the year. The ruffed grouse is a species of the hardwood forest and woodlots, living on various seeds and insects. The wild turkey is found chiefly in the East associated with hardwood cover as well as the pine woods. The sage hen is associated with the sagebrush cover of the semi-arid West, and feeds chiefly on sage and succulent vegetation. The prairie chicken ranges the north central states on prairie areas, dry marshes, or pine barrens, living chiefly on seeds, insects, and berries. The ptarmigan (Arctic grouse) of the Alaska tundra is still an important food source for natives. Ptarmigan occupy rocky areas, living on buds, twigs, insects, and berries. Southward in the West, ptarmigan are found above the timberline on rocky habitat.

WATERFOWL

Waterfowl are certainly one of the important facets of wildlife, both from the aesthetic and from the recreational hunting points of view. The recreational significance of waterfowl is indicated by the fact that about sixteen and one half million are estimated to be bagged by about two million hunters annually. Waterfowl include many species which wildlife specialists group into six classes: (1)The swans and geese, of which the Canada goose is the chief objective of the hunter and more important than all other geese (Figure 9 – 5); (2) the tree ducks, of which the fulvous tree duck, or squealer, found in the rice fields and marshes of the South, is the only important species; (3) the spiny-tailed ducks, represented chiefly by the small ruddy duck;[5] (4) the mergansers, which are not particularly prized for food because their diet consists of fish; (5) the diving ducks, often called bay or sea ducks, (some frequent fresh water), which congregate in huge flocks, or rafts, making them especially vulnerable to hunting (among this group are the scaups, eiders, scoters, redhead, and canvasback, the latter two species being the most abundant and most often hunted game birds); (6) the surface feeders, or dabbling ducks, who feed by tipping in relatively shallow water, the most common species of this group being the mallard which is found throughout the northern hemisphere, and is the most popular game duck.

THE FLYWAYS. For the most part, waterfowl migrate with the seasons. From their breeding grounds in the northern United States and Canada, they migrate to the wintering grounds along the Gulf Coast and southern Atlantic and Pacific littorals. These annual migrations from the breeding grounds to the wintering areas and back follow more or less fixed routes which are known as *flyways*. There are four generally recognized flyways: (1) The Pacific, located mainly west of the Rocky Mountains; (2) the Central, which is over the High Plains; (3) the Mississippi; and (4) the Atlantic (see Figure 9 – 6 and Table 9–5). It is estimated that 85 per cent of the waterfowl nest in Canada, but that 75 per cent of the hunting is done in the United States. This international characteristic of the wildfowl has led to management treaties between the United States and Canada on the north and Mexico on the south.

FISH RESOURCE FOR SPORT FISHING

Fish resource in lakes, streams, reservoirs, and coastal waters is the basis of a traditional recreational activity of the American scene. It is estimated that at least a third of the total population over twelve years of age participate in recreational fishing, and that the number of fishermen is increasing faster than the total population (Figure 9 – 7).[6]

The fishing area is estimated to be about fifty million acres of inland and coastal waters not counting the area of the Great Lakes or Alaskan waters (Table 9–6). Especially significant is the major role of man-made water bodies in providing fishing area in all parts of the nation (Table 9–7).

The Sports Fishing Institute estimates that fish population may range from fifteen pounds per acre to as much as one thousand pounds,

9 – 5 Canada goose on a marshland nest. Marshes provides much of the needed habitat for waterfowl. Drainage of such marshes threatens migratory wild fowl. This Canada goose has utilized a muskrat house as a nesting site in Blackwater Refuge, Maryland. (Fish and Wildlife Service)

TABLE 9–5

WATERFOWL KILL BY FLYWAYS, 1963–64 SEASON*

	Pacific	Central	Mississippi thousands of birds	Atlantic	Total
Ducks	3,033	1,610	4,457	1,242	10,342
Coots	165	41	371	92	669
Geese	354	301	291	184	1,130
TOTAL	3,552	1,952	5,119	1,518	12,141

* "Waterfowl Status Report 1965," Special Scientific Report—*Wildlife No. 90* (Washington: Fish and Wildlife Service Special Scientific Report, 1966), 110 pp.

with an average for the nation of one hundred and fifty pounds per acre.[7] Fishing pressure varies greatly with accessibility, stocking, and

TABLE 9–6

RECREATIONAL FISHING AREA*

Area	Million acres
Great Lakes	35.8
Inland waters	25.3
Reservoirs	(10.2)
Lakes	(9.3)
Streams	(5.8)
Coastal waters	25.0
Alaskan waters	12.4
TOTAL AREA	98.5

* "National Survey of Fishing and Hunting 1965," *Resource Publication No.* 27 (Washington: Bureau of Sports Fisheries and Wildlife, 1966), 76 pp.

TABLE 9–7

TYPE OF FRESH WATER FISHING AREA BY SECTIONS IN PER CENT*

Section	Man-made	Natural Lakes	Rivers and Streams
New England	13	59	29
Middle Atlantic	20	31	49
East North Central	23	52	25
West North Central	31	35	34
South Atlantic	45	18	37
East South Central	53	8	38
West South Central	59	15	26
Mountain States	46	10	44
Pacific	26	27	47
TOTAL UNITED STATES	35	30	35

* "National Survey of Fishing and Hunting 1965," *op. cit.*

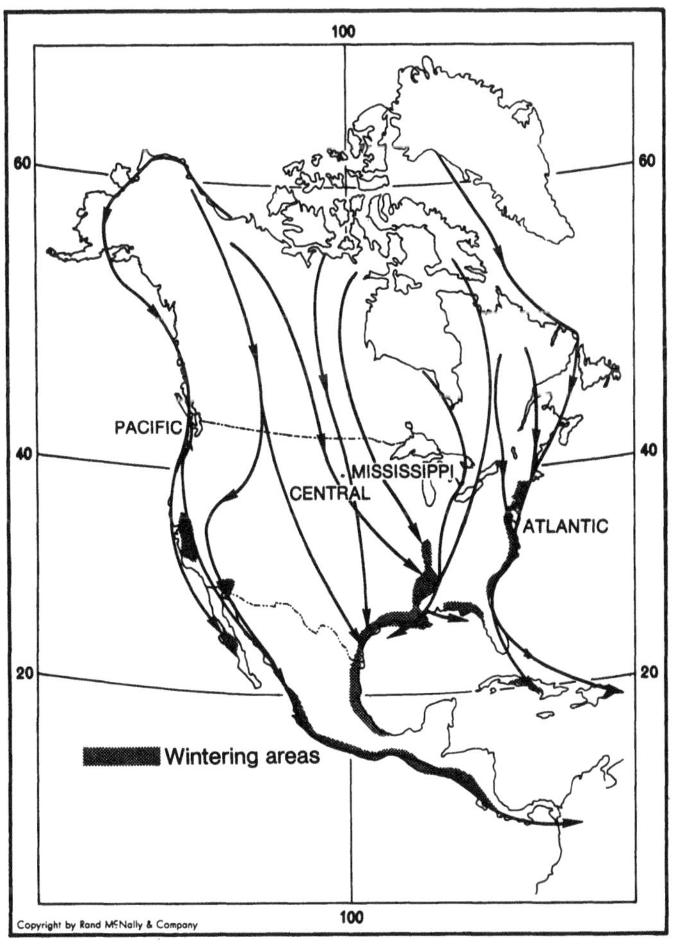

9–6 Principal waterfowl flyways and wintering areas. **(After Frederick C. Lincoln, "Migration of Birds," Circular No. 16 [Washington: Fish and Wildlife Service, 1950]. 102 pp.)**

size. Thus a small pond near Los Angeles is reported to support 682 man-days of angling, while a reservoir in Tennessee is utilized for only five man-days of fishing. The 1965 National Survey of Fishing and Hunting reported angler pressure and success as shown in Table 9–8. These average pressures are, of course, magnified by the limited seasons on most waters and the concentration of fishing days on week ends (Figure 9 – 8).

The harvest by sports fishermen is large. Data gathered in the 1965 survey indicate that fresh water fishermen caught 555 million pounds, and coastal salt water anglers about 661 million pounds for a total of 1.2 billion pounds (Table 9–8). The corresponding "eating fish" catch by commercial fishermen is about 2.5 billion pounds.

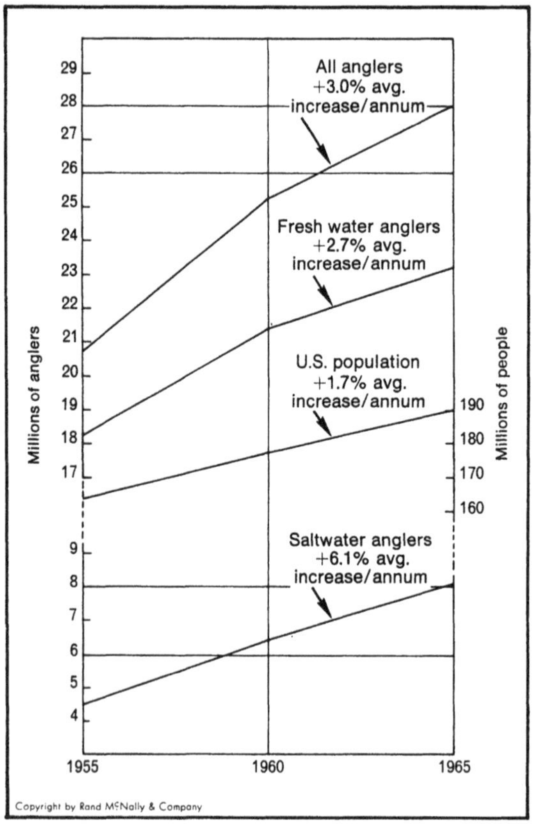

9 – 7 Increase in number of recreational fishermen. **Data from the 1955, 1960, and 1965 National Surveys of Fishing indicate that participation in recreational fishing is growing nearly twice as fast as total population. (After Sport Fishing Institute)**

Ownership of Wildlife

The question of ownership of wildlife has always been a major conservation problem. It is a legal question having direct bearing upon regulation and other management authority. The courts of law have repeatedly ruled that wildlife is a public resource, and that no person may claim ownership of living wild things in nature. From this tenet, Federal, state, and local governments derive authority to regulate hunting and otherwise manage wildlife in the interest of the public. Among the implications of wildlife law is the concept that landowners may not claim ownership of wild animals just because they inhabit private land, nor is ownership of game or other wildlife acquired through capture, except as allowed by the hunting regulations. Moreover, the legal concept of *ferae naturae* has been specifically interpreted to mean that citizens may not raise native animals in captivity and thereby gain ownership. For example, if a person were to raise deer in cap-

TABLE 9–8
ANGLER PRESSURE AND CATCH 1965*

	Fishing pressure in angler days		Catch in pounds		
	Millions	Per acre	Millions	Per angler day	Per acre
Fresh water	425	17	552	1.3	22
Coastal waters	94	10	661	7.0	26
TOTAL	519	10	1,213	2.3	23

* From data in "National Survey of Fishing and Hunting 1965," *op. cit.*

tivity on his own private land (fenced in) the offspring would be public property. In like manner, the raising of wild birds, such as pheasants, in captivity does not grant the right of private ownership, except under authority granted and licensed by the state. Thus wildlife, more than any other resource, is managed for and by the people, and regulation is the sole responsibility of government.

Under the United States system, primary authority over wildlife within their boundaries is vested in the states. State ownership of wildlife is evident in the common requirement of special licenses for out-of-state hunters, and states may limit or prohibit out-of-state shipment of game even though it was legally bagged. State ownership of wildlife is also evident in the fact that hunting and fishing on Federal lands is regulated by laws of the state, and no Federal regulation may be in conflict with state law, although hunting and fishing may be prohibited on Federal land through the right of land ownership. A private landowner, too, has the right to prohibit hunting on his

9 – 8 ***Intense angler pressure on a fishing resource.*** Streams and lakes close to great urban populations are often under heavy angler pressures as illustrated here. Attention to conservation of existing resource opportunities and creation of new waters are called for to meet growing demands. (California Fish and Game Commission)

land, but must obey the hunting laws.[8] With the great increase in the number of hunters and fishermen, private landowners more and more are posting property against hunting. In general, the law of ownership of fish is similar; however, it is a common rule that private waters are exempt from fish laws. Definition of private waters is sometimes difficult, but it usually means isolated bodies of water such as man-made ponds or lakes. Fishing on navigable waters is almost always a public right. Thus the hunting and fishing license issued by a state is a legal document authorizing a citizen to convert public game to private use, but it does not grant permission to hunt or fish upon the private lands of others.

Value of Wildlife

The wildlife resource has a many-faceted value. It is one major foundation of the growing recreational "industry." Wildlife still has commercial value in furs and supplies subsistence food in many areas (commercial fisheries are discussed in Chapter 13). Aesthetic values are widely recognized. Finally, wildlife is valuable to man in control of harmful rodent populations.

RECREATIONAL HUNTING AND FISHING

Recreational hunting and fishing, from the materialistic point of view, is the most important value derived from wildlife. The 1965 survey estimated that twenty-eight million people fished, fourteen million hunted, and nine million did both (Table 9–9).

TABLE 9–9
NUMBER OF RECREATIONAL HUNTERS AND FISHERMEN BY NUMBER OF PARTICIPATIONS*

	Fished	Hunted
	In thousands	
Total	28,348	13,583
Once only	3,559	1,586
2–5 times	8,385	4,415
6–15 times	8,416	4,358
16–50 times	6,277	2,736
Over 50	1,711	488

* "National Survey of Fishing and Hunting 1965," *op. cit.*

The 1965 National Survey estimated that thirty-three million persons devoted 700 million participant days to recreational hunting or fishing. In total they travelled about thirty billion passenger miles (fourteen billion car miles), and spent four billion dollars. The data indicate that fishermen account for about three-fourths of the expenditures (about one hundred dollars a year per person). The distribution of expenditures (Table 9–10) shows that fishing gear accounted for only 10 per cent of money spent; transportation, food, and lodging together total 30 per cent; auxiliary equipment such as boats and motors total about 27 per cent; bait, guides, and other trip costs total a surprising 28 per cent.

Hunters account for about one-quarter of the total expenditures (about eighty dollars per person). The distribution of expenditures, however, is different; hunters spent one-third of the total outlay on hunting equipment, 13 per cent on dogs, 11 per cent on auxiliary equipment, and about 28 per cent on transportation, food, and lodging (Table 9–11).

AESTHETIC AND BIOLOGICAL VALUES

The aesthetic values of wildlife are beyond material calculation, but are of great importance to all. Great numbers of people enjoy bird watching and photographing wildlife in their native habitats. The natural beauty of a graceful heron fishing in the marsh, the color and song of birds in city and country are important to our society, and none of us would wish to lose this part of the wildlife heritage.

Wildlife perform some service to man in control of insects and pests. Hawks and owls are known to feed on rodents such as mice. Other animals such as foxes and weasels feed on injurious rodents. It is, of course, true that wild animals occasionally prey on domestic animals and damage crops and other property, especially in periods of natural food shortage. In balance, however, the values of wildlife outweigh the damage.[9]

TABLE 9–10

EXPENDITURES OF FISHERMEN, 1965*

Items	Fishermen (thousands)	Total spent (millions of dollars)	Average spent (dollars)
Food	16,292	340	12.01
Lodging	3,837	104	3.67
Transportation			
Automobile	23,826	400	14.10
Air, rail, bus	433	29	1.03
Equipment			
Boats and motors	1,936	626	22.08
Fishing	17,111	324	11.41
General	5,473	161	5.67
Licenses and permits	16,018	68	2.40
Bait, guide fees and other trip expenses	20,975	755	26.62
Privilege fees			
Daily entrance fees	3,439	52	1.83
Annual leases, etc.	749	11	.40
Boat launching	1,491	16	.55
Other	4,700	40	1.42
TOTAL	27,445	$2,925	$103.19

* "National Survey of Fishing and Hunting 1965," *op. cit.*

TABLE 9–11

SELECTED EXPENDITURES OF HUNTERS, 1965*

Items	Hunters (thousands)	Total spent (millions of dollars)	Average spent (dollars)
Food	6,998	116	8.53
Lodging	1,005	21	1.58
Transportation			
Automobile	11,720	157	11.53
Air, rail, bus	207	12	0.86
Hunting equipment	11,061	397	29.25
Auxiliary equipment	11		
Boats	111	29	2.16
General	3,529	96	7.09
Licenses and permits	10,371	68	5.02
Duck stamps	1,268	4	0.28
Dogs	2,021	146	10.78
Privilege fees, etc.			
Annual leases	528	15	1.12
Entrance fees (private)	322	9	.68
Entrance fees (wildland)	415	10	.70
Other	1,766	17	1.25
TOTAL	13,402	$1,121	$82.54

* "National Survey of Fishing and Hunting 1965," *op. cit.*

Need for Conservation

The national population is expected to double in the next half century. As a result there will be more reduction in wildlife habitat as the urban area spreads, perhaps more pollution, and certainly greatly increased hunting and fishing pressure. Man now has the power to exterminate wildlife; in fact, a number of species have already been reduced to rare status, and some are extinct.

In 1965 Congress appropriated funds to implement a program of research on birds and mammals considered to be in danger of being lost unless conservation measures are adopted to protect or improve their habitats. Wildlife specialists indicate that some forty-five species already are rare, and at least eighty-one species are endangered (Table 9-12).

The increasing number of fishermen has already created great pressure on the fish producing capabilities of streams and lakes. Moreover, urbanization and industry continue to cause major pollution damage in many streams and lakes. The number of hunters also is increasing, while the area open to them is decreasing. It is thus clear that if the nation would have wildlife in variety and quantity to sustain future recreational hunting and fishing at levels comparable to the present, more self-restraint, rational regulation, and other conservational management practices are essential in all fifty states. The following section reviews, first, the rise of wildlife conservation, and second, major conservation needs and practices.

TABLE 9–12
WILDLIFE EXTINCT OR THREATENED*

Extinct species		Threatened species
Species	Extinction date	
Michigan grayling	1930	Great white heron
Great auk	1853	Trumpeter swan
Pallas cormorant	1852	California condor
Heath hen	1931	Mississippi kite
Passenger pigeon (a specimen was kept in captivity until 1914)	1898	Red bellied hawk
		Sage hen
		Whooping crane
		Ivory billed woodpecker
Carolina paraquet	1904	
California grizzly	1900	Grizzly bear
Sea mink	1860	Sea otter
Arizona elk	1901	Wolf and cougar
		Sierra bighorn
		Rock sturgeon

* Leonard W. Wing, *Practices of Wildlife Conservation* (New York: Wiley, 1951), p. 134.

Evolution of Wildlife Conservation

Need for wildlife conservation was recognized in colonial times, but the majority of the population saw only immediate values and continued to slaughter wildlife. Important actions to regulate hunting and fishing and to begin practices to conserve wildlife, therefore, were slow in evolving.[10]

Hunting regulations in various forms began in the early years of the American colonies, and state regulation was well established by 1850. Resident hunting licenses were required in Michigan and North Dakota as early as 1895.[11] Federal activity began with the appointment of a commissioner of fisheries in 1871, and in 1873 funds were provided for the propagation and distribution of fish. In 1903 these activities were consolidated into the Bureau of Fisheries. In 1885 Congress appropriated money for the study of birds and mammals, and in 1896 it recognized the work by creating the Division of Biological Survey, which in 1906 became the Bureau of Biological Survey. These two bureaus (fisheries and biological survey) were combined in 1940 to form the Fish and Wildlife Service. In 1886 the first Audubon Society was organized and became an important factor in arousing public interest in birds and other wildlife. In 1894 the Yellowstone Protection Act prohibited public hunting in the Park.

The League of American Sportsmen was organized in 1898, and the American Game Protective Association was organized in 1900. Both of these groups did important work for conservation legislation. The Izaak Walton League was organized in 1922.

The Federal government became involved in the regulation of wildlife through the Lacey Act of 1900 which prohibited the transportation of game taken in violation of state laws, thus supporting the authority of states over hunting of wildlife. The first national wildlife refuge, Pelican Island Refuge in Indian River, Florida, was created in 1903. Others were soon established, and before President Roosevelt left office in 1909 the Federal wildlife refuge concept was an established principle. In 1909 killing or disturbing birds and animals on Federal refuges was prohibited. In 1913 the Weeks-McLean Act gave the Bureau of Biological Survey the mandate to protect migratory birds, and it was implemented by the enactment of the Migratory Bird Treaty of 1916. These acts were for the purpose of providing protection, through international regulation, to game birds passing from breeding grounds in Canada to wintering grounds in Mexico. In 1929 the Migratory Bird Conservation Act provided for acquisition of refuge areas in support of the treaty. Mexico became part of the cooperative international agreement in 1937.

Major advances in Federal activities took place in the 1930's. The Fish and Wildlife Coordination Act of 1934 was for the purpose of fostering coordination and integration of wildlife activities in the several bureaus of Federal government. In the same year, the National Forest Fish and Game Sanctuary Act authorized the President to establish refuges by proclamation within national forests. The Cooperative Wildlife Research Unit Program was authorized in 1935. The Migratory Bird Hunting Stamp Act (Duck Stamp Act) was enacted in 1934—it effected the first Federal game license, because all duck hunters were required to obtain a duck stamp sold by post offices. The income derived from the sale of stamps is used for sanctuaries and enforcement of the treaty.

The Pittman-Robertson Act of 1937 provided financial aid to states for wildlife restoration projects through funds derived from a 10 per cent tax levied on arms and ammunition. The Act requires states to furnish 25 per cent of the cost of approved projects. The Dingell-Johnson Law of 1952 provides the counterpart of the Pittman-Robertson Act for fisheries through an excise tax on the sale of fishing tackle, and makes funds available to states for fisheries research and restoration. Increased recognition of wildlife is evident in the Fish and Wildlife Act of 1956, which reorganized the Fish and Wildlife Service. It created an Assistant Secretary of the Interior for Fish and Wildlife and divided the service into two bureaus, the Bureau of Commercial Fisheries and the Bureau of Sport Fisheries and Wildlife. The Federal Water Pollution Control Act of 1948, amended in 1956, to assist cities and industries to control sewage and toxic materials has important potentials for preservation of aquatic wildlife habitat. In 1958 the Fish and Wildlife Coordination Act was amended to require that wildlife be given equal consideration in all water resource projects such as watershed and river basin developments.

Agencies and Organizations

Many agencies and organizations now have important functions in the conservation of wildlife. As noted, states hold regulatory authority, and consequently most action programs are implemented by them. Federal regulations, management, and aid in general, supplement the primary role of the state. A third important facet is the work of the many private organizations serving to alert the public and lawmakers to needs, and thus provide impetus for desirable legislation.

STATE ACTIVITIES

All the fifty states have departments concerned with wildlife which regulate the hunting season, the size and kind of bag and equipment, and which even prohibit certain types of lures, nets, ammunition, and guns. Research scientists maintained by these agencies are continually striving to extend knowledge of wildlife and interrelationships of wildlife with habitat so essential to sound regulations and management programs. Fish and game commissions are increasingly being called upon to evaluate developments—such as hydro dams, reclamation projects, or industrial pollution potentials—which might conflict with wildlife interest. Most states maintain fish hatcheries, and many operate game farms. Funds for operation of state wildlife research and management are supplied by income from license fees and Federal aid funds such as those of the Dingell-Johnson and Pittman-Robertson acts. The amount of income from fishing licenses is over sixty million dollars annually. Federal aid to states for sports fisheries restoration under the Dingell-Johnson Act amounted to six million dollars in 1965. In all, since 1952 over sixty-five million dollars have been made available to states. Several hundred fishery scientists are working on investigations supported by the funds. Under the Pittman-Robertson Act in 1965, 16.5 million dollars were made available to the states for wildlife restoration projects. In total since 1937, 246 million dollars have been made available in support of wildlife conservation.[12]

FEDERAL AGENCIES

The Federal government has entered wildlife regulation and other conservation practices in the national public interest, especially through regulation of interstate commerce, levying taxes for general welfare, and regulation of Federal land. From the early beginnings there has developed extensive and highly significant activity on the part of the Federal government for the benefit of wildlife. The major responsibility is delegated to the agencies of the Department of the Interior, particularly the Fish and Wildlife Service. Other agencies which by their nature necessarily have interest in wildlife (even though it is not their main role) include the National Park Service, Forest Service, Soil Conservation Service, Tennessee Valley Authority, Public Health Service, Corps of Engineers, and the Bureau of Reclamation.

THE UNITED STATES FISH AND WILDLIFE SERVICE. The Fish and Wildlife Service, created in 1940 by consolidating work already in progress, now functions through two bureaus, one for commercial fisheries and the other for sports fisheries and wildlife. The Bureau of Sport Fisheries and Wildlife has many duties. It is responsible for administration of migratory bird treaties, management practices essential to obligations established by treaties such as establishment of refuges along flyways, research and management of marine and inland fisheries, administration of the Pittman-Robertson funds, and it approves state projects for support, the Dingell-Johnson funds for fishery rehabilitation projects, and Duck Stamp funds for purchase and operating of national waterfowl refuges and public hunting grounds (Figure 9 – 9).

The Bureau of Sport Fisheries and Wildlife operates cooperative wildlife research units usually established in cooperation with the nation's land grant universities under the Cooperative Wildlife Research Unit Program (eighteen in 1965). Fish culture stations (ninety-one) are maintained in forty-two states. Under the authority of the Migratory Bird Conservation Act of 1929, and with support from Duck Stamp funds, the Bureau has developed a nationwide system of national wildlife refuges. In 1964 the system included 297 units, mostly on land controlled by the Bureau (Table 9–13). The revenue-sharing provision of the Land and Water Conservation Fund Act provides that counties in which refuges are located receive payment of ¾ of 1 per cent of the value of the land or 25 per cent of the refuge receipts.[13]

OTHER FEDERAL AGENCIES. The Soil Conservation Service through cooperation in farm plans and watershed management contributes much to wildlife conservation. The wildlife watershed management program authorized in 1954 has done much to improve habitat. Hundreds

TABLE 9–13

AREA AND CLASSIFICATION OF NATIONAL WILDLIFE REFUGES

Type	Number	Acres
Migratory bird refuges—waterfowl	229	3,612,430
General	46	3,708,590
Big game refuges	14	5,190,995
Game ranges	5	4,604,258
Alaska wildlife ranges	3	11,185,016
TOTAL	297	28,301,288

of thousands of farm ponds have been created, many stocked with fish and providing habitat for waterfowl. By the end of 1962 about 2.2 million farm ponds were known to exist, of which one million had been designed for fish production, and about five hundred small watershed projects had been developed or planned to include fish and wildlife. The Tennessee Valley Authority developments have made notable contributions to the nation's wildlife habitat. About half a million acres of impounded water area have been created by the system of multiple-purpose reservoirs, supporting an annual catch of ten to fifteen million pounds of fish and generating many millions of dollars worth of business from sports fishing alone. Since 1945 these waters have been open to year-round fishing. In addition, wildfowl and animal life have been greatly benefited.

The Forest Service is another major agency in fostering wildlife conservation. From its beginning in 1905 the Forest Service has implemented an active policy of multiple use of forest lands, and wildlife has been of major concern. The extensive national forests provide considerable wildlife habitat for big game. The National Park Service plays a major role in nurturing certain species because parks comprise the nation's principal system of sanctuaries for game. Fishing is permitted in parks, but hunting is prohibited. The ranger-naturalist program of the Park Service performs a significant service in wildlife education.

The Corps of Engineers and the Bureau of Reclamation both have direct concern for wildlife. The Corps of Engineers recently has devoted large amounts of funds to fishery research, especially in river basin studies, and both agencies are involved in provision for passage of anadromous fish. In addition, the major reservoirs of more stabilized water levels provide extensive habitats for fish and wildlife.[14]

PRIVATE AGENCIES

Numerous private organizations play an important role, especially by arousing public interest in conservation needs, providing basic information, and bringing pressure to bear on law-making groups.[15] One of the earliest and

9–9 *Fishery scientist at work.* Wildlife scientists are increasingly extending knowledge as a foundation for improved management. Here a fishery scientist is taking a stream bottom sample with a square-foot sampler as part of a trout stream food study. (Fish and Wildlife Service)

largest of these groups is the National Audubon Society, created in 1905. The Society issues a number of publications including a bimonthly magazine, and has local chapters and clubs throughout the nation. Ducks Unlimited was organized in 1937 to foster practices which would increase and perpetuate the supply of ducks. The National Wildlife Federation, established in 1936, is comprised of conservation leagues and wildlife confederations throughout the United States. The Federation publishes *National Wildlife*, a monthly magazine, and numerous booklets and pamphlets. The Izaak Walton League, organized in 1922, has nationwide membership and local chapters throughout the United States working for sound land and water management. The Wildlife Management Institution is a nonprofit private organization, incorporated in 1946. The Institute works with government officials, provides factual information, and publishes numerous reports, including the annual "Transactions of the North American Wildlife Conference." The Conservation Foundation was established in 1948 to aid in nonpartisan research projects and provide educational material. The Foundation has contributed notably to studies of wildlife, forest, soil, water, and conservation education. The Sport Fishing Institute was organized in 1949 to promote interest in conservation of the national recreational fishery resources.

Conservation Practices

There can be no doubt that wildlife resources need serious consideration in almost every resource development project, for wildlife is fundamentally interrelated with the total mosaic of environment. Nor can there be any question of the desirability of wildlife conservation for society as a whole. Fortunately, the need for conservation is being recognized, as evidenced by much of the preceding review of conservation evolution and the activities of agencies and organizations. Increasingly, land owners and managers are including consideration for wildlife in their management plans. There is also growing recognition of the essential need for interstate and international cooperation.

Many conservation practices have been devised by wildlife specialists to assist in assuring adequate wildlife for the future, but most can be grouped in three classes: (1) Practices concerned with preservation or improvement of habitat; (2) practices concerned with regulation of harvest by laws which control hunting and fishing; and (3) practices designed to increase natural stocks by artificial means.

MAINTENANCE OF HABITAT

Habitat maintenance, development, and preservation is a most important avenue toward wildlife conservation. Here man is cooperating with nature in keeping with the basic principle that man must devise his plans to get along with and assist nature. Many aspects of this cooperation could be noted, but only the role of the farmer, the public habitat, and a few problem areas will be reviewed.

IMPORTANCE OF THE FARMER. The farmer plays the major role in habitat provision, because it is estimated that 75 per cent of all wildlife find cover on farm areas. What the farmer does to provide cover and food for wildlife is a major factor in conservation.[16] Many farmers have a true conservation spirit. Some leave a portion of grain unharvested for wildlife, and winter feeding stations are often part of the farm activity. Land unsuited to cropping—marshes, rough lands, farm woodlot areas—can be developed for wildlife habitat (Figure 9 – 10). Desirable practices of this sort are being encouraged by Federal aid through soil conservation programs, by the Conservation Reserve provision for aid to farmers for tree planting, and by the small Watershed Development Program. At the end of 1963, about two million acres of trees had been planted on farms, three hundred thousand acres improved as wildlife cover, and over nine hundred thousand ponds provided on farms for waterfowl as well as fish. Property line hedgerows, perhaps incompatible with modern intensified agriculture, are nevertheless major ecotones for wildlife. Mortality

9 – 10 *Planted wildlife habitat.* Many farmers are providing wildlife habitat as part of total land management. This planting near Allen, Nebraska, contains blackberry, American plum, Russian olive, multiflora rose, and Nanking cherry. (Soil Conservation Service)

of valuable game birds, such as pheasant, nesting in grain and hay fields has been high. Now many farmers cooperate in research and in using techniques such as chains and flushing bars designed to reduce mortality during harvest by modern machinery.[17]

INCREASING NON-FARM HABITAT AREA. The farmer under pressure of commercial agriculture obviously cannot be called upon to provide all needed habitats. Growing population and spread of urban developments need to be offset through creation of additional habitat areas. Public wildlife habitat areas have been established in many states, some as sanctuaries for wildlife in which hunting is prohibited. The most notable of these areas are the national parks, in which hunting is prohibited, but fishing is generally allowed, and the national forests, in which hunting and fishing are both allowed. In some states, especially Pennsylvania and Michigan, large public shooting grounds have been set aside. Another approach is the private hunting and fishing club with leased, rented, or owned lands set aside for hunting. Many object to this procedure, however, because persons with more than average wealth thus are thought to gain unfair hunting privileges. States' conservation agencies frequently create new fish waters (Figure 9 – 11).

In connection with habitat need, it is worth noting that the widespread drainage of marshes, especially under the Swamplands Act of 1950, has had a significant undesirable effect for wildlife. Many feel that the drained lands too often did not achieve the objective of creating new farmlands or desired mosquito control, but only destroyed wildlife habitats to the detriment of society. Such problems highlight the need for full analysis with concern for *all* aspects of conservation before starting developments.

THE CONFLICT OF FISH VERSUS DAMS. Major changes in fish habitats result from the building of dams which block access and alter conditions of stream flow. Not all effects of dams are detrimental, for they provide major areas of new habitat. The most serious problem is with the anadromous fish such as the salmon. At present the attempts at solution have been the construction of means for the fish to pass the dam (fish ladders, which are stairsteps of flowing water with about a twelve-inch rise or conveyors of some type), or the capturing of fish and transporting them to release areas above the dam. The river basin development and multiple-use concepts have been helpful in gaining financial support for the intensive fishery research now underway. What is unfortunate is that too little fishery research was accomplished decades ago to supply answers now badly needed. For example, in the case of anadromous salmon, if fishery scientists can learn why the fish behave the way they do and, as a result, how to guide their travels, conservation solutions may be achieved that will be satisfactory to both fishery and power interests. Another avenue of fish conservation is the designation of certain streams as inviolate fish spawning areas. Perhaps this approach is little used because it is difficult to keep them so designated when industry and other uses increase demands on the water resource.

CONTROLLING POLLUTION. A significant facet of habitat maintenance, especially for fish, is control of pollution. Occasional pollutants are directly toxic to fish; but more important, oxidation of industrial and urban waste can and does cause oxygen content of stream water to drop below tolerance levels for fish life. In some cases oxygen requirements of pollutants are so great that low-oxygen barrier zones develop and prevent fish from ascending the stream. Siltation at times results in damage to spawning grounds and to shellfish culture grounds. A most important action program began with the Federal Water Pollution Act of 1956, which provides for regulation and for Federal financial aid for states and municipalities for construction of treatment plants. Pollution control programs need the widespread support of the public and should be carried out with definite regional and national standards. Local areas individually cannot effectively or adequately control pollution and clean up rivers and lakes and tidal areas.

INSECTICIDE USES AND WILDLIFE. A new threat to wildlife is posed by the widespread use of insecticides on farm and forest land, especially where aerial spraying techniques are employed over large areas in repeated treatments. If not carefully controlled by experts and used with proper regard for wildlife, widespread harm to fish and game can occur. The problem has received the serious attention both of foresters interested in control of endemic tree insect population, and of scientists concerned with the welfare of wildlife. Research is being carried out to determine maximum strength of insecticides and best application techniques to minimize undesired harm to wildlife.[18]

CONSERVATION BY HARVEST REGULATION

Wildlife populations were regulated by inexorable processes of nature before man became so numerous. Now, with hunting and fishing pressures intense in some places and light in others, a conscious management effort through regulation of the harvest is often required. Every range has a maximum capacity, and, when population grows beyond this

9-11 *Fertilizing a fish pond.* Fish ponds such as this one are becoming increasingly popular, especially in the plains portions of the nation. Here fertilizer is being applied as part of the management practice. (Fish and Wildlife Service)

capacity, either the range is damaged by overgrazing or animals starve. Fishing waters likewise have biological limits. Hunting and fishing accomplishes an orderly harvest of the biological surplus—a concept similar to the forester's allowable cut.

A notable example of the need for a realistic approach to harvest and conservation in the cultural environment set up by man's influence is the Kaibab deer episode. In 1906 President Roosevelt established the area in the Arizona Grand Canyon region as a game refuge. Predators such as the wolf were hunted down and domestic grazing prohibited. With the environmental resistance, especially competitors, removed, the deer multiplied, and in a few years their number passed the capacity of the range; even the conifers were browsed to as high as the deer could reach. The Forest Service urged a harvest, and struggled unsuccessfully with the State of Arizona for legal authority. The deer increased from about four thousand to one hundred thousand. Then in 1924–1930 at least eighty thousand were estimated to have starved because there were more animals than the land could support.[19] The population was reduced to about ten thousand.

All the states have regulations for hunting and fishing. The rules are based on the best available knowledge supported by field research. The principle involved is illustrated by Figure 9 – 12 which concerns deer, but is applicable to all wildlife. Hunting or fishing should harvest the biological surplus, but ought not to exceed the annual increase except under unusual conditions.

Fish and Game Commissions in most states set the date and length of time of hunting and fishing, number of the bag, sex in the case of deer and pheasant, etc., and other regulations as deemed desirable for the particular wildlife area. The regulations are in the interest of the

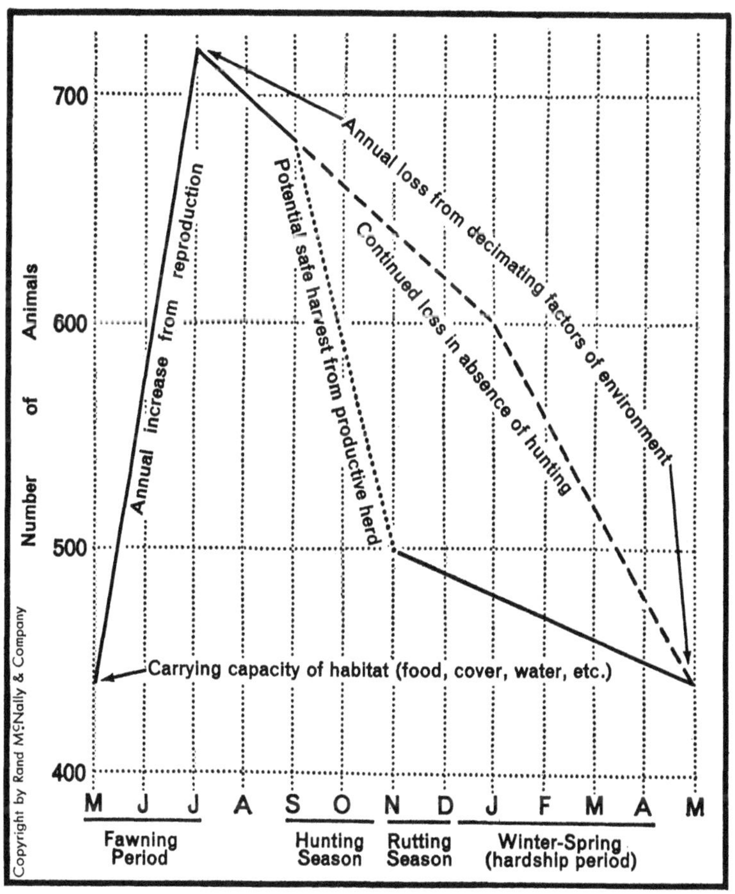

9 – 12 *The annual fluctuation in a deer population and the shootable surplus.* (From, with permission of, Raymond F. Dasmann, *Environmental Conservation* [New York: John Wiley and Sons, Inc., 1959], p. 220.)

wildlife as well as the public, and all citizens need to recognize their value and obey the hunting and fishing laws.

INCREASING WILDLIFE BY ARTIFICIAL TECHNIQUES

Popularity of recreational fishing has led nearly all states to initiate extensive programs for artificial propagation of game fish and the stocking of streams and lakes with considerable success. In 1962 the states devoted about thirty-two million dollars to management out of a total expenditure of sixty-five million dollars for fisheries functions. Table 9–14 shows how the fishery management funds were spent in the nation.

Nearly all of the states now carry on rehabilitation work to improve fish habitat when undesirable fish have overrun a lake or stream. In such cases the usual conservation practice is to kill all fish chemically and then plant the water with desired species.

INTRODUCED SPECIES AND PROPAGATION. Introduction of wildlife from foreign areas has long been practiced—sometimes inadvertently, sometimes with desirable results, sometimes not beneficially. A few notable examples include the introduction of the English sparrow and the European starling, both of which have become liabilities rather than assets. The carp introduced as a food fish has become a problem in many waters. Two of the most successful introductions are the ring-necked pheasant and the Hungarian partridge. The moral is consider all aspects before introducing exotics.

Recent evidence indicates that artificial propagation as the basis for management is less highly regarded than it was in the early stages of wildlife management. According to Dale, "one of the least profitable game management measures is the release of hand-reared birds to build up wild stock where the species are already established."[20]

Conclusion

This brief review of the nation's wildlife resource makes it clear that wildlife forms a facet of the total environment without which man would be poorer in aesthetic values, recreation, and economic benefits. If one accepts the conclusion that wildlife is a desirable facet of the total environment, then the question is: To what extent and how can wildlife be maintained or managed in the best interests of the nation? It is, of course, self-evident that conditions of natural wilderness that once typified most of the nation's territory cannot be retained

TABLE 9–14
EXPENDITURES FOR MANAGEMENT PER FISH CONSERVATION DOLLAR*

Activity	West 13 states	North Central 12 states	South 13 states	Northeast 12 states
Coldwater fish stocking trout-salmon	29¢	6¢	4¢	32¢
Warmwater fish stocking plus salvage and transfer	1¢	11¢	8¢	4¢
Land acquisition and access development	2¢	3¢	1¢	4¢
Construction of new fishing lakes	7¢	13¢	10¢	4¢
Rehabilitation of old fishing waters	2¢	1¢	2¢	2¢
Fish population control	1¢	8¢	2¢	—
Habitat improvement	5¢	4¢	7¢	5¢
All other forms of fish management	4¢	5¢	7¢	4¢
TOTAL	51¢	51¢	41¢	55¢

* "Fish Conservation Highlights 1960–62" (Washington: Sports Fishing Institute, 1963), p. 20.

except in comparatively small protected areas, and even in such preserves the conditions are modified. It has been noted that wildlife has been and still is abundant in the nation; nevertheless, modifications in and reduction of habitat, increased numbers and greater efficiency of hunters led early to regulations prohibiting or regulating hunting in efforts to protect species that were observed to be declining in population. Such protection is now recognized as only a small part of the conservational program required. Although it is clear that more research is needed for full understanding of the problem of wildlife conservation, the following five conclusions are drawn from the review of programs and management.

1. Because a given habitat is generally agreed to have maximum carrying capacity, wildlife scientists and managers recognize control of wildlife population as a major task of wildlife conservation.

2. Management of wildlife to provide for the wants of recreational hunters and fishermen is only a part of wildlife conservation. There is ample reason to believe that there should be preservation of at least remnants of wilderness wildlife such as the buffalo which are not adaptable to the man-dominated environment. There is need also to include environmental management for songbirds and other wildlife for their aesthetic values.

3. Conservation of wildlife needs an ecological approach that considers the complexities of the food web in which wildlife function in the total environment.

4. Wildlife management programs need to be part of and compatible with agricultural resource utilization, forestry, or watershed management, for wildlife is inevitably involved.

5. Research is making clear that a viable wildlife population can be preserved only through maintenance or provision of habitat that is adequate in quantity and quality.

Much is now being done in management of wildlife by many agencies, federal and state, as well as by dedicated private organizations. Their research and rational programs of management need to be continued and in some cases, no doubt, expanded, if the nation desires to maintain an abundance and variety of wildlife.

FOOTNOTES

[1]Helmut K. Buechner, "The Bighorn Sheep in the United States: Its Past, Present, and Future," *Wildlife Monograph, no. 4* (Chestertown, Md.: Wildlife Society, May, 1960), 174 pp.

[2] "Fur Catch in the United States 1964," *Wildlife Leaflet* no. 471 (Washington: Fish and Wildlife Service, Department of the Interior, August, 1959), 4 pp.

[3] For example see Chester E. Kebbe, "Oregon's Beaver Story," *Oregon State Game Commission Bulletin*, vol. 15, no. 2 (February, 1960), p. 306.

[4] "The Seals, Sea Lion and Sea Otter of the Pacific Coast," *Wildlife Leaflet No. 344* (Washington: Fish and Wildlife Service, Department of the Interior, February, 1953), 28 pp.

[5] Leonard W. Wing, *Practices of Wildlife Conservation* (New York: John Wiley & Sons, Inc., 1951), p. 250. Wing considers only five groups and includes the ruddy duck in the diving class.

[6] The study made for the Outdoor Recreation Resources Review Commission on sports fishing gives projections for increases to sixty-three million anglers and 1.3 billion angler days of fishing by the year 2000. See "Sports Fishing Today and Tomorrow," *Special Report No.* 7 (Washington: Bureau of Outdoor Recreation, Department of the Interior, 1962), 127 pp.

[7] Richard H. Stroud and Robert M. Jenkins, "Fish Conservation Highlights 1957–1959" (Washington: Sports Fishing Institute, 1960), pp. 7–10.

[8] This concept has interesting, even serious, implications. For example, if a farmer's crop or orchard is being heavily browsed by deer during a time of meager normal food, the farmer, in most states, must inform the state game agency of his intent or action in killing animals to protect his crop. Then he may not possess the meat.

[9] See Alfred Stefferud and Arnold Nelson, eds., *Birds in Our Lives* (Washington: Bureau of Sports Fisheries and Wildlife, 1966), 372 pp.

[10] For a review of evolution of wildlife conservation and basic concepts see John D. Black, *Biological Conservation* (New York: McGraw-Hill Book Company, Inc., 1954), 328 pp.

[11] See "Hunting in the U.S.: Its Present and Future Role," *Special Report No. 6* (Washington: Outdoor Recreation Review Commission, 1962), p. 117.

[12] "Federal Aid in Fish and Wildlife Restoration 1965," (Washington: Wildlife Management Institute and Sport Fishing Institute, 1966), 88 pp.

[13] In 1963 there were 12.3 million recreational visitor-days recorded at these national refuges, mostly for purposes other than hunting or fishing, which, nonetheless, testify to the value of the refuges to society.

[14] Reservoirs created by the Corps of Engineers have added at least three million new acres of fishing water; nineteen million man-days of angling were provided in 1959, and a catch of 32.4 million pounds was made. "Fish Conservation Highlights," (Washington: Wildlife Management Institute and Sports Fishing Institute, 1960), p. 38.

[15] A valuable listing of organizations of all types and much useful information is found in *The Conservation Directory* (annual) (Washington: National Wildlife Federation, 1966), 123 pp.

[16] For example see Carl V. Swanson and Charles F. Yocom, "Upland Game Bird Population in Relation to Cover in Southeastern Washington," *Transactions of the Twenty-Third North American Wildlife Conference* (Washington: Wildlife Management Institute, 1958), pp. 277–90.

[17] For example see Eugene D. Kloghan, Russell L. Robbins, and Bromfield L. Ridky, "Evaluation of Effectiveness of Pheasant Flushing Bars in Iowa Hayfields," *Proceedings of the Iowa Academy of Science*, vol. 66 (Des Moines: The Academy of Science, December, 1959), pp. 534–52.

[18] For a review of the problem and a good bibliography see Daniel L. Leedy, "Pesticide-Wildlife Problems and Research Needs," *Transactions of the Twenty-Fourth North American Wildlife Conference* (Washington: Wildlife Management Institute, 1959), pp. 150–65; and "Effects of Pesticides on Fish and Wildlife," *Circular No. 226* (Washington: Fish and Wildlife Service, August, 1965), p. 77.

[19] D. I. Rasmussen, "Biotic Communities of the Kaibab Plateau," *Ecology Monograph, No. 3* (Durham, N. C.: Duke Univ. Press, 1941), pp. 229–75; also discussed in Raymond F. Dasmann, *Environmental Conservation* (New York: John Wiley & Sons, Inc., 1959), 307 pp.

[20] Fred R. Dale, "Management of the Ring-Necked Pheasant," *Wildlife Leaflet No. 412* (Washington: Fish and Wildlife Service, Department of the Interior, 1960), p. 3.

10

Recreation Lands

PROVISION FOR OUTDOOR RECREATION has become one of the major considerations in the nation's allocation of resources. As a result, new and difficult conservation problems, which require thorough study and answers from the nation's citizens and from legislative representatives, have been posed. Among the major problems are such questions as: What kinds of resource environments need to be provided for outdoor recreation? How much ought to be provided by public investment? Where is provision and development needed? How much access and development is appropriate? What agencies—Federal, state, local, and private—should be responsible, and for what? Perhaps the most important question for conservation is: How can the nation *now* best reserve adequate and suitable resources for outdoor recreation needs of future citizens?

Societal Conditions and Outdoor Recreation

Outdoor recreation as a means of "re-creation" of the body and spirit of man is far from new. There are, however, a number of significant evolutionary changes that have led to the current public concern for conservation of outdoor recreation resources, and also to some compelling needs. The complexity of urban living, with its tension-producing demands, has encouraged a popular turning to the out-of-doors for recreational activities which is being magnified by the increase in population. Even more important are three conditions that have combined to further increase pressures on available resources for outdoor recreation. These three conditions are: (1) Increased discretionary time, (2) increased discretionary income, and (3) expanded and improved transportation facilities.

INCREASED DISCRETIONARY TIME

A major variable, essential to consideration of the nation's need to allocate resources for outdoor recreation is the decreasing number of hours which people must dedicate to earning a living and accomplishing daily chores of the home. The average work week was at least 70 hours in 1850; it is currently 40 hours, and there is reason to believe that it will be reduced to about 30 hours in the 1970's. Moreover, in our homes machines of all sorts, together with other aspects of modern technology such as commercial food processing and packaging, have reduced time and labor required of the homemaker. The result is liberation from many hours of toil, and thus more time during which people may *choose* what to do. This time is designated as *discretionary time*, and it is part of this time that people can and do devote to outdoor recreation.

Perhaps even more significant than the reduction in the work week is the increased number of days available as holidays and paid vacations. Paid holidays increased from an average of 4 days in 1946 to 8 days in the 1960's, and paid vacations have doubled. The United States Bureau of Labor estimates that 80 per cent of the nation's industrial workers are employed by firms offering three weeks or more of paid vacation. Thus, the average American worker now has about 125 days per year not devoted to earning a living. In addition, the "three-day weekend" has become common, increasing the number of multiple-day vacation periods available to American families. A provocative analysis of the nation's "time budget" for 1950 and projection for the year 2000 is illustrated in Figure 10 – 1 and Table 10–1. The implications of large amounts of discretionary time for conservation of resources for outdoor recreation are compelling.

INCREASING DISCRETIONARY PURCHASING POWER

Along with increasing leisure, the people of the United States are enjoying increasing incomes. Especially important is the portion of income above requirements of the normal cost of living. This income—or purchasing power—is designated as *discretionary purchasing power*. A series of data developed by the National Industrial Conference Board is significant (Figure 10 – 2). The chart reveals that there has been an indicative increase in discretionary purchasing power expressed in an indicative increase in discretionary purchasing power expressed in 1958 standard dollars, rising from 93 billion dollars in 1949 to 196 billion dollars in 1966. On a per capita basis, discretionary purchasing power rose from 623 to 997 dollars,

TABLE 10–1

THE NATION'S TIME BUDGET, 1950 AND 2000*

Activity	1950	2000
	Billion hours[1]	
Sleep	514	1,131
Work	132	206
School	32	90
Housekeeping	68	93
Preschool nonsleeping	56	110
Personal care	74	164
Discretionary	453	1,112
Daily	189	375
Weekend	179	483
Vacation	35	182
Retired	24	56
Other	26	16
TOTAL TIME	1,329	2,906

* Mary A. Holman, "A National Time-Budget for the Year 2000" *Sociology and Social Research*, vol. 46, no. 1 (1961), p. 19.

[1] Based on the 24 hour day for a population of 152 million in 1950 and projection to 322 million in year 2000.

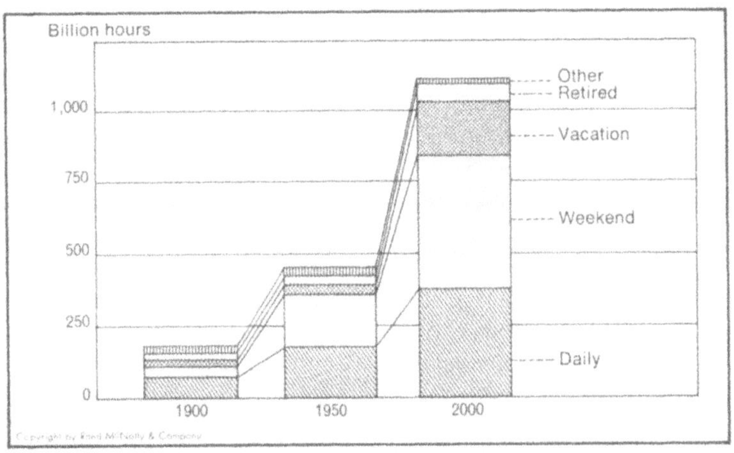

10 – 1 *Division of the nation's discretionary time budget.* The projected increase in weekend and vacation leisure time for the year 2000 has major implications for the type and location of outdoor recreation needs. (Data from Mary A. Holman, "A National Time Budget for the Year 2000," *Sociology and Social Research*, vol. 46, no. 1 [1961], p. 19.)

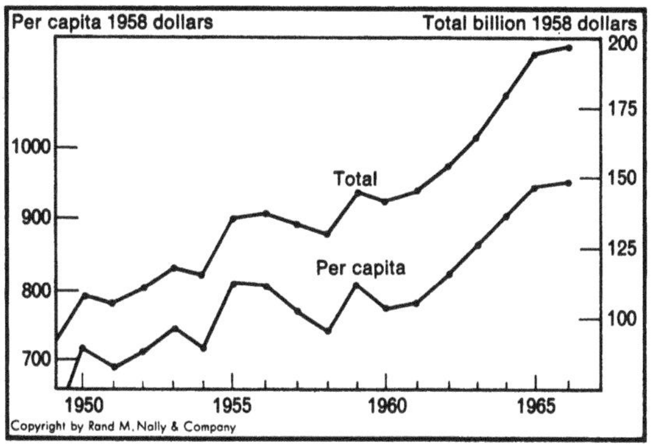

10 – 2 ***Increase in discretionary purchasing power.*** **Since 1949 discretionary purchasing power has increased significantly. (Based on data from National Industrial Conference Board, *Technical Bulletin No. 17* and the Conference Board Record.)**

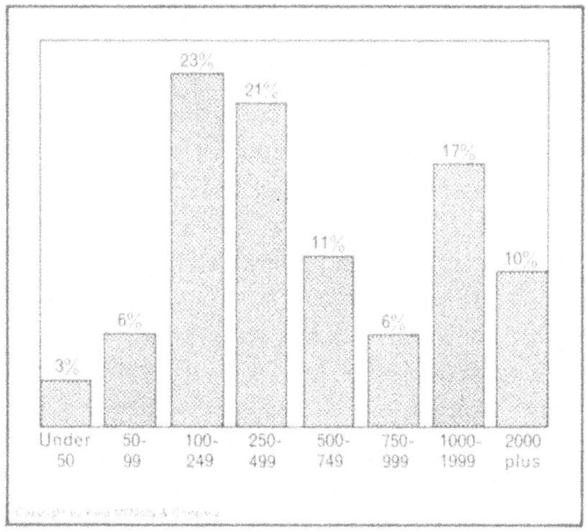

10 – 3 ***Distance traveled one way on vacation trips.*** **The data reveal how great is the range of vacation travel, and reflect the impact of modern transport in making distant scenic and recreational attractions accessible. (Based on data in "Participation in Outdoor Recreation, Report 20" [Washington: The Outdoor Recreation Resources Review Commission, 1962], p. 43. The sample included 1,186 cases.)**

for a 44 per cent increase (Figure 10 – 2). The trend suggests that discretionary purchasing power will continue to increase, and it is significant that this trend also applies to the world community. Out of discretionary purchasing power, Americans are buying recreation equipment, participating more in recreational activities, and traveling farther each year.

THE IMPACT OF MODERN TRANSPORTATION

The question of how much allocation of resources for outdoor recreation (and where) requires consideration of the impact of modern transportation. The airplane, for example, has internationalized the recreational range of the more affluent citizen for hunting, fishing, skiing, and sightseeing. Better automobiles and better roads, combined with discretionary time and money, have greatly increased recreation opportunities for American families. A sample study in 1960 showed that over half of vacation trips are 1,000 miles or more in length (Figure 10 – 3). A different study revealed that urban residents travel farther than rural residents on vacations and recreational outings (Table 10–2).

TABLE 10–2

DAYS OF VACATION TRIPS AND MILES TRAVELLED*
JUNE 1960–MAY 1961

Place of residence	Vacation days	Trip days	Miles travelled on Vacation	Trips	Outings	Total
Standard metropolitan areas						
Over a million	11.5	2.4	974	192	335	1,501
Under a million	10.0	2.6	831	208	332	1,371
Rural	8.8	2.4	833	266	399	1,498
Urban outside SMA	12.0	2.5	765	156	276	1,197
Nonfarm rural	10.4	2.6	589	178	305	1,072
Rural farm	9.5	2.3	410	90	227	727

* Based on data from "National Recreation Survey" special report to the Outdoor Recreation Review Commission, report number 19 (Washington: 1962), pp. 368, 371, 372.

THE PROBLEM OF THE MOTORIZED OUTDOOR RECREATION SEEKER

Concomitant with more discretionary time and money, a significant number of the people seeking outdoor recreation have developed a desire to do so in comparative comfort. One indication is the rapid increase in the number of motorized camping units. A recent survey showed that there were over a million recreational vehicles in use in 1965, and that owners on the average, travel 5,244 miles a year with their trailers or motorized camper units. Table 10–3 indicates the use of travel trailers by outdoor recreation categories. Such units require suitable space for parking and special facilities such as electricity, water, and provision for the removal of sewage. To what extent ought public investment to provide such facilities, and at what charge? Whatever the answer, the need to provide for the thousands of motorized camping units and trailers that arrive at local, state and national parks has already required significant reevaluation of resource need (see Figure 10 – 4).

Another aspect of the problem of the motorized outdoor recreation seeker is the great expansion in boating, especially motorboating, as an outdoor recreation activity. Canoeing and sailing have long been part of the outdoor recreation panorama, and the yacht club a feature of the nation's coastline and major lakes. Until recently, only a small number of families owned boats; but in 1965 there were about seven million recreational boats in use. Thus, just as the motorized camper has added new dimensions to the outdoor recreation needs, the mass ownership of small boats has added new problems ranging from provision of launching facilities to needs for safety regulations.

The result of this evolution has been to expand outdoor recreation activities so that they have become a major resource utilization in numbers of participants, time involved, and dollars expended. Especially in the United States, discretionary time and money have become characteristics of the entire social structure, and participation in outdoor recreation has become a major part of the American mass consumption pattern. As a result, the questions of what and how much ought to be provided by public allocation of resources have become major conservation problems.

TABLE 10–3
MAJOR USES OF TRAVEL TRAILERS*

	Per cent
Travel and sightseeing	42
Camping and hunting	36
Camping–no hunting & fishing	13
Camping and travelling	6

* Merril D. Ormes, "Skyrocketing Recreational Vehicle Sales Spur Need for Parking Areas", *Trends*, vol. 3, no. 2 (April, 1966), p. 14–19. Travel trailers are defined as units less than eight feet wide and less than thirty-two feet long that are handled by the family car without special permits (costing 1,100 to 8,000 dollars, averaging 2,260 dollars). A pick-up coach is a structure mounted on a pick-up truck body which costs 1,100 to 5,500 dollars. A camping trailer is a canvas cover on wheels that folds out and costs 350 to 1,200 dollars. A motorized home is a portable dwelling built as an integral part of a self-propelled vehicle costing 5,000 to 30,000 dollars. In 1965, sales were estimated at 250,000 units annually.

The Nature of Resources for Outdoor Recreation

Space for outdoor recreation gains its resource quality or value status from the combination of several variables. The most significant qualifications of an area for outdoor recreation are scenic attractiveness, accessibility, adequate size, and development for use in keeping with the type of outdoor recreation for which the area will be used. The wide variety of environments which needs to be included in provision for outdoor recreation activities is especially notable. These environments range from seashore areas to high alpine peaks, from roadside picnic spots to spacious national parks, and from intensively developed activity areas such as playgrounds, tennis and golf courses, to large preserved wilderness areas.

Unlike most aspects of the nation's resource foundation, allocation of land for outdoor recreation is not wholly identifiable with the profit motive of the nation's economic system. There are, of course, outdoor recreation resources such as private golf courses and camps which are established as profit-making enterprises in serving public desire for outdoor recreation. This chapter is concerned with provision of outdoor recreation resources by public allocation and investment for enjoyment by the nation's citizens.

EVALUATION OF RESOURCES FOR OUTDOOR RECREATION AND USE

The question of what kind and how much environmental space ought to be allocated to public outdoor recreation use is one of the basic problems of conservation. The variety of kinds of outdoor recreation participated in by the nation's citizens indicates that there is need to provide for a wide range of activities, including local picnic, playground, and swimming facilities as well as large national parks and wilderness areas. Table 10–4, based on a study of a sample of the national population in 1960, gives an indication of the variety and amount of participation in outdoor recreation. As would be expected, the most general participation is in such activities as picnicking, swimming, driving, and sightseeing. The significance of the data, however, is in the consistent indications of

10–4. *Trailer facility area in Yellowstone National Park.* Increasing use of recreational vehicles, such as trailer homes, camper pickups, and fold-out trailers, which provide sleeping accommodation is a phenomenon of concern to managers of outdoor recreation areas. For the National Parks especially, the trend is making necessary increased provision of space and major capital investments to provide facilities such as electricity, water, and sewage hookups. In general these facilities developments are operated by concessionaires under supervision of the National Park Service and special fees are charged. Most parks record overflow use during the summer, especially on Saturdays. With capacity reported as 3.7 million, in 1967 the National Park Service recorded 4.6 million camper days by persons using such recreational vehicles. For comparison 4.7 million camper days were recorded by persons using tents. Yellowstone National Park recorded 675,330 camper days in trailers during 1967—nearly double any other single park. (Courtesy National Park Service)

TABLE 10–4

PARTICIPATION IN OUTDOOR RECREATION

	U.S. population (over 12 years) per cent participation[1] 1960	2000	Days per participant 1960[2]
Driving for pleasure	52	60	12.7
Picnics	53	61	4.0
Swimming	45	63	11.5
Sightseeing	42	52	5.2
Walking for pleasure	33	43	13.1
Playing outdoor sports	30	42	12.3
Fishing	29	36	6.8
Attending outdoor sports	24	30	5.5
Boating	22	38	5.5
Nature walks	14	19	5.2
Hunting[3]	13	14	5.6
Bicycling	9	14	19.4
Attending outdoor concerts	9	17	2.4
Camping	8	14	5.7
Hiking	6	12	4.4
Water skiing	6	12	0.1
Horseback riding	6	12	7.5
Skiing[4]	3	?	?
Canoeing	2	—	3.0
Sailing	2	—	3.0
Mountain climbing	1	—	3.7

[1] "Prospective Demand for Outdoor Recreation" report to the Outdoor Recreation Review Commission, number 26 (Washington: 1962), p. 27.

[2] "National Recreation Survey" *op. cit.*

[3] Fall, 1960.

[4] Skiing was not included in the survey. It is estimated that there are about five million skiers (about 3 per cent of the population over 12 years ago).

expected increased participation. The study indicates that in the future there will be not only a larger population but also a higher percentage of general participation and more days of participation. It is reasonable to suggest that all of these activities should be provided by public investment and allocation of resources.

Closely related to the need for balance in providing for the total variety of outdoor recreation activities is the need for quality of resources. Space for outdoor recreation ought not to be just some of the leftover lands that have little or no commodity production value. Outdoor recreation ought to be ranked near the top of the scale of alternate land use values whether in the urban, rural, or wilderness environment. There is a need to conserve a reasonable portion of the nation's high quality lands for outdoor recreation in strategic locations accessible to urban and rural families, as well as to conserve major areas of national, scenic, and historic quality.

Although there is a rationale for considering the benefits of outdoor recreation to be beyond normal economic appraisal, nevertheless it is proper to confront the question of economic appraisal in considering allocation and development of resources for outdoor recreation. It is desirable to appraise proposed resource allocation for outdoor recreation in terms of alternative commodity production and dollar values that society would forego. It is then proper and desirable to estimate and compare potential economic benefits of recreational development. The calculating of benefits from outdoor recreation, however, is even more complicated than for alternate uses, and there are several opinions on what should be measured and how.[1] Entrance fees and other direct income at developed areas are readily calculated, but obviously do not reveal the full benefit to a community or to the nation.

Dedicating areas to outdoor recreation commonly does involve foregoing some commodity production or economic use; however various studies indicate that development of major areas for outdoor recreation commonly results in economic benefits at least as great as the alternative uses foregone.[2] Secretary of the Interior Stewart L. Udall has expressed this point with reference to national parks as follows: "It has become abundantly clear that national parks are not only sound social investments, but sound use of public funds as well. Time and time again citizens adjacent to new parks have bemoaned the loss of revenues from resources "locked up" inside a new reserve—lost taxes, uncut timber, undiscovered minerals, unharvested game—only to learn later that the income from providing services to visitor-tourists has equalled or surpassed whatever sums might have been gained exploiting these park resources."[3]

In the final analysis, however, decisions by Congress and other responsible agencies relating to conservation of outdoor recreation resources ought not to be based primarily on the probability of economic benefit either to the local area or to the national economy. Nor should such decisions be based solely on the needs of the present population. The decisions should be based on outdoor recreation needs of future citizens as well as the present population, and on integrated evaluation of the resource quality and suitability to provide for the need.

MEASURING ADEQUACY OF SPACE FOR OUTDOOR RECREATION

A rational answer to the basic question of how much space should be allocated for outdoor recreation to satisfy the reasonable needs of the nation's future population requires that there be some quantitative measures for determining what is adequate. Establishment of measures of adequacy is complicated by the variety of activities that need to be satisfied, by the spatial variation in the nation's population density, by the seasonality of the pressures on outdoor recreation areas, and by difficulties of knowing what the population in the future will desire. Nevertheless, several methods of measuring what is adequate have been suggested; the most common is that of acres per one thousand population. Table 10–5 is a proposed composite standard that is useful as a model with which to compare national, regional, and local provision for outdoor recreation.

TABLE 10–5
PROPOSED STANDARD OF OPEN SPACE FOR ULTIMATE POPULATION*

Type	Acres per thousand population
Urban government	10
Neighborhood parks	2.5
District parks	2.5
Large urban parks	5.0
State and local nonurban	80.0
Large extraurban parks	15
State parks	65
National government	100
TOTAL	190

* *Outdoor Recreation Space Standards* (New York: National Recreation Association, 1965), p. 55.

Cities have commonly utilized ten acres per one thousand population as a standard for urban open space and recreational area. The simplicity of the standard no doubt has fostered its common use; however the limitations and inadequacies are readily apparent. All sizes and types of cities cannot equally be served by the same index acreage, and moreover, the average does not consider the distribution of open spaces for access within urban areas. Various modifications have been proposed by city planners. For example, it is suggested by some that cities under 10,000 population ought to reserve fifty to seventy-five acres per thousand people; that cities over 500,000 need only five acres; and that for cities over one million population three acres per one thousand population may be adequate. Others have suggested that at least 10 per cent of a city's area should be in open space and recreational areas.

In the past, cities have been concerned only with their corporate area, but in recent years the magnitude and sprawl of urbanization and increased use of the automobile have led to recognition of a broader need and responsibility. The need today is for cities and other local governments to expand conservational planning to include coordinated development of total environments within a range of two hours driving time as a minimum.

Comparison of the proposed standard with the realities of dedicated space for recreation is instructive. The fifty largest cities in the United States indicated an average park and recreational area of seven acres per thousand population, and a quarter of them meet one of the common standards. The implication is that the proposed standard is reasonable, but at the same time that there is need to reserve more recreational areas to provide for the needs of the future. Comparison of the standards for state parks with the realities shown in Table 10–12 is also instructive. Five of the fifty states do in fact have over sixty-five acres per thousand population dedicated to state parks. It is also significant that were the proposed standards to be met for a population of three hundred million, the total space devoted to recreational parks at the state and local level would then be only about 3 per cent of the land area of the fifty states.

Suitable provision of space for outdoor recreation would not likely be accomplished by considering a standard only in terms of total areas. There is a need for evaluation in terms of *where* to provide space, especially with consideration for easy accessibility to population concentrations for daily use as well as for longer-period vacation usage. The Outdoor Recreation Resources Review Commission properly emphasized this need in urging conservation consideration for six classes of outdoor recreation areas as follows:

Class I—*High-density recreation areas*—areas intensively developed and managed for mass use.

Class II—*General outdoor recreation areas*—areas subject to substantial development for a wide variety of specific recreation uses.

Class III—*Natural environment areas*—various types of areas that are suitable for recreation in a natural environment and usually in combination with other uses.

Class IV—*Unique natural areas*—areas of outstanding scenic splendor, natural wonder, or scientific importance.

Class V — *Primitive areas* — undisturbed roadless areas, characterized by natural, wild conditions, including "wilderness areas."

Class VI—*Historic and cultural sites*—sites of major historic or cultural significance, either local, regional, or national.

MULTIPLE USE AND OUTDOOR RECREATION

Analysis of the problem of providing adequate quantity and quality of space for outdoor recreation also ought to include consideration of several ways in which such provision may be developed. It is in the public interest that some areas of quality and scenic significance be reserved and developed primarily for outdoor recreation, but in addition there are many possibilities to develop outdoor recreation space and facilities as part of multiple-use developments of land and water resource complexes. It is doubtful that the nation could rationally provide for the projected magnitude of outdoor recreation by space designated chiefly for recreation.

Areas dedicated to outdoor recreation as the primary use include playgrounds, golf courses, parks, wild rivers, and wilderness areas. Such areas may in a sense be termed "single purpose," yet one should reflect that in reality outdoor environments inevitably have multiple resource values. A golf course, park, or wilderness area, for example, designated primarily for outdoor recreation, still serves as watershed and habitat for wildlife (See Figure 10 – 5).

It appears that in the future an increased share of space and facilities for outdoor recreation will need to be provided as part of multiple-use developments of commercial forests and water resources by private as well as public enterprise. It is helpful to note in this connection that multiple use does not commonly mean that all parts of a resource development complex are equally managed for outdoor recreation activities. In a commercial forest, for example, timber production, wildlife habitat, and watershed management are intermingled throughout the area, and the forest area may be open to hunting and fishing, but camping and picnic sites are specifically designated areas allocated to outdoor recreation.

In recent years development of recreation facilities has become recognized as part of public and private natural resource development and management. There are many notable examples. Reservoirs constructed by the Corps of Engineers and Bureau of Reclamation usually have recreational values that are being developed.[4] Wildlife refuges have been opened to outdoor recreation, and national forests for decades have provided picnic and campgrounds as well as many miles of forest trails. The Tennessee Valley is an outstanding example of an area in which recreation benefits have grown to major significance in multiple use development.

Multiple use development and management with provision for recreation is not confined to public agencies. Throughout the nation, commercial forest lands of private industry are being opened to public outdoor recreation, and increasingly developed picnic and camping areas are being provided. As noted in Chapter 9, most private forest industry lands are open for hunting and fishing, and thus serve as recreational lands. Electric utilities commonly develop park sites for picnicking and camping, as well as water sports, in association with their reservoirs (Figure 10 – 6). In the future, provision for recreation as part of private resource developments devoted primarily to commodity production very likely will provide an even more significant part of the space and facilities needed for outdoor recreation activities.

The Conservation Problem

There is no doubt now that society and legislative leaders realize the need and the desirability of outdoor recreation. Providing for outdoor recreation, however, is not so much a problem of the present as it is a problem of *reserving adequate space* and of *preserving quality natural scenic endowments* for enjoyment by citizens of the *future*. Some of the difficulties and questions involved in considering conservation of resources for outdoor recreation have already been noted. The questions are not easy to answer, in part because our knowledge is incomplete. The psychology of outdoor recreation is not fully understood, and thus it is not wholly clear what the population of the future will want. Not the least of the problems is that of reconciling competitions and conflicts between alternate resource uses. To what extent can recreation be provided for as part of multiple use on lands primarily dedicated to commodity or other economic production? How much space needs to be dedicated to recreation as the primary use? It appears certain that there

10 – 5 *A public golf course.* The golf course is a ubiquitous element of the modern urban scene. Lands such as this, although dedicated to single-purpose recreation, also provide cover for wildlife and open space in the city, helping to create a more desirable total environment. (New York City Park Department)

is need for greater provision for recreational land use than at present. It is equally clear that conservation of open space and recreational land resource involves difficult problems in resolving claims of competing land use. How can one measure the worth of preserving a scenic area? How can it be objectively decided that a portion of commercial forest land should be dedicated to recreation and the harvest of timber foregone? How can the pressure of urbanization and short-term profit motives be kept from making automobile parking lots out of urban open spaces?[5]

THE PROBLEM OF USER FEES

One of the perplexing problems resulting from allocation of public resources to preservation and development for outdoor recreation is the question of user fees. Should those who use the outdoor recreation areas pay in greater portion than the non-using taxpayer? And if so, how much?

The Land and Water Conservation Fund Act of 1965 is especially significant because it may be assumed that it establishes the principle that users of facilities provided by public investment should pay a user fee. Apparently the first user fee at a national park was the automobile fee at Mount Rainier National Park, first assessed in 1908 for use of the road. Since then most National Parks have had entrance fees of varying amounts, generally associated with the automobile entrance. The Land and Water Conservation Fund Act, however, expanded the user fee concept to include designated areas under nearly all Federal agencies where investment of public funds has been made to provide facilities for outdoor recreation. There are a number of fee schedules depending upon the length of use, but it is the annual family use permit, the so-called Golden Eagle, that no doubt will be best known and most widely used. The Golden Eagle permit in 1966 cost seven dollars and admitted the holder and his family as well as occupants of the car driven by the permit holder to any and all designated facilities.[6]

THE PROBLEM OF USE-INTENSITY VERSUS QUALITY

Another perplexing problem in conservation of resources for outdoor recreation is that of maintaining quality of the resource and of the recreational activity. How many people can

10–6 ***Public recreation area at a private utility dam.*** **Throughout the nation many private enterprise resource developments are now making such provision. This picnic ground and swimming area at Merwin Dam on the Lewis river in southwestern Washington is readily accessible to population of the Portland metropolitan area. (Courtesy Pacific Power & Light Company)**

a given area provide for without lessening the individual's recreational experience or without damaging the resource itself? Both questions deserve study, and neither will be easy to answer definitively. The answer to how many visitors ought to be accomodated at any given time or within a specific area in large part will be a personalized answer, and also will vary with the activity. The reality is that some people enjoy crowds, and even prefer to have many tents and trailers close by, while other outdoor recreation seekers become distinctly unhappy if there is even a single person within sight or sound. Yet it is clear that for the average American there is some degree of crowding—whether at the seashore, the picnic grounds, or in the wilderness area—beyond which the area begins to lose its appeal. Thus development of easy access to outdoor recreation areas may result in spoiling the very experience that the access was developed to provide.

Unfortunately, too many people are lacking in public responsibility, with the result that accessible facilities commonly become littered, vandalized, and otherwise downgraded—streams become polluted; the natural environment becomes humanized; and the quiet sounds of nature are shattered by automobile horns, radios, and even TV. In addition, there is considerable evidence that the treading of the ground by many people significantly alters the ecological environment, leading to a change in vegetation.

The question of use-intensity level is a problem that the National Park Service in particular has faced for many years. How much access can be tolerated without too much downgrading of the significant resource quality that the parks were established to preserve? In general, it is probable that there is an inverse correlation between quality conservation of an outdoor recreation resource and the intensity of its use through provision of access and facilities. The question that needs to be answered for each type of outdoor recreation activity and specific area is that of what degree of quality in outdoor recreation resource is rational and appropriate to maintain in the public interest. Some students of the problem believe that quality may best be regulated—or maintained—by managing the intensity of use through varying the amount of the user fee. Others believe that the desired quality of recreational resource and the associated experience should be maintained through provision of sufficient area and facilities. Whatever the final answer, it is clear that projected population increases and a greater degree of participation in outdoor recreation activities make the question of the appropriate level of quality and how to maintain it a fundamental one for conservation.

AVAILABILITY AND USE

In the mid 1960's the nation had about 570 million acres of public lands designated as available in some form for outdoor recreation—about 2.8 acres per capita. In addition, private forests total 366 million acres, of which about sixty million acres of industrial forest lands are known to be open to the public in some measure; certainly large areas of other private forest lands sometimes are used by the public for outdoor recreation. Perhaps 180 million acres of private wildland thus may also be considered as part of the outdoor recreation space, making a total of about 3.7 acres per capita (Table 10–6). For consideration of national needs it is pertinent to note that the available lands and public facilities are unevenly distributed with reference to population density, with most of the population in the East but with most outdoor recreation space in the West. (See Table 10–7). An estimate of the developed area and future requirements is summarized in Table 10–8 and Figure 10 – 7.

Evidence accumulated by the Outdoor Recreation Resource Review Commission (1960 to 1962) indicates major increases in Outdoor Recreation by 1976 and at least doubled participation in most activities by the year 2000. The projected increases in space-requiring activities such as camping, boating, and hiking are especially significant (Table 10–9). To provide for the expected number of visits and use certainly will require a doubling or tripling of facilities and would appear to call for additional reservation of space by national, state, and local agencies.

TABLE 10–6
PUBLIC RECREATION AREA AND VISITS*

	1960 total area	Designated[1] open for recreation	Visits (millions) 1950	1960	1965
	Million acres				
National forests	186	165	27	93	150
Bureau of Land Management	464	330			27
National parks	26.5	26.5	33	72	112
Indian lands	4.7	0.3			N.A.
Wildlife refuges	17.2	9.6	4.2	10	12.5 (1963)
TVA reservoir areas	1.0	0.6	16.6	42	49.4
Corps of Engineers			16	106	168
Water surface	4.7	3.9			
Land		0.25			
Bureau of Reclamation areas			6.3	24.3	32
Water surface	1.4	1.4			
Land	7.6				
State parks	5.8	3.6	114	259	N.A.
Other state lands	48.1	32.0			N.A.
Municipal and county parks	3.5	3.5		over 1,000	
Industrial forest lands	67.0	60.0			N.A.

* Based on various departmental annual reports—forty-eight conterminous states only.
[1] In considering the acreages it should be noted that the actual accessibility of lands varies. For example, Forest Service lands, Bureau of Land Management lands, and other state and industrial forest lands, although designated open for outdoor recreation, are only partly accessible even by trail, and the facilities areas are but a small portion compared with state parks. In reality, perhaps fifty million acres of public land can be considered readily available.

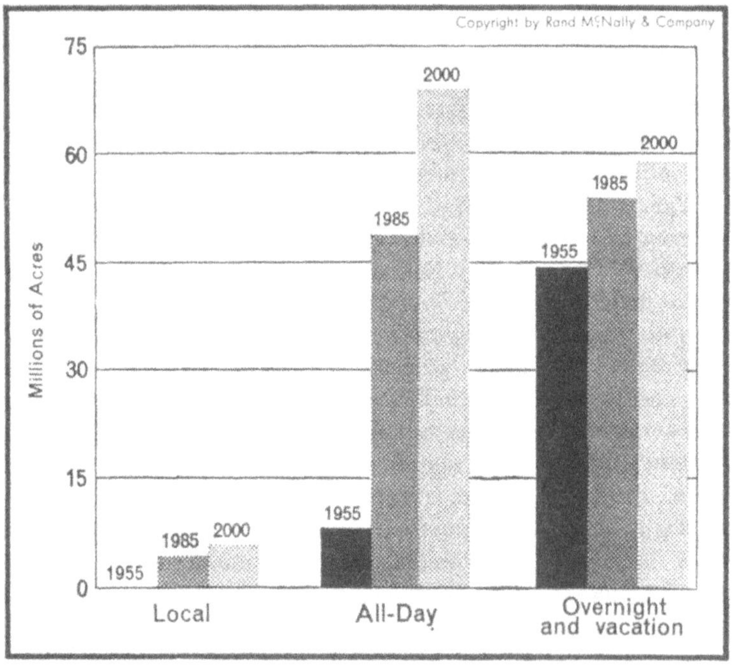

10–7 Area of recreation lands by type in 1955; projected land needs for 1985 and 2000.

TABLE 10–7

CAMPGROUND AND TRAILER AREAS AND CAPACITY BY REGIONS, 1960*

Region	Campgrounds No.	Campgrounds People capacity 1,000's	Campgrounds Per cent	Trailer camps No.	Trailer camps People capacity 1,000's	Trailer camps Per cent
Northeast	293	100	13	39	5	7
North central	1,359	153	20	93	10	14
South	1,009	107	14	237	16	22
Mountain West	1,744	126	17	67	6	8
Pacific	2,206	270	36	119	36	49
TOTAL	6,612	756	100	555	73	100

* Based on data in "Public Outdoor Recreation Areas—Acreage, Use, Potentials" (Washington: Report no. 1 to *Outdoor Recreation Resource Review Commission*, 1962), p. 29, 32.

TABLE 10–8

RECREATION USE AND AREA IN THE UNITED STATES*

	1955 Actual area (million acres)	Visits million	1985 Estimate of needed area and visits (million acres)	1985 (visits)	2000 (million acres)	2000 (visits)
Municipal and county parks, area	0.7		4		5	
Million visits		1,000		2,800		3,750
State parks and reservoirs, area	9.0		50		70	
Million visits		312		2,500		5,000
National areas	45		55		60	
(Parks, forests, wildlife refuges)						
Million visits		116		3,400		5,000

* 1955 and 1985 data based on Marion Clawson, *The Dynamics of Park Demand* (New York: Regional Plan Association, 1960), p. 37, Table 4.

The evidence of the studies indicates that conservation of recreational resources needs to be implemented and integrated on at least three levels with attention to quality, variety, and also strategic location to serve the growing urban population of the nation. First, and perhaps most important, is conservation at the local level in provision of city parks and open spaces to serve short period activities. Second is conservation at the state level for provision of regional facilities outside of the cities but within a travel time distance of one or two hours from urban areas; such facilities should provide for camping as well as for picnic grounds, trails, and beaches. Third is conservation at the Federal level to provide large recreation areas; such areas in general should be of national significance.

CONSERVING OUTDOOR RECREATION RESOURCES AT THE NATIONAL LEVEL

With the exception of the National Park Service, direct involvement of Federal agencies in outdoor recreation was secondary and a by-product of other primary responsibilities until about 1950. Then began an evolution leading to major assistance programs and direct involvement of federal agencies in outdoor recreation. The following selected Acts of Congress are indicative of the recognition of need for conservation of resources for outdoor recreation at the national level.

1952. The Cape Hatteras National Seashore was established by the Congress as authorized in 1937. This development is significant as the first

TABLE 10–9

GROWTH PROJECTION FOR SELECTED ACTIVITIES IN NUMBER OF OCCASIONS OF PARTICIPATION (PERSONS 12 YEARS OLD OR OVER)*

Sport activity	Millions 1960	1976[1]	2000[1]
Camping	60	113	235
Swimming	672	1,182	2,307
Playing outdoor games	474	825	1,666
Picnicking	279	418	700
Fishing	260	350	521
Boating	159	285	557
Nature walks	98	153	263
Hunting	95	123	174
Water skiing	39	84	189
Hiking	34	63	125
TOTAL (17 major activities)	4,377	6,926	12,449

* Based on data given in "Prospective Demand for Outdoor Recreation" (Washington: Report 26 to the *Outdoor Recreation Resources Review Commission*, 1962). p. 22.

[1] Based on present level of opportunity. With increased development (opportunity) the projection would be higher.

Federal reservation of seashore for outdoor recreation use.

1954. The Recreation and Public Purposes Act amended the General Recreation Act of 1926 to authorize sale of tracts of public land up to 640 acres to state and local governments or nonprofit associations for outdoor recreation. In 1959 the Act was extended to allow larger sized sales for state parks, and included the O and C Lands.

1956. The "Mission 66" plan of the National Park Service proposed major developments in the national parks to provide facilities required by the projected usage envisioned for the 1960's.

1957. The "Operation Outdoors" plan of the United States Forest Service proposed greatly enlarged outdoor recreation facilities to provide for the needs of the projected millions of recreation visits to national forests.

1958. The Outdoor Recreation Resources Review Commission was authorized and directed to carry out a major analysis of the nation's needs with respect to outdoor recreation. Its report, submitted in 1962, led to the creation of the Bureau of Outdoor Recreation and provided a major document on the nation's needs in space and facilities for outdoor recreation.

1961. The Open Space Program of the Housing and Home Finance Agency authorized financial assistance to local agencies for purchase of land to provide space and other outdoor recreation needs.

1962. The Federal Recreation Advisory Council was formed by Executive order to advise the President (now the President's Council on Recreation and Natural Beauty). The Council is comprised of six cabinet members and six other federal officials.

1962. The Citizen's Advisory Committee on Recreation and Natural Beauty was formed to advise the President's Council and the President. This Council is comprised of twelve private citizens.

1962. The Rivers and Harbors Act authorized the Army Corps of Engineers to restore and protect publicly owned beaches and shore areas for development of outdoor recreation.

1962. The Bureau of Outdoor Recreation was created by the Congress as a step toward focusing and intensifying efforts to assure adequate resources for outdoor recreation. The Bureau was assigned the responsibility for coordination of all Federal programs in outdoor recreation, for conducting surveys and establishing needs, and for developing a program of assistance to the states. The Bureau was also directed to prepare a National Outdoor Recreation Plan to serve as a guide to public policy and public investment in providing outdoor recreation opportunities in quality and quantity desired by the nation's citizens and in preserving the nation's natural beauty.

1964. The Wilderness Act authorized the creation of a National Wilderness Preservation System to provide a reasonable amount of enduring wilderness for future generations.

1964. The Land and Water Conservation Fund Act provided for a twenty-five year program of Federal grants to state and local areas. The Act provided that funds from the sale of surplus Federal real property, from a tax on motor boat fuel, and from user fees at federal recreation areas where facilities are provided by federal investment, be designated for assistance to outdoor recreation. Sixty per cent of the proceeds from user fees are designated for assistance to states on a fund matching basis.

1965. The Federal Water Projects Recreation Act elevated outdoor recreation potentials at Federal water projects from the status of by-product by directing that outdoor recreation, including fish and wildlife, be given equal treatment with other resource values in all Federal water projects. The Act also established the policy that non-federal agencies be encouraged to assume operation of recreational facilities, and provided for payments of up to one-half of the cost of such developments.

1965. A Wild Rivers study by a Committee representing the Secretaries of the Interior and of Agriculture was completed. Recommendations were later made to Congress for the establishment of a National Wild Rivers System.

1965. The 1965 Housing Act provided matching funds, under the Housing and Urban Development Program, for state and local areas for purchase of open spaces for parks, outdoor recreation, and for preservation of scenic and historical areas.

1965. The Food and Agriculture Act expanded the authority of the Secretary of Agriculture to permit five to ten year programs of assistance to private farmland owners for development of outdoor recreation use areas and for payments of forty per cent of the cost. The Act also authorized grants to state and local governments for acquisition of farmlands for parks and recreational purposes.

New Concepts for National Recreation Areas

In the 1960's the President's Advisory Council on Recreation recommended the establishment of a system of *National Recreation Areas* which would include appropriate areas already developed and gradually to be expanded, to meet growing needs. The term "National Recreation Areas" designates large areas that have scenic or other natural endowments of greater than local significance in quality and recreation appeal, but do not meet the standards for national park status, and are to be developed for outdoor recreation use. Such areas for Federal investment in facilities for outdoor recreation are to include national seashores, national lakesides, national riverways, and national reclamation dam areas. Various Federal agencies would be involved in their administration and management. Some of the facilities come under management by the National Park Service, some will be established with jurisdiction by the National Forest Service,[7] a few will be jointly managed, and it is probable that other agencies will become involved.

THE NATIONAL SEASHORE DEVELOPMENT

A significant movement, begun in the 1930's by conservation-minded citizens in cooperation with the National Park Service, is the effort to reserve a portion of the nation's seashore as natural areas for noncommercial public recreational use. In 1936 the National Park Service, with the aid of donated funds, began studies to identify outstanding seashore areas that ought to be conserved for public enjoyment along the Atlantic coast. A similar inventory was conducted on the Pacific coast to identify areas of national significance and to recommend a few for establishment as national seashores.[8]

In view of the obvious pressure of commercialization on the nation's seashores as a result of expanding urbanism and private shore developments, the preservation of a number of significant seashore areas for recreational use with a minimum of disturbance of the natural environment seems to be reasonable and in the national interest (See Figure 10 – 8). In this context it is significant that the Select Committee on National Water Resources, United States Senate, 86th Congress, recommended that 15 per cent of the ocean and major inland water shoreline be acquired by various levels of government for public recreation purposes.[9]

It is evident that it is already too late to reserve major areas of quality seashore without conflict and necessity of compromise between competing use values. Further delays in reservation of seashores and other outdoor recreation areas only increase the difficulties and greatly raise the cost because competing uses of land become increasingly established. The Point Reyes National Seashore acquisition is a current lesson in escalation of cost resulting from delay. The original studies of land needs and appraisals indicated acquisition costs of ten to fourteen million dollars. When Congress finally approved the seashore, fourteen million dollars was authorized for land acquisition. In the intervening time, however, land values soared, and as a result a supplementary appropriation of 44.5 million dollars was necessary. It is unlikely that the appraisers so greatly underestimated land values. The larger question involved is how much increment in land value is reasonable at the public expense when it is the result of a decision to establish an outdoor recreation area or other development with public investment?

THE NATIONAL TRAILS PROPOSAL

In 1966, a study submitted to Congress recommended the creation of a nationwide system of trails to further provide for recreational enjoyment of activities such as hiking and horseback riding. The system would be accomplished by improving, expanding, and connecting existing trails, especially in areas administered by the Secretaries of the Interior and of Agriculture. If the System were approved, Federal funds such as those from the Land and Water Conservation Act would be available to assist in constructing and improving trails. It is proposed that the System be composed of four classes of trails: (1) National scenic trails, those which are of national significance either for historic reasons, such as the Lewis and Clark Trail, or for their interregional extent, such as the Appalachian Trail or the Pacific Crest Trail; (2) Federal parks and forest trails; (3) state parks and forest trails; and (4) metropolitan trails. Wherever possible, the access trails would

10–8 ***Cape Cod National Seashore, looking south at Coast Guard Beach area.*** **This area of beaches, dunes, woodlands and marshes was authorized by Congress in 1961 to be preserved for public recreational use as part of the system of national seashores. (Courtesy National Park Service)**

lead directly from urban areas (See Figure 10 – 9).

THE WILD RIVER CONCEPT

A long-discussed proposal to preserve selected rivers or parts of rivers as wild rivers received major impetus in the 1960's. In his 1965 address to the Congress, President Lyndon B. Johnson stated, "The time has come to identify and preserve free-flowing stretches of our great scenic rivers before growth and development make the beauty of the unspoiled waterway a memory." Various study teams have evaluated many rivers, of which the following have been recommended for preservation:[10]

- Rogue River in Oregon
- Salmon and Middle Fork of the Clearwater in Idaho
- Rio Grande in New Mexico
- Eleven Point in Missouri
- Shenandoah in West Virginia
- Cacapon in West Virginia
- Alagash in Maine

THE WILDERNESS PRESERVATION SYSTEM

A category of resource for outdoor recreation has come to be known as "wilderness" to signify large areas comparatively unmodified by human occupance. In part, the establishment of the National Parks was predicated on recognition of the values and desirability of preserving wilderness, even though facilities for visitors modify large portions of them. In the National Forests, preservation of wilderness areas began in the 1920's with the setting aside of an area in the Gila National Forest as a result of the efforts of Aldo Leopold, who was then a member of the Forest Service Staff. Additional stimulus for wilderness preservation was given in 1935 by the formation of the Wilderness Society under the leadership of Robert Marshall. Although primitive areas existed in the National Parks, and the National Forest Service for many years had identified large areas for management as wilderness under the administrative terms, "primitive area," "wild area," or "wilderness," advocates of wilderness continued to press for greater assurance of preservation. In the late 1940's and especially in the 1950's, Congress seriously discussed the need for wilderness preservation, but it was not until the 88th Congress in 1964 that a bill was written that gained approval and became law (Figures 10 – 10 and 10 – 11).

One of the significant outdoor recreation accomplishments of the 1960's was the authorization of the Wilderness Preservation System under the Wilderness Act. The rationale for wilderness preservation is made clear by the following excerpts from the Act: ". . . to assure that an increasing population . . . not occupy and modify all areas . . . leaving no lands designated for preservation and protection in their natural condition, it is hereby declared to be the policy of the Congress to secure for the American people of present and future generations the benefits of an enduring resource of wilderness . . . [T]hese shall be administered for the use and enjoyment of the American people in such manner as will leave them unimpaired for future use and enjoyment as wilderness and so as to provide for the protection of the areas, the preservation of their wilderness character . . ." The Act defined wilderness to be "an area where the earth and its community of life are untrammeled by man, where man himself is a visitor who does not remain." The Act further defines a wilderness area as being at least five thousand acres, or of sufficient size to be effectively preserved and to provide an outstanding opportunity for solitude or a primitive and unconfined type of recreation.

The Wilderness Act prohibits all forms of motorized equipment, roads, and structures. Hunting and fishing are permitted according to state regulations. In concessions to commercial resource interests the Act allows grazing under regulations designed to protect the wilderness. Mineral exploration and mining both are authorized until 1984 with the restriction that the activity be carried on "in a manner compatible with the preservation of the wilderness environment." Proponents of wilderness preservation argue that mining and wilderness are not compatible under any circumstances. Only time will tell what the provision means. After January 1, 1984, all wilderness areas are withdrawn from

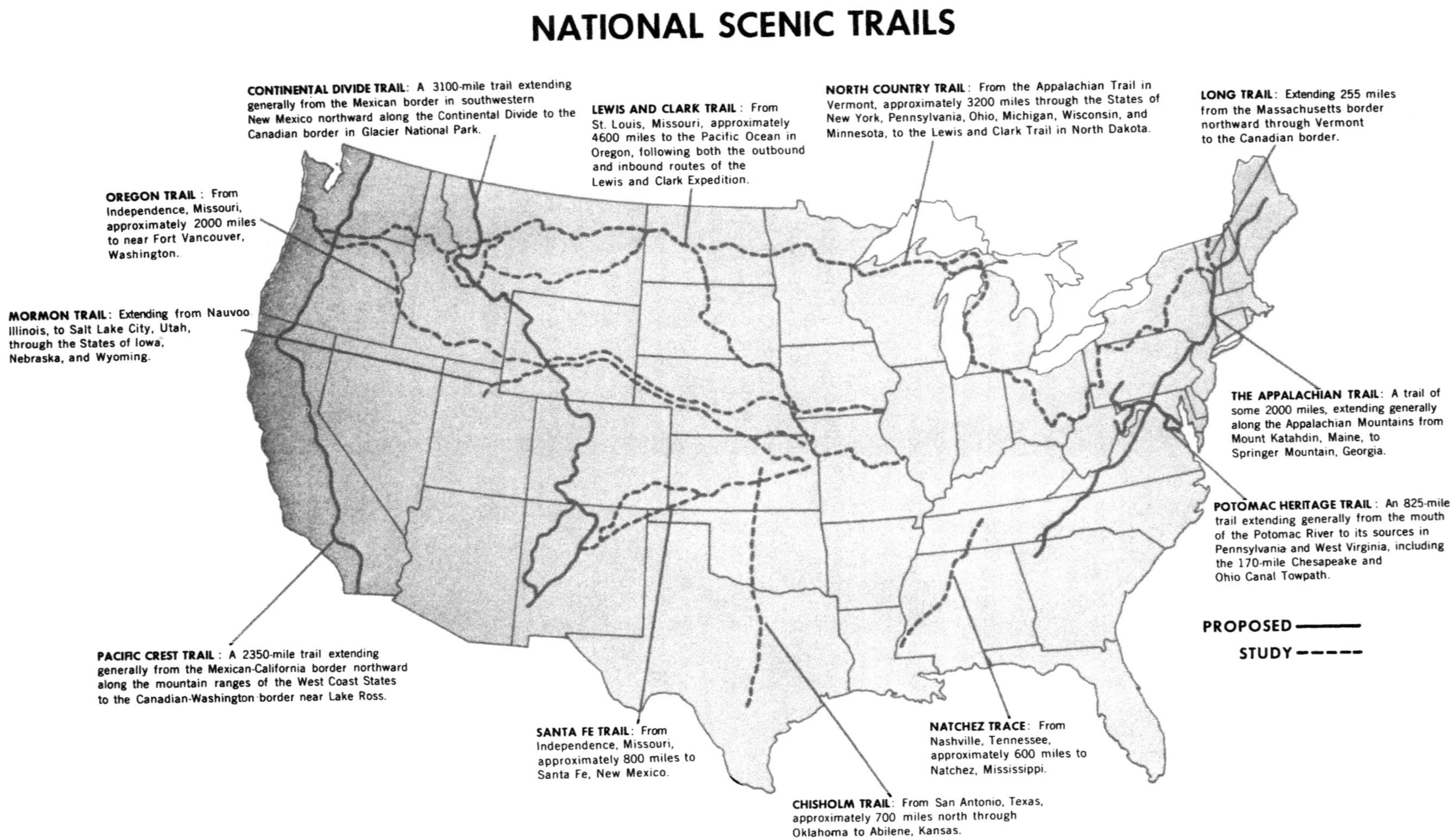

10–9 ***Proposed National Trails System.*** **To provide for recreational hiking, improvement of and integration with existing trails is proposed. Map indicates major units under consideration.**

DEPARTMENT OF THE INTERIOR Bureau of Outdoor Recreation March 1967 entry for minerals exploration, however, existing valid claims and patents will continue and mining from them is authorized. The Act modifies the mining law in that patents approved in such wilderness areas will not convey surface rights.

The Wilderness Act immediately established a National Wilderness Preservation System by designating fifty-four areas of National Forest lands totaling 9.1 million acres as wilderness, and by authorizing specifically the inclusion of the Boundary Waters Canoe Area in Superior National Forest. In addition, the Act directed that fifty-two million acres be evaluated for possible inclusion into the Wilderness System at a future time, such inclusions, however, to be authorized only with approval of the Congress (See Table 10–10).

Outdoor Recreation and National Agencies

A number of agencies have long been involved in providing outdoor recreation facilities and opportunities for the public. The National Park Service and the Forest Service were actively involved by the 1920's.[11] In the 1960's these agencies are among the most significant Federal agencies, and they have been

10 – 10 ***The enduring wilderness.*** **The conservation of wilderness areas is designed to preserve unspoiled areas such as this one in the Three Sisters Wilderness of the Oregon Cascades. The entire nation benefits from their existence as wildlife habitats, watersheds, and scientific laboratories for ecological study, as well as for recreation use by the comparatively few who can seek out the solitude of the wilderness. (Forest Service)**

10 – 11 ***National forest and the wilderness preservation system, 1965.***

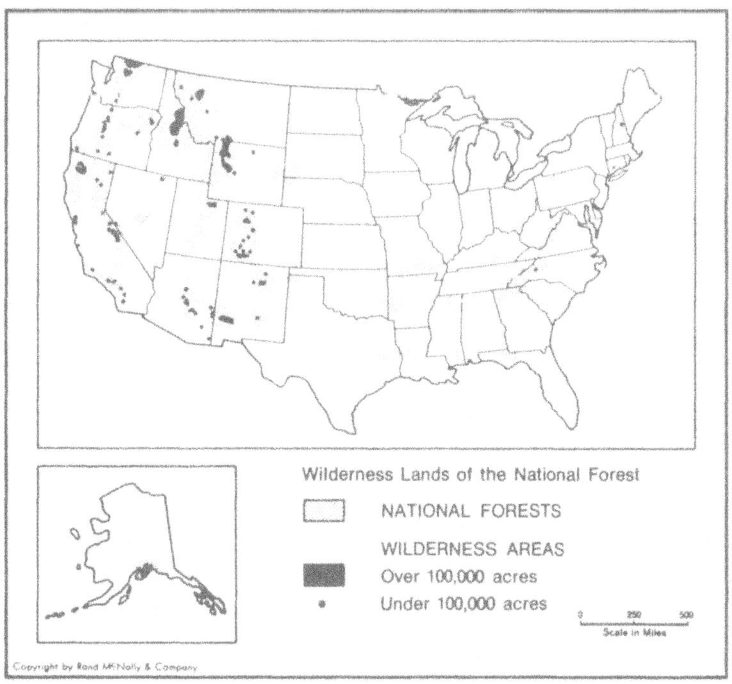

TABLE 10–10

THE ORIGINAL WILDERNESS PRESERVATION SYSTEM AND AREAS DESIGNATED FOR REVIEW*

Forest service	No.	Designated 1964	To be reviewed
Arizona	5	420,707	250,936
California	13	1,256,884	560,399
Colorado	5	280,104	554,283
Idaho	1	987,910	1,642,388
Minnesota	1	886,673	–
Montana	5	1,482,567	417,140
Nevada	1	64,667	–
New Hampshire	1	5,552	–
New Mexico	5	679,398	335,424
North Carolina	2	20,975	–
Oregon	9	665,062	86,700
Utah	–	–	240,717
Washington	3	576,589	801,000
Wyoming	4	1,780,994	586,000
TOTAL	54	9,108,082	5,474,987
National parks			22,200,000
Federal wildlife range			24,400,000
TOTAL		9,108,082	52,074,987

* Based on data from *Report of the Chief of Forest Service*, 1965 (Washington: Forest Service, Department of Agriculture, 1966), p. 8.

joined by many others in the effort for outdoor recreation.

NATIONAL PARK SERVICE

Specific participation by the Federal Government in conservation of space for outdoor recreation began in 1864 when Congress authorized a land grant including Yosemite Valley to the State of California (later included in Yosemite National Park in 1890). The Yellowstone area was withdrawn from the public domain in 1872 to become the first national park. In 1916 the National Park Service was established in the Department of the Interior with the mandate "to conserve the scenery and the natural and historic objects and the wildlife therein and to provide for the enjoyment of the same in such manner and by such means as will leave them unimpaired for future generations."

The early leaders whose vision and dedication gave character to the National Park Service included Stephen Mather, the first administrator, who is often called the father of the National Park Service.[12] Horace Albright, who served with Mather from the beginning, became the second administrator. The Park Service was assigned broader responsibilities in 1933 when museums, military parks, battlefields, and the like were added to its responsibilities. In 1936 the Recreation Act established a desirable pattern of cooperation by authorizing the National Park Service to undertake recreation planning surveys in cooperation with states. In 1962 such cooperation and coordination was delegated to the Bureau of Outdoor Recreation.

THE NATIONAL PARKS SYSTEM. The National Park Service today is responsible for 214 units, of which the most important for recreation are the thirty-two parks containing over thirteen million acres, the seventy-seven monuments with nearly nine million acres, and the eleven recreation areas containing thirty-five million acres (Figure 10 – 12). As part of its professional staff serving the public, the Park Service has nearly two hundred naturalists and archaeologists. They perform important services in the education of the public for better understanding of the many-faceted resources of the national parks through preparation of booklets, illustrated lectures, and guided tours offered at the major developed areas of the parks.

The mandate from Congress in establishing the national parks states clearly that the national parks will be held forever free from commercial commodity production, and that development will be limited to that which is needed to make the areas accessible without destroying the natural scenic attractions. Therefore, with few exceptions, the parks and monuments are not open to commercial commodity production such as mining, forestry, or grazing. Roadways within the parks and public facilities are established by the federal service. Within the parks, development of lodging facilities, food service, gasoline stations and the like are typically provided by private enterprise under concession or lease from the National Park Service. Major provisions for overnight lodging and related commercial enterprises are commonly developed adjacent to but outside of the park. Within the park, camping has long been of major importance and many areas have been developed to provide for tent campers and more recently for trailers and other motorized camping units.

INCREASING USE—PRESSURE AND NEEDS. The national parks and other areas managed by the National Park Service had small patronage in the early decades of the twentieth century. Then, with increased leisure, general ownership of automobiles, and improvement in highways, visits increased at a rate for which the Park Service was unable to prepare. In the first decade of the century, annual visits averaged less than one hundred thousand. In 1950 visits to National parks totalled fourteen million, in 1960 about twenty-seven million, and in 1965 over thirty-six million. "Mission 66" of the National Park Service, begun in 1956 and developed with 1966 as the target year (a year that fittingly marked the Golden Anniversary of the Park Service), was a major effort to expand facilities to serve increased numbers of visitors. More development will be needed, especially for overnight stays, including provision for trailers and other motorized camp units.

The increasing use-pressure on National Parks raises a serious question for conservation. How much access and development can there be without seriously impairing the basic values for which the parks were authorized? Part of the solution may well be to encourage a major share of commercial accommodation to become established adjacent to but outside of the park area. The Wilderness Act very likely will assist the Park Service in its desire to preserve some major areas, and wilderness areas in the parks may be designated. The National Parks Service has recognized the problem and is endeavoring to plan development within a hierarchy of access and use intensity. The National Parks Service Space Use Classification is as follows:

Class I and II lands—intensive development —lands reserved for administrative facilities, campgrounds, and lodges.

Class III lands—wilderness threshold—natural environment areas to be provided with moderate access and minimum development.

Class IV lands—unique features—outstanding natural areas to be preserved.

Class V lands—primitive areas—wild areas to be preserved; access limited to trails.

Class VI lands—historical and cultural areas —sites preserved for historical or cultural significance.

NATIONAL FOREST SERVICE

The National Forest Service is the other great provider for the outdoor recreation needs of the nation. The National Forests include 186 million acres, nearly all open to recreation usage by trails for hunting, fishing, camping, and hiking. Included in the forests are many well developed recreation areas,[13] and large areas withdrawn from commercial timber production to be managed as wilderness.

The greatly increased pressure of people seeking outdoor recreation in the national forests points to a need for expansion of developed facilities to handle the recreational visitors (Table 10–11). There is great need for more picnic and camp grounds, as well as for access roads and for development of more winter sports areas (Figures 10 – 13, 10 – 14). It has been esti-

TABLE 10–11
USE OF NATIONAL FORESTS FOR RECREATION*

	Number of visits		
	1950	1960	1964
General enjoyment	7,969	30,181	49,145
Picnicking	6,326	19,497	20,664
Fishing	4,885	14,535	19,358
Hunting	2,285	7,591	10,817
Camping	1,534	6,597	10,420
Winter sports	1,517	4,499	7,773
Swimming	902	2,801	3,749
Hiking and riding	635	2,081	3,263
Other	1,315	4,894	8,573
TOTAL	27,368	92,595	133,762

* Data based on several editions of *Statistical Abstract of the United States.* (Washington: U.S. Bureau of the Census, Forest Service, Department of Agriculture, annual).

mated that two-thirds of the nation's skiing is on National Forest lands managed by about two hundred private development concessions.[14] Not only is there a need to provide facilities; there is also a need to regulate recreational activities for safety, for fire prevention, for pollution control, and to appropriately accommodate outdoor recreation activities and usages in the multiple-use management of commercial forest lands.

BUREAU OF LAND MANAGEMENT

Another major land resource agency, the Bureau of Land Management, through multiple use management responsibility for 484 million acres of public forest land range has also become involved in a major expansion to serve outdoor recreation. As early as 1927 the Bureau (then the General Land Office) became involved when the General Recreation Act au-

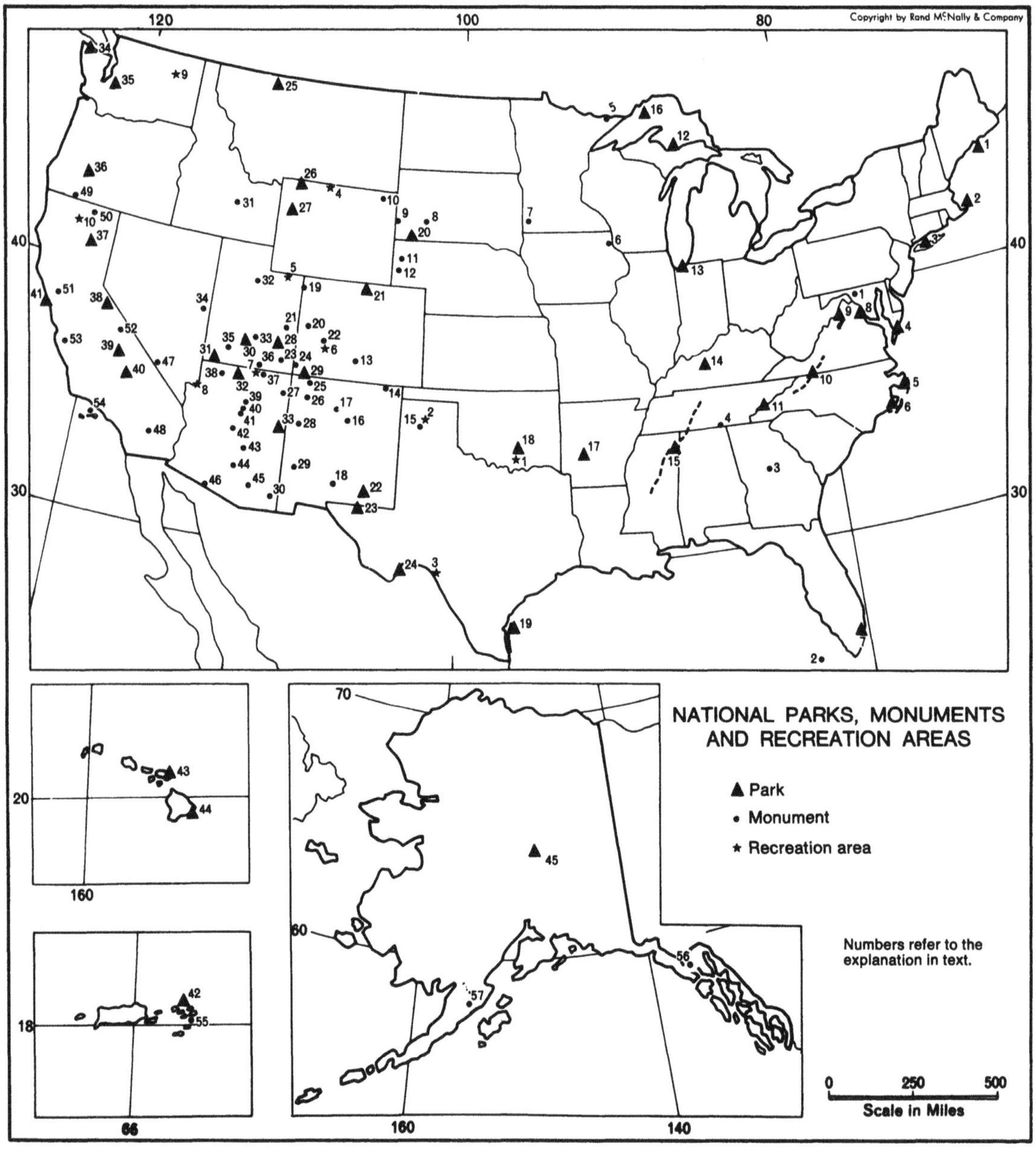

10–12 *The National Parks System,* selected monuments

National Parks (N.P.), National Seashores (N.S.), National Lakeshores (N.L.), and National Parkways

1. Acadia, N.P.
2. Cape Cod, N.S.
3. Fire Island, N.S.
4. Assateaque Island, N.S.
5. Cape Hatteras, N.S.
6. Cape Lookout, N.S.
7. Everglades, N.P.
8. George Washington, Memorial Parkway
9. Shenandoah, N.P.
10. Blue Ridge Parkway
11. Great Smoky Mountains, N.P.
12. Pictured Rocks, N.L.
13. Indiana Dunes, N.L.
14. Mammoth Cave, N.P.
15. Natches Trace Parkway
16. Isle Royale, N.P.
17. Hot Springs, N.P.
18. Platt, N.P.
19. Padre Island, N.S.
20. Wind Cave, N.P.
21. Rocky Mountain, N.P.
22. Carlsbad Caverns, N.P.
23. Guadalupe Mountains, N.P.
24. Big Bend, N.P.
25. Glacier, N.P.
26. Yellowstone, N.P.
27. Grand Teton, N.P.
28. Canyonlands, N.P.
29. Mesa Verde, N.P.
30. Bryce Canyon, N.P.
31. Zion, N.P.
32. Grand Canyon, N.P.
33. Petrified Forest, N.P.
34. Olympic, N.P.
35. Mount Ranier, N.P.
36. Crater Lake, N.P.
37. Lassen Volcanic, N.P.
38. Yosemite, N.P.
39. Kings Canyon, N.P.
40. Sequoia, N.P.
41. Point Reyes, N.S.
42. Virgin Islands, N.P.
43. Haleakala, N.P.
44. Hawaii Volcanoes, N.P.
45. Mount McKinley, N.P.

National Monuments.

1. Chesapeake and Ohio Canal
2. Fort Jefferson
3. Ocmulgee
4. Russell Cave
5. Grand Portage
6. Effigy Mounds
7. Pipestone
8. Badlands
9. Jewel Cave
10. Devils Tower
11. Agate Fossil Beds
12. Scotts Bluff
13. Great Sand Dunes
14. Capulin Mountain
15. Alibates Flint Quarries and Texas Panhandle Pueblo Culture
16. Pecos
17. Bandelier
18. White Sands
19. Dinosar
20. Colorado
21. Arches
22. Black Canyon of the Gunnison
23. Natural Bridges
24. Hovenweep
25. Aztec Ruins
26. Chaco Canyon
27. Canyon de Chelly
28. El Morro
29. Gila Cliff Dwellings
30. Chiricahua
31. Craters of the Moon
32. Timpanogos Cave
33. Capitol Reef
34. Lehman Caves
35. Cedar Breaks
36. Rainbow Bridge
37. Navajo
38. Grand Canyon
39. Wupatki
40. Sunset Crater
41. Walnut Canyon
42. Montezuma Castle
43. Tonto
44. Casa Grande Ruins
45. Saquaro
46. Organ Pipe Cactus
47. Death Valley
48. Joshua Tree
49. Oregon Caves
50. Lava Beds
51. Muir Woods
52. Devils Postpile
53. Pinnacles
54. Channel Islands
55. Buck Island Reef
56. Glacier Bay
57. Katmai

National Recreation Areas

1. Arbuckle
2. Sanford
3. Amistad
4. Bighorn Canyon
5. Flaming Gorge
6. Curecanti
7. Glen Canyon
8. Lake Mead
9. Coulee Dam
10. Whiskeytown-Shasta-Trinity

thorized sale of public domain lands to states and counties for public outdoor recreation use. Various other acts involved the Bureau of Land Management in conservation of wildlife and development of the range to enhance outdoor recreation for hunting and fishing.

The Bureau of Land Management began development of major facilities for outdoor recreation in Alaska in 1959, but since Alaskan statehood most of the facilities have been transferred to state administration. In 1960 a program of facilities development for outdoor recreation was begun in Oregon, and since 1965 regular appropriation for facilities in support of outdoor recreation have been made.[15]

FEDERAL WATER DEVELOPMENT AGENCIES IN RECREATION

As a result of the many outdoor recreational activities and their relationship with a variety of environments, it was inevitable that many other federal agencies would become involved in provision for outdoor recreation. The most notable, perhaps, are agencies that deal primarily with water resource development: The U.S. Army Corps of Engineers, the Bureau of Reclamation, and the Tennessee Valley Authority.

The *Corps of Engineers* became directly involved with outdoor recreation at reservoirs built for flood control and power developments, as well as in beach control projects. Its first specific authority for investment in recreation came in 1944 under the Flood Control Act, but recreation then was only a minor consideration. By 1960, however, the Corps had invested about twelve million dollars directly in outdoor recreation supporting facilities, and it was estimated that private enterprise had invested about ten times as much in associated developments at Corps of Engineers reservoirs. As early as 1961 the Chief of the Corps of Engineers directed that recreation facilities and use be given full consideration.[16]

The *Bureau of Reclamation* also has become directly involved through reclamation reservoirs which are open for a variety of water activities.[17] Because the Bureau of Reclamation is not primarily concerned with outdoor recreation, the development and management of facilities for camping, picnicking, and water access have typically been delegated to other agencies, notably the National Park Service and the Forest Service. Outstanding examples where recreation responsibility has been delegated include the recreation facilities at Lake Mead, Lake Powell, and Flaming Gorge on the Colorado; Grand Coulee on the Columbia; and the Whiskeytown-Shasta-Trinity National Area.

The *Tennessee Valley Authority* is another example of a major outdoor recreation resource through multiple-use development by public investment. TVA dams have created about 600,000 acres of water surface with about 10,000 miles of shoreline. Access to the waters is provided by several hundred boat docks and nearly a hundred parks. In 1965 there were 11,000 private residences on lake frontages. Nearly two hundred million dollars had been invested by non-TVA enterprises related to the waters, about a third in boats and facilities. In the mid 1960's TVA began development of a major demonstration center for national recreation and conservation education in a 170,000

10–13. *Winter sports area developed in a national forest.* **Skiing, by nearly five million people, has become an important recreational activity requiring greatly expanded development of facilities including tows and chair lifts, lodgings, and automobile parking areas.**

acre area to be known as the "Land Between the Lakes" (so named because the area is in fact between Lake Barkley on the Cumberland and Kentucky Lake on the Tennessee). Heavy usage is reflected by the 49.4 million recreational visits in 1965.[18] (Figure 10 – 15).

OTHER AGENCIES

A number of other agencies have a lesser or less direct involvement, but are increasingly assisting in provision for outdoor recreation. The *Bureau of Indian Affairs* through administration of Indian lands has long been concerned because lands have been open to hunting and fishing. Some notable outdoor recreation developments have been made on Indian lands. The *Bureau of Sports Fisheries and Wildlife* is directly involved because its twenty-eight million acres of wildlife refuges have become popular outdoor recreation areas, and increasing provision is being made for picnicking, hiking, and camping. Other agencies involved less directly include the *Federal Power Commission* through licensing development projects, *the Rural Community Development Service* through aid to local area development including provision for outdoor recreation, the *Soil Conservation Service* through technical assistance in the small watershed program which provides half cost payment for recreation facilities development, the *Economic Research Service* and the *Cooperative State Research Service* through assistance in basic research to stimulate outdoor recreation development in rural areas. The *Federal Cooperative Extension Services* increasingly are developing and expanding educational programs involving outdoor recreation, and many states now provide one or more outdoor recreation specialists.

THE BUREAU OF OUTDOOR RECREATION. The creation of the Bureau of Outdoor Recreation was a significant outgrowth of the 1962 report of the Outdoor Recreation Resources Review Commission, and no doubt in part of the multiplicity of involvements in outdoor recreation by Federal, state, and local agencies. Unlike most of the agencies already noted, the Bureau of Outdoor Recreation is not an operating agency, but rather is concerned with planning, advising, and coordinating. Its specific task is to evaluate the nation's needs for outdoor recreation facilities and to develop a plan for adequate conservation through coordinated effort and investment by local, state, and Federal agencies. The Bureau also is charged with the task of managing the Land and Water Conservation Act Funds, which requires review and approval of state programs before fund matching support is authorized.

State Provision of Regional Facilities

It is generally accepted that the states have responsibility for the provision and develop-

10 – 14 *Camping in the forest.* Scenes such as this are common at Forest Service campgrounds and in national and state parks. In 1965 it was estimated that total camper-days for all areas was about 40 million. (Forest Service)

ment of outdoor recreation facilities of a regional service character.[19] These may be defined in general as scenic and recreation areas of less than national significance which serve the needs of metropolitan population centers for day and weekend outdoor activities. Such facilities include roadside parks and picnic grounds, ocean and lake beaches, and in most states major camping areas. (In 1962 state parks in the nation recorded a total of 284 million camper days out of a total of 400 million visits.)

EXISTING STATE PARKS

Since the beginning of state parks over a century ago, the responsibility for providing regional outdoor recreation facilities has increasingly been accepted by the states.[20] Today there are about 2,500 state parks which include 5.7 million acres of land (Table 10–12). Maintenance and supervision of these parks employ nearly eight thousand workers year-round, and about ten thousand more in the summer season; annual state park expenditures are about 110 million dollars. During the past decade there has been major development, yet nearly all state facilities are overtaxed during the main season of use. (See Figure 10 – 16 for the pattern of state investment in outdoor recreation.)

NEED FOR ADDITIONAL STATE PARKS

If outdoor recreation activities expand as projected, there is reason to believe that states need to plan for additional areas and facilities development. At present the states vary in provision, with nine providing more than the average thirty acres per thousand population, and twenty-three providing less than ten acres per thousand population (Table 10–12).

In reality, the common need appears to be for reservation of quality outdoor recreation space strategically located for access by urban populations to provide for day and weekend outings, and for more facilities for trailers, camping, picnicking, and related activities (Figure 10 – 17).

10 – 15 ***Launching a pleasure boat, TVA Kentucky Lake facility.*** **Kentucky Lake, 184 miles long, is one of the nation's large man-made lakes that help provide for water oriented outdoor recreation. Over 100 such boat launching ramps have been provided on this lake alone, and in 1966 were used by about 12,500 boats. In 1965 it was estimated that American families owned about seven million recreational boats. (Photo courtesy of Tennessee Valley Authority)**

Local Provision of Recreation Land

The local recreation facilities may be defined as parks, playgrounds, refuges, and beaches within city boundaries. Their provision is thus the responsibility of the municipal government. It is self-evident that such recreation areas are used more intensively, by more people, and generally through a longer season than either regional or national areas. New York's 840-acre Central Park is an example of forethought and of the desirability of preservation of open space in cities. In 1844 William Cullen Bryant, editor of the *New York Evening Post*, conceived and promoted the idea which became Central Park. The city paid about seven million dollars for the land. Today its assessed value is more than five hundred million dollars and, had the land not been set aside over a century ago, no amount of money could provide it now. As Stuart Jones says, "The abiding miracle of Central Park is the simple fact that it exists... Serenely it offers harassed millions its one great

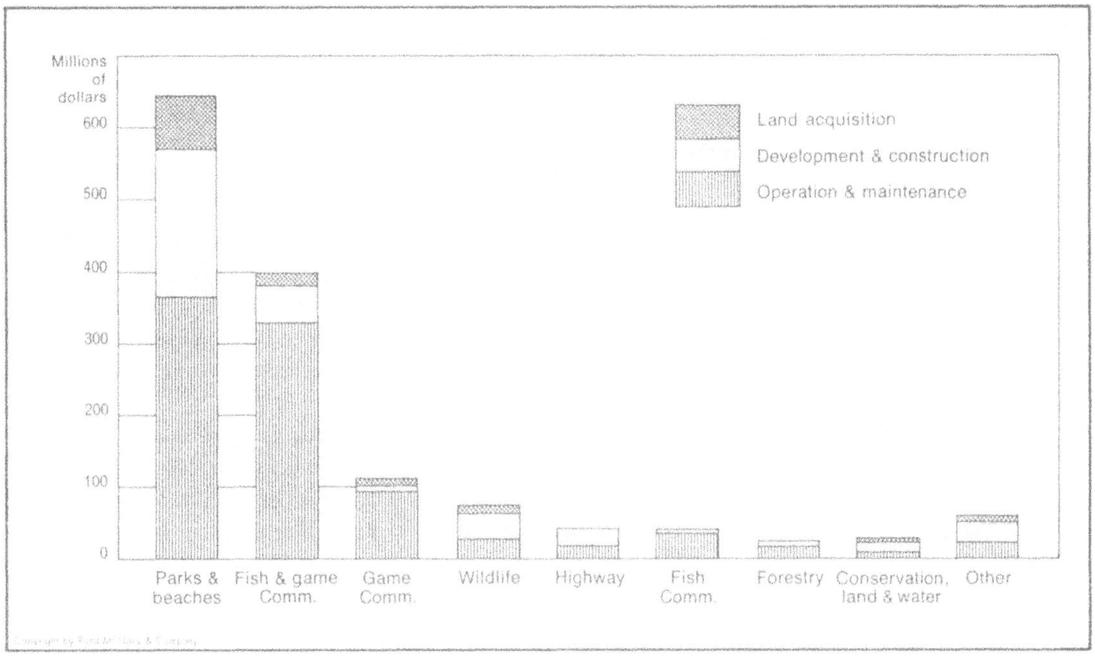

10 – 16. ***State financing of outdoor recreation by categories, 1951-60.*** **("Public Expenditures For Outdoor Recreation," [Washington: *Report* 25 to the Outdoor Recreation Resources Review Commission 1962] p. 15.)**

10.– 17 *Roadside picnic site.* In most states the motoring weekend or vacation tourist may enjoy picnic facilities. They range in size from the small roadside resting spot and single picnic table to large areas with camping facilities for hundreds. (Oregon Highway Department)

gift—room to breathe, refuge from the pressures and tensions of big city living" (Figure 10 - 18).[21]

EXISTING LOCAL RECREATION SPACE

A century ago cities seldom provided parks or open space for recreation, although the town square as open space was characteristic of many eastern cities. Urbanization, as it is known now, is recent, and with the growth of cities the need for more adequate open space and park provision has intensified. In 1880 only about half of the cities of one hundred thousand or more had even one park. Now all have parks. In total, there are recorded over 10,000 city parks totaling nearly 400 thousand acres.

Data assembled for the fifty largest cities of the nation indicated that only half meet the proposed standard of either 10 per cent of the city area or of ten acres per thousand population (Table 10–13). It should be kept in mind that total acres and averages cannot fully evaluate the adequacy of recreational space either in reflecting the quality of facilities or their distribution with reference to population density patterns. Moreover, areas and populations of corporate cities fail to reflect the realities of modern urban centers, because most centers are conurbations.

City administrators and planners fortunately have become increasingly sensitive to the need for outdoor recreation facilities accessible by walking or by public transportation and to the

TABLE 10–12

ACREAGE OF STATE OWNED PARKS AND RECREATION AREAS, 1962*

State	Parks	Area per 1,000 population	Annual visits: Total	Annual visits: Per acre
	Acres	Acres	1,000	Number
States with over 65 acres per 1,000 population				
Wyoming	161,664	479.7	1,043	6.5
Maine	212,423	216.3	788	3.7
New York	2,589,122	146.2	34,201	13.2
South Dakota	74,884	101.6	5,235	69.8
Tennessee	250,000	67.7	4,623	18.5
States with over 30 acres but less than 65 acres per 1,000 population				
California	707,566	40.2	27,810	39.3
Idaho	25,561	35.8	829	32.4
Oregon	64,689	35.4	12,458	192.5
Nevada	11,123	30.2	272	24.5
Minnesota	100,864	28.8	3,287	32.6
Oklahoma	66,857	26.9	10,250	153.2
Washington	79,959	26.2	9,135	114.2
West Virginia	44,532	25.0	1,922	43.2
Vermont	9,631	24.6	865	90.1
Michigan	188,754	23.2	14,125	74.8
Florida	121,776	21.5	3,911	32.1
South Carolina	48,663	19.6	3,255	66.8
Kentucky	60,285	19.5	8,532	141.5
Missouri	75,381	17.4	8,097	107.4
Pennsylvania	169,793	14.9	23,533	138.6
Montana	10,526	14.9	441	42.0
Indiana	54,502	11.6	3,082	56.5
Georgia	46,119	11.1	3,665	79.5
Alabama	36,184	10.8	3,136	86.6
Arkansas	20,120	10.8	4,141	206.0
Iowa	28,710	10.3	7,578	264.0

TABLE 10–12 (Continued)
ACREAGE OF STATE OWNED PARKS AND RECREATION AREAS, 1962*

State	Parks	Area per 1,000 population	Annual visits	
			Total	Per acre
	Acres	Acres	1,000	Number
State with under 10 acres per 1,000 population				
Rhode Island	8,500	9.6	3,021	355.4
Ohio	94,601	9.3	22,789	240.9
Hawaii	6,420	9.2	501	78.3
Connecticut	22,222	8.3	4,848	218.4
North Carolina	38,427	8.1	1,829	47.6
Maryland	25,664	7.8	6,695	260.5
Colorado	14,535	7.4	1,276	87.8
Delaware	3,495	7.3	24	6.8
Virginia	30,890	7.1	1,267	41.0
North Dakota	4,426	7.0	570	129.5
Nebraska	10,095	6.9	4,166	412.5
Massachusetts	34,343	6.6	1,757	51.2
Texas	66,616	6.4	7,116	106.8
Mississippi	14,400	6.3	2,024	140.5
Wisconsin	22,204	5.5	5,323	239.8
Illinois	42,936	4.7	9,574	223.0
New Mexico	4,591	4.5	1,492	324.3
Kansas	9,592	4.3	1,989	207.2
New Jersey	23,802	3.7	6,536	274.5
Arizona	5,801	3.7	246	42.4
Louisiana	12,240	3.6	1,573	128.9
New Hampshire	1,336	2.1	2,438	1,824.9
Utah	1,320	1.3	949	719.0
TOTAL	5,763,142	32.0	284,795	49.4

* Computed from data in State Outdoor Recreation Statistics–1962 (Washington: Bureau of Outdoor Recreation, Department of the Interior, 1963).

desirability of making city environments attractive. Many modern industrial plants are setting good examples through landscaping open spaces around their facilities.

NEED FOR ADDITIONAL LOCAL AREAS

The combination of increasing population, increasing participation in outdoor activity, and continuing growth of urban complexes certainly indicates the need to conserve open spaces and recreation areas in cities. In 1960 cities of ten thousand population or more accounted for one hundred million people, indicating a need for one million acres of park and open space, but data on city parks record only 400,000 acres. The indication is that city parks and recreation areas are inadequate now, not to mention the needs of the future (Figure 10 – 19).

To assure parks, playgrounds, and other open spaces in the future, it would appear that every city ought to have regulations requiring minimum open spaces and recreation areas in newly developing urban areas. It is not just quantity that is needed; there must also be wisdom in selection of site to best serve the growing neighborhoods. A most favorable sign for the future is the widespread acceptance by city and county governments and by the citizenry of the desirability and need for professional planning. The day is coming when every city and county will have its planning department, and there will be continual reappraisals guiding our growth. Zoning, which implements planning, also is being increasingly accepted as a means by which people are benefited through assuring better living environments. In the context of conservation of open space for recreation, it is notable that planners today believe the only way to preserve open spaces effectively is for cities to buy the desired land outright. An alternative to land purchase, which is now receiving considerable attention, is the practice of acquiring easements to assure conservation of open space. The increasing use of Federal-city cooperation in urban renewal projects to replace degraded areas with planned neighborhoods, including parks and other open spaces, is another development that will improve our cities.

The city is the apex of its region, an expression of the genius of man, and utilitarian in serving the needs of modern society. Cities will continue, and they almost certainly will grow larger, but with foresight and especially with planning for parks, playgrounds, and other open spaces, cities can be and ought to be attractive.

Conclusion

All who enjoy the nation's areas of unspoiled natural scenery, its parks and open spaces, are the beneficiaries of the foresight and effort of men such as Theodore Roosevelt, John Muir, Stephen Mather, Gifford Pinchot, and Horace Albright. Much has been accomplished since the establishment of Yellowstone National Park. The national park system has been expanded, and the "Mission 66" program implemented. The Forest Service, through its

10–18 ***Central Park, New York.*** **An outstanding example of foresight. (New York City Park Department)**

"Operation Outdoors," has intensified efforts to provide recreational facilities of all sorts in keeping with its multiple-purpose mandate. Recreation facilities at Federal and private reservoirs have become common and are heavily used. The Bureau of Outdoor Recreation has been created, and recreational potentials are being given equal consideration in Federal projects. State and municipal parks are recognized as essential, and are gaining local tax support. There has been notable expansion of public recreation facilities provided by private utilities and forest industry. Indeed, the record in conservation of resources for outdoor recreation is one of outstanding accomplishment—yet more is needed.

The philosophy and rationale for conservation of open space and of quality scenic and wilderness environments were concisely expressed in 1959 by General of the Army Omar Bradley when he said "Each of us has need to escape occasionally from the noisy world which surrounds us and find refreshment in the grandeur of nature. Yet, year after year, our scenic treasures are being plundered by what we call an advancing civilization. If we are not careful, we shall leave our children a legacy of billion dollar roads leading nowhere except to other congested places like those they left behind. As the pressures of civilization mount, it would seem to me that we probably have as much need for part-time Thoreaus as we have for full-time nuclear scientists. Since the beginning of mankind, people have always drawn great strength from their nearness and kinship to nature. If we

TABLE 10–13

URBAN RECREATION AREA IN SELECTED LARGE CITIES, 1960*

City	Acres	Acres per 1,000 population	Per cent of city area
Over 10 acres per 1,000 population			
Phoenix	22,757	51.8	19.0
Denver	15,927	32.2	35.1
Fort Worth	9,586	26.9	10.7
Portland, Oregon	7,109	19.1	16.5
San Diego	8,054	14.1	6.5
Tulsa	3,634	13.9	11.9
Dallas	8,808	13.0	4.9
Omaha	3,540	11.7	10.8
Minneapolis	5,533	11.5	15.3
Louisville	4,100	10.5	11.2
Cincinnati	5,059	10.1	10.2
Less than 10 acres per 1,000 population but over 10 per cent of city area			
Washington, D.C.	7,531	9.9	19.2
New York	36,663	4.7	18.2
Baltimore	7,052	7.5	13.9
San Francisco	3,774	5.1	12.4
St. Louis	4,692	6.3	12.0
Boston	3,455	5.0	11.3
Philadelphia	9,378	4.7	11.5

TABLE 10–13 (Continued)

URBAN RECREATION AREA IN SELECTED LARGE CITIES, 1960*

City	Acres	Acres per 1,000 population	Per cent of city area
Below 10 acres per 1,000 and less than 10 per cent of city area			
Honolulu, Hawaii	2,914	9.9	5.4
Columbus	4,242	9.0	7.4
Oklahoma City	2,882	8.9	1.4
Kansas City	4,100	8.6	4.9
Los Angeles	19,856	8.0	7.7
Oakland	2,928	8.0	8.6
Indianapolis	3,762	7.9	8.3
Memphis	3,948	7.9	4.8
Toledo	2,362	7.4	7.7
Dayton	1,903	7.3	8.8
Long Beach	2,461	7.2	8.4
Rochester	2,103	6.6	9.0
Seattle	3,500	6.3	6.2
St. Paul	1,907	6.1	5.7
Atlanta	2,500	5.1	3.0
Cleveland	4,500	5.1	8.7
Houston	4,604	4.9	2.2
San Antonio	2,827	4.8	2.8
Norfolk	1,427	4.7	4.5
Pittsburgh	2,274	3.8	6.6
Birmingham	1,256	3.7	2.6
Detroit	6,106	3.7	6.8
Miami	900	3.1	4.1
New Orleans	1,708	2.7	1.3
Akron	600	2.1	1.7
Chicago	7,627	2.1	5.3
Buffalo	1,069	2.0	4.2
Tampa	249	0.9	0.5
TOTAL	263,167	7.0	7.8

* Excerpted from *Recreation* Vol. 58, No. 1 (National Recreation Association, 1965), p. 21.

close off this source of strength by plowing under our scenic resources, we may soon find ourselves so baffled by the pressures of urbanization, that we risk damage to our character as people and therefore to our institutions."[22]

The future portends needs for recreational space which are difficult to comprehend because of the projected magnitude. The need is not only for space in quantity, but is also for preserving variety and quality. The need is to provide opportunity for outdoor recreation in the interests of maintaining high-level national health and character and of fostering appreciation of nature as exemplified in the nation's outstanding areas of natural phenomena and scenic resources. The need is to reserve strategically located quality areas of seashore, sports areas, picnic and camping areas at the regional level, and playgrounds and open spaces in cities.

The conservation of adequate variety and quantity of recreational land is a matter of national, state, and local concern requiring action now; for every year of delay, the conflicts with competing land-uses grow and provision of resources for outdoor recreation and open spaces becomes more difficult and expensive.

10 – 19 *Kensington Metropolitan Park.* This outstanding multiple use facility was developed to serve the Detroit urban area. The 4,300 acre site provides for swimming, boating, fishing, golfing, nature trails, picnicking, and winter sports. Martindale Beach is in the foreground and Maple Beach on the far side of the 1,200 acre lake formed on the Huron River. Intense use is indicated by the photograph. On a single day 55,000 visits have been recorded; in 1965 over two million visits were recorded. As of 1966 about six million dollars had been invested in the development. Other than for golfing no fees are charged. (Courtesy of the Huron-Clinton Metropolitan Authority)

FOOTNOTES

[1] For reviews of measurement techniques and some of the problems see the following: Marion Clawson, "Methods of Measuring the Demand for and Value of Outdoor Recreation," *Resources for the Future*, Reprint no. 10 (Washington: Resources for the Future, 1959), 36 pp.; Andrew N. Trice and Samuel Wood, "Measurement of Recreation Benefits," *Land Economics*, vol. 34, no. 3 (1958), pp. 195-208. (This contains a good review of the problem and offers some suggestions for methodology); Phillip O. Foss, "Problems in Federal Management of Natural Resources for Recreation," *Natural Resources Journal*, vol. 5, no. 1 (May, 1965), p. 6394; Robert K. Davis, "Recreation Planning as an Economic Problem," *Natural Resources Journal*, vol. 3, no. 2 (October, 1963), pp. 239–49; Marion Clawson and Jack L. Knetsch, "Outdoor Recreation Research: Some Concepts and Suggested Areas of Study," *Ibid.*, pp. 250–75.

[2] For example see Floyd K. Harnston, Richard E. Lun, and J. Richard Williams, *A Study of the Resources, People and Economy of Teton County, Wyoming.* (Laramie: Univ. of Wyoming, 1959), 97 pp. This is an illuminating study on the impact of developing an area for recreation.

[3] *First World Conference on National Parks.* (Washington: National Park Service, Department of the Interior, 1962), p. 471.

[4] The Select Committee on National Water Resources of the United States Senate, 86th Cong., made the following recommendation: "That Congress adopt the policy that the recreational potentialities of all Federal multipurpose reservoirs are to be developed for public use." "Water Resources Activities in the United States" *Committee Print No. 17* (86th Cong., 2nd Sess.), (Washington: Government Printing Office, 1960), p. 2. In 1965 the Federal Water Projects Recreation Act implemented this recommendation.

[5] It is significant that the 1965 Housing Act provides strong support for assistance to state and local governments in acquiring open space for outdoor recreation uses. In addition, voters of a number of states have authorized substantial bond issues for purchase of lands to be preserved as open space; for example, New York authorized fifty billion dollars and Maryland, one billion dollars.

[6] Some two thousand areas were designated by the National Forest Service, and over one million dollars collected in user fees during the summer of 1965. In total, by the end of 1965, eighty-three million dollars had been received into the fund; twenty-eight million dollars from motor boat fuel tax, 6.8 million dollars from user fees and forty-eight million dollars from sales of surplus Federal property. Over seven thousand areas had been designated for 1966.

[7] For example, in 1965 the first all Forest Service area, the Spruce Knob-Seneca Rocks National Recreation Area, including 100,000 acres, was established by action of Congress. In 1966 the Mount Rogers National Recreation Area in Virginia was authorized within Forest Service administration.

[8] See "Our Vanishing Shore Line" (Washington: National Park Service, Department of the Interior, 1955), 36 pp., and "Pacific Coast Recreation Area Survey" (Washington: National Park Service, Department of the Interior, 1959), 207 pp.

[9] "Water Resources Activities in the United States." *loc. cit.*

[10] A Wild River Bill was introduced into the 88th Congress, but was not acted upon.

[11] For valuable historical background on federal involvement in outdoor recreation see "Recreational Resources of Federal Lands" (Washington: National Conference on Outdoor Recreation, 1928), 141 pp.

[12] Robert Shankland, *Steve Mather of the National Parks* (New York: Alfred A. Knopf, Inc., 1951), 326 pp.

[13] In 1965 the Forest Service reported 7,229 developed sites occupying 41,000 acres and capable of safely providing for 425,000 persons (255,000 tent camps, 41,000 trailer campers, and 125,000 picnickers).

[14] In 1965 national forests reported that there were 199 developed areas covering 46,000 acres, providing over 600 tows, and capable of handling nearly 300,000 skiers at one time. *Report of the Chief of the Forest Service, 1966.* (Washington: Forest Service, Department of Agriculture, 1967), p. 16.

[15] As of 1965 the Bureau of Land Management had about 150 sites in nine states with eleven thousand acres dedicated to outdoor recreation facilities, plus over 100,000 acres on lease for recreation use. Visits were estimated at twenty-seven million visitor days. *Public Land Statistics 1965*, (Washington: Bureau of Land Management, Department of the Interior, 1965), p. 71.

[16] In 1965 there were 350 Corps of Engineer reservoirs in forty-four states providing a total of four million acres of water surface available for recreation uses; four hundred state and local parks and six hundred swimming beaches occupy a quarter million acres of reservoir shoreland.

[17] As of 1964, the Bureau of Reclamation reported 1.4 million acres of water surface and ten thousand miles of shore with over two hundred developed acres and facilities sites. Over thirty-two million visits were recorded.

[18] Annual Report of the Tennessee Valley Authority, 1965 (Knoxville: Tennessee Valley Authority, 1965), 245 pp.

[19] It is significant that the Outdoor Recreation Resources Review Commission reports (and later the Federal assistance program administered by the Bureau of Outdoor Recreation) strongly supported the concept of state responsibility. Technical and financial assistance by Federal agencies, however, has been authorized in unprecedented amounts.

[20] The development of state park systems probably began in 1865 in California with Mariposa Park which later became part of Yosemite National Park. The second notable state park was Niagara Falls State Park, established in 1885.

[21] Stuart E. Jones, "Central Park, Manhattan's Big Outdoors," *National Geographic*, vol. 18, no. 6 (December, 1960), pp. 781–811.

[22] Quoted in *First World Conference on National Parks*, (Washington: National Park Service, Department of the Interior, 1962), p. 106.

11

Fresh Water Resources

"THE WATER REQUIRED for daily living in the United States weighs more than 100 times as much as all other materials consumed including food, fuel, metals, plastics, lumber, sand, gravel, and stone."[1] The myriad ways in which water directly and indirectly enters our daily lives are attested to by the above quotation. The Greeks considered it one of the fundamental elements; centuries later, early scientists considered it to be the basic substance from which all matter was made. It is certain that life itself as we know it is dependent upon water (Figure 11 – 1). It is basic to the formation of the protein molecule and is essential to photosynthesis; without it no seed will sprout.[2] Most living things include large amounts of water; 71 per cent of man's body weight is water. The very origin of life on this planet is believed to have occurred in the sea; eons later nature's highest form of life, man, still reflects this origin in the saline character of his body fluids.

Water is unique in its physical character. Despite its very real significance in the gaseous state as a part of the atmosphere, it is dominantly present on earth as a liquid, its most usable physical state. This is the result, in part, of its heat of vaporization, which is distinctly higher than that for most liquids with similarly simple molecules.[3] Its solid state, ice, is lighter than the liquid state; hence, lakes, rivers, ponds, etc. freeze from the top down. Almost no other substance has this property, yet its significance to life in water is paramount. The role of water in the atmosphere and as a temperature regulator needs no elucidation here, but does deserve mention. The cohesive and adhesive properties

11 – 1 *Water: essential to all life.* The Stehekin River in the northern Cascade Mountains illustrates that water has beauty as well as utility. (David Simons)

of water are notable and important. Its ability to wet surfaces and to rise by capillarity is basic to the retention of water by the upper levels of soils in which plants grow, and to the use of water by plants in general.

Water is more intricately interwoven with the several facets of the total environment than any other natural resource. The effect of significant variation in the supply of water in a given area is felt by many components of the environment, often in a chain reaction fashion. Variation in available water can reduce the type of protective vegetation covering watersheds, thereby increasing runoff which in turn increases the possibility of erosion, thus affecting soils, stream wildlife, landforms, and the water supply itself. Where man seeks to utilize water for a given purpose, he finds that a multitude of additional purposes to which it is already being put are also affected. Hence conservation of water resources is perhaps the most complex conservation problem of all.

The Nature of the Water Resource

The United States is well endowed with water resources (Figure 11 – 2). The national annual average precipitation of thirty inches represents some 4,300 billion gallons a day.[4] Although somewhat more than two-thirds of this amount soon is returned to the atmosphere (Figure 11 – 3), the annual runoff of nearly 1,200 billion gallons per day still is over four times the annual withdrawal of 270 billion gallons per day.[5] Withdrawn water is here considered to be any water which is taken from its natural setting except for water which is only momentarily diverted as in passing through hydroelectric installations. In 1960 a little less than one-fourth of the water withdrawn was consumed (*i.e.*, eliminated from immediate reuse),[6] a fact which further indicates the adequacy of this basic resource. Overall averages and summary statements are of considerable value for planning, but they leave much

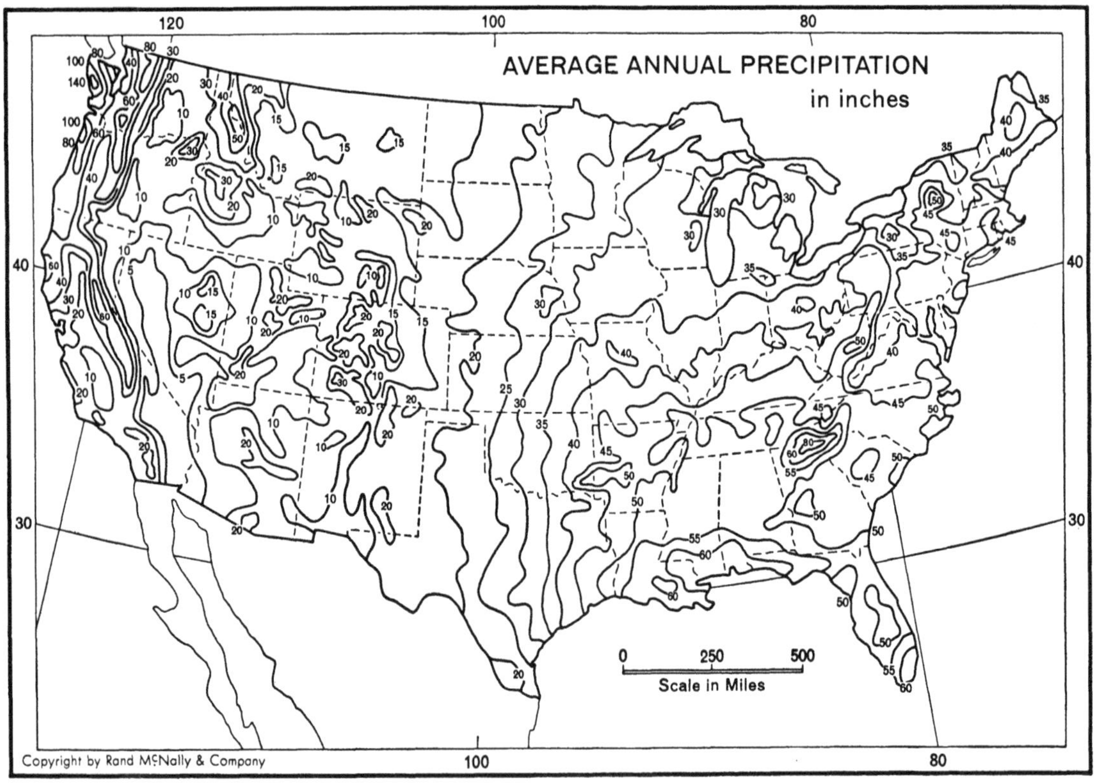

11 – 2 ***Average annual precipitation in the United States.***

to be desired in consideration of local or seasonal situations. The United States has locally experienced water shortages within the past decade on a number of occasions. Most streams have a variable flow from season to season. For the nation as a whole, for example, an estimated two-thirds of the annual runoff passes rapidly back to sea under flood or near-flood conditions. There are, therefore, times during the year when stream flow is significantly below average. Then too, the "dry years" play their part in making up the overall averages and must be considered along with the "wet years." Indeed, the usual situation is one of above or below average annual precipitation on the local level. The early and mid-1960's reminded easterners, especially residents of the largest northeastern cities, that drought still occurs. The near return to dust bowl conditions in the south-central states during the early and mid-1950's and the floods on the west coast in 1964 are indicative of the inconstancy of the water supply.

Aside from natural water shortages, man, as a result of his insistence on settling in semi-arid and arid regions, has created water shortages where they need not have occurred. Unlimited withdrawal from ground water supplies with limited recharge has caused wells to go dry, water costs to rise, streams to dry up, saline waters to invade coastal locations, and a host of other problems (Figure 11 – 4). Literally mining the ground water of the High Plains in Texas, man is faced with the possibility of having an impressive investment in land and equipment become disastrously reduced in value within a few decades as water supplies are exhausted. Man-made water shortages are not restricted to the arid and semi-arid parts of the nation, however. As urban centers grow with expanding national population, water supply facilities become insufficient and urban areas in crowded parts of the East find their efforts to extend their supply lines meeting opposition from neighboring communities with similar

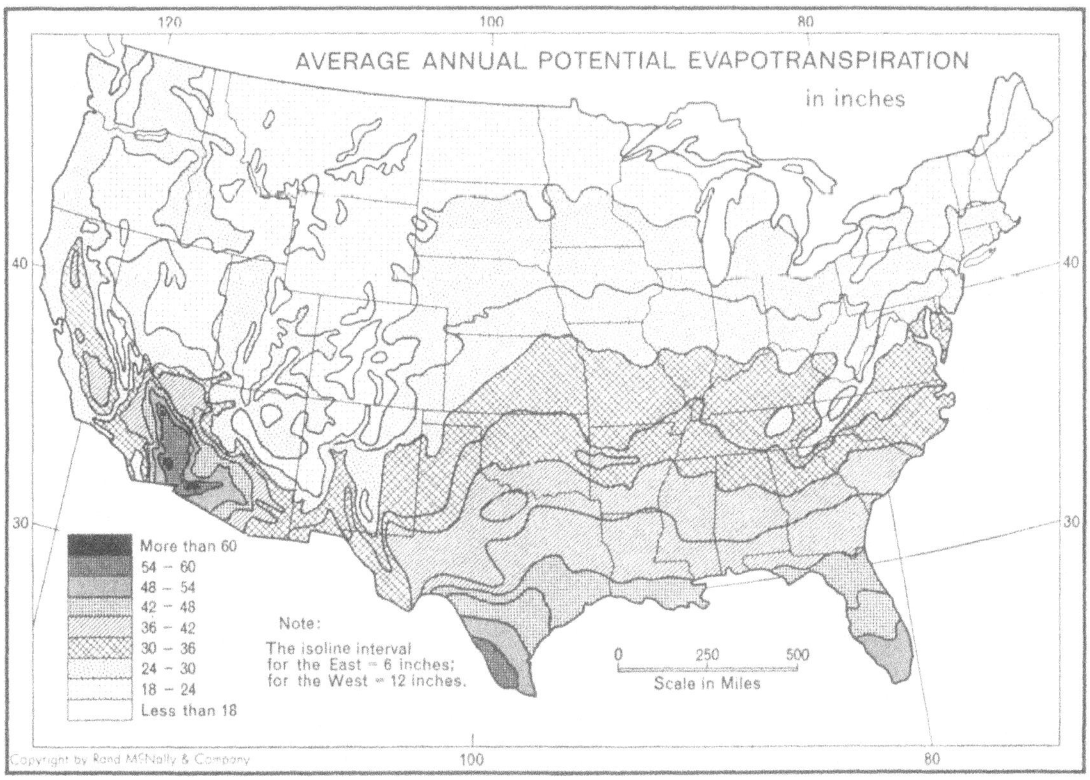

11 – 3 ***Average annual potential evapotranspiration in the United States.*** **Compare this map, which indicates water need, with Figure 11 – 2 and with Figure 11 – 10. (After Thornthwaite)**

problems. Inadequate planning is often responsible for needless water problems.

There is not a national water shortage currently; the total supply is more than adequate. The distribution of water over the United States is uneven, however, and locally critical water shortages exist. As the prospect for a doubled population in less than fifty years appears real, the need for more careful assessment of the nation's water resources and planning for the most efficient use of these resources is self-evident.

THE HYDROLOGIC CYCLE

The water resources of the earth's surface may be divided into two general categories: Salt water found predominantly in the oceans, and fresh water which is a resource of the continents. Man, a land dweller, makes an overwhelmingly greater direct use of the water on the land. Yet the two are not finally separable because the fresh water has been, and will be again, salt water. Through the action of energy from the sun and gravity, the earth's water moves endlessly from the sea to the land and then back to the sea again (Figure 11 – 5). Utilizing the transportation medium of the atmosphere, water evaporated from the natural reservoirs of the earth (the oceans) moves over the land where it is released by condensation and precipitation. Having left its salt behind at the point of evaporation, and having changed from gas back to liquid during condensation, the water is in its most usable form for man. Although man may succeed in trapping some of it temporarily for the increasing variety of his uses, water inexorably makes it way back to the sea, some by rivers, but most by the same medium which brought it, the atmosphere. This restless movement of water has neither beginning nor end; it is a complete cycle and is called the *hydrologic cycle*.

An estimated eighty thousand cubic miles of water are evaporated from the oceans, and somewhat more than fifteen thousand cubic miles are evaporated from the land and its associated water bodies annually.[7] Although the bulk of the moisture taken into the atmosphere is returned directly to the sea by subsequent precipitation, an estimated twenty-four thousand cubic miles fall on the land surfaces. Of the water released over the land as a result of cooling due to lifting, a portion does not reach

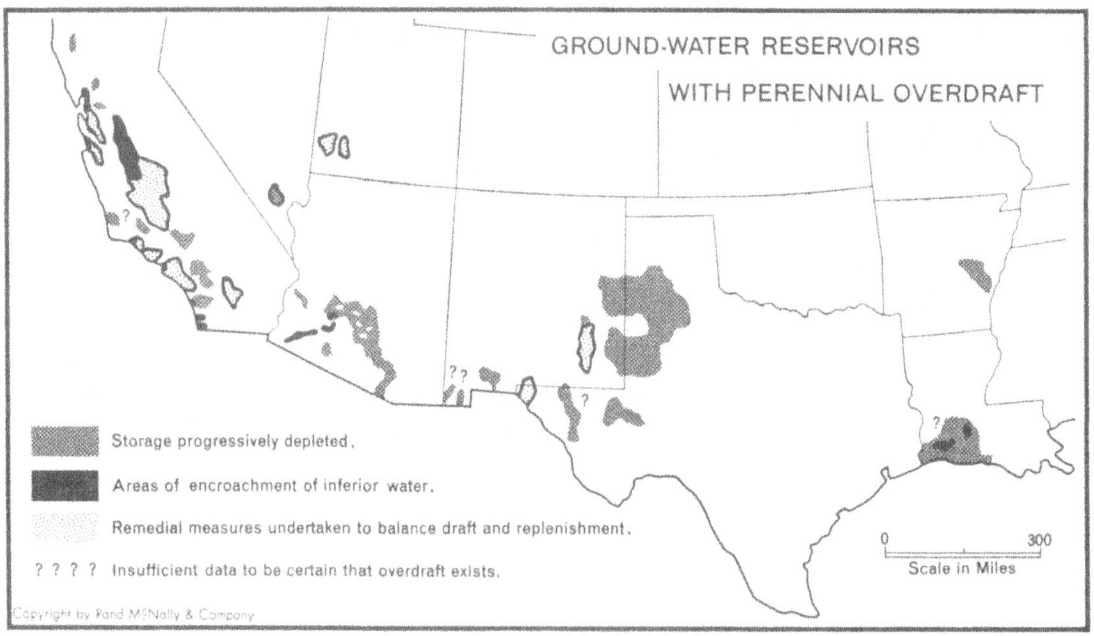

11 – 4 Areas of perennial overdraft of ground water. (*Water, the Yearbook of Agriculture*, [Washington: Department of Agriculture, 1955.] p. 70.)

the ground. Some returns to the vapor state in falling and some is intercepted by trees and evaporates from foliage. Interception has been measured in several instances, and the loss is estimated to have represented between 5 and 15 per cent of the annual precipitation.[8] Other estimates are much higher, the amount being quite variable depending upon vegetation type, intensity of precipitation, and the like. In amounts that are increasingly being recognized as significant, drip from fog-shrouded foliage and dew adds to the basic source of water, precipitation.[9]

Almost as though reluctant to stay, the water reaching the earth's surface immediately begins its return to the sea in one or more ways. Some precipitation falls into bodies of standing or running water and may return to the atmosphere through evaporation or to the sea as stream discharge. Of that portion reaching the ground, some may be shed by the surface if precipitation intensity and surface character warrant, but some will penetrate the soil, entering the spaces between soil particles and becoming subsurface water. The soil is wet progressively deeper as water continues to enter the soil surface. The uppermost layer of the soil, that portion from which plants are directly supplied, is called the "soil water zone." As this topmost zone becomes filled, some water begins to move downward into the intermediate zone, held away from the saturated zone below by dominance of molecular attraction over gravity. The intermediate zone is the connecting link by which water leaving the saturated soil water zone may add to the ground water supply. Addition to the top of the intermediate zone results in release of water from the bottom of it

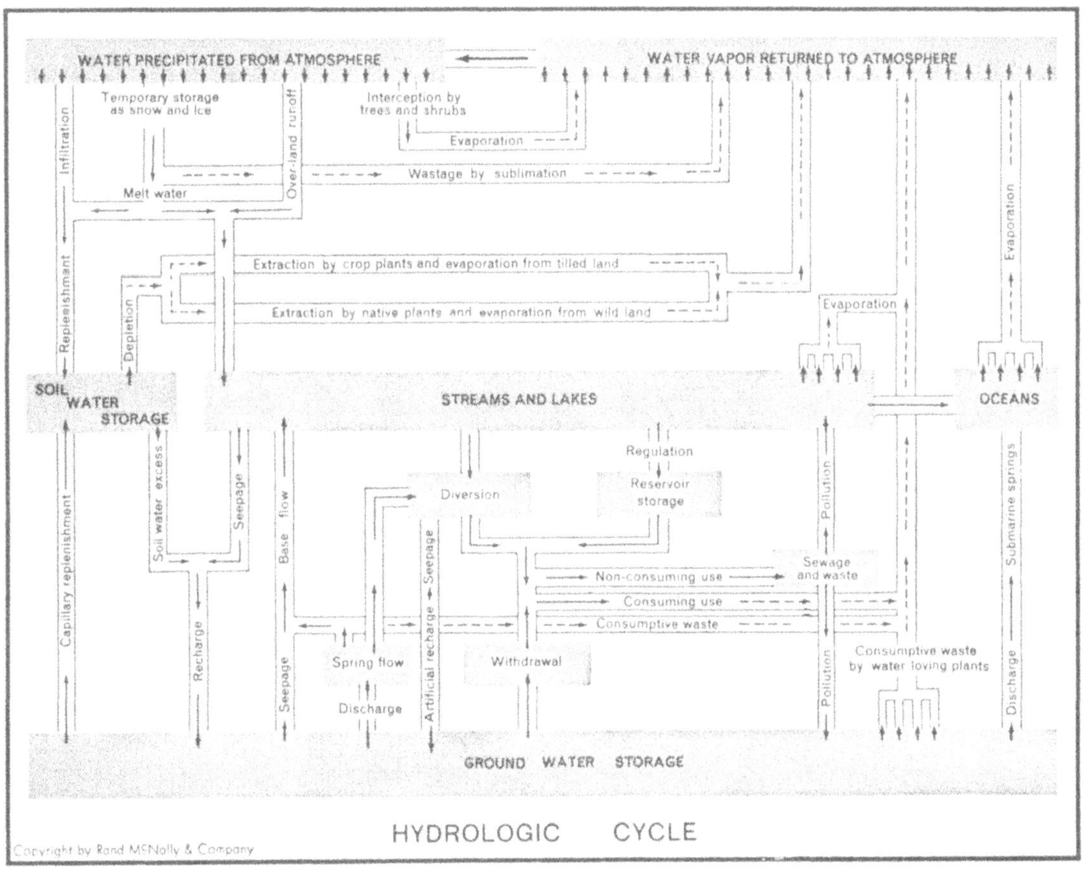

11–5 *The hydrologic cycle.* **Solid-flow lines indicate water movement as liquid; dashed lines, movement as vapor. (House Committee on Interior and Insular Affairs, 1952)**

to the capillary fringe (water being held above the zone of saturation by capillarity).

Thus precipitation at last adds to the vast reservoir underlying most of the land in amounts which, in the United States, have been estimated at as high as thirty-five times the average annual national runoff.[10] This figure, even if correct, is misleading from the resource point of view in that much of the ground water is unobtainable. Even so, this great subsurface body represents the natural storage which supplies all of our wells and maintains the flow of humid-area streams and springs during dry periods. The upper surface of the saturated zone, the water table, is at variable depth, but in humid areas approximates the contours of the land surface. The saturated zone extends downward into the rocky crust as far as there is pore space or a crack for the water to occupy, in some instances over a mile.

Of the several zones mentioned, it is the uppermost and lowermost from which water moves directly back toward the sea. Water from the soil water zone evaporates and also is put back into the atmosphere by transpiration. Transpiration and evaporation from freshwater bodies and the land collectively are believed to return to the atmosphere some 70 per cent of all the precipitation falling on the United States (Figure 11 – 6).[11] Permanent streams and springs indicate places where the water table intersects the earth's surface. Here the natural reservoir of ground water supplies the stream as water moves down the slope of the water table to the lowest local level, the stream bed. Stream flow, although the most obvious mode of return of water to the ocean, actually carries only about one-fourth of the annual precipitation.

Thus the hydrologic cycle makes fresh water available to the several forms of land life and at the same time maintains the level of the seas. Man has caused minor changes in the cycle, some deliberately and to his benefit, others un-

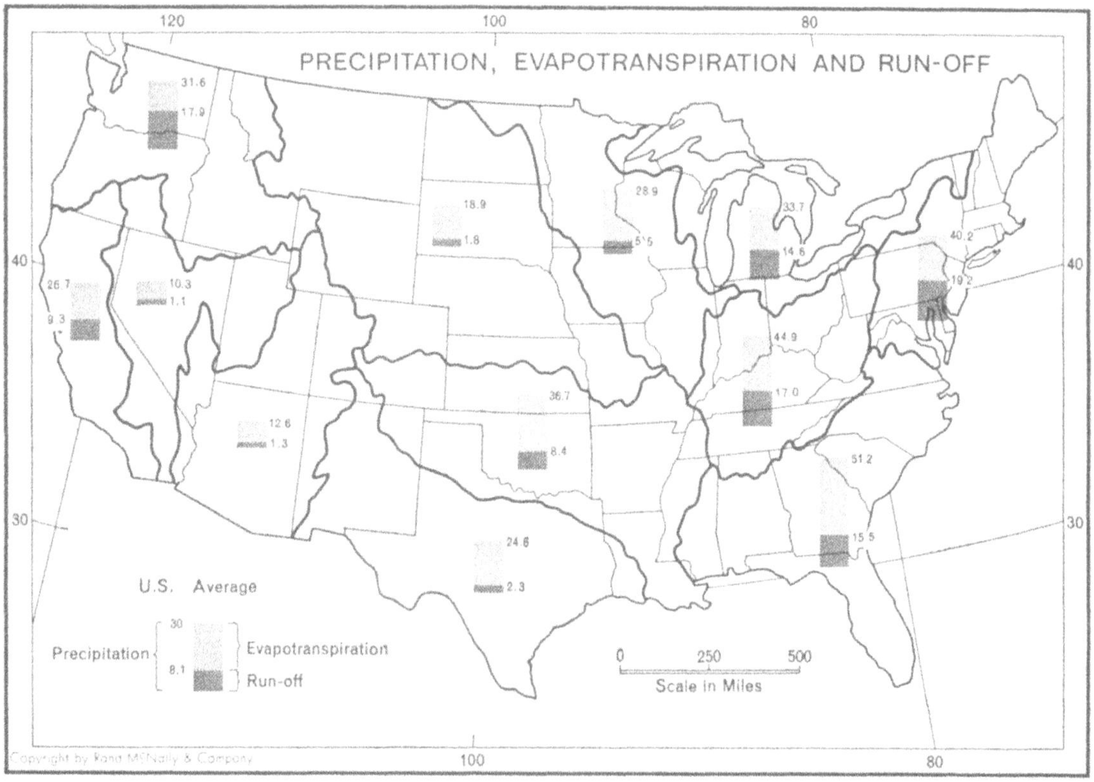

11 – 6 *Precipitation, evapotranspiration, and runoff in the United States by major drainage basins.* Contrast the West with the East. (House Committee on Interior and Insular Affairs, 1952)

wittingly and distinctly to his detriment. Locally he has slowed the cycle by decreasing the rate of runoff with dams; in other places agriculture and forestry have removed the protective mantle of vegetation so helpful in retarding the movement of surface water. In the latter case not only is more rapid runoff abetted and thus erosion also, but the opportunity for the surface water to become subsurface water is lessened. Through irrigation in areas of marginal precipitation he has hastened the return of ground water to streams and to the atmosphere, removing the water from natural storage at rates more rapid than recharge can compensate. By terracing, contouring, planting shelter belts, cloud seeding, and many other diverse schemes he has sought to hold the water on the land a bit longer or make the amount a little greater. His successes have been modest, and his failures have been only temporarily significant; the natural hydrologic cycle has been little affected. But certainly its complete understanding is essential for a conservational adjustment.

WATER RESOURCE TYPES

SURFACE WATER. The water in lakes, streams, and reservoirs comprises the surface water supply. Of the 270 billion gallons withdrawn per day during 1960, 220 billion gallons per day (over 80 per cent) were taken from surface water supplies.[12] All of the several major categories of water use (public supplies, industrial, irrigation, power, and transportation) are served chiefly by surface waters. Industrial water supply comes overwhelmingly from surface sources. During 1960, 95 per cent of the self-supplied water used by industry was surface water. This figure represents an increase of 4 per cent since 1955. Such supplies are especially desirable where considerable volume is required at a rapid rate of flow. Lakes, reservoirs, and large rivers thus are sought by industry as well as for metropolitan water supply systems. Surface water supplies generally have lower mineral content than ground water supplies in the same area, an important factor for many industrial processes as well as for irrigation and many other uses.

Surface waters, nevertheless, vary considerably in their mineral content. Some parts of the United States are noted for the softness of the water and thus are preferred by certain industries. In general, the central part of the country has most of the surface water areas with the greatest amount of dissolved salts (notably calcium), and the East coast and Pacific Northwest are notable as areas which have unusually soft surface water (Figure 11 – 7).

The proportion of surface water to ground water used for all purposes is lowest in the drier parts of the nation. Ground water rises in importance in the southern Great Plains and the Southwest. Here supplies of surface water are

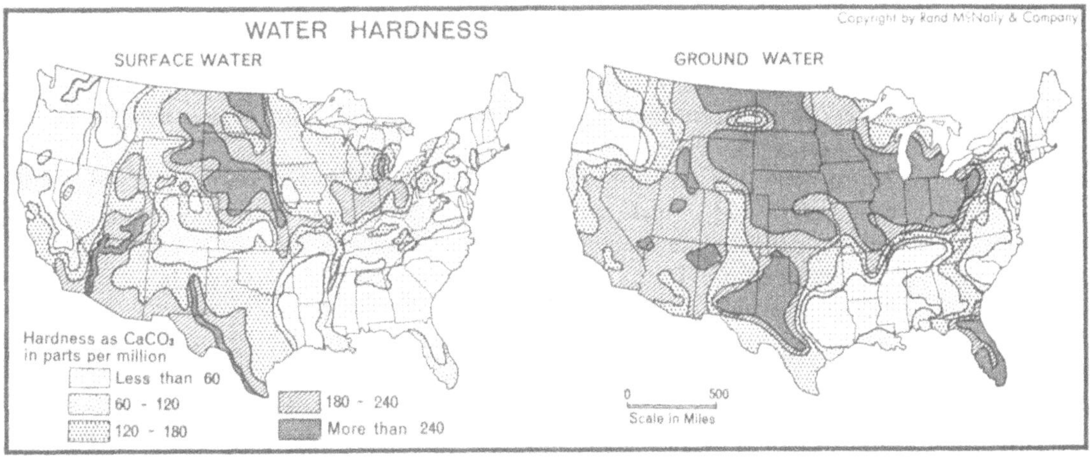

11 – 7 ***Water hardness in the United States.*** **Note the general similarity of pattern but greater concentration of salts in ground water. (After Ackerman)**

limited, and ground water must supplement or replace surface water as a source of supply. Despite this fact, California, which annually withdraws more water for purposes other than water power than any other state, used some twenty billion gallons per day of surface water in 1960. An additional eleven billion gallons per day or more were supplied from ground water sources. Texas, the third-ranking water-use state, is a notable example of a state in which ground water use dominates, the figures for surface and ground water withdrawal in 1960 being 5.6 and 9.1 billion gallons per day respectively. In the East, New York led in water use with a total withdrawal of fourteen billion gallons per day in 1960, of which thirteen billion gallons per day were derived from surface supplies.[13]

Surface waters have associated natural problems other than dissolved mineral content, notable among these being irregularity of volume or flow and sedimentation. Although man is responsible for much of the suspended material the streams transport and deposit by reason of unwise land-use practices, he is by no means responsible for all. Erosion is a geologic process which goes on in some degree with or without man's aid. Aside from creating problems in reservoirs and hydroelectric installations, undue concentrations of suspended materials may render surface water supplies useless for many purposes or make their restoration inordinately expensive. Even the finest-textured particles can damage the intricate machinery common to the industry of today.

Some irregularity of flow or volume is to be expected from natural water sources; however, the extremes of irregularity may constitute literally a matter of life or death. In areas which suffer from a dearth of natural water supplies, crops, wildlife, and domesticated animals may pay the supreme price for inadequate preparation to meet the vagaries of nature. In the case of the opposite extreme (floods), man may also pay with his life. An analogy has been suggested between floods and weeds: A flood is an occurrence of water out of place. In point of fact the waters are not so much out of place as the people who, for what seem to be sound economic reasons, insist on living next to the mean position of the stream borders. The maximum extent of the waters during peak flow periods causes great damage. It was pointed out earlier that an estimated two-thirds of the annual runoff returns to the sea during flood periods. The conservationist seeks to play the middle against both ends—to control floods by storage of excess water and utilize the excess to alleviate shortages.

GROUND WATER. That water which fills the cracks and pore spaces below the water table found at variable distances beneath the surface is known as ground water. It contributes about 17 per cent of the water withdrawn for public, rural, self-supplied industrial, and irrigation uses, amounting to forty-seven billion gallons per day during 1960. Of the several categories of use, ground water supplies are proportionately most important to rural use. Rural use is here considered to be use in homes not served by public water supply systems. Ground waters supplied over three times the amount furnished by surface waters for rural use in 1960, the amounts being 2.8 billion and 850 million gallons per day respectively.[14] From the standpoint of volume, however, use for irrigation ranks highest. As mentioned earlier, drier parts of the country dominate in this respect. Ground waters are typically free of significant amounts of suspended materials, but dissolved materials may be even more prevalent than in surface waters. The area having hard water in the United States is significantly larger for ground water than for surface water. Although the pattern is similar to that for surface water, each of the several categories of increasing water hardness occupy greater areas in the case of ground water.

The ground water resource of the United States constitutes an enormous natural reservoir far exceeding the annual use in quantity on a national scale (Figure 11 – 8). In many parts of the country, however, withdrawals exceed recharge, and shortages of ground water already exist locally. In certain areas, awareness of the situation has resulted in some effort to restrain unlimited withdrawal; but the fact must be faced that with continuous use natural recharge can never restore the ground water supply to anything resembling its original status. A considerable proportion of the ground water underlying the nation is not obtainable economically or physically with present methods. As water use in general increases, so will ground water use. In the future much more attention will need to be given to the management of the supply to assure that expansion of settlement and industry will have a firm water resource base upon which to rely.

SOIL MOISTURE. There is an additional water supply category which occupies an intermediate position between ground and surface water. Soil moisture, existing as it does in the pore spaces of the soil, is beneath the surface and thus not surface water. Because it is above the water table, neither is it properly termed ground water. This category of water resource is not comparable to either of the other two in terms of volume, but its importance to both natural and cultivated plants far outweighs its volume status. Because of its special importance to cultivated plants, soil moisture is considered in Chapters 4, 5, and 6.

Water Use in the United States

Water use in the United States during 1960 involved an estimated 2,270 billion gallons per day which were withdrawn or were temporarily diverted from streams or lakes.[15] The fact that this figure exceeds the supply which was available (assumed to be equal to the average annual runoff of nearly 1,200 billion gallons per day) is understood when it is recognized that, of the total, two thousand billion gallons per day were used for hydroelectric generation. Obviously much of this water served other purposes before it returned to the sea or atmosphere. The figure would be much higher if it were possible to compute and add transporta-

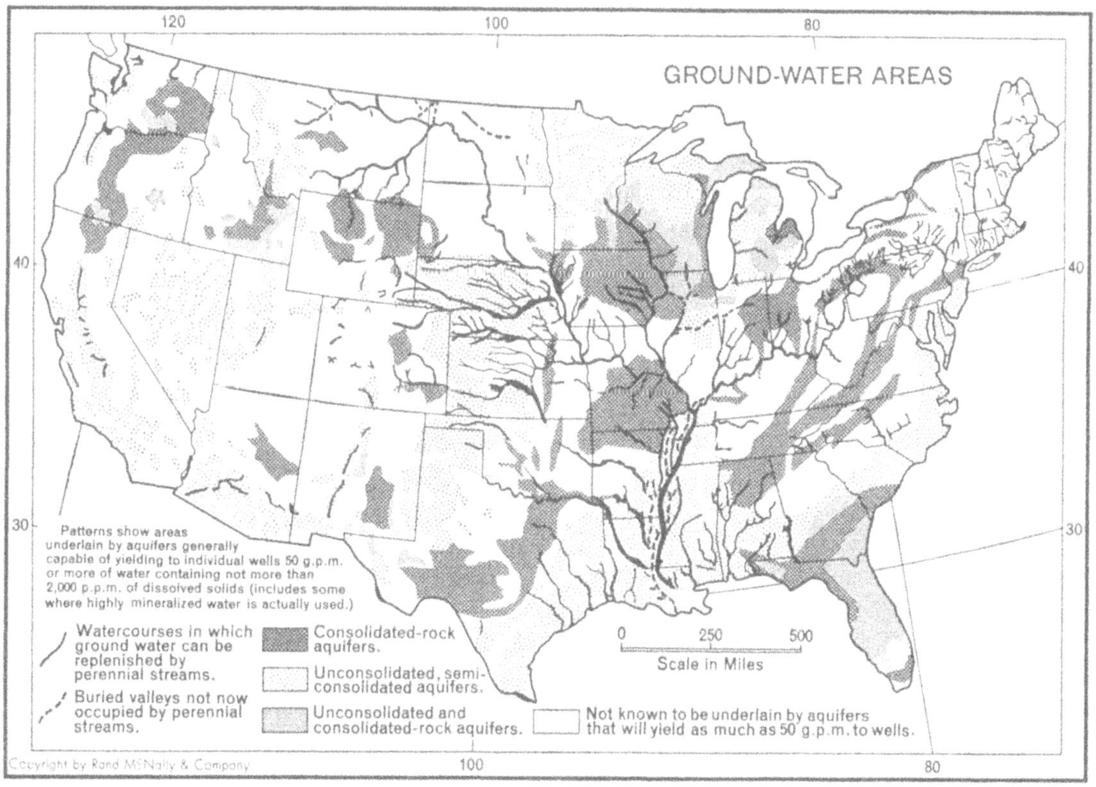

11 – 8 *Ground water supplies in the United States.* The area west of the Rocky Mountains, an outstanding user of ground water, has the least impressive supplies. (After Thomas)

tation and recreational uses. Furthermore, the withdrawal figure of 270 billion gallons per day does not represent the consumptive use of water, some of the water having been re-used and thus counted more than once. Water moving along a river from which many cities draw their supplies commonly is reused several times before it completes the return trip to the sea or becomes water vapor. Consumed water, returned to the atmosphere during use or incorporated into products, is estimated at sixty billion gallons per day during 1960, of which 85 per cent results from irrigation.

TABLE 11–1
WITHDRAWAL WATER USE, 1960*

(Billion gallons per day)	
Public supplies	21.0
Rural	3.6
Irrigation (including conveyance loss)	110.0
Industry (self-supplied)	140.0

* After Kenneth A. MacKichan and J. C. Kammerer, "Estimated Use of Water in the United States, 1960," Geological Survey Circular 456 (Washington: Government Printing Office, 1961), p. 39.

DOMESTIC USE. Food, shelter, and clothing are commonly listed as the three basic needs of man; but unless water is considered food, the list is incomplete. The water need is more immediate than the need for food. Despite the loss of half of its protein and nearly all of its fat, an animal can continue to live; loss of one-tenth of its water results in death.[16] Thus domestic water supply receives high priority attention; water use from public water supply systems in the United States during 1960 averaged 151 gallons per person per day and served about 136 million people.[17] This is a misleading figure, however, because it includes a host of commercial and industrial uses with which the individual himself has no direct contact. The figure of twenty to eighty gallons per capita per day for household use by people living in homes with running water is more appropriate, even though it includes such nonessentials as lawn watering.[18] Quantitatively, public water supply is one of the least significant categories of water use despite its fundamental importance (Table 11–1).

URBAN USE. About 17 per cent of the water withdrawn for public supply is consumed. A major consumptive use is for the watering of lawns and gardens wherein evapotranspiration eliminates much of the water from re-use. Urban dwellers today must have collectively an annual water bill of well over five hundred million dollars.[19] The estimated one gallon per day drinking water for the average individual is but an infinitesimal part of this. A much larger proportion is used for cleaning and waste disposal. A study of the use of hot water alone by an average family of five revealed the following monthly use totals: 2,562 gallons in the dishwasher, 1,908 in the laundry machine, 4,830 for bathing and 3,208 in the sink. Hot water in this instance represented only one-fourth of all water used.[20]

RURAL USE. The rural category of withdrawal water use is the smallest category primarily because so few people are involved. Also, an estimated one-fourth of the rural homes are still without running water;[21] water use per capita in such homes is quite low. It may be surprising that stock watering does not boost the total substantially; water requirements of livestock, however, are modest in comparison to the demands placed upon the nation's water resources by the activities associated with the other categories of water use. Milk cows and beef cattle have some of the highest use rates, ranging from ten to thirty gallons per day per head. Even so, domestic use of water is believed to have slightly exceeded that used by livestock in rural areas in 1960. Indeed, a study in one state indicates that over four times as much water evaporates from stock watering ponds alone as the livestock of the state consume.[22]

PROBLEMS. There are numerous problems being encountered daily by the municipal agencies charged with the responsibility of maintaining adequate public water supplies, many of which are caused by pollution and mineral content. In 1963 there were 5,831 communities, serving nearly thirty-six million

Americans, which discharged untreated sewage, discharged inadequately treated sewage, or which had no sewage collection or treatment system at all.[23] In addition, private waste outlets of industries pour large amounts of impurities into streams and lakes, much of which is untreated or inadequately treated. This is especially disconcerting when one considers the fact that over two-thirds of the United States population is served by a public water supply which obtains over three-fourths of its water from surface water supplies. The factor of time provides an additional problem in this connection. Some studies suggest that unless the present rate of sewage treatment plant construction is accelerated, the municipal sewage discharge in 1980 will be equivalent to the untreated waste that a population of 114 million would produce.[24]

The most basic problem already being faced by many communities and destined to be faced by most others is simply that of the provision of adequate quantities of water. About 40 per cent of the cities of the United States in 1960 were estimated to have inadequate municipal supply systems.[25] This is the result of several factors, notable among which are the growth in the proportion of urban dwellers, increased population, and greater use of water per capita. The population of the United States has more than doubled since 1900 but the rate of water use in that time has more than quadrupled. If one but reflects on the use of water around the home today, a number of practices, appliances, and comforts readily come to mind that were not a part of everyday life in 1900. As the cities grow ever larger, the concentration of demand is overtaxing the local water supply area. Thus cities such as Los Angeles are having to extend their facilities great distances to procure adequate supplies.

INDUSTRY

The largest increase in water use between 1955 and 1960 was in self-supplied industrial water. This increase amounted to 27 per cent. Industry is typically an urban phenomenon, and industrial use of the public water supply is common. Between one-third and one-fourth of the public water supply was so used in 1960. The overwhelming majority of industrial water (140 billion gallons per day or about 98 per cent), however, was self-supplied. The major need of industry for water is in thermal generation of electric power, amounting to over twice that of other industrial uses. Condenser cooling employs virtually all of this water. Because this results in little change other than in temperature, most is returned to the source.

Water for cooling is important to many industries; a majority of the self-supplied water of industry is used for cooling in some way or another. The twenty-four thousand gallons used, on the average, in making a ton of steel are mainly for cooling. Water has properties, however, which make it uniquely suited to a multitude of other industrial uses, for example as a solvent, a diluent, a waste carrier, a raw material, and for flotation and energy transfer. Impressive figures commonly cited, such as the 248 million gallons per day intake of fresh water by the automobile industry alone, 1,741 gallons to process a barrel of crude oil, and 57,000 gallons to produce a ton of paper pulp[26] reflect the very real importance of industry as a water user. Actually neither these figures nor the total figure for industry mentioned earlier correctly indicate the magnitude of this water use because they do not indicate recycling within plants. A 1959 survey indicated substantial increases in re-use of water by most of the plants surveyed and an average use of two and one-half times the water used in all such plants. A majority of the paper and pulp plants had an average re-use figure of 180 per cent, while the 182 refineries surveyed had an average water re-use figure of three and one-half times. One refinery decreased its daily requirements from six hundred million gallons to forty million gallons, a 92 per cent reduction.[27]

PROBLEMS. The eastern half of the United States includes the overwhelming majority of the nation's industry, with the major concentration located in the northeast quarter (Figure 11 – 9). Although some industrial plants treat their waste water before discharging it, many do not. Unfortunately, however, the effectiveness of treatment is too often modest at best,

TABLE 11–2

HYDROELECTRIC CAPACITY AND GENERATION IN THE UNITED STATES, JANUARY 1, 1964*

Major drainage division	Installed Capacity—kw developed	Installed Capacity—kw undeveloped	Average annual generation 1,000 kw hours
Hudson Bay	14,325	0	69,200
Great Lakes-St. Lawrence	4,041,797	1,414,305	24,861,300
North Atlantic	2,499,630	7,902,576	10,715,270
South Atlantic	2,771,340	5,436,423	6,816,150
Eastern Gulf	1,339,213	3,372,600	4,798,800
Western Gulf	345,618	1,154,085	1,046,700
Lower Mississippi River	1,031,610	3,247,000	2,785,900
Upper Mississippi River	609,427	577,660	3,125,825
Ohio River	4,563,292	5,694,883	19,582,700
Missouri River	2,618,826	9,003,225	12,200,000
Great Basin	483,821	330,750	1,720,200
North Pacific	13,589,081	38,519,985	85,851,700
South Pacific	4,219,367	9,904,056	20,407,300
Colorado River	1,994,495	7,284,450	7,503,380

*After *Hydroelectric Power Resources of the United States, Developed and Undeveloped* (Washington: The Federal Power Commission, 1964), Summary Table A.

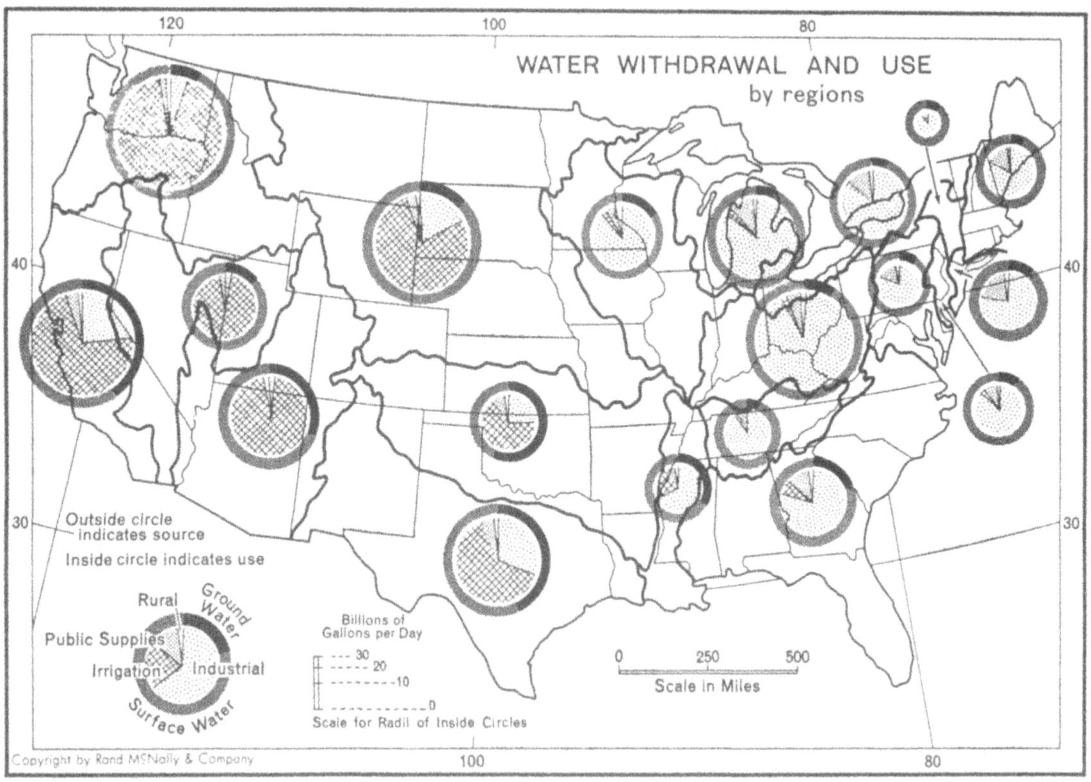

11–9 *Water withdrawal and use by regions in 1955.* Note the importance of irrigation in the West and industry in the East. (United States Geological Survey)

with a strong reliance upon "solution by dilution" still predominating.

In the case of many small establishments, discharge is into public sewers, and treatment is left to the municipality. It is the northeastern portion of the nation also which has the greatest population density. The combination of inadequately treated municipal sewage and industrial waste constitutes one of the most important problems of conservation. Enough data have been collected on the undesirable effects of known substances on crops, wildlife, natural vegetation, and on the health of human beings to point to the pressing need for more regulation. When the variety and nature of industrial wastes with imperfectly known toxic effects are considered, the problem assumes critical proportions.

IRRIGATION

Irrigation water use is a major category nearly comparable to self-supplied industrial water use in total volume. Actually there is a considerable amount of water withdrawn for this use which is lost by seepage and evaporation from transportation arteries. Of the total water withdrawn for irrigation in 1960, 21 per cent is considered to have been lost during conveyance to the farms.[28] Water which is seemingly lost by seepage, however, may well become a part of the ground water obtained from wells and also used for irrigation. Although the dominant type for irrigation is surface water from streams and lakes, about 36 per cent comes from wells and springs; this figure represents a substantial upward trend in ground water use for irrigation over the previous five years. An interesting but relatively unimportant source of irrigation water is that reclaimed from sewage, primarily in the southernmost states in the western half of the nation. Utilization of this source, however, is significant to conservation on two counts: Pollution is reduced and re-use is achieved.

The location of the major area of water withdrawal for use in irrigation is in the low precipitation areas of the West where water deficits are large (Figure 11 – 10). The seventeen western states which received only about 25 per cent of the annual national precipitation have 94 per cent of the nation's irrigated land. California, clearly the leader with an annual withdrawal of some twenty-five million acre-feet (including conveyance losses), is followed by Idaho, Texas, and Colorado in that order, each of which had a total withdrawal of about one-half that of California in 1960.[29]

The eastern thirty-one states use somewhat more water annually than the seventeen western states, but less than 5 per cent of it is used for irrigation. Use of water for supplementary irrigation in the East is on the increase, however, and is encountering competition even at this beginning stage.

PROBLEMS. The importance of irrigation as a water user and the crop production it makes possible have been commented on in this and other chapters, but this use is not without its problems. "No other use of water requires basic data to as great a degree as that for irrigation . . . irrigation farming is exceedingly complex, perhaps more so than any other kind of water project. Chances for error are innumerable and any one of them could ruin a project."[30] Unwise dependence upon limited ground water supplies, overconcentrations of salts in soils, soil water-logging and inadequate drainage, water table alteration, and insufficiently protected watersheds and resultant reservoir silting are examples of irrigation water-use problems.

WATER POWER

Use of the water resource for the generation of power involves a greater volume of water than all other computable uses combined (Figure 11 – 11). It should be recalled that this is a nonconsumptive use. During 1960, 2,000 billion gallons per day were so used, an increase of 33 per cent over the 1955 figure. Despite the impressive size of this figure, falling water remains far less important than the other sources of energy such as petroleum, coal, and natural gas, furnishing less than 5 per cent of the nation's energy. Neither does this form of energy dominate the electrical energy scene, because it furnishes less than one-fifth of the total electrical energy generated. As of December 31, 1962,

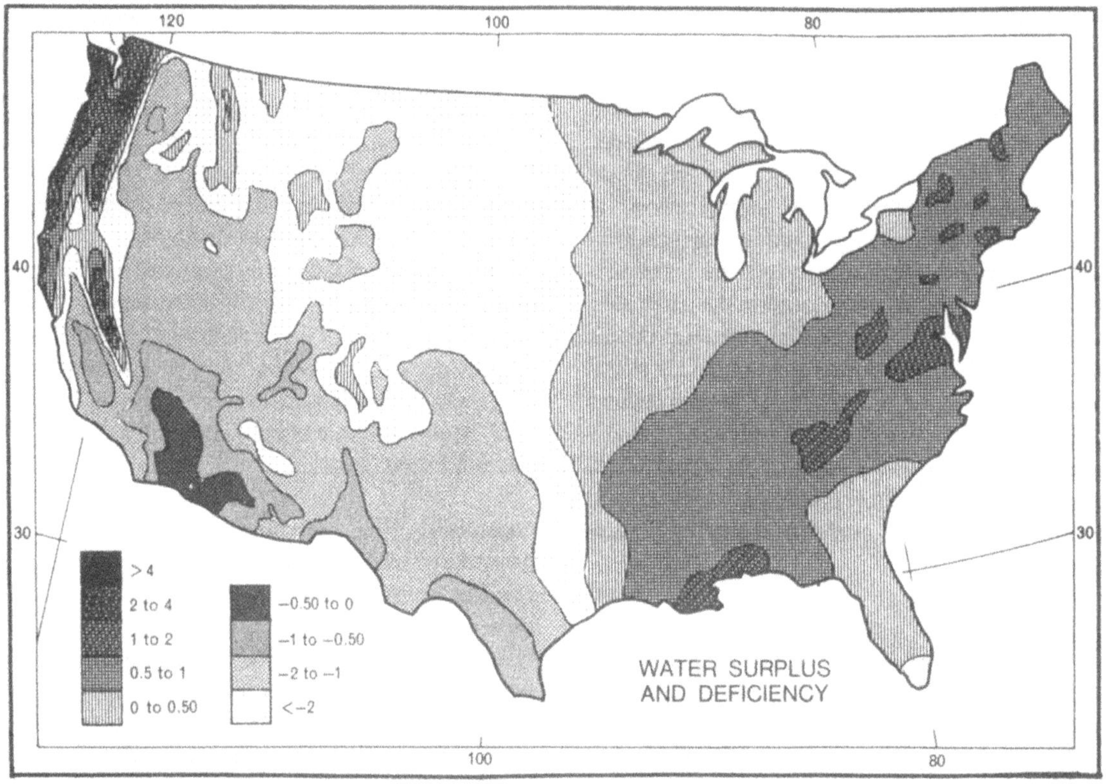

11 – 10 ***Water surplus and water deficiency in the conterminous United States*** **(Plate 2 in "Has the U. S. Enough Water?"** ***Geological Survey, Water Supply paper 1797*** **[Washington: Department of the Interior, 1965.])**

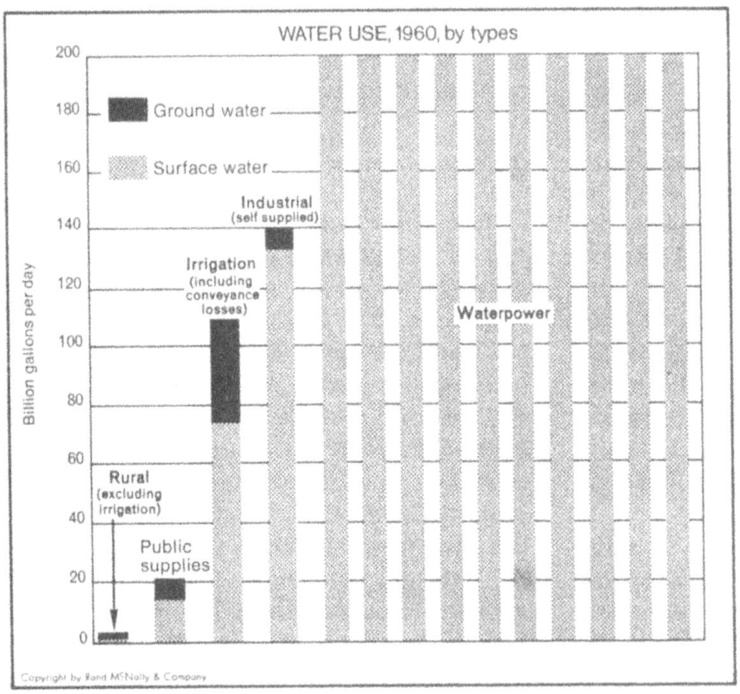

11 – 11 ***Water use in the United States by types.*** **(United States Geological Survey)**

the United States had a developed installed hydroelectric capacity of 38,600,000 kilowatts.[31] In some regions, water power has major importance; for example in the Pacific Northwest over 95 per cent of the electrical energy is supplied by hydroelectric installations.

Water power development is not concentrated in any one section of the United States, although there are several areas worth specific mention. Among these the Pacific Northwest with its great Federally sponsored dams on the Columbia River stands out. Four of the six first ranking dams in the United States, which incidentally rank among the world's greatest, are all located in this region—Grand Coulee, The Dalles, Chief Joseph, and McNary. Large Federally sponsored dams are not the only important producers of hydroelectric power; about one-half of the nation's hydroelectric power capacity is in non-Federal installations. In the eastern part of the country, the many smaller private utility dams taken as a whole produce a significant proportion of the hydroelectric power of the nation.

PROBLEMS RELATED TO RESERVOIRS AND DAMS. The rather large body of water and the dam required for a hydroelectric plant create problems. Many of these are encountered in any case where a dam to impound a large body of water is created; but the conditions accompanying use of the water for power generation cause some of them to be aggravated. This is due to size, height, turbines, and fluctuations of water release.

The conflict of dams with wildlife, notably anadromous fish, is one which has so far defied a completely successful solution. Although fish ladders, trapping and moving fish in tanks, and other devices have partly alleviated the problem, many fish are still being killed trying to get to or from spawning grounds (Chapter 13). The numbers killed are sufficiently large to constitute significant waste and research still goes on in pursuit of a better solution.

Impounding large quantities of water behind a dam necessarily requires that a considerable proportion of the land adjacent to the dam site be eliminated from any other use. In areas of considerable relief the area covered by the water often represents the best agricultural land; with large dams the amount of land may be a significant proportion of that locally available. Displacement of settlements, railroads, and highways has occurred in a number of places. Even in the case of water transportation, which may be aided by the resulting rise in water level, the dam itself presents a barrier. Although it is a surmountable barrier when locks are added to the structure, the building, maintenance, and operation of the locks are an additional expense. The reservoir, moreover, may not improve the scenic quality of the area, particularly if there are large fluctuations of water level exposing unsightly floors. As population increases and spreads, the likelihood of such conflicts with other land uses grows more acute.

Creation of a large body of water at a higher level than the original stream has an effect on the water table of the surrounding area. Raising the water table would seem to be a desirable feature, but in some instances the results have been unfavorable. Additionally, areas some distance downstream from the dam may experience a lowered water table as a result of reduced stream flow and the recharge it would afford.

Storage of large volumes of water in reservoirs inevitably results in considerable surface area being exposed to evaporation loss. The larger lakes and reservoirs of the United States, those with a capacity of more than five thousand acre-feet, experience an evaporation loss which approximates the total public water supply use. Loss from deep narrow reservoirs is the least, the great expanse of water exposed in a wide, shallow reservoir of equal volume contributing to the greater loss.

A major problem affecting the reservoir itself is that of silting or filling by sediments carried by the streams which feed it. The reduction in velocity the stream experiences as it enters the body of standing water is accompanied by a reduction in carrying capacity and resultant deposition of suspended sediment. Such silting commonly progresses from the upstream end of the reservoir toward the dam. Some silting is inevitable, and allowance for a reasonable amount is usually included in the planning of a reservoir. A properly maintained watershed will usually make this factor a minor problem; sedimentation at the present rate in the TVA reservoirs will allow the reservoirs to function properly for centuries. For too many reservoirs, however, the outlook is not this bright; some had lost a significant proportion of their storage capacity before remedial action was undertaken, and they will render less service than was expected when they were constructed. Others, fortunately few in number, have not only ceased to serve any useful purpose but, in addition, have made useless a section of the stream valley. Physical removal of the accumulated sediment by dredging is an expensive process, estimated in some instances to be more costly than the original investment in the installation. Silting continues to rob the nation's reservoirs of several hundred thousand acre-feet of capacity annually.[32]

There are additional problems occasioned by dams such as the inevitable conflicts in water management which arise with multiple-purpose use. Discussion of this concept and attendant problems is found in the chapter following.

The advantages which accrue from dams and the water they store—maintenance of constant flow, a power source, flood control, recreation, and the like—are sufficient to insure that they will continue to be built. The above-mentioned problems are examples of factors which are receiving attention, but which still need greater consideration in future planning for dams.

TRANSPORTATION

The inland waterways of the United States may be thought of as having three major component parts, although these are actually interconnected (Figure 11 – 12). The Great Lakes form the finest and most heavily used inland waterway in the world. The sprawling Mississippi-Missouri-Ohio system interconnects the several parts of the Midwest as well as linking the area to the sea in Louisiana. The Intracoastal Waterway constitutes the third major component and affords a protected water route

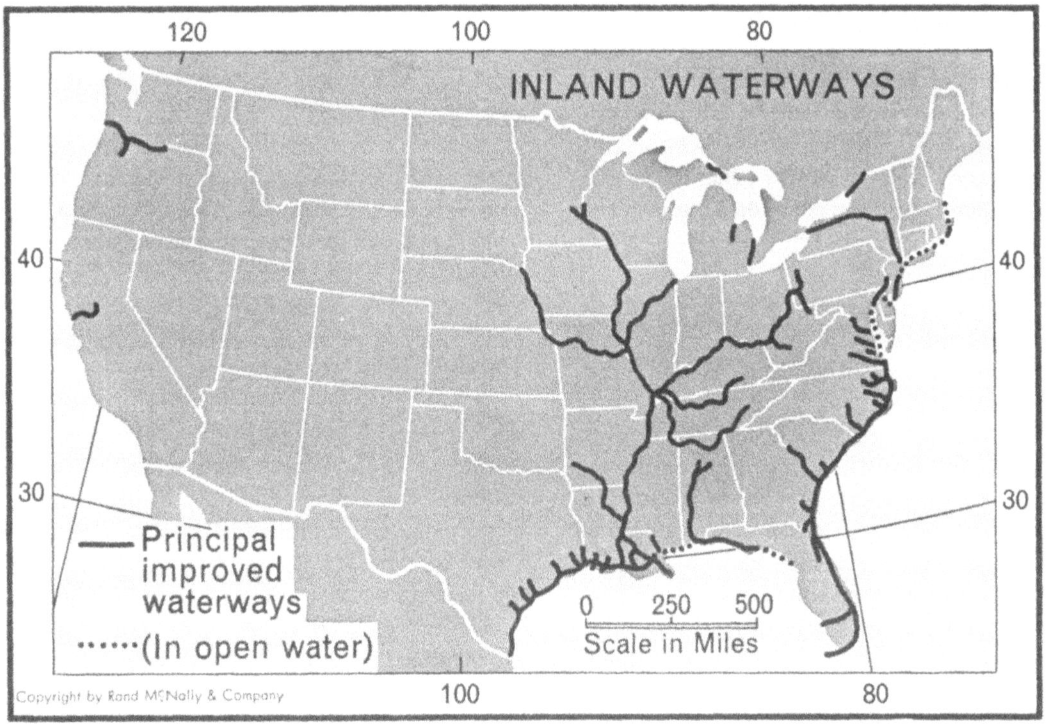

11 – 12 ***The inland waterways of the United States.*** **(Corps of Engineers)**

extending from Brownsville, Texas, to Trenton, New Jersey, which is interrupted by open water stretches or connecting waterways only locally.

The use of the nation's water resources for transportation is less readily evaluated than the uses considered thus far. Although it may occupy a status of relatively lesser importance to the total transportation situation than it does in many countries, water transportation in the United States is undergoing a slow but steady increase. During the decade 1948 through 1957, the total water-borne commerce, domestic and foreign, increased by 338,400,000 short tons[33] to a total of 1,131,400,000 tons in 1957. The increase has continued; the figure for 1964 is 1,238,000,000.

The greatest increase was in traffic between the United States and foreign nations. Domestic traffic, while experiencing a lesser increase, nevertheless comprises two-thirds of the total. Of the several categories of domestic water-borne commerce, the greatest increase was experienced in internal shipments on rivers. This category also accounted for the greatest tonnage in 1964.

The increased importance of inland waterways, other than the Great Lakes, is revealed in the rise of ton-miles of freight carried on inland waterways. In 1948 the Great Lakes system accounted for 119 billion ton-miles of freight, while other inland waterways handled only forty-three billion. In 1964 the positions were reversed, 106 billion ton-miles having been carried on the Great Lakes system and 144 billion on the other inland waterways.[34] It may thus be seen that, although the Great Lakes system has not substantially decreased in absolute importance, the other inland waterways have experienced a significant increase in relative importance.

Of the several waterways, the Mississippi River system ranks closest to the Great Lakes in ton-mileage of freight carried. This system, which includes the Ohio River and other tributaries, accounted for over 60 per cent of the ton-mileage carried during 1964 on inland waterways other than the Great Lakes. The importance of other waterway systems is indicated in Table 11–3.

TABLE 11–3

TON-MILEAGE OF FREIGHT CARRIED ON INLAND WATERWAYS OF THE UNITED STATES IN 1964*

System	1957
Great Lakes system	105,912,000
Mississippi River system, including Ohio River and tributaries	89,348,000
Atlantic Coast waterways	27,861,000
Gulf Coast waterways	20,646,000
Pacific Coast waterways	6,347,000
Other waterways	11,600

* *Waterborne Commerce of the United States, Calendar Year 1964*, part 5. National Summaries (Washington: Department of the Army, Corps of Engineers), p. 40.

In 1964 about 38 per cent of the domestic water-borne commerce of the United States was comprised of petroleum and petroleum products, 20 per cent coal and coke, 13 per cent iron ore and iron and steel, and 12 per cent sand, gravel, and stone.[35] Water transportation is notably cheaper than other means, and this factor is responsible for the dominance of bulky low-unit-value-commodities so moved. Some types of high unit-value items are being moved by water, but in total they represent less than 10 per cent of the domestic commerce.

PROBLEMS. The nation's inland waterways pose a number of problems, and their interrelatedness is typical of water resources in general. A permanent waterway should have a constant minimum depth, yet withdrawal of water during a dry year or season adversely affects this depth. Excessive watershed erosion may necessitate dredging operations in the channels choked with sediment. The maintenance of minimum channel depth or elimination of rapids may require dams and locks; problems occasioned by dams were mentioned earlier. In addition to problems of this sort there is the controversial nature of Federally sponsored inland waterways. The long-debated questions—Why should the whole nation finance the building of waterways for the use of only a few? Why shouldn't the users of the waterways pay for their use and maintenance as do the railroad users and users of the highways?—have never been answered to the satisfaction of all, nor are they apt to be. Nevertheless, the present and indicated future importance of the nation's waterways suggests that this use of the water resource must be given its share of attention in future planning.

Conclusion

The attempt has been made to assess the water resource of the United States for character, quantity, importance, type, use, and major associated problems. Coverage is by no means complete; the attempt to make it so would be both impractical and impossible for this volume. In addition, other chapters such as those on recreation, wildlife, and land resources deal incidentally with water resources. The purpose of this chapter is to provide a base from which to consider the conservation of this all-important natural resource. As one contemplates the water requirements for an estimated doubling of the population within a half-century or less, the need for water conservation planning becomes apparent.

FOOTNOTES

[1] Robert T. MacMillan, *Mineral Facts and Problems*, 1960 edition (Washington: Bureau of Mines, Department of the Interior, 1960), p. 957.

[2] For an excellent summary statement on the importance of water see Edward A. Ackerman, "Water Resources in the United States," lecture for the Department of Agriculture Graduate School given at Washington, D. C., February 13, 1957 (Washington: Resources for the Future Inc., *Reprint No. 6*, March, 1958), p. 2.

[3] Sterling B. Hendricks, "Necessary, Convenient, Commonplace," *Water, the Yearbook of Agriculture, 1955* (Washington: Department of Agriculture, 1955), p. 10.

[4] MacMillan, *loc. cit.*

[5] Kenneth A. MacKichan and John C. Kammerer, "Estimated Use of Water in the United States, 1960," *Geological Survey Circular 456 (1961)*, p. 1.

[6] *Ibid*

[7] William C. Ackermann, E. A. Colman, and Harold O. Ogrosky, "From Ocean to Sky to Land to Ocean," *Water, the Yearbook of Agriculture, 1955, op. cit.*, p. 41.

[8] *Ibid.*, p. 45.

[9] See F. W. Went, "Fog, Mist, Dew and Other Sources," *Water, the Yearbook of Agriculture, 1955, op. cit.*, pp. 103–9.

[10] MacMillan, *op. cit.*, p. 3.

[11] *Ibid.*, p. 1.

[12] MacKichan and Kammerer, *loc. cit.*

[13] *Ibid.*, p. 34.

[14] *Ibid.*, p. 11. For a nontechnical introduction to ground water see Helene L. Baldwin and Charles L. McGuinnes, "A Primer on Ground Water," (Washington: Government Printing Office, 1963), 22 pp.

[15] MacKichan and Kammerer, *op. cit.*, p. 1.

[16] Ackerman, *op. cit.*, p. 2.

[17] MacKichan and Kammerer, *op. cit.*, p. 4.

[18] Luna B. Leopold and Walter B. Langbein, "A Primer on Water," *Geological Survey* (Washington: Department of the Interior, 1960), p. 32.

[19] Bernard Frank, "The Story of Water as the Story of Man," *Water, the Yearbook of Agriculture, 1955, op. cit.*, p. 7.

[20] Harry E. Jordan, "The Problems that Face our Cities," *Ibid.*, p. 652.

[21] MacKichan and Kammerer, *op. cit.*, p. 4.

[22] *Ibid.*

[23] "A Study of Pollution-Water," A Staff Report to the Committee on Public Works, United States Senate (Washington: Government Printing Office, 1963), pp. 85–88.

[24] *Ibid.*, p. 8.

[25] MacMillan, *op. cit.*, p. 10.

[26] *Water in Industry* (New York and Washington: The National Association of Manufacturers and The Chamber of Commerce of the United States, January, 1965), p. 49.

[27] *Ibid.*, section C, pp. 50–81, and "Water Resource Activities in the United States. Present and Prospective Means for Improved Re-use of Water," Select Committee on National Water Resources, United States Senate, *Committee Print No. 30* (Washington: Government Printing Office, 1960), p. 35.

[28] MacKichan and Kammerer, *op. cit.*, p. 17.

[29] *Ibid.*, p. 14.

[30] Walter B. Langbein and William Hoyt, *Water Facts for the Nation's Future* (New York: The Ronald Press Company, 1959), p. 143. See also Roger Revelle, "Water," *Scientific American*, vol. 209 (September, 1963), pp. 93–108.

[31] Lloyd L. Young, "Summary of Developed and Potential Waterpower of the United States and Other Countries of the World, 1955–62," *Geological Survey Circular 483* (1964), p. 2.

[32] Peveril Meigs, "Water Problems in the United States," *The Geographical Review*, vol. 42 (July, 1952), p. 359.

[33] *Waterborne Commerce of the United States, Calendar Year 1964*, part 5, National Summaries (Washington: Department of the Army, Corps of Engineers 1965), p. 13.

[34] *Ibid.*, p. 41.

[35] *Ibid.*, p. 21.

12

Water Resources Conservation

Although the more impressive water conservation activities in the United States are of relatively recent development, early attention to water arose as an adjunct to erosion and soil conservation problems. The interrelatedness of water with other resources led conservationists to recognize that many problems were water problems in part. Hence, many of the measures adopted and practices employed in the conservation of other resources were, and are, water conservation practices also. In addition, because of the diversity of uses made of this resource—power, irrigation, transportation, public supply, industry, etc.—water conservation measures per se are quite numerous. The quantitative extremes of the resource, drought and flood, further contribute to the variety of management practices applied to water. It is not possible, therefore, to attempt here anything approaching a complete survey of the evolution of water conservation. An effort is made instead to point to some of the more notable developments pertaining specifically to water resources which have contributed to the present status of water conservation in the United States. Brief reference to the legal aspects of water conservation, however, should precede such discussion.

Water Law

As pointed out in earlier chapters, any attempt to effect conservation of a resource involves consideration of ownership. In the case of water, the situation is somewhat different than with most resources in that several levels of control may be in existence simultaneously. Landowners whose properties adjoin a river in the eastern United States hold equal legal rights to the water in the river; the state in which they live, however, is directly involved in the regulation of the use of that stream's waters. Should the importance of the stream extend beyond the limits of the state boundary, in the instance that the stream is a part of the inland waterways of the nation, the Federal government is involved also. The case of water ownership is further complicated in that water use, the purpose of ownership, is not always consumptive, and that which is "owned" or used by one person will shortly become the property of someone downstream. Thus for conservation a consideration of water rights rather than ownership is more appropriate.

The water laws of the several states for consumptive use of surface water have developed from two distinct legal foundations, the doctrine of *riparian* water rights and the doctrine of *prior appropriation*. Ground water law developed from a third concept, *the rule of capture*.

THE DOCTRINE OF RIPARIAN WATER RIGHTS

The doctrine of riparian water rights concerns only surface water and grew out of the

English Common Law. In essence it provided that all landowners had equal right to use of water and to unimpaired flow of streams on or contiguous to their land. Thus water rights under the riparian doctrine are based on land ownership. The concept of riparian rights was introduced into the well-watered eastern United States and is the basis of the water law in most of the eastern half of the nation; it was satisfactory as long as water was available for all. As populations have grown and per capita use of water has come to press on supplies, however, there have been modifications of riparian water rights toward limiting individual claims in the interest of the public welfare, and additional ones are still needed. As the doctrine of riparian water rights was extended to areas of less ample water supply, various alterations were made. Nevertheless, the concept of a riparian water right as an inalienable right of land ownership continues to be important as the foundation and legal base of many water contests, especially where Federal lands are concerned. For example, in the West the water rights of the Federal government have in several recent cases been established in the courts on the basis of land ownership.

THE DOCTRINE OF PRIOR APPROPRIATION

The doctrine of prior appropriation as a water law evolved from common practices in the early West where the water resource was generally limited and often inadequate. Riparian concepts obviously were unsuitable. The mining camps and range areas developed the concept of "first in use, first in right" which is the doctrine of prior appropriation. Unlike the riparian rights, the prior appropriation doctrine has little or no relationship to ownership of land —the water is considered public property subject to claim. Thus a gold miner having used water beneficially for sluicing operations, by the unwritten law came to have the right to use this needed water over and above the needs of those who might come later. Later on, formal systems of filing claims for water were established, usually through the state engineer's office.

In 1866 Congress confirmed these local practices of the West. Even water rights on public land came to be subject to state rather than Federal law. State laws came to be the prime regulators of the water resource. In recent years the Federal government has asserted stronger claims to water, often basing its claim on the doctrine of riparian rights and associating these claims with Federal ownership of land.

As the West has become more occupied and water problems increasingly complex, there have been numerous modifications. Most states now require specific filing of water rights with the state engineer. Many states have or are initiating legal steps to make the needs of the public interest superior to the doctrine of prior rights. To maintain the water right, a provision is increasingly being made that there must be continuous beneficial use. A major problem of water rights in much of the West is overappropriation. In many cases water rights which are not actually used are still valid and, were all of the rights on a given stream to be utilized, the amount of water would be quite insufficient.

GROUND WATER

Ground water is a special case for which regulation and law have only recently developed. Originally the underground water resources of the land came under the legal concept of the rule of capture. The water belonged to the person bringing it to the surface. Thus an owner of land had the right to pump ground water without regard for the fact that the water might be draining from ground water reservoirs under a neighboring property. With the populating of the West, water needs have made this situation generally intolerable. Therefore, most states today have developed modifications requiring permits for the pumping of ground water and reasonable consideration for the rights of others. For example, in California ground water is generally regulated under the concept of correlative rights—not unlike the basic concept of the riparian rights of the East which in effect held that all had equal right. The problem of ground water regulation is still very great and the legal base is still evolving as it is with all water.

GOVERNMENT RIGHTS

While state and Federal governments as land owners have rights analogous to those of individuals, as government bodies they have other roles to fulfill with respect to the nation's water resources. These roles have developed along with the nation itself, and are reflected in part by some of the significant legislation of the past century or so. It should be noted at the outset that control of the majority of the nation's water resources traditionally has been and still remains in the hands of the several state governments. With the exception of interference with navigational waters or waters on Federal property, state jurisdiction prevails over the water within the state boundaries. The increasing interest and activity of the Federal government in water resources development, however, is an important consideration from the conservation standpoint. In some instances conflicts have occurred because of the expanding role assumed by the Federal government.[1] Thus it may be useful to consider in summary fashion the basis for and development of this role.

The authority of the Federal government stems primarily from the supreme law of the land, the Constitution of the United States, and is limited to that which is expressly delegated or reasonably may be inferred.[2] Although at first glance this may seem to restrict significantly the Federal government's authority, in point of fact final limits have not yet been established and probably never will be. For example, Congress is empowered by the Constitution to regulate commerce with foreign nations and among the several states. To do so it becomes concerned with navigation; Congress thus has jurisdiction over all navigable waters of the United States. In addition, its jurisdiction is extended to the non-navigable upper portions of streams which may affect the navigable capacity of a stream being used for commerce. Included in the authority to regulate commerce are flood protection and watershed development, implying an authority as broad as the needs of commerce themselves. Nor is the authority of Congress with respect to water resources limited to the regulation of commerce and all that it implies. The Property Clause of the Constitution grants Congress the power to "dispose of and make needful regulations respecting government property."[3] Hence, although the several states may establish the character of water rights pertaining to streams within their boundaries, they may not infringe upon the right of the United States to the continued flow of those streams for whatever uses beneficial to affected Federal government property are deemed necessary. In this respect, it may be noted that the water rights mentioned earlier are in fact rights to the use of water, not to the corpus of the water itself. In similar fashion, other powers granted the Congress under the Constitution extend its present and potential authority considerably.

It will readily be seen that considerable variation in the interpretation of the Constitution is possible. Many have felt that the Federal government should have used a wider interpretation of its authority and played a more extensive role in the regulation and development of the nation's water resources. Many others feel that, especially in the past several decades, the Federal government has played too dominant a role, specifically, that it became involved at the state and local level where it should not have. It is not our purpose to deal with this question other than to point out that the primary role of the Federal government is to act on water resource problems of national scope. The decision as to which future water problems are of this dimension and which are not will have to be made in the light of existing situations.

EVOLUTION OF FEDERAL ACTIVITY. The Federal government entered the water resource picture rather early in the nation's history. As early as 1789 the establishment of navigational aids along the coast was aided by making available to the coastal states operational and maintenance funds from Federal sources.[4] In fact, most of the Federal government's early interest in water resources was from the navigational aspect. Significant improvement of navigable waterways began in 1824, and from that time to the present the Federal government has played an increasingly significant role in the development of water resources.

The role of the government has evolved in a rather irregular fashion, however, with the emphasis shifting and additional types of activity being introduced as a result of the exigencies of the moment. The several major phases of activity—navigation, irrigation, flood control, and power—developed independently and often with only incidental relatedness. Major activity in the field of reclamation and irrigation began with the Reclamation Act of 1902. In 1920 the Federal Water Power Act served to mark the advent of extensive participation by the Federal government in the development of water power. Although a number of activities concerned directly or incidentally with flood control were undertaken prior to its time, the Omnibus Flood Control Act of 1936 marked the beginning of consideration of flood control as Federal responsibility on a broad scale. Many other aspects of water conservation have in similarly irregular fashion evolved from Federal activity, for example the Federal Water Pollution acts of 1948 and 1956. As pointed out earlier, however, the most notable aspect of Federal activity in the field of water resource development is not so much the extent of this activity as the recently accelerated rate. Of the 16.3 billion dollars disbursed by the government between 1824 and 1955 for water resource and power projects, 88 per cent was spent after 1930, 71 per cent after 1940, and the bulk of it after the end of World War II.[5]

An increasingly prominent role for the Federal government seems assured in view of the events of the past few years; several examples will illustrate. In April, 1959 The Senate Select Committee on National Water Resources was formed to consider present and future extent, character, and use of the nation's water resources. The Federal Water Pollution Control Act of 1956, the first really comprehensive water pollution control legislation, was amended in July, 1961 to provide a more effective program of enforcement, support and assistance, and research. Amendments to the Flood Control Act have recently authorized regulation of river flow by Corps of Engineers' dams for municipal water storage and stream flow augmentation to offset pollution. Authorization of funds for additional investigation of water resource problems was provided by the Water Resources Research Act of 1964. In 1965 the Water Resources Planning Act with its basin-wide planning emphasis and provision for organization of river basin commissions was complemented by the 1965 Water Quality Act which established the Federal Water Pollution Control Administration. Indicative of the pace of change in the role of the Federal government is the fact that in 1966 the Clean Waters Restoration Act became law, extensively amending the Water Quality Act passed the year before.

In view of the prospects for increased demand on the water resource, much attention must be given to the development of a policy which will employ the efforts of all regulating bodies—Federal, state, and local—each in its proper and most effective role. To this end, each citizen as an individual or collective user of the resource has the responsibility of keeping himself informed and making proper use of the information.

Notable Concepts and Developments in Water Conservation

Water and land are inextricably related; that which affects one will have related effects on the other, directly or indirectly. Similarly, significant events in one section of a drainage system will be reflected in resultant effects in downstream areas. Recognition of these two facts was essential to the planning for and achievement of the present status of water conservation in the United States.

THE RIVER BASIN AS A NATURAL UNIT

Early efforts to use or alter portions of river systems were of a minor nature (or were ineffective enough) so that the remainder of the system was little disturbed. Ultimately, the projects began to assume proportions, however, which resulted in the downstream portions of the drainage system being significantly affected, often undesirably. Similarly, unsuccessful efforts to control floods in the lower portions of river systems by concentration of effort on creation of protection devices inexorably led to the realization that effective downstream flood control requires upstream control. From this state of awareness it was only a short additional step to the recognition that integrated control efforts in all parts of the drainage system held the greatest promise for the most complete use and control of water resources. The river basin has thus come to be recognized as a natural planning unit especially adaptable to water resource systems considerations.

It should be noted that there are some water resource planning problems that do not lend themselves to restriction to the river basin as the basic unit. Already it is apparent that some developments would benefit from interbasin transfers of stream flow. Also, many river basins are too small to make feasible plans for local changes in the hydrologic cycle without consideration for areas both to the windward and leeward. Still, the advantages seem to outweigh the disadvantages, and the recognition of the basin as a viable planning unit is implicit in its employment in numerous interstate compacts, interagency committees, and the like.[6]

As the idea of the unity of a river basin developed, so did an awareness of the interrelatedness of water with the whole of the natural complex. Although early limited efforts to make use of water had little effect on related resources, man's increasing numbers and per capita use of water resulted in significant alteration of the natural setting. It became evident that the aforementioned goal of integrated control in all parts of a drainage system required consideration of man and his activities as well as of the total physical complex. Although to a few an awareness of the complexity of water conservation came early, the awareness came more slowly to the majority. The relatively recent recognition of the role of subsurface water in the total picture has brought us still closer to the realistic appraisal of the nation's water resources which is a necessary prelude to the establishment of the kind of national water policy the future demands.

MULTIPURPOSE PROJECTS. Multipurpose water development today is designed primarily to derive the maximum benefit from the available resource. Early accomplishments in multipurpose development were occasionally inadvertant, resulting from efforts to avoid reducing the utility of the resource for one purpose while in the process of increasing its utility for another purpose. Increasing size of projects with increasing costs and growth in conflicting uses finally brought the government into leadership in the interest of maximum multiple-use development. In government-sponsored programs, the cost of certain benefits (nonreimbursable costs: recreation, transportation, pollution abatement) are paid by the total populace while costs of hydro and irrigation benefits are paid by the beneficiaries.

The beginning of multipurpose dam projects in this country occurred in 1879 when the army, authorized to lease water power, developed the Federal navigation dam on the Mississippi River.[7] At about the same time the relatedness between navigation projects and flood control began to receive serious attention. In 1888 flood control and irrigation were linked in a statute which was a forerunner of the more comprehensive acts of a similar nature which subsequently developed. Theodore Roosevelt's Inland Waterways Commission of 1907 urged that adequate consideration for all purposes be the objective of future river basin development,[8] and since that time attention has been given to the possibility of more complete use of water resources.

Multipurpose projects are not without their problem aspects. It is readily apparent that the most effective use of a reservoir for flood control does not go hand in hand with storage of water for irrigation or water power. For power, the reservoir should be full most of the time, for flood control, nearly empty. Similarly, withdrawal of water for irrigation during a dry period may drop water levels, and so interfere with water transportation or water storage for continued hydroelectric power production. No multipurpose development can serve all purposes with optimum effectiveness; yet to serve one purpose effectively and render some service to other needs simultaneously is generally better than to have the lesser needs receive no consideration. The many examples of conflict that could be cited in the face of continued development of multipurpose projects collectively imply that the benefits derived are sufficient to make them desirable. In actuality, where several similar neighboring developments exist or additional auxiliary facilities are available, it is possible to make the system serve each of several purposes with a high degree of effectiveness (Figure 12 – 1). True multipurpose development of a river basin requires the correlation of the functions of the integral parts of the entire system. Co-ordinated effort based on study of the likely need for flood storage space as opposed to volume requirements for waterpower or irrigation allows for balance. Power pools provide more efficient utilization and distribution through the exchange of power from one area to another. Within an entire basin development some dams may function chiefly for a single purpose within the total multiple-use design.

It will be appreciated that for multipurpose use where needs are opposed, knowledge of precipitation character and amount, runoff, evapotranspiration, and the like cannot be too complete. Here, in fact, is one of the great stumbling blocks to truly efficient harnessing and use of the water resource: The lack of sufficient good basic data.[9] Despite the advanced position of this nation compared to most others with respect to basic data coverage, knowledge is woefully inadequate for the kind of planning needed for large integrated water resource development programs. The forested areas, the mountains, and other sparsely populated sections of the country have few or no first class weather data collection installations, or have acquired them only recently. The need has been recognized, however, and the movement in the right direction has begun, albeit halting and occasionally sporadic. With more complete data comes the opportunity to choose between alternatives, to plan the most efficient structures, to properly delegate responsibility, to integrate related conservation programs more effectively —in short to approach the ultimate use of the water resource, the theoretical goal of multiple use.

TVA. As the multipurpose and total basin concepts became accepted and began to be implemented, the problems presented by independent and variable land-use practices within a given river system made conclusive proof of and experimentation with these concepts difficult. "Congress accorded unique recognition to the need for coordinating multiple uses of water under comprehensive river-basin development when it established the Tennessee Valley Authority in 1933."[10]

The Tennessee Valley Authority development is a unique and controversial Federal government water resource project. Many were opposed to the original use of Federal tax money for this project, and continue to be opposed to additional use of such funds for the project. National and international recognition of the worth and uniqueness of the project, however, is evidenced by the continual stream of visitors from all parts of this nation and the world who come to see an entire river system under one central management where current conservation and multiple-use concepts are being applied in an effort to secure the ultimate development of the area. Created with an original admonition to construct such dams and reservoirs as needed to promote navigation on the Tennessee River and its tributaries, the regulation of stream flow thereby was primarily to serve the needs of navigation (Figure 12 – 2)

12 – 1 ***McNary Dam on the Columbia River, a multipurpose dam.*** **The navigation lock is in the foreground, and the power installation is on the other side. (Corps of Engineers)**

12 – 2 *Barges on the Tennessee River.* This navigation lock at Wilson Dam has one of the world's highest single lifts: one hundred feet. (American Waterways Operators, Inc.)

and flood control. Authority to operate its projects for power generation was granted also, so long as interference with the first-named objectives was avoided, and along with the resultant generation of power came studies and experiments on wider and better use of power for agricultural, domestic, and small local business purposes. Additional tasks set for TVA included consideration of proper use of marginal land, proper methods of reforestation of all parts of the drainage basin of the Tennessee River suitable for such treatment, and the bringing about of the social and economic well-being of the inhabitants. Many other specified or implied items are included in the activities of this organization, such as soil conservation, fertilizer manufacture and experimentation, and the like.

Aside from the unusually broad multipurpose nature of TVA, the interesting feature is the physical area itself. Comprised of parts of eight states and including over forty thousand square miles, TVA is concerned with the entire drainage basin of a major river—a complete natural unit. (Figure 4 – 22). In cooperating with state and local authorities, TVA can effect regulation of all activities which would either benefit or deter the many and interrelated uses of the area and its resources. Moreover, such regulation makes possible the elimination of many variables which in other areas effectively hinder experimentation. Thus the Tennessee Basin has served as the proving ground for many water resource conservation developments.

WATERSHED MANAGEMENT CONTROL. As watershed management came to be considered the cornerstone of water resource conservation, organization on the major river basin level and the need for the Federal government to play an increasingly important role seemed to be indicated. Yet the idea of large river systems being administratively subdivided into smaller natural units, such as tributary watersheds, has merit. In such instances, organization and control of the unit could be in the hands of local or state authorities with the Federal agency occupying only a supervisory or coordinating role at the most. Because it is unlikely that all major river basins will be developed under as broad an authority as TVA, an alternative way of achieving complete cooperation in the use of the resources of the Basin is desirable. It has been suggested that such cooperation would be more readily forthcoming if the extent of the administrative unit were held to a size such that the individual could appreciate his part in the whole.[11] Additional possible benefits include administration by local or state personnel perhaps more attentive to local problems, greater local participation in decisions of the nature and extent of expenditures, and the like. Public Law 566, The Small Watershed Act established in 1954, is a step in this direction.

As implied earlier, there are divergent views on the question of Federal versus state and local control; indeed this is one of the major political issues of the time. Lest one become discouraged that there seem to be so many complexities to water resource conservation, it bears repeating that having reached the point where the population is aware enough of these complexities to have soundly divergent views is an achievement in itself. It is in the framework of just such debate that the most useful planning is often developed.

Conservation Practices

A majority of the advances in water conservation in the United States have been made relatively recently. The overall concept of conservation itself has made its greatest advances only since the turn of the century, and most of the actions in the first half of the century which have involved the water resource have done so only incidentally or as a means to the conservation of related resources. With the exception of the dry parts of this nation where some degree of "home grown" conservation of water has been necessary from the beginning of settlement, Americans by and large have tended to regard water as inexhaustible, as the great forests once seemed to the pioneers. Beginning about mid-century an awareness that such was not the case began slowly to emerge. To be sure, there were a few who knew earlier, but for most Americans the past decade has been an awakening period during which the nation's water resource has experienced a decided change of image. Many factors have contributed to this change. As the research begun in the forties and fifties by both government and private organizations began to bear fruit, a clearer picture of the facts and a more solid basis for prognostication emerged. The rapid rates of increase of both population and water use began creating problems which affected large numbers of people who had been able to ignore them in the past. Almost as though by design, nature helped focus attention on the situation by contributing periods of below-average precipitation during the 1950's and 1960's which aggravated these problems. Certainly the determined program of information and education sponsored by public and private agencies alike must be given credit for much of the change in the viewpoint of the adequacy of the water resource. There are still too many who do not yet feel the personal need for conservation of this resource, but the size of the body of informed citizens and the number of organizations directly concerned with seeking solutions are now more nearly commensurate with the task. Although we are just on the threshold of sound and lasting water conservation, significant advances already have been made in research and implementation. The following discussion provides only a sample of some of the presently applied and suggested procedures.

CONSERVATION AND DOMESTIC USE

One of the most striking examples of mistreatment of a natural resource is the despoilment of many of the nation's streams and lakes resulting from the pouring of sewage into them by urban centers. Even water bodies the size of the Great Lakes are susceptible, as the seriously polluted condition of Lake Erie clearly shows. Although not all cities discharge untreated or partly treated sewage into natural water bodies, many still do. At the beginning of 1962 an estimated 63 per cent of the total U. S. population was served by municipal sewers. Of the more than 11,000 sewered communities, about 80 per cent treat the sewage in some fashion before discharging it.[12] Unfortunately, the extent of treatment varies from rather effective systems employing both primary and secondary treatment down to extremely simple procedures which do little more than allow for settling. In too many cases the decision of what the condition of the water that is returned to the stream after use should be is based on the poor quality of the water when it was withdrawn. Aside from the fact that this line of thought leads into a vicious circle, the worst feature of its tacit acceptance is the effect it has on areas between use points. Insufficient or complete lack of attention to this problem has allowed some streams to become little better than open sewers befouling and rendering undesirable the areas through which they flow. If at least enough treatment is given to sewage before it is discharged that it does not destroy the natural condition of the river, some purification of the water will occur of its own accord. If no treatment is afforded, however, waste in sufficient quantity will destroy the forms of life which assist in this natural purification, and only putrefying bacteria and undesirable life forms will find such an environment satisfactory.[13] Fish and other forms of desirable animal and plant life in or along the stream will disappear, and as the whole balance is upset additional problems begin to occur. Completely aside from the elimination of the stream for other uses such as fishing, recreation, irrigation, and the like, a stream reaching such a condition will become at best a more expensive water source for the very cities contributing to its depreciation.

Treatment of sewage has features to recommend it other than the elimination of the above-mentioned problems. Fertilizer manufacture from sewage recovery products has already aided in the amortization of the cost of some treatment plants, and the use rather than misuse of this waste is sound conservation in itself (Figure 12 – 3). With a unified program requiring all persons to purify discharged materials to an acceptable minimum standard, the reduced cost of purification equipment used for incoming water could assist in paying for treatment of outgoing water. Unforeseen benefits accruing from greater future use of the streams thus improved may well prove the axiom that sound conservation practices return the investment manyfold from related effects, whereas flagrant mistreatment of one resource will typically result in additional losses in associated resources.

Aside from quality, the problem of the quantity of available water is an ever growing one demanding the attention of many American cities. The nation's largest city, despite its location in an area of distinctly humid climate adjacent to an ocean, continues to experience water shortages and associated problems during dry years. New York City derives the bulk of its water supply from over a thousand streams in seven watersheds, with a combined area half again as large as Rhode Island,[14] yet additional facilities are being built. As is true in most humid areas, reservoir facilities cannot realistically be built large enough to contain the runoff and precipitation of the wettest periods; hence, water that could be useful is allowed to escape. Even so, six thousand or more permanent residents from more than twenty villages and hundreds of farms have been displaced from the now inundated areas which comprise the existing reservoirs,[15] and as the population increases the need for greater reservoir space will conflict more and more with the larger numbers of persons to be displaced.

12 – 3 *Orzan A, a product recovered from waste liquors from a paper mill. Instead of befouling the river, this lignin product is recovered and sold as a binder, dispersant, and wetting agent. (Crown Zellerbach Corporation)*

The phenomenon of water shortage in a humid area is, as suggested earlier, due in considerable part to the concentration of demand and withdrawal. In the case of that proportion of water which comes from subsurface supplies, disastrous side effects are liable to accompany the primary problem of insufficient supply. New York City found that unrestricted withdrawal of ground water resulted in invasion by saline water threatening permanent loss of this source of supply. Insistence by law on limited withdrawal and artificial recharge along with expanded surface supplies has lessened this threat. A significant consideration in this regard is the decrease in surface recharge area in large cities resulting from the large amount of surface area paved or covered by various kinds of structures. Of course ground water can and will move into these areas from all sides, but if one side adjoins the sea the opportunity for saline invasion is thus enhanced.

As the population of an urban unit increases, so must the amount of its usable water source. In the case of ground water, reduction in demand upon a limited volume of aquifer should minimize depression cone draw-down and contribute to maintenance of a more steady water table level. Surface water supplies may be increased by acquiring additional watershed areas and extending supply lines greater distances and to more sources; but competition from neighboring cities in densely populated sections of the nation is making this increasingly difficult. A more immediate source is closer at hand. Estimates of water loss through leaks in the New York City supply system exceed thirty million gallons per day. An inadequate metering system (covering only about 25 per cent of the domestic users) encourages deliberate or unconscious waste. It is suggested that installation of water meters in all residences and adoption of a realistic pricing policy could reduce domestic water use by as much as two hundred million gallons per day.[16] To be sure, the achievement of such measures would not be possible without encountering problems, not the least of which are political considerations. It is not likely that the task will become easier in the future, however, and alternate solutions are becoming increasingly difficult.

CONSERVATION AND INDUSTRIAL USE

As the fastest growing user of the water resource, industry needs to give special attention to conservation. Two factors are worth noting at the outset in this connection. First, the bulk of the industry is, and probably will continue to be, in the humid eastern half of the country where the water supply is most plentiful. Secondly, most of the water used by industry is used for cooling purposes, a use which consumes little. Rather than suggesting that there is no need for water conservation by industry, these two points are made to indicate that there is no need for industrial water problems on a national scale if sound conservation policies are formulated and implemented. The basic use pattern of water by industry thus is amenable to conservation. That problems do exist, however, has already been noted.

To a large extent, the problem of water quality is of industry's own making. In crowded industrial sections where numerous installations use the same river, upstream waste discharge locally has caused neighboring industries to seek other water sources or install expensive treatment equipment. The nature of some of the ingredients in industrial effluents today poses as great a problem as the amount. A few industries, realizing the unknown potentials of the combinations of their waste and that already present in streams, have accepted the responsibility of thorough treatment of their own waste (Figure 12 – 4). A large eastern chemical firm which manufactures over eight hundred chemical products has constructed a treatment plant capable of handling seventeen million gallons of waste water per day. Another firm has joined in cooperative effort with three neighboring towns to build a twenty-one million gallon per day facility to handle both industrial wastes and sewage.[17]

It is evident from the passage of the Federal Water Quality Act of 1965 that water quality standards are going to become a feature of each state's conservation program, because the act authorizes the Federal government to set standards if the states does not. Standards are necessary to the orderly development of pollution control, and will make feasible the monitoring that is a fundamental part of pollution control. Examples of the fact that projects for control can be cooperatively and successfully developed include the Ruhr region in Germany and the Ohio River valley in the United States.[18]

The Ohio River Valley Water Sanitation Commission (ORSANCO) was established in 1948 through the cooperative efforts of eight adjoining states, and it has raised the proportion of the local population having sewers served by treatment plants from 38 per cent in 1948 to 90 per cent in 1966. Thousands of industries were discharging treated or untreated waste into the river with little effective control being possible in 1948; in 1966 four-fifths of the industrial establishments were operating waste control and treatment facilities rated adequate by the signatory states.[19]

Industry is aware of the necessity for water pollution control; indeed, a significant proportion of the nation's manufacturers have had pollution abatement procedures in operation for many years. A portion of the manufacturers of the nation were surveyed on the costs of treatment of final waste waters before discharge; an estimated one hundred million dollars were spent for this purpose in 1959 by the firms surveyed.[20] This figure represents only a fraction of the cost of water pollution to the industries in question, because most will have treated the water while it moved through the plant at least once, and many will have had to treat it before it could even begin to go through the plant. Additionally, this figure represents the cost of *operation* of the pollution abatement facilities, not their installation. Yet it is clear that industry must look forward to markedly greater and more universal effort as the complexity of its processing with its diversity of effluents and the sheer volume of its operations continue to increase.

Additional practices designed to reduce the quantities of water needed per plant should complement efforts to preserve or improve quality. Data on water re-use were cited earlier as an example of what could be done in this respect. One of the most frequently cited exam-

12–4 ***A storage basin for waste liquid chemicals.*** The forty-six acre seepage-proof basin outlined in white is on an unused island in the Columbia River near Camas, Washington. (Crown Zellerbach Corporation)

ples of industrial water-use conservation is at the Kaiser steel plant at Fontana, California. Here the production of one ton of steel requires an intake of a fraction of the water used in conventional plants. With an organization of the water-use system to fit an area notable for its dearth of water, the plant employs many devices to conserve the water such as having subsequent uses of it be those which can employ water of somewhat higher temperature.[21] Other suggestions include cooling with fans, reducing pressure in piping systems, using separate pipes for different "grades" of water, extensive metering, more insulating of pipes, seeking water substitutes, and using lower grades of water. In the last case both salt water and sewage effluent have been successfully incorporated into industrial operations. The Bethlehem Steel Company's Sparrows Point plant uses 590 million gallons of water daily for cooling purposes. While some of this water is fresh water, and some is saline from Chesapeake Bay, almost a fifth of it is sewage effluent from the nearby urban centers' treatment plants. An average of 110 million gallons per day was used in 1960, and 150 millions per day from this source were to be used by 1965.[22]

Technological and scientific improvements can and must improve the conditions of industrial water use if the quantity and quality of this resource is to suffice for demands which will be put to it in the future.

CONSERVATION AND IRRIGATION USE

The impressive stature of irrigation as a user of water is accompanied by an equally impressive host of conservation problems. Work on all of these problems is in progress, but successes in most instances have been modest, and the need for much more research has become evident.

The great loss incurred in moving water from the source to the fields has received considerable attention, and numerous suggestions for its reduction have been advanced. Unlined irrigation ditches may lose more than 10 per cent per mile through zones of porous soils.[23] Conveyance ditches lined with materials such as clay, plastic, and concrete to restrict loss by seepage, and completely enclosed pipes above or below ground are in use locally and have proved effective. Although many minor problems remain unsolved, the chief problem is economic rather than technical. The proper precautions for seepage reduction are somewhat expensive despite the availability of some subsidies. While he may make more water available for others, a user with adequate water rights would not increase the amount of water for his own use by lining his canals and laterals, nor is he likely to be fully reimbursed for having done so. Here is an opportunity for legislation which will encourage conservation.

In addition to loss of water by seepage, great losses are incurred through evaporation and useless transpiration. Atmospheric conditions in areas where irrigation is necessary are ideal for extensive evaporation and, aside from completely enclosed conveyance systems, elimination is virtually impossible. Reduction of such loss in storage areas is possible, however, by restriction of surface area of the water body through creation of deep narrow reservoirs. Use of and further research on substances such as *hexadecanol* which forms a monomolecular film that retards evaporation without significantly affecting the usefulness of the water also shows progress. Estimates indicate possible national annual reduction in reservoir evaporation losses of as much as two million acre-feet.[24]

Substantial waste of carefully collected water by *phreatophytes* is another problem receiving attention. Literally translated, the term means "well plant," and refers to plants whose roots extend down into the capillary fringe or the saturated zone itself.[25] In the dry lands they grow along stream beds. Cottonwoods, saltgrass, willows, and greasewood are examples. Some grow in the water itself, and, in the case of salt cedar (tamarisk), consume more water than an equal area of irrigated crops.[26] Undesirable phreatophytes (some have use and value) should be eliminated or replaced by plants with lower water consumption rates and greater usefulness. In 1960 the seventeen western states were estimated to have had over sixteen million acres occupied by nonbeneficial phreatophytes, which are believed to have used between twenty and twenty-five million acre-feet of water annually.[27] Although programs of control are underway, the spread of these plants has increased at an alarming rate, especially in the southwest, and their elimination provides a task of monumental proportions.

Of the numerous other problems accompanying irrigation water use, many simply require extensive planning for their solution. The balance of supply and demand of water for irrigation, present and future, is essential if problems like that of excessive withdrawal in the Southwest are to be avoided (Figure 12 – 5). Readily available surface water supplies are already largely committed in the seventeen western states,[28] and further significant expansion of irrigation there should be preceded by development of additional supplies or improved use of supplies or both. Similarly, careful assessment of the likely effects of alteration of the water table in areas to be irrigated should precede any major projects which would affect it. The ounce or more of prevention may serve related resources in as important a way as the water or land itself.

FLOOD CONTROL

Large dams represent man's most obvious and spectacular effort to restrain the headlong rush of water back to the sea, sometimes for the purpose of making it serve him, sometimes to restrict its destructiveness, and often for both of these reasons. Dams are, however, but a part of the vast array of physical facilities and measures that are necessary for effective control. Moreover, all of the many activities included in the management of upstream watershed areas

12 – 5 ***The product of misuse is waste.*** **(Soil Conservation Service)**

(controlled logging and grazing, reforestation and re-establishment of healthy and useful cover in grasslands, the myriad conservation practices associated with tilling the soil designed to slow down the rate of water movement and thereby control erosion as well as to further "soak-in," the small dams and ponds in the areas of steeper slopes which seek to frustrate the role of gravity before it can create unmanageable velocities), and many more are necessary for the continued function and usefulness of the large dams (Figure 12 – 6). Without the assistance of upstream measures, large multipurpose dams, which commonly have only a limited amount of storage space available for flood waters, would make only the briefest contribution to flood control. Moreover, their usefulness would be eliminated rather quickly through sedimentation.

The upstream conservation activities have an equally important additional role. Floods occur in headwater areas as well as the lower courses of streams, and the upstream measures and structures are at least as much for control of such floods as for those downstream. The enactment of the Flood Prevention Act of 1954, commonly referred to as Public Law 566, was a recognition of the fact that even floods on small streams required engineering works for effective control.

The best efforts to create a watershed which will give up its water reluctantly are insufficient, however, in the case of the weather extremes. Even a pristine wilderness cannot prevent a flood when widespread high-intensity rainfall persists, as evidenced by present-day floods in such areas and records of floods in the journals of early explorers of America. In this instance numerous correctly spaced reservoirs, each capable of absorbing a significant proportion of the water; little-used areas where overflow can be allowed; straightened and reinforced por-

12 – 6 ***Small watershed erosion and flood control.*** **Note the contours, terraces, grassed waterways, and dams, all designed to slow down runoff. (Soil Conservation Service)**

tions of the stream course; and dikes and levees are all needed to control these occasional floods. The role of the large dam and its associated reservoir is thus a complementary one, no more and no less.

Efforts of the Federal government toward flood control in areas of private ownership may be thought of as being mainly in the hands of two agencies.[29] The Corps of Engineers is, among other things, concerned with downstream structures and measures for flood control, while the Department of Agriculture, incidental to its concern with the nation's land resources, has the task of dealing with upstream areas. While large structures were originally considered to belong in downstream areas, advancement of the government further and further into the flood control field has resulted in expanded activities in upstream areas with an accompanying increase in the size of structures there.[30] Regardless of where upstream ends and downstream beings, or which Federal agency is concerned, flood control requires coordinated effort throughout the watershed and not only by Federal agencies. State agencies, local organizations, and individual efforts are as essential as those of the Federal government. The best conservation policy would result from each of these serving in that capacity in which it would be most effective and for which it is best suited. The achievement of such a cooperative program is as important as the solution of the many scientific and technological problems receiving so much attention.

As mentioned earlier, there is one cause of flood damage that has little to do with natural factors, namely the increased settlement and industrial expansion on floodplains. While there may be perfectly sound reasons for wanting to be immediately adjacent to a transportation artery or water source, the fact remains that from the present viewpoint flood control does not mean flood prevention. In those instances when prolonged widespread rains fill the best natural and man-made reservoirs and still continue, or spring "snowmelt," frozen ground, and high intensity rains combine their effects, floods will occur.

From a projection of continued use and development of flood plains it has been estimated that the current rate of Federal expenditures will just about enable flood protection to keep up with increases in flood damage until 1980. Even a substantial increase in engineering works, it is estimated, will succeed only in holding damage to about a half billion dollars a year, on the average.[31] If more complete education of the populace on this situation is ineffective, it seems evident that some form of restrictive legislation on the state or local level will be necessary. Restriction of these areas to uses which would not involve extensive loss of life and property during extreme though infrequent flood conditions has been suggested.[32] Section 206 of the Federal Flood Control Act of 1960 points to the need for regulation of floodplain use and development by states and municipalities, and authorizes the Corps of Engineers to provide them with requested information on floods, flood damage, and engineering procedures.[33]

CONSERVATION AND WATERPOWER, TRANSPORTATION, AND OTHER USES

WATERPOWER. Use of the water resource for the generation of electric power has several favorable aspects from the conservation standpoint. Such use only momentarily diverts the water and consumes virtually none of it. It serves also to help prolong the supplies of other energy sources, especially the depletable mineral fuels. It is an endlessly renewable resource which, with proper management of the installation and watershed, can continue to provide a source of power through several generations of users. The reservoirs created to assure a constant source of water for power generation often make a contribution to local recreation also, as camping, fishing, boating, and swimming facilities may be established by the management organization, public or private. The facilities provided along the shores of the several lakes created by Pacific Power and Light Company dams along the Lewis River in southwestern Washington provide a case in point. Merwin Picnic Park adjacent to Merwin Dam, Speelyai Bay and Yale boat launching sites downstream and upstream respectively from Yale Dam, and Cougar Creek campground, a few miles downstream from massive Swift Dam are examples of these facilities. Cooperation between the Company and state game and fishery departments has resulted in trout stocking programs for the River and reservoirs and research devices and methods for getting fingerling salmon downstream past dams.

Although the United States ranks above most nations in terms of developed waterpower as compared to potential (Figure 12 – 7), there remains still a greater potential waterpower resource. The dams required for development and the waters they impound still create problems of displacement of settlement and of land coverage, however, and such factors have to be balanced against the need for service and that which would be rendered by the hydroelectric installation. While the "fuel" to run the generators may seem to be free, the cost of the installation and the retirement of the land from other use are expenses that require careful consideration. Development of the water resource for power in those areas suited for and needing hydroelectricity in such fashion as to maintain its reasonable integrity for other uses can and should be the goal, but more research on the basic data needed for the establishment of these criteria is under way.

TRANSPORTATION. As mentioned earlier, use of the waterways of the United States has been increasing substantially over the past two decades. During the eleven-year period, 1946 to 1957, for example, total inland waterways traffic increased to four times the 1946 figure (Figure 12 – 8). The amount of increased use has been likened to the additional transport facility that would be provided the nation by the addition of another transcontinental railroad for each of those eleven years.[34] A major factor contributing to this increase was the development of modern highly efficient "towboats." Diesel powered and designed specifically to fit the task, a 4,000 horsepower towboat can handle over 20,000 tons of freight. While the more than 4,000 towboats presently in use on American waterways average between 1,000 and 4,000 horsepower, the trend is toward more powerful boats like the 8,500 horsepower *United States* (Figure 12 – 9). It is believed that the use of

12 – 7 *Grand Coulee Dam.* The pumping plant, discharge pipes, and feeder canal serve the Upper Grand Coulee Reservoir. (Bureau of Reclamation)

the waterways will increase markedly in the future; intercity freight traffic, for example, is visualized as experiencing a several-fold increase in the next several decades (Figure 12 – 10). Inland waterways play an important part in the total transportation picture of the nation; many of the low unit-value items thus transported would probably move only short distances or not at all in the absence of this form of transportation. In the case of strategic materials the delivered price would rise significantly. The economical aspect of water transportation is a conservation factor in itself, even if considered only in terms of fuel requirements per ton-mile. A factor occasionally overlooked in attempts to assess the dollar value returned from expenditures to improve inland waterways facilities is their use by sportsmen and vacationers and for weekend family outings. With the number of families owning boats increasing, along with the increase in time away from work, recreational use of the waterways is clearly destined to increase. For these and many other reasons the maintenance and necessary improvement of water transportation facilities should continue to advance as needs dictate. A way will be found to resolve the question of how such expenditure should be financed if sufficient effort is devoted to it.

The problems referred to in the discussion of this use of the resource can and will be solved as research and data collection show the way. Steps already taken imply a growing awareness that maintenance of the waterways is interwoven with, depends upon, and contributes to conservation of other resources. As ability to utilize the water resource more fully advances, water transportation as a logical and economical use of the water will come closer to a position commensurate with its early importance to this and most nations.

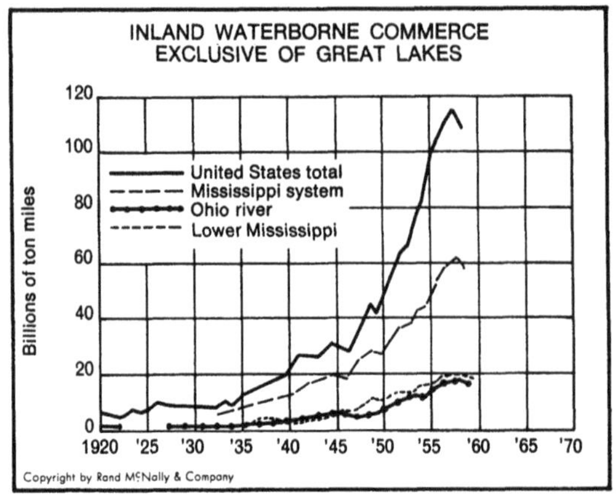

12 – 8 ***Inland waterborne commerce, exclusive of Great Lakes.*** **(From "Water Resource Activities in the United States, Future Needs for Navigation," Select Committee on National Water Resources,** ***Senate Committee Print No. 11*** **[Washington: Government Printing Office, 1960.])**

12 – 9 The M/V United States on the right with a tow of 21,651 tons, enroute from the Gulf Coast to Great Lakes and river ports. **It is passing the M/V** ***America*** **with a south bound tow of 26,170 tons between the same ports. (Federal Barge Lines Inc.)**

Outlook

It was pointed out early in the discussion of water resources that the increased use of water was due to two factors: greater use per capita and more people. These same two factors will continue to cause ever greater demands to be levied against the available water supplies. Estimates of the future population distribution in the United States along with the land and water resources are shown in Table 12–1. The unequal distribution of people and water is notable.

The question of the adequacy of the nation's water resource, which would have seemed utterly pointless only a generation ago, is being debated at length by many writers today.[85] Numerous approaches to the question have been followed, and the conclusions drawn are variable. There are several good reasons for this lack of agreement; perhaps the foremost is that the figures we have to work with are still only approximations of an average or hypothetical situation. Much basic research is necessary before a greater degree of reliability can be attached to the quantification of the actual hydrologic cycle and the role of ground water in it. Also, the spectre of dry cycles requires consideration of below-average supply periods and all of the interrelated problems they would raise; the basic supply is inconstant. Finally, estimates of demand must be predicated upon

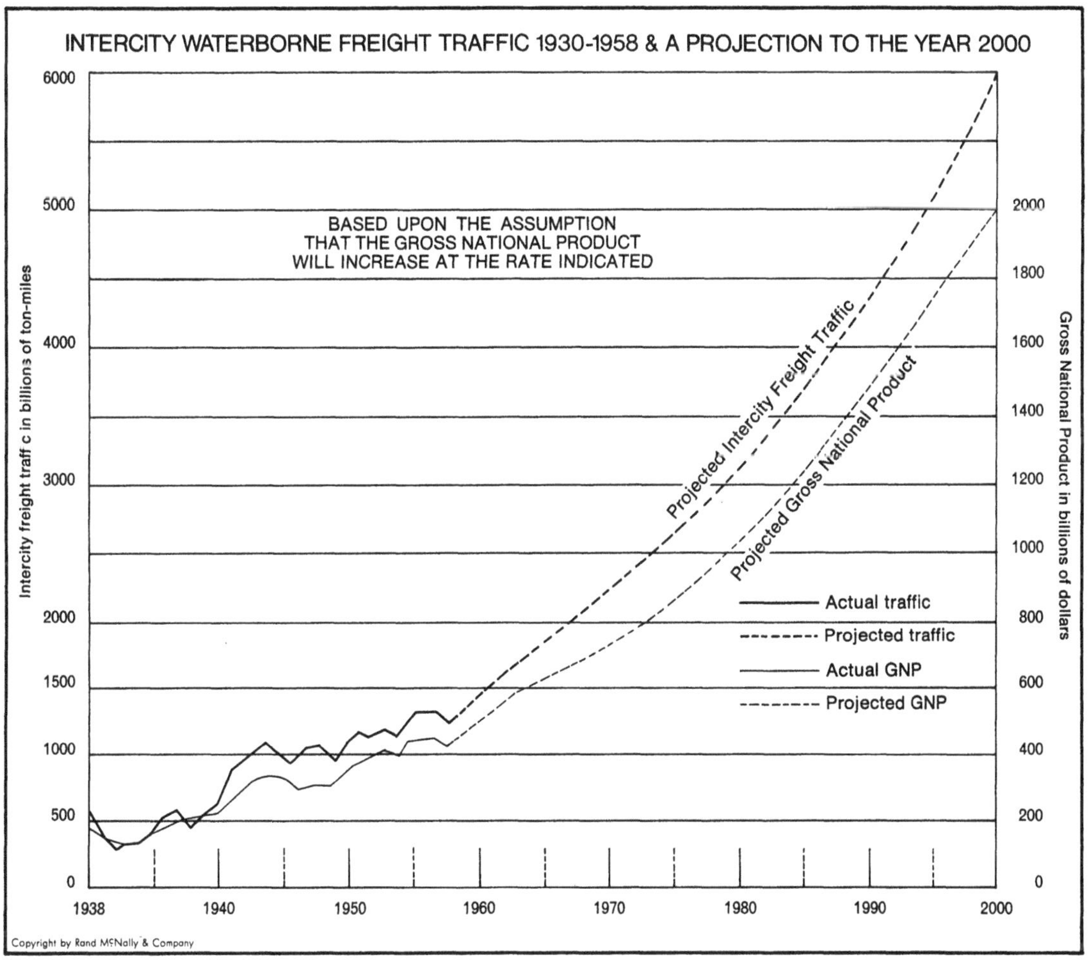

12 – 10 ***Intercity waterborne freight traffic 1930-1958 and a projection to the year 2000.*** **(From "Water Resources Activities in the United States, Future Needs for Navigation,"** ***op. cit.,*** **p. 6.)**

TABLE 12–1
REGIONAL DISTRIBUTION OF POPULATION (2000 A.D.), LAND AREA, AND RUNOFF*

Region	Population (2000 A.D.)	Land area	Runoff (water supply)
	per cent	per cent	per cent
Northwestern 8 states	5–6	21	18
Southwestern 9 states	23–24	38	14
Northeastern 22 states	53–54	26	38
Southeastern 9 states	17–18	15	30

* "Water Resource Activities in the United States, The Impact of New Techniques on Integrated Multiple-Purpose Water Development," Select Committee on National Water Resources, United States Senate, Committee Print No. 31 (Washington: Government Printing Office, 1960), p. 7.

"projections" of population increase, industrial expansion, urban settlement growth, and developments in irrigation agriculture, each of which is a complex variable in itself.

Despite the vexation of having to work with so many variable causes, work with them we must. Too many sober appraisals, carefully developed to reduce the variables as effectively as possible, point to the same conclusion: A decided change of attitude toward and treatment of water resources is essential to the future adequacy of the resource. Water must be cleaned more effectively and more thoroughly before it is released after use. Water must be reused more frequently than at present. More care must be taken to reduce waste and loss through evaporation. Additional practices and structures designed to hold the water on the land must be employed where appropriate. Substitution of lower grades of water for use in less demanding situations should become a more common practice. Only through the employment of these and similar actions can we be assured of making the most effective use of the supply of water which is naturally furnished and readily available.

In addition to the numerous known measures advocated for making better use of what exists, there are efforts to increase the

12–11 *A spherical weather experiment laboratory.* Raindrop size, rate of growth, temperature, and electrical properties are studied under controlled conditions in this sixty-foot sphere. (Weather Bureau)

basic amount of water. The most familiar of these efforts are weather control and conversion of salt water to fresh. In 1952 the Office of Saline Water in the Department of the Interior was established by Congress to advance and coordinate research on the conversion problem. Economical conversion of salt water to fresh is believed inevitable, although the solution will come grudgingly and with presently indeterminate delay. In New York, it is planned to build a nuclear-powered multipurpose plant at Riverhead which would offset the cost of fresh water produced through the sale of radioisotopes and the generation of electricity. Some success with brackish waters has been achieved, but economical desalinization of sea water is still an elusive prize. In any case, this possible source would be limited in use area by the cost of transportation. The Weather Bureau and the National Science Foundation are similarly concerned with weather control. While knowledge has been advanced significantly in the relatively short time since the idea gained prominence, results of experiments have been, by and large, inconclusive to date. In view of the complexity of atmospheric processes, it seems unwise to put much dependence upon its becoming a significant source of fresh water in the near future (Figure 12 – 11).

Aside from efforts to make greater use of the readily available natural supply and to obtain additional supplies from the atmosphere and the sea, there remains yet another way to offset local water shortages: long distance transfer. Numerous examples of extant water transfer may be pointed out, ranging from simple extension of supply lines to neighboring sources to rather more sophisticated systems

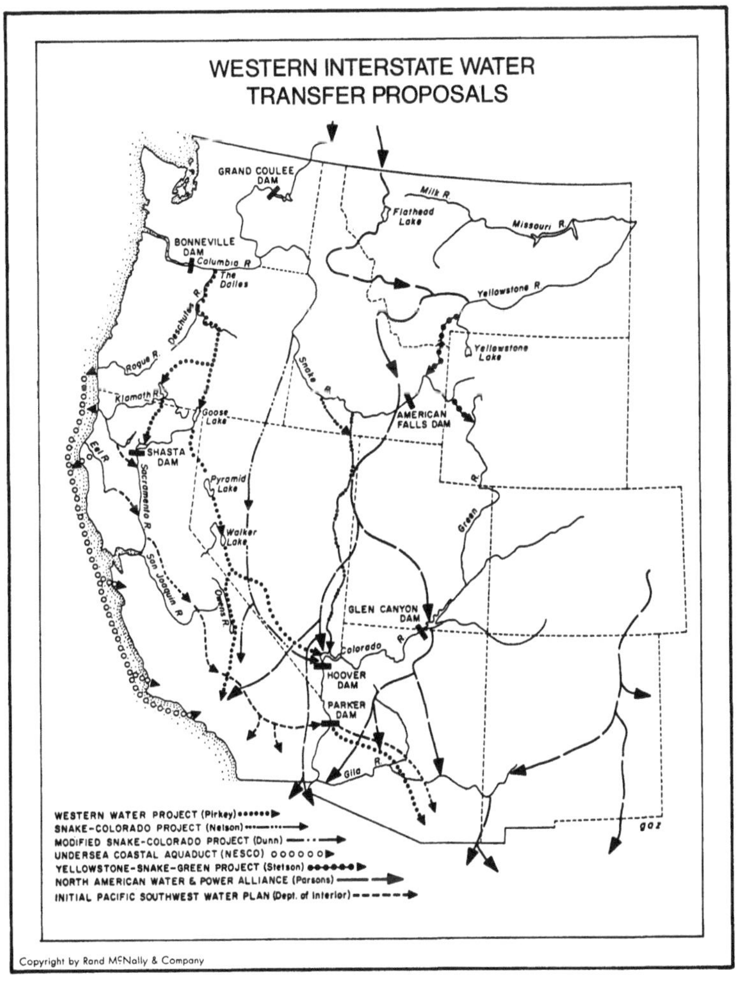

12 – 12 *Western interstate water transfer proposals.* (From Ernest A. Englebert, ed., *Strategies for Western Regional Water Development,* Proceedings of the Western Interstate Water Conference, Corvallis, Oregon, 1965. [Los Angeles, University of California Printing Dept., 1966] 195 pp.)

covering more area and distance and involving complex exchanges and balances such as are operative in the present California program.[36] Long distance transfer as used here is meant to imply substantially greater distances than any present system involves; not only interstate transfers, but transfer on a continental scale is envisioned. A number of schemes have been proposed for the movement of water from the water-rich Pacific Northwest states to the southwestern states (Figure 12 – 12). Impressive and controversial as such schemes may be, they are diminished in stature by the North American Water and Power Alliance proposal which would redistribute water from water-rich areas on the continent to water deficient areas in Canada, The United States, and Mexico. Indeed, in recognition of the international stature that water management problems have achieved, the period January 1, 1965, through December 31, 1974, has been designated the International Hydrological Decade, a period of research and of collection and exchange of data and ideas among nations.

Redistribution proposals which would transfer water across political boundaries raise a host of questions. In addition to political, economic, and legal considerations,[37] there is the question of what effect the transfer of water in such quantities would have on the natural environment of both the recipient and source areas. The number, complexity, and scope of the questions raised suggest that their investigation has not begun any too soon, nor is the amount of effort thus expended likely to be too much.

Although research should and will continue, the most promising way of alleviating existing water shortages and preparing for the immediate future is simply through the nationwide implementation of known conservation practices. With the wholehearted support of an informed citizenry and properly coordinated effort at the several levels of government, the quantity and quality of the present water resource can be improved so that it will serve the needs of the immediate future.

To do this much is to do no more than should be expected of the people; to do less is to invite hardship, needless expense, and waste less excusable than that blamed on our predecessors, for the people now have the knowledge and means available to avoid it.

FOOTNOTES

[1] For a notable and far-reaching example see the Pelton Dam Case, *United States* vs. *State of Oregon*, in which the Federal government asserted that its right to regulation of water resources was superior to that of the state on the grounds that the water flowed from, and contiguous to federal Indian reserve land.

[2] *A Water Policy for the American People, the Report of the President's Water Resources Commission*, vol. 1 (Washington: Government Printing Office, 1959), p. 277. Chapter 19 of this volume is a summary of the more extensive coverage of water law in the companion volume, *Water Resources Law, The Report of the President's Water Resources Policy Commission*, vol. 3 (Washington: Government Printing Office, 1950), 777 pp.

[3] *Water Resources Law, op. cit.*, p. 279.

[4] Ben Moreell, "Federal Water Policies—Past and Present," lecture 1 of the series on *Our Nation's Water Resources—Policies and Politics* (Chicago: Univ. of Chicago Press, April 30, 1956), p. 12.

[5] *Ibid.*, p. 9.

[6] Abel Wolman, "Water Resources, A Report to the Committee on Natural Resources," *Publication 1000–B*, (Washington: National Academy of Sciences—National Research Council, 1962), p. 12.

[7] *A Water Policy for the American People, op. cit.*, p. 289.

[8] "Water Resource Activities in the United States, The Impact of New Techniques on Integrated Multiple-Purpose Water Development," Select Committee on National Water Resources, United States Senate, *Committee Print, No. 31* (Washington: Government Printing Office, 1960), p. III.

[9] For a treatment of the many facets of this point see Walter B. Langbein and William Hoyt, *Water Facts for the Nation's Future*, (New York: The Ronald Press Company, 1959).

[10] *Water Resources Law, op. cit.*, p. 306.

[11] For a discussion of this point see Charles H. Callison, *America's Natural Resources* (New York: The Ronald Press Company, 1957), p. 52.

[12] Andrew C. Glass and Kenneth H. Jenkins. *Statistical Summary of 1962 Inventory, Municipal Waste Facilities in the United States* (Washington: Public Health Service, Department of Health, Education and Welfare, 1964), p. 20.

[13] For a depiction of the effect raw domestic sewage has on stream biota see Alfred F. Bartsch and William M. Ingram, "Stream Life and the Pollution Environment," *Public Works*, vol. 90 (July, 1959), p. 104. An excellent data summary and bibliography of biological water pollution investigation is found in William M. Ingram, Kenneth F. Mackenthun, and Alfred F. Bartsch, *Biological Field Investigative Data for Water Pollution Surveys* (Washington: Department of the Interior, Federal Water Pollution Control Administration, 1966).

[14] Anastasio Van Burkalow, "The Geography of New York City's Water Supply: A Study of Interactions," *The Geographical Review*, vol. 49, no. 3 (1959), pp. 369–86.

[15] *Ibid.*

[16] R. W. Derrick Sewell, "The New York Water Crisis," *Journal of Geography*, vol. 65, no. 8 (1966), pp. 384–89.

[17] "Water Resource Activities in the United States, Present and Prospective Means for Improved Reuse of Water," Select Committee on National Water Resources, United States Senate, *Committee Print No. 30* (Washington: Government Printing Office, 1960), p. 5.

[18] See "Water," a special report in *Power*, vol. 110, no. 6 (1966), 48 pp.

[19] "Water in Industry" (New York and Washington: The National Association of Manufacturers and the Chamber of Commerce of the United States, January, 1965), p. 39.

[20] *Ibid.*

[21] Harry E. Jordan, "The Increasing Use of Water by Industry," *Water, the Yearbook of Agriculture, 1955*, (Washington, Department of Agriculture, 1955), p. 654.

[22] "Water Resource Activities in the United States, Present and Prospective Means for Improved Reuse of Water," *op. cit.*, p. 8.

[23] Wayne D. Criddle, "Reducing Water Losses from Storage and Conveyance," *Proceedings of the National Water Research Symposium* (Washington: Government Printing Office, 1961), p. 144.

[24] *Ibid.*

[25] H. C. Fletcher and Harold B. Elmendorf, "Phreatophyte—A Serious Problem in the West," *Water, the Yearbook of Agriculture, 1955, op. cit.*, p. 423.

[26] Walter B. Langbein and William Hoyt, *op. cit.*, p. 151.

[27] "Water Resource Activities in the United States, Evapo-Transpiration Reduction," Select Committee on National Water Resources, United States Senate, *Committee Print No. 21* (Washington: Government Printing Office, 1960), 42 pp.

[28] Langbein and Hoyt, *op. cit.*, p. 145.

[29] For a historical summary of flood control legislation up to 1960 see Section III in "Water Resource Activities in the United States, Floods and Flood Control," Select Committee on National Water Resources, United States Senate, *Committee Print No. 15* (Washington: Government Printing Office, 1960), pp. 10–15.

[30] For an interesting treatment of this topic see Luna B. Leopold and Thomas Maddock, Jr., *The Flood Control Controversy* (New York: The Ronald Press Company, 1954), 278 pp.

[31] "Water Resource Activities in the United States, Floods and Flood Control," *op. cit.*, pp. 48, 49.

[32] For a consideration of this point see Francis C. Murphy, "Regulating Floodplain Development," *Department of Geography Research Paper No. 56* (Chicago: The Univ. of Chicago Press, November, 1958), 204 pp.

[33] *Ibid.*, p. 13.

[34] Water Resource Activities in the United States, Future Needs for Navigation," Select Committee on National Water Resources, United States Senate, *Committee Print No. 11* (Washington: Government Printing Office, 1960), p. 5.

[35] For example see Arthur M. Piper, "Has the United States Enough Water," *Geological Survey Water-Supply Paper 1797* (Washington: Government Printing Office, 1965), 22 pp.; John C. Maxwell, "Will There Be Enough Water," *American Scientist*, vol. 53 (March, 1965), pp. 97–103; and "A Study of Pollution–Water," *staff report to the Committee on Public Works*, United States Senate (Washington: Government Printing Office, June, 1963), 100 pp.

[36] For a diagram of this system see "Water," *Power*, *op. cit.*, pp. 518–19.

[37] For an interesting comment on this factor see Irving K. Fox, "Water–Supply, Demand, and the Law" reprint no. 15, (Washington: Resources for the Future, January, 1960), 7 pp.

13

Resources of the Salt Water Margins

THE OCEANS OF THE WORLD, covering nearly three-fourths of the surface of the globe, provide a resource which serves the continents in many and varied ways. The most important of these is the one most often overlooked—the oceans are the primary source of atmospheric moisture which supplies the continents with the fresh water so necessary to life. As regulators of temperature, both world-wide and along continental margins, the oceans perform a vital function in making the continents more habitable. Were the areal extent of land masses and oceans to be reversed, the climates of the resultant land masses would be strikingly different from those which exist.

The oceans have long been a source of food. Although the total world catch of fish does not approach in value the annual harvest of any one of the major grains, millions of people are provided with more food, a better balanced diet, and a means of earning a livelihood through fishing. Nor are the fish caught restricted to those which can be used for food, as evidenced by the overwhelming dominance in the total volume of the menhaden catch, almost none of which is used for human food. In 1964 only 55 per cent of the catch in the United States was used for human food.[1]

Since the time of the Phoenicians man has used the seas increasingly for transportation and commerce. The seas linked the infant United States to the Old World, where markets for her products supported an occasionally faltering economy. Although international trade has been a proportionately small factor in the United States economy in modern times, the actual amount is significant. As interdependence of the world's nations increases, the United States, with three coasts, will utilize the seas for transportation to greater extent.

The sea has long served as an inexhaustible source of at least one mineral, salt. While its continued importance in this respect is probably assured, the first steps taken to recover minerals present in lesser concentrations (such as magnesium, iodine, and bromine) afford a glimpse into the future when the estimated fifty million billion tons of dissolved materials in the sea will become an increasingly important resource.[2] As indicated in Chapter 12, sea water *per se* is being used in ever greater amounts for industrial tasks traditionally allotted to fresh water. Although little actual development has occurred to date, the first steps have been taken to harness some of the energy of the restless seas through employment of tidal energy also.

One of the most familiar uses of the seas, especially their borders, is for recreation. In the United States and many other nations, the increasing recognition of the importance of recreational facilities implies an increased future importance for the therapeutic values afforded by the scenery, climate, and natural sports facilities of the seashore.

Thus man has long made some use of the sea and its resources, but his attention and best efforts have been turned chiefly to the land, his natural habitat and chief provider.

The Limited Use of the Sea

Although men have sailed its waters for centuries, fought over its control, fished many and varied parts of it, and marveled at its changeable and diverse character, relatively little is known about the sea. Man is most familiar with its surface and some of the shallower sections adjacent to the continents; the great bulk of the sea remains essentially unexplored and unused. The surface area of the earth approximates two hundred million square miles, but almost three-fourths of it is sea surface. Even so, the remaining fifty-eight million square miles have proved sufficiently large to support man (albeit often marginally) throughout his existence on earth, despite the large proportion which is either unused or presently unusable. As world population has increased and the most readily usable cropland has been surveyed and resurveyed, recognition of the need for increased acreage has become widespread, and planning to increase the acreage and productivity whenever and wherever possible is under way. Prospects for the next several decades seem favorable in this and some other parts of the world, but the future beyond seems somewhat less assured. The rate at which population is increasing is alone impressive, but when this rate is applied to the numbers presently in existence, the specter of a population explosion looms ever more real. With the world ratio of cultivated land to population at about an acre per capita[3] now, it seems possible that increasing acreage and productivity may do little more than keep up with population increase. But whether the accomplishment on the land is this much or more, there remains a vast, little-used frontier which should supplement the land, the sea.

The story of man s development is one of increasing ability to use resources to shelter, clothe, and feed himself. His early status as a gatherer and hunter supplementing his finds with a rudimentary agricultural effort is a far cry from his ability to produce food on the land today. The suggestion is made, however, that his present use of the sea resembles rather closely his earlier use of the land—he gathers wild shellfish, hunts at sea, and practices a little rudimentary farming of oysters here and there.[4] Future methods of using and managing the sea and its resources offer the possibility of enormously greater production "per acre" as well as the additional use of portions presently little used or not used at all.

Reference to the inexhaustibility of the sea is based on a striking contrast with the land resource—the mobility of the sea. Although continued cropping of a piece of land without fertilization will rob the land of its store of nutrients, it is essentially impossible to remove the nutrients permanently from a portion of the ocean. And it is these same nutrients which are utilized by the *phytoplankton* which serve as food for the chain of animal-life forms present (ranging from zooplankton to whales), some of the larger members of which constitute the "crop" to be harvested from the sea's acres. Similarly, separation of minerals from sea water does not result in "mined-out" sections of the sea; by its own devices the sea continues to offer up, ever fresh, "ore" of essentially the same "tenor" year in, year out. Moreover, the waters running off the land that are often the despair of the soil conservationist are busily replacing many times over the miniscule withdrawals represented by today's proud harvest of the sea's mineral wealth. That which is lost to the land is gained by the sea, but man has not yet learned to use it.

No one knows the quantity of life the sea contains. Enough is known about the sections which have become accepted as the presently known best fishing areas to indicate that the numbers and variety present beggar description. Yet knowledge of the depths has only recently included the truly deep portions of the seas, and this in the form of relatively few and widely scattered glimpses. Limited though they are, they reveal that life is found even in the unimaginable pressures of the greatest depths and may therefore be presumed to exist at all depths. Even if distribution is highly variable, the immensity of the volume of space involved implies a resource which is largely unused at present.

To say that the total resources of the sea are essentially inexhaustible is one thing; to say this about individual forms of sea life is something entirely different. Man has repeatedly proved his ability to decimate certain species of sea life almost as effectively as some forms of animal life on the land. While man is in the hunting and gathering stage, the area he inhabits cannot support a very great number of hunters. As his numbers increase, the more desirable forms of game decrease. It is this same selectivity that presents one of the greatest dangers to the resources of the sea, especially when coupled with the lack of sufficient knowledge of the habits of the species sought. Unlimited catches of a few species with little knowledge of or consideration for such a factor as spawning time can hardly result in anything short of those species eventually becoming unobtainable. Man has significantly increased his ability to take fish in larger quantity, but this is not the same as having improved his ability to use the resources of the sea. The latter requires that he expand his selection of species to relieve the pressure on a few and, more importantly, that he manage and harvest the produce on at least a sustained-yield basis. Indeed, the possibility of increased yields based on far more intimate knowledge of the sea and its life, and the application of that knowledge in sound management practices should be no more unimaginable than what can be done now in conservation of forests, minerals, and land. The chief requirements are increased knowledge, the formulation of conservation practices suited to the sea, and international legal agreements. Probably the present limited use of the sea is a factor in man's favor. If management practices can be formulated and sufficient knowledge accumulated before the sea comes into its own as the last great earthly frontier, perhaps man can avoid many of the kinds of mistakes he made in the use of his natural habitat, the land. The development and application of the finest management practices by only some of the users of the sea, however, will not suffice. Effective conservation of the resources of the sea will become increasingly impossible without international legislation and cooperation.

United States Fisheries

FOOD FROM THE SEA

Although he may be using only a fraction of its potential, man is now, and long has been using the sea as a source of food. The overwhelming majority of food taken from the sea is comprised of forms of animal life—fish, crustaceans, mollusks, and the like. From the two most fertile sources of such food, the fishing grounds of the north Atlantic and north Pacific oceans, tens of billions of pounds are taken annually, representing about three-fourths of the world total.[5] The United States, whose fishing fleet utilizes parts of both of these areas, recorded a catch of nearly six billion pounds in 1964, being exceeded by Peru, Japan, mainland China, and the USSR.[6]

The United States has always been a fishing nation. The economy of early New England was significantly dependent upon fishing, salted codfish being one of the important items exported to Europe. So important was the codfish to Massachusetts that the Commonwealth Legislature in 1784 had an image of the cod hung in its meeting place, where it may be seen today. Atlantic cod no longer occupy this prominent a place in United States fishing, partly because of an actual decline in the annual cod catch, but also because of the rise in importance of other fishing areas and other species of fish.

VOLUME OF THE CATCH. The commercial fisheries of the United States produced a total catch of 4.5 billion pounds during 1964.[7] Although this figure includes the catch from domestic fresh-water sources, the overwhelming dominance of marine fisheries is evidenced by the fact that about 4.4 billion pounds, over 96 per cent of the total catch, came from marine waters and coastal rivers.

One species clearly dominates the total catch figure: Over 1.5 billion pounds of menhaden, which is used primarily for poultry and other livestock feeds, were landed in 1964. Although this figure is somewhat below that of previous years, the 33 per cent of the total catch which menhaden represents is typical. This fish is caught chiefly in middle and south Atlantic and Gulf coastal waters.

The three species next in rank as to volume were relatively close in 1964. The figure for salmon was 352.3 million pounds; for tuna, 305.8 million; and for crabs, 270.4 million.

Salmon are predominantly from Alaskan fisheries (Figure 13 – 1), although other West Coast states contribute; tuna (Figure 13 – 2) are chiefly the product of Pacific fisheries. Crabs are taken from Atlantic, Gulf, and Pacific fisheries. Although their relative importance may vary from year to year (Table 13–1), these species typically rank high.

These species include representatives from all three primary marine fish environments: Upper levels of shallow seas, lower levels, and estuarine or bay areas. Representatives from the upper levels of the shallow seas, *pelagic* fish, include herring, menhaden, and sardines. Those which live along the bottom of the shallower parts of the sea, *demersal* fish, include haddock and flounders. *Crustaceans* and *mollusks*, such as shrimp and crab, are common to estuaries, bays, and in general the shallower areas close to shore. The salmon is probably the best known example of *anadromous* fish which are born and spawn in fresh water, but spend their lives at sea. The variety of environments and types of fishing methods used along with the international use of the sea and its resources pose a real challenge for conservationists, present and future.

VALUE OF THE CATCH. The value of the 1964

13 – 1 ***A purse seine boat in operation.*** **After a set has been made the net is pursed and the fish brailed from it. These boats are fishing in southeastern Alaskan waters. (Fish and Wildlife Service, F. A. Davidson)**

catch by commercial fisheries in the United States was 389,498,000 dollars. Although many of the species ranking high in the volume of catch also rank high in value (for example, salmon and tuna), there are changes both in species and relative position of species. As shown by Table 13–2 shrimp led all species in value of the catch, although ranking fifth in volume. Gulf waters furnish over nine times the value of shrimp obtained from the next ranking region, the south Atlantic. Ranking third in importance are the little known Pacific and

TABLE 13–1

THE TEN TOP RANKING FISH AND SHELLFISH (VOLUME) CAUGHT IN 1964*

Species	Quantity (1,000 pounds)	Per cent of total	Record catch Year	Record catch 1,000 pounds
Menhaden	1,569,965	34.6	1962	2,347,944
Salmon	352,321	7.8	1936	780,884
Tuna	305,829	6.7	1950	391,454
Crabs	270,442	6.0	1964	270,442
Shrimp	211,821	4.7	1954	268,334
Flounders	176,351	3.9	1963	176,798
Haddock	133,498	2.9	1929	273,809
Sea herring	115,929	2.5	—	—
Whiting	94,233	2.1	1957	133,041
Jack mackerel	89,985	2.0	1952	146,522

* After C. H. Lyles, "Fishery Statistics of the United States, Fish and Wildlife Service, 1964," *Statistical Digest*, No. 58 (Washington: Department of the Interior, 1966), p. 28.

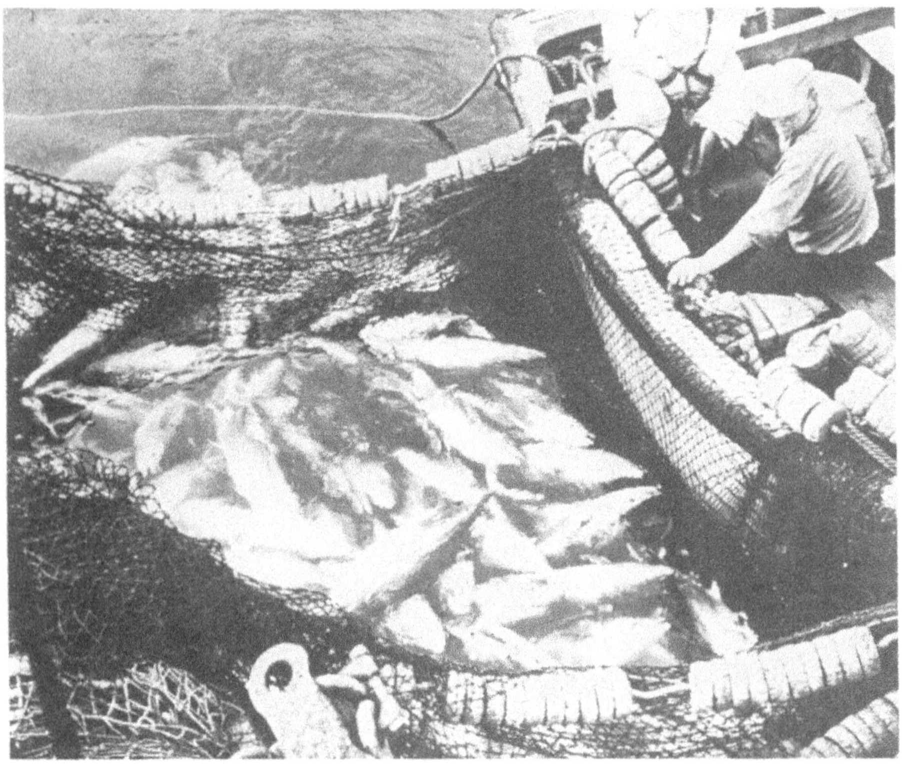

13–2 *A catch of tuna.* These fish have long ranked high among the species caught both in value and volume. Note the coarseness of the net, which allows smaller fish to slip through it. (Fish and Wildlife Service)

TABLE 13–2
THE TEN TOP RANKING FISH AND SHELLFISH (VALUE) CAUGHT IN 1964*

Species	Value ($1,000.00)	Per cent of total	Record value	
			Year	$1,000.00
Shrimp	70,376	18.1	1953	76,641
Salmon	55,995	14.4	1962	56,353
Tuna	39,398	10.1	1950	61,342
Oysters	27,926	7.2	1961	33,204
Crabs	23,813	6.1	1964	23,813
Menhaden	22,190	5.7	1956	28,425
Northern lobsters	19,856	5.1	1964	19,856
Flounders	15,006	3.9	1963	15,411
Clams	14,890	3.8	1964	14,890
Haddock	11,845	3.0	1946	13,043

* After C. H. Lyles, *op. cit.*, p. 29.

Alaskan shrimp waters which furnish small cocktail shrimp. The Chesapeake Bay fisheries lead the next ranking source of oysters, the Gulf region, by more than two to one. Northern lobster, the primary commercial variety, comes overwhelmingly from New England, with Maine alone harvesting over 70 per cent of the total United States catch by value.

NONFOOD PRODUCTS

An estimated 45 per cent of the total United States fish catch in 1964 was used for products other than human food. Among these products were fish meal for chickens, furbearing animals, and pets, fish oil for industrial use or export, and buttons manufactured from shells.

Although menhaden figures most importantly in the nonfood products derived from United States fisheries, other species of fish also are so used. The 1964 catch of "industrial fish" (unclassified mixed species used for bait, reduction to meal or oil, and animal food) totaled 262 million pounds, thus collectively ranking in the top ten categories with regard to volume. In addition, cannery by-products add significantly to these uses.

Because of the very considerable amount of imported nonedible products of fisheries (131 million dollars worth in 1964) from other nations, the data which follow on manufactured fishery products should not be related to earlier listed domestic catch figures; they are presented as an indication of the diversity and importance of nonfood (human) fishery products in the United States.

Almost five hundred million pounds of fish meal and scrap for chicken feed, fertilizer manufacture, and the like were produced in 1964. An additional 350 million pounds of fish products went into over forty-three million dollars worth of animal food. The value of canned salmon eggs for bait alone exceeded one million dollars in 1964. Oil produced chiefly from fish, but in part also from seals and whales, amounted to 180 million pounds, while the single biggest item with regard to volume was the 725 million pounds of lime and poultry grit made from oyster shells. Although relatively small in volume, over one million dollars worth of buttons were made from marine pearl shells and mussel shells. A variety of miscellaneous products which were collectively evaluated at over sixteen million dollars in 1964 includes kelp (a type of seaweed) products such as iodine, agar-agar (also from seaweed), irish moss extract, and fish feed pellets. Although they do not complete the list, fur from seals plus sponges and pearls serve to round out this summary of the major nonfood products of United States fisheries.

The purpose of this summary of the nation's fishing activity is to provide an indication of its status as a base for the consideration of some of the conservation problems which are already present, as well as those which will demand attention as man seeks to utilize the resources of the sea to a far greater extent. Although only 127,875 men were employed as fishermen in the United States during 1964, the number of additional people directly involved in the fishing industry was considerably greater. This is immediately evident when one considers those who construct and repair the ships, boats, and fishing gear, those who process and market the harvest, and so on. The sea can and will be made to support and feed vastly greater numbers, however, in this and other countries if man will but strive to have wise management accompany more intensive use.

Conservation and Fisheries

Conservation of the fishery resource involves consideration not only of the fish themselves but of their habitat as well. Indeed, elimination or despoilment of the habitat will result in the disappearance of the fish as effectively as will their unrestricted and wasteful harvest. Fishery habitats may be thought of in three categories: (1) Inland fresh-water streams and lakes: (2) the salt-water oceans; and (3) the meeting place of the two, the estuarine and coastal waters. This discussion of fishery conservation will be concerned chiefly with consideration of estuarine and coastal waters.

SIGNIFICANCE OF CONSERVATION IN THE SALT-WATER BORDERS

There are several sound reasons for special consideration of the estuarine and coastal waters. These areas are of outstanding significance to the species which comprise at least 60 per cent of the total combined sport and commercial fish catch.[8] The zones of brackish waters found here are the home of mollusks like the oyster and clam; they are the spawning ground for fish species such as the striped bass; here the shrimp develops and lives until time to return to the open sea to spawn; and their role in the lives of anadromous fish, while as yet incompletely determined, is recognized as an important one.[9]

Equally important is the consideration of this particular habitat as the most readily usable for present and future "sea farming." The protection afforded locally by coastal landforms, the proximity to the land, and the shallowness of the sea help contribute to the potential utility of such areas. Moreover, "this transitional area . . . is one of the most interesting of all marine ecosystems, and perhaps from a fishery viewpoint potentially the most valuable. In some places it is richer than the richest farm country, for it is lavishly fertilized with inorganic nutrients which the land is continually pouring into it."[10]

Coastal locations have special utility for man. In the United States the most productive of these estuarine areas are adjacent to some of the most densely populated and heavily industrialized parts of the country. They are thus subject to the effect of man's activities to a high degree. These factors suggest that nowhere are man's best efforts at conservation of fishery resources more immediately necessary or apt to yield greater returns.

POLLUTION. To the objections raised in earlier chapters concerning sewage and industrial waste being poured into streams must be added the effect of these contaminants in estuarine and coastal waters. Despite the greater volume of water into which these materials are poured, and the mixing effect of currents, direct toxic effects and reduction of dissolved oxygen in the water is taking its toll in bays, river mouths, and along the shore as well as in the inland water bodies. Nor is the source of pollutants restricted to the land, for that matter; it is along coastal margins and in the harbors that ship concentrations are greatest, and where much of their "housecleaning" is done.

It is in fact suggested that pollution is probably the single biggest factor causing habitat loss in coastal areas.[11] Along with industrial waste and sewage, the effect of chemicals sprayed on coastal areas for the control of vegetation or insect life should be included. There are too many instances of accidental "fish kill" from this and similar episodes on record already. Pollutant discharge into the sea can be at least as costly as in inland areas, and potentially even more so with regard to the total fishery resource. It is the sea and its margins that furnish the majority of the fishery products; moreover, the future importance of these areas will be greater, not less. True, the volume of the sea dilutes the discharge being poured into it to harmless concentrations, but before this dilution has been achieved the pollutants must pass through many of the most productive fishing areas.

EFFECT OF OTHER UPSTREAM ACTIVITIES. About half the total commercial fisheries' catch is made up of species dependent upon fresh water during part or all of their lifetime. Although in general those practices which effect conservation of fresh-water fish may contribute in some way or another to marine fisheries, there are some which have a more direct effect than others. Probably the best known example is the effort made to avoid impediments to the passage of migratory fish. One of the most important species in value and volume, the salmon, has been and continues to be the cause of much research of this type. The disappearance of Atlantic salmon as a commercial species and the decline in shad and alewife runs in New England are considered to be directly attributable to damming of many streams without provision for these anadromous species to reach spawning areas. Fish ladders, traps below the dam to collect adults to be taken upstream, traps above the dam to collect fingerlings to be taken downstream, artificial completion of the cycle in hatcheries, and other devices seek to circumvent the problem created by dams. Nor is the physical barrier the only problem created for anadromous fish by damming. Lowered water levels downstream may dry up spawning pools or excessive depth of impounded still waters may make formerly used areas above the dam undesirable. The silt-trap nature of the impounded waters may cause eggs laid in the shallow water near the upper end of the reservoir to become buried, or the draw-down of waters to leave them "high and dry." Variable release rates of water can smother or bury estuarine marine life or the food it is dependent upon under the silt load brought in during a period of high rate of water flow. Here, conversely, is one way dams may be beneficial—by alleviating natural variation of flow. If these important and desirable species are to continue to augment our marine fisheries production, future planning for area development must consider them.

The silt-laden stream emerging from overcut, overgrazed, or poorly farmed lands bids fair to add to the nation's loss in soil a loss in breeding grounds for many types of estuarine marine life, notably shellfish. The dredging of stream bottoms to improve navigation is often accomplished at the cost of fish habitats, in addition to increasing the silt-carrying capacity of the stream through the induced velocity increase. Release of water used by industry for cooling may alter the temperature in estuaries or shallow coastal areas enough to upset the ecological balance critically.

FULFILLING THE NEEDS OF CONSERVATION

The foregoing discussion has sought to indicate the present importance of the salt-water borders' fisheries and to cite examples of some of the conservation problems faced by such areas. The need for conservation has been recognized at both state and Federal government levels. Each state has a fish and game department or commission having primary jurisdiction over the fishery resources of the state. Management and regulation of the harvest of this resource is their task. The Fish and Wildlife Act of 1956 marks the increasing recognition by the Federal government of the need for conservation of these resources. The Bureau of Commercial Fisheries, responsible for Federal programs relating primarily to commercial fisheries, was established by this Act.[12] The opening lines of the 1956 Act indicate the awareness of the Federal government of the importance of the fish and wildlife resources of the nation: "The Congress hereby declares that the fish, shellfish, and wildlife resources of the Nation make a material contribution to our national economy and food supply, as well as a material contribution to the health, recreation, and well-being of our citizens; that such resources are a living, renewable form of national wealth that is capable of being maintained and greatly increased with proper management, but equally capable of destruction if neglected or unwisely exploited. . . ."

Although the future will undoubtedly reveal management practices capable of vastly increasing the use of the seas, the people have at hand currently the ways and means for substantial improvement of their use of marine resources. They are aware that proper management of watersheds can provide clear streams with less erratic flow, and the fact that watershed management and fishery conservation methods coincide only adds to the insistence that they be implemented. Elimination of pollution in coastal areas and streams and habitat improvement for fishery conservation, rather than being an additional task, becomes instead an additional reason for water conservation. Recognition that land-based conservation alone is inadequate to the task is implicit in the provisions of the 1954 International Convention for the Prevention of the Pollution of the Sea by Oil.

Restrictions on the catch have been and are being devised which consider size or age of the fish, type of gear used, time in the reproductive cycle of the species, quantity of a given species, and the like. These restrictions should have two effects:To increase the variety of the catch and to assure sustained yield. To the future is relegated the task of significantly increasing numbers or size or use of the several species. Much of the total catch now landed is wasted in processing; more efficient storage, marketing, and utilization of the present catch can provide additional products and jobs. Although there are many examples of coordination between equal levels (interstate and international), there is room for considerable improvement to insure that the good accomplished in one location is not wasted or undone by conflicting legislation elsewhere. Long recognized as the foremost need, the collection of data and search for knowledge about marine life through experimentation continues to provide ever more sound bases for conservation measures. Whatever successes there have been to date are the result of these two activities, and future advances are dependent upon them. Hence they must be included in, if not placed at the head of any list of conservation efforts.

INTERNATIONAL COOPERATION

Conservation of ocean resources, notably fish, requires a measure of international cooperation that is not necessary for most other resources.[13] Fish move freely from waters of one nation to waters of another. For example, anadromous fish of the north Pacific spawn in Canadian and United States waters, yet they may be caught on the open seas. There would be little value in the United States' regulating the catch of salmon on its west coast if the Japanese or Canadians did not have similar regulations. A large portion of the problem involves the question of the international character of the seas.

There are many examples of international agreements for conservational regulation of fishing. Probably the first international attempt at conserving and rebuilding a marine fishery was embodied in the treaty between Canada and the United States, signed in 1923 as an outgrowth of the Convention for the Preservation of the Halibut Fishery of the North Pacific Ocean.[14] The International Convention for the Northwest Atlantic Fisheries, signed in 1949 by representatives of eleven countries, and the Convention for the Regulation of Whaling, attended in 1946 by fifteen nations, are additional examples of the numerous efforts that have been made from time to time to exert some form of control over the harvest of individual species. Unfortunately such examples of international agreement are easily matched with examples of lack of agreement. With the improvements in fishing equipment and sophistication of processing techniques (including processing at sea in a factory ship), fishing fleets ranged farther and farther afield, coming into contact with fleets of other nations and often on what have been, in terms of usage, the home grounds of those nations. Although it is common to consider the "three-mile limit" of territorial jurisdiction over waters surrounding a nation's borders as applying to fisheries as well as other activities, such a limit is emphatically rejected as being too limited by many nations.

In fact, there is considerable lack of agreement as to what such a limit should be—or, in the case of some, whether there should be a limit. In 1956 the International Law Commission was forced to admit that delimitation of the territorial seas by the several nations was not uniform, and, in the same year, the Council of Jurists of the Organization of American States maintained that a three-mile limit of territorial seas did not constitute a general rule of international law.[15] In 1958 the United Nations General Assembly convened the International Conference on the Law of the Sea with eighty-six nations represented by delegates in attendance. While success on some matters was achieved, agreement on the breadth of territorial seas or fishing limits was not achieved. Two years later, a second conference had little more success with regard to these two matters. Indeed, after the 1960 Geneva convention came to its unsuccessful conclusion there seemed to be an implied question of whether any universal territorial limit could be assumed to be in force, and many nations made statements reaffirming old or establishing new limits.

Perhaps it is not essential that there be a universal breadth of territorial jurisdiction over coastal waters; it is, however, becoming increasingly essential that nations find bases for agreement on this and other matters pertaining to the utilization of the sea's resources. Although the present use of these resources is relatively limited, it is clear that the picture must change. Indeed, if recent trends are indicative, the change has already begun. The world total catch of fish, mollusks, crustaceans, and aquatic animals and plants has more than doubled since 1950. As the search for additional sources of food for expanding populations urges greater use of marine resources, and as technology makes it increasingly possible to realize larger harvests, the necessity for effective conservation will grow apace. And conservation of the sea's resources is not possible without international cooperation.

Other Uses of the Sea

To be sure, the sea is used other than for its fishery resources. Foremost among its additional uses are transportation and recreation, with minerals recovery and water supply probably ranking much lower, but destined for future greatness.

TRANSPORTATION AND RECREATION

It is difficult to assign concise values to the first two uses. Tonnages hauled, value of cargoes, ships and men employed, tourist spending, sports equipment purchased—all of these criteria furnish fairly accurate indications of certain aspects of the value of the sea and have been alluded to in Chapters 9, 10, and 11. Other criteria of comparable or greater importance, however, elude evaluation. What measure of worth should be assigned to the role of the sea as a naturally created and maintained highway which makes possible the intercontinental exchange of materials in significant volume and at a cost such that areas not as well equipped to produce them need not be forced to do so? How shall one evaluate the results of time spent in relaxation and pleasure, or the scenic beauty afforded by the endless variety of seacapes? The seas will continue to serve man in these ways to some extent whether, in recognition of their incalculable values, he employs conservation to this end or not. But they will serve him more effectively if he does. Perhaps there is neither reason nor method at present for attempting conservation of the high seas themselves, but the salt-water borders afford a real opportunity. Just as pollution and sediment rob the interior water bodies of transport facility and recreation opportunity, so do these same undesirable elements lessen the value of the sea margins. To note that conservation measures applied in the interior of the continent are important to its margins only serves to re-emphasize the interrelatedness of water resources, fresh and salt.

MINERALS

The 165 million short tons of dissolved salts in a cubic mile of sea water are chiefly in the form of eight ions: Chloride, 54.8 per cent; sodium, 30.4 per cent; sulphate, 7.5 per cent;

magnesium, 3.7 per cent: calcium, 1.2 per cent; potassium, 1.1 per cent; carbonate, 0.3 per cent; and bromide, 0.2 per cent.[16] Although every element is believed to be present in at least trace quantities, the above-named ions represent over 99 per cent of the dissolved salts. To date, man has recovered only one of these dissolved salts in quantity—sodium chloride, which is dominant and easiest to separate (Figure 13 – 3).

Some six million tons of salt are recovered from sea water annually throughout the world by solar distillation.[17] This salt has been obtained from sea water since several thousand years before Christ. Although numerous earlier efforts to recover other elements from sea water were successful, the extraction of bromine directly from sea water by the Ethyl-Dow Chemical Company in the 1930's was significant in that no previous concentration of the source material was needed for the operation. Since 1941 magnesium has been recovered from sea water at the Dow Chemical Company plant at Freeport, Texas,[18] and this eighteen-million-pound-per-year plant was so successful that a second plant was constructed a year later (Figure 13 – 4). Although, as noted earlier, other minerals are taken from the sea in limited amounts, what has been accomplished thus far can hardly be considered more than a beginning.[19] In recognition of this, the U. S. Bureau of Mines initiated its marine minerals technology program. First becoming operational in 1965, its purpose is to develop systems for the mining of mineral substances on and below the sea floor.[20] As research and technology point the way, the sea will come to serve man more significantly as a mineral source complementing the more familiar sources on the land.

13 – 3 ***Salt from the sea.*** **Solar energy is utilized to evaporate sea water, leaving the salt. (Leslie Salt Company)**

Despite the difficulties caused by dissolved salts, brackish and salt waters are being increasingly used for purposes more commonly served by fresh water. Water for cooling in industry is an example. Engineering advances will undoubtedly increase the amount of sea water so used, but the dream of economical conversion of sea water to fresh water has long taxed man's imagination. The word "economical" is the key word, of course; desalination of sea water on board ship has been going on for over one hundred years. The way has not yet been found to produce fresh water from sea water in quantity and at low enough cost, despite the variety of approaches that have been tried.

Two processes which employ membranes to produce fresh water from salt water are *electrodialysis* and *reverse osmosis*. Electrodialysis utilizes two types of membranes, one permeable only to cations, and the other only to anions. Salt water between a pair of such membranes will, under the influence of an electromotive force, lose sodium ions through one membrane and chloride ions through the other to enrichable surrounding salt water, causing the water between the membranes to be depleted of its salt. Coalinga, California, was the first town in the United States to obtain its drinking water from brackish well water by this process.[21] The cost of drinking water for the city was cut substantially thereby. In reverse osmosis, pressure greater than osmotic pressure is exerted on sea water, causing it to move through an ion-restraining membrane toward a supply of fresh water, leaving the salt ions behind.

The most popular processes presently in use, however, are the several forms of distillation. The *long tube vertical* process employs the heat released by condensing vapor to boil sea water falling through long tubes in a reduced pressure environment. *Multistage flash distillation* introduces heated sea water into large chambers in which the pressure is just below the boiling point of heated sea water; some of the liquid boils (flashes) immediately upon introduction, and the vapor is later condensed.

13–4 ***Separating magnesium from sea water.*** **In these ponds a mixture of sea water and lime is producing magnesium hydroxide which will be converted to magnesium chloride and then electrolyzed to produce magnesium. (The Dow Chemical Company)**

Freezing sea water and extracting the salt-free or salt-poor ice is yet another possibility. One of the most interesting lines of research would utilize algae which are known to absorb and retain sodium chloride. Use of insoluble organic liquids which absorb large quantities of water affords another possibility for extraction of the water without the salts. One of the simplest methods is one of the oldest, solar distillation of sea water. Although solar-powered stills have sought to avoid the cost of fuel required to accomplish evaporation, this method has its obvious shortcomings in terms of time.

The tremendous potential reward incumbent to success, as much as local water shortages in coastal areas, causes the search to be continued. As water use increases, however, water-use costs are rising to meet the descending cost of salt-water desalination. An intriguing possibility is the production of fresh water as a by-product of a future seaside mineral extraction plant bent on recovery of all dissolved materials for one purpose or another.

Outlook

Man's use of the sea seems destined to increase significantly in this and other nations in the near future. It must be made to produce more to supply at least its proportion of the needs of the population of the future, and should actually increase in importance proportionately. This will not happen without the establishment of sound and integrated conservation measures coupled with improvement in techniques to enable man to advance from the hunting and gathering stage of harvesting the sea's resources. A chief deterrent has been and still is lack of knowledge of the sea and the life it contains. There is probably no subject area of research on earth which offers greater proportionate opportunity.

FOOTNOTES

[1] Charles H. Lyles, *Fishery Statistics of the United States, 1964*, Statistical Digest no. 58 (Washington: Fish and Wildlife Service, Department of the Interior, 1966), p. 4.

[2] C. M. Shigley, "Minerals From the Sea," *Journal of Metals*, vol. 3 (January, 1951), p. 25.

[3] Richard M. Highsmith, Jr. and J. Granville Jensen, *Geography of Commodity Production* (Philadelphia: J. B. Lippincott Company, 1963), p. 14.

[4] Hawthorne Daniel and Francis Minot, *The Inexhaustible Sea* (New York: Dodd, Mead & Company, Inc., 1954), p. 3.

[5] Lionel A. Walford, *Living Resources of the Sea* (New York: The Ronald Press Company, 1958), p. 6. Note also the interesting series of world maps (pp. 15–37) dealing with the geography of the seas, their resources and use.

[6] Unless otherwise indicated, data quoted are from Charles H. Lyles, *op. cit.*

[7] The nearly 1.5 billion pounds difference between this figure and the one comparing the United States catch to that of other nations is chiefly due to the fact that the weight of mollusk shells is not included in this figure. The earlier-mentioned comparative figures are on a live weight basis, the standard form in which United Nations Food and Agricultural Organization data are published.

[8] "Water Resource Activities in the United States—Fish and Wildlife and Water Resources," Select Committee on National Water Resources, United States Senate *Print No. 18* (Washington: Government Printing Office, April, 1960), p. 12.

[9] Walford, *op. cit.*, p. 134.

[10] *Ibid.*, p. 133.

[11] The remainder of this section is principally based on *Committee Print No. 18*, Senate Select Committee on National Water Resources.

[12] For a more extended coverage of the government's role see Chapter 9.

[13] For a detailed consideration of conservation and international law see Douglas M. Johnston, *International Law of Fisheries* (New Haven: Yale Univ. Press, 1965), 554 pp.

[14] *Ibid.*, p. 372.

[15] Shigeru Oda, *International Control of Sea Resources* (Leyden, Netherlands: A. W. Sythoff, 1963), p. 15.

[16] Shigley, *op. cit.*, p. 25.

[17] John L. Mero, *The Mineral Resources of the Sea* (New York: Elsevier, 1965), p. 27.

[18] *Ibid.*, p. 26.

[19] For a discussion of ocean mining methods see Mero, *op. cit.*, Ch. 7.

[20] "Interagency Conference on Continental Shelf Research," *U. S. Army Coastal Engineering Research Center Miscellaneous Paper No. 1-66* (January, 1966), p. 28.

[21] "Water Resource Activities in the United States—Saline Water Conversion," Select Committee on National Water Resources, United States Senate, *Committee Print No. 26* (Washington: Government Printing Office, 1960), p. 21. For a partial listing of the larger water desalting plants of the world see "Saline Water Conversion Report for 1965," (Washington: Department of the Interior, Office of Saline Water; Government Printing Office, 1966), Appendix E.

14

Mineral Resources

THIS NATION'S SUPERLATIVE MATERIAL STANDARD of living is in a large measure based on utilization of enormous quantities of inanimate energy resources harnessed through machines constructed of metals. Only a century ago the muscles of man and animal accounted for at least two-thirds of the nation's total work output. Now, nearly all commercial energy is derived from mineral fuels (Figure 14-1). Moreover, almost all material things are either made of metals or processed or manufactured by machines of metal powered by mineral energy sources. The spectacular increase in the nation's agricultural productivity has been achieved in large part by the use of farm machines powered by petroleum fuels, by widespread application of mineral fertilizers, and by the use of mineral-derived chemical pest controls. The importance of minerals in the American economy is thus evident.

The United States' consumption of minerals has multiplied in response to increasing demands of the economy and increasing population. The population grew from 76 million in 1900 to about 195 million in 1965. This is an increase of about two-and-a-half times, but mineral consumption multiplied six-and-a-half times. Some provocative indications of the growing pressure on minerals are revealed in Table 14-1 and Figure 14-2. Moreover, all projections for future needs predict increasing per capita consumption by increasing numbers of people. At 1965 growth rates there are three million new consumers each year—8,200 each day—in the United States. Thus demands on mineral sources will continue to increase in the foreseeable future. Such enormous consumption of mineral resources raises profound questions concerning how long such demands can be met.

Rational conclusions in studies of mineral resources conservation must be predicated on knowledge and understanding of the national economy as well as on comprehension of the implications of the special characteristics of mineral resources. This first chapter dealing with mineral resources, therefore, presents the characteristics of mineral resources and realities of ownership which have implications for conservation analysis. The second chapter on minerals reviews the mineral resource base of the United States in terms of adequacy as a basis for conservation analysis. The third chapter is concerned with conservation needs, practices, and possibilities.

Characteristics of Mineral Resources

Mineral resources, being nonrenewable, are not subject to sustained management by man as are forests, wildlife, agricultural lands, or fisheries. Moreover, mineral resources have a number of characteristics not common to other resources which need to be understood if one would comprehend the mining industry and realistically appraise mineral resources from a conservation point of view.

FORMATION OF MINERAL DEPOSITS

The mineral deposit is a concentration in a limited area of one or more earth elements to a degree that the desirable minerals are rich enough and large enough to encourage exploration and possibly to support mining. *Ore* is defined as earth material which under given economic conditions and technology can be economically mined. Various geologic processes have resulted in mineral deposits. Many types of metal ores formed as lodes in cracks and cavities when mineral-rich fluids rose from magmas, others formed by enrichment of host rocks around rising magmas. Placer deposits, such as gold, resulted from erosion of primary lodes followed by accumulation in stream gravels. Other ores resulted from concentration by leaching or by evaporation. Coal, oil, and gas fields formed from vast quantities of organic materials.

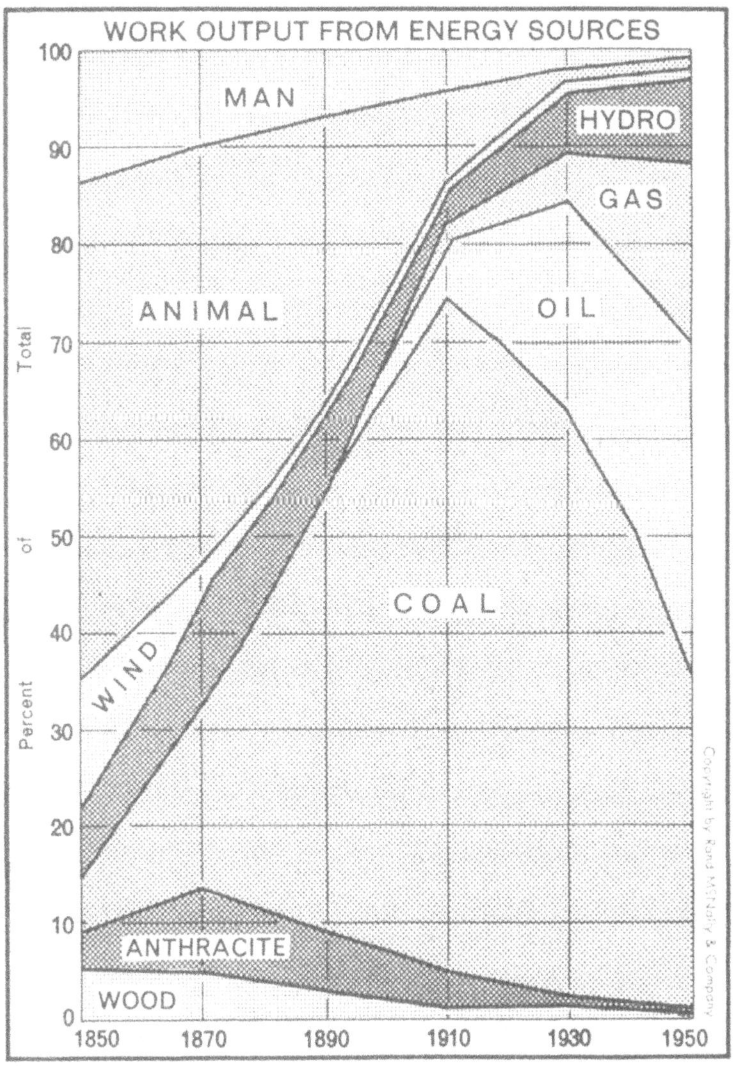

14 – 1 ***The changing pattern of work output from energy sources.*** **(Based on a graph in** ***America's Needs and Resources: A New Survey,*** **by J. Frederic Dewhurst and Associates. [New York: The Twentieth Century Fund, 1955], p. 907.)**

TABLE 14–1

CONSUMPTION OF SELECTED MINERALS 1900–1965[1]

	(Thousand tons unless otherwise stated)		
	1900[2]	1965[3]	Approximate increase
Coal (bituminous)[4]	223,782	459,164	2
Petroleum (million barrels)	63	3,310	51
Natural gas (billion cubic feet)	180	16,033	90
Iron ore	28,399	131,888	5
Manganese ore	268	2,866	11
Copper (refined)	357	2,005	6
Bauxite	31	13,534	450
Lead	269	1,241	5
Zinc (slab zinc)	99	1,354	14
Phosphate rock[5]	1,491	19,500	13
Sulphur	408	7,959	20
Salt	3,114	36,409	12
Cement (thousand barrels)	20,573	389,159	19

[1] National net consumption including import-export balance.

[2] *Mineral Resources of the United States, 1901* (Washington: United States Geological Survey, Department of the Interior, 1901).

[3] *Minerals Yearbook, 1965* (Washington: Bureau of Mines, Department of the Interior, 1965).

[4] The peak production year for coal was 1947 when 630.6 million tons were mined. Mine production of coal in 1965 was 512 million tons.

[5] Mine production of phosphate rock in 1965 was 26.4 million tons.

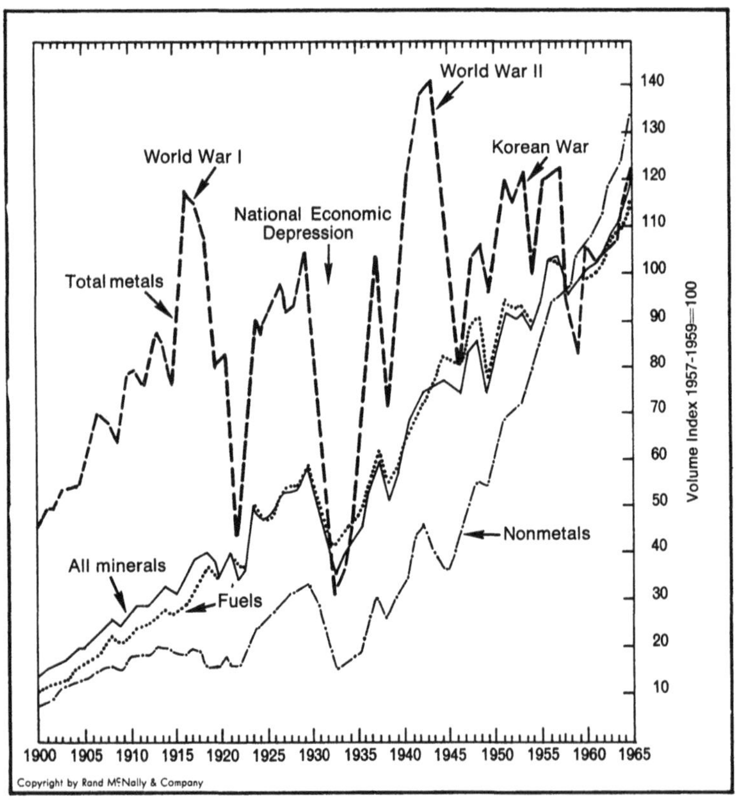

14 – 2 ***Growth of physical volume of mineral production in the United States, 1900–1964.*** **(After graph in Edward E. Johnson, "Index Numbers for the Mineral Industries,"** ***Bureau of Mines Information Circular 8275.*** **[Washington: Department of Mines, 1965], p. 3.)**

The geologic processes which result in formation of mineral deposits are only partially understood, but earth scientists are intensifying research toward more complete understanding, especially in geochemistry and geophysics.[1] It is probable that mineral deposits are being formed even now, but the rate of formation is so very slow that this factor may not be considered as significant for conservation study. Thus the mineral resources must be considered as a heritage of geologic time—a capital or stock resource.

CHARACTERISTICS

The circumstances of formation are responsible for characteristics of mineral deposits that are important to conservation appraisal. The most significant conditions and implications include the following:

DEPOSITS ARE SCATTERED AND OCCUPY SMALL AREAS. In contrast with forest or agricultural lands, mineral deposits are scattered and occupy small areas. It has been estimated that "90 per cent of our metallic wealth has come from a scant thousand square miles of our domain."[2] The great molybdenum ore body at Climax, Colorado, occupies less than a square mile, yet this mineral concentration area produces about 60 per cent of the world's supply. Coal and oil fields are more extensive, but even so the mining areas do not compare in extent with forest growing areas (Figure 14 – 3).

A number of implications are worthy of note: (1) Control of exploitation by relatively few monopolistic companies is fostered; (2) major regional differences in mineral endowments exist; and (3) no region, not even any single nation, has within its borders economic deposits of all of the seventy-five or more minerals required by modern industrial economies. International trade is, therefore, highly desirable to supplement domestic mineral supplies for the United States and the world economy.

DEPOSITS ARE HIDDEN FROM VIEW. Mineral deposits are nearly always partially or completely hidden from view. The implications are perhaps obvious, but they are profound. Large expenditures of capital are generally required to determine whether a surface indication represents a deposit of sufficient size and richness to support mining, and it is not unusual to prove the deposit to be uneconomical—hence the often noted gamble in mining ventures. In petroleum exploration, drilling cost averages 15 dollars per foot of drilling depth, and chances are one in fifteen of finding a small new field, only one in two hundred of finding a medium field, and one in one thousand of finding a major new field.[3] Modern geochemistry and geophysical exploration, providing man with the means of "seeing" into the earth, may in the future find important mineral deposits which are not now in any way visible. The fact that deposits are hidden from view creates difficulties in knowing exactly the quantity of the resource, a major factor that will be discussed later.

14 – 3 *The Berkeley pit.* This outstanding copper mine is a good example of the localization of minerals in relatively small areas. In recent years this one pit has produced about seven million tons of copper ore annually, accounting for nearly half of the copper mined in Montana. (Anaconda Copper Company)

MINERAL DEPOSITS ARE NONRENEWABLE. Mineral resources are esentially nonrenewable and not subject to sustained-yield management. As noted earlier, formation of minerals, including petroleum, well may be continuing, but the rate of formation is insignificant compared to the rate of depletion through mining. Thus every mine is faced with gradual depletion of its capital resource, leading ultimately to closure. This fact is recognized in the tax laws which allow the mineral industry to amortize the resource through the depletion allowance and which provide various incentives to encourage and aid exploration.

COMMODITIES ARE DURABLE. With a few exceptions, mineral commodities can be stockpiled without deterioration, and in some cases, reused. Stockpiling can be accomplished as ore concentrate or as primary metal. Moreover, many of the metals have high percentage reuse possibilities through scrap recovery, and these in-use metals form an in-use stockpile. The influence of this reserve is partly economic, because the price and availability of scrap metal influence mining. In addition, it is clear that intensification of reuse of mineral commodities will be an important conservational measure of the future.

MINES FACE INCREASING COST AND RISK OF NEW DISCOVERY. Two important characteristics of mining are the tendency toward increasing costs of operation, and the ever-present possibility of a richer, more favored deposit being discovered. Most mines, especially underground mines, tend to become more costly to operate with increasing depth of working areas and length of haulage. Moreover, there is frequently a decreasing quality of ore as time goes on, further emphasizing the risk of new discovery of a richer deposit which may be able to produce at lower cost. Few mines literally are "mined out"; rather they close because cost of operation comes to exceed the value of the production. It is the operation of these characteristics that leads to ghost towns. Virginia City, site of one of the richest deposits in America, the Comstock Lode, is today a ghost-town tourist center.

MINERAL CONSERVATION PRACTICES ARE ESPECIALLY VULNERABLE TO PRICE FLUCTUATION. From the conservation point of view, it is important to recognize that major mines involve large capital investment and that maintenance is costly. Thus mines operate under considerable economic risk and are generally less flexible in management possibilities than the production facilities of most industries. It follows that conservational practices tend to be fostered by long periods of stability of mineral markets. In periods of depressed prices there is a tendency to "high grade" a mine to stay open rather than close down. The result is often irretrievable loss of low-grade ores. Underground mines in particular are not readily closed and reopened to match fluctuating prices because working areas, timbering, and machinery deteriorate, with resultant loss of invested capital. These are factors that appear to justify government assistance to regulate imports and maintain prices of mineral commodities.

DETERMINATION OF RESERVE IS DIFFICULT. In the appraisal of mineral resource adequacy and conservation needs, the most difficult problem is that of adequately defining the available quantity. Ore has been defined to mean material of such quality and quantity that it can be economically mined. The term *reserve* is used to indicate quantity of ore available. The problem of appraising mineral resource adequacy for the future then has two major facets: First, defining the minimum quality that is minable at a profit; and second, measuring the existing quantity. Both factors present major difficulties.[4]

In estimating quantity of marketable material, miners have long used the terms *proved ore* meaning ore actually exposed on three sides, *probable ore* meaning exposed on two sides and therefore uncertain of continuity, and *possible ore* meaning only that the geologic environment suggests its presence. As an approach to objective statements of mineral reserves, the United States Geological Survey uses the following terminology to connote degrees of physical certainty.

Measured reserves are those for which tonnage is computed from dimensions revealed in outcrops, trenches, workings, and drill holes for which the grade is computed from the results of detailed sampling. The sites for inspection, sampling, and measurement are spaced so closely and the geologic character is so well defined that size, shape, and mineral content are well established. The computed tonnage and

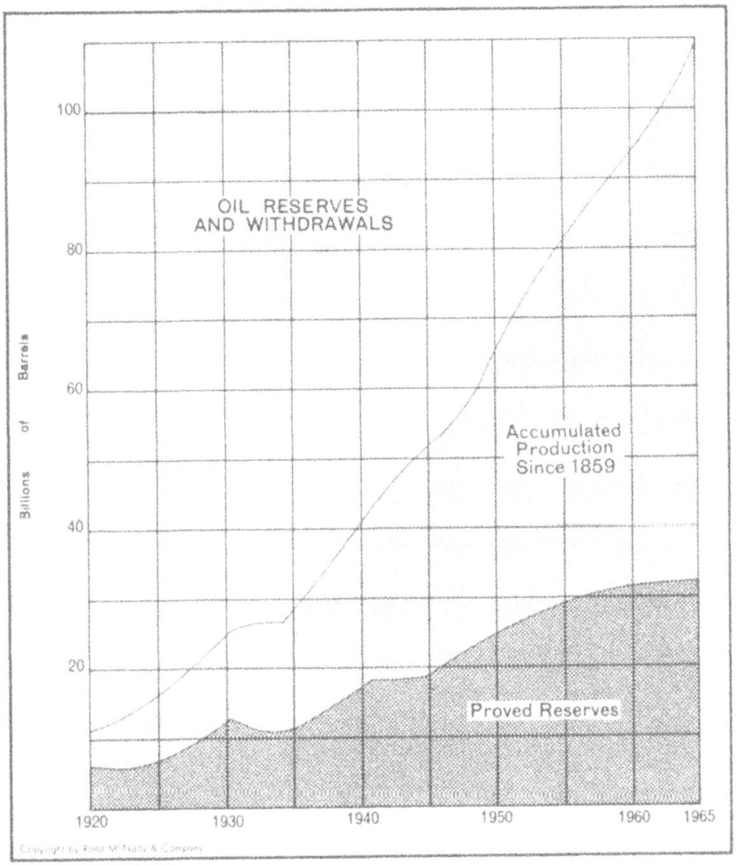

14–4 The trend in United States petroleum reserve.

grade are judged to be accurate within limits which are stated, and no such limit is judged to be different from the computed tonnage or grade by more than 20 per cent.

Indicated reserves are those for which tonnage and grade are computed partly from specific measurements, samples, or production data, and partly from projection for a reasonable distance on geologic evidence. The sites available for inspection, measurement, and sampling are too widely or otherwise inappropriately spaced to permit the mineral bodies to be outlined completely or the grade established throughout the area.

Inferred reserves are those for which quantitative estimates are based largely on broad knowledge of the geologic character of the deposits and for which there are few, if any, samples or measurement. The estimates are based on an assumed continuity or repetition of which there is geologic evidence; this evidence may include comparison with deposits of similar types. Bodies that are completely concealed may be included if there is specific geologic evidence of their presence. Estimates of inferred reserves should include a statement of the specific limits within which the inferred material may lie.

The economic realities, especially the costs of exploration, mine development, extraction, and transportation, in relationship to market values are explicit in the definition of reserves. Thus, as technology improves, market values change, more detailed measurements are made, and new deposits are discovered, the quantity estimates of reserves change. For example, United States coal reserves were estimated in 1928 as 3.2 trillion tons, but in 1953 as only 1.9 trillion. The history of steadily increasing petroleum reserves is an outstanding example of the need for understanding the meaning of proved or measured reserves (Figure 14–4).

The definition of the lower limits of quality of ore that can be mined economically is the major variable which makes prediction of future adequacy difficult. In a large measure the question is a function of technology and the willingness of the economy to pay the cost of production from lower grade materials. For example, United States workable manganese reserves are only one million tons; but if improved technology or increased price would permit use of lower grade material, then over one hundred millions tons are known to exist.[5]

TABLE 14–2

THE ABUNDANCE OF TWENTY SELECTED ELEMENTS IN EARTH CRUST*

Element	Units of element per million parts	Concentration required to be considered ore	Units per million in sea water
Silicon (Si)	277,200		4.0
Aluminum (Al)	81,300	4 times	1.9
Iron (Fe)	50,000	6 times	0.02
Calcium (Ca)	36,300		400.0
Sodium (Na)	28,300		10,561.0
Potassium (K)	25,900		380.0
Magnesium (Mg)	20,900		1,272.0
Titanium (Ti)	4,400		
Phosphorus (P)	1,180		0.01
Manganese (Mn)	1,000	350 times	0.01
Sulphur (S)	520		884.0
Zirconium (Zr)	220		0.014
Chromium (Cr)	200	1,500 times	
Zinc (Zn)	132	300 times	0.014
Vanadium (V)	110		0.0003
Nickel (Ni)	80	125 times	0.0005
Copper (Cu)	70	140 times	0.01
Tin (Sn)	40	250 times	0.003
Lead (Pb)	16	2,500 times	0.005
Uranium (U)	4	500 times	0.0016

* Kalervo Rankama and T. G. Sahama, *Geochemistry* (Chicago: Univ. of Chicago Press, 1950).

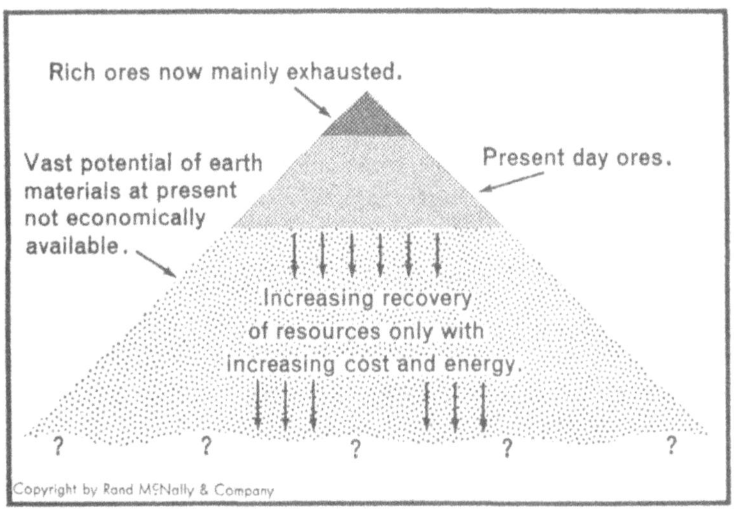

14 – 5 *A diagrammatic concept of the term "mineral reserve."* It is not intended that a specific mathematical relationship be expressed between the amount of energy or capital expended and the increase in reserve.

Figure 14 – 5 illustrates this concept of the mineral reserve as a variable. The ultimate of mineral reserve is the entire earth crust because a mineral deposit in fact is only a special category of earth material. A few long-range observers are suggesting that man will in time be processing earth crust material and sea water en masse for all contained elements. It well may be significant for the future resource potentials to consider the relative abundance of elements in the total earth crust, but practical consideration of this view is remote (Table 14–2).

The concept *reserve* then is seen to be a complex function of cost factors in production and processing—physical conditions for mining, richness and character of the ore, technology, transport costs, and the market value at any given time. Published quantitative statements of mineral reserves are seldom as definitive as they appear to be, and they must be understood for what they are—usually no more than estimates subject to major expansion or contraction as knowledge increases or economic and technical conditions change.

Ownership of Mineral Resources

Just as it is necessary to understand the special characteristics of mineral resources, so it is essential to know their ownership and legal status as a foundation for understanding conservation needs and possibilities. The question of who owns mineral deposits formed eons ago by geologic processes has ethical ramifications with reference to the interest of the whole society, as well as legal aspects. The discussion here is limited to the legal status.

In the United States the right of private ownership and the right to personal gain from whatever a man can produce by his work and enterprise has always been fundamental in the economic philosophy. At the same time there always has been a measure of appreciation of and respect for the rights of others, including the obligation of the individual resource-developer to consider the welfare of the community and nation above personal gain. As a result there has evolved a body of mineral resource law based on recognition of private ownership, the welfare of the public, and stewardship for future generations.

In the broadest sense, three specific major principles are recognized and implemented by the mineral laws. The first principle is the concept that the finder of mineral wealth on public land is entitled to its ownership and to whatever profit he can gain from it. The general Mining Laws (1872) which concern metallic minerals embody this concept.[6] The second principle is that the Federal government in the interest of the present and future welfare of the nation has the inherent right to permanent ownership and management of mineral resources within the public domain lands. This concept is implemented chiefly in the Mineral Leasing Act (1920), and the Outer Continental Shelf Act (1953). The third principle is the concept that mineral rights and exploitation on public land ought not to inhibit seriously or prohibit development of other resource values. This philosophy is embodied in the Multiple Surface Use Act (1955), the Multiple Use Mining Law (1955), and specific modifications of the mining laws such as are in the Wilderness Act of 1964. Understandably, the realities and implications of the laws affecting mining and mineral resources are complicated (in fact the first two principles are clearly in conflict). The following discussion, therefore, reviews only selected highlights together with comments germane to understanding mineral resources in the United States and their conservation prospects.[7]

EMERGENCE OF THE MINING LAWS

The question of mineral rights and mineral resources was of little concern during the early decades of the acquisition and disposal of the public domain. In 1785 Congress recognized mineral resources in its policies of land disposal by providing for withdrawal of mineral lands from sale under the agricultural land disposal laws. In addition, Congress also provided that a third of all gold, silver, copper, and lead mined in the then "Western Territory" be given as a royalty to the Federal government. In 1807 the first mineral land-leasing ordinance provided for leasing of lead deposits to private individuals and corporations. But long before mineral raw materials came to have major importance in the

TABLE 14–3

MINERAL LEASES AND RECEIPTS TO UNITED STATES TREASURY UNDER MINERALS LEASING ACT BY STATES*

State	Receipts		Leases and area, 1965	
	1920–1965 accumulated (million dollars)	1965	Number	Thousand acres
Wyoming	541.6	40.3	33,942	18,480
New Mexico	241.2	24.3	12,607	8,104
California	196.6	5.9	3,253	1,388
Colorado	148.6	8.6	8,635	6,329
Utah	101.8	9.8	19,917	11,351
Montana	65.7	5.7	5,429	4,368
Alaska	53.3	7.5	5,507	10,056
Nevada	11.8	1.5	6,931	2,263
Idaho	6.7	1.1	171	263
Arizona	5.6	0.4	886	707
North Dakota	5.2	0.5	611	272
Louisiana	4.8	0.6	453	53
Kansas	3.7	0.4	153	47
South Dakota	3.5	0.3	811	9
Oklahoma	1.8	0.3	915	208
Arkansas	1.2	0.1	400	331
Alabama	0.3	0.02	58	14
Nebraska	0.3	0.006	77	13
Mississippi	0.2	0.03	249	21
Washington	0.2	0.01	16	23
Michigan	0.1	0.008	27	19
Florida	—	—	12	2
TOTAL	1,394.7	107.3[1]	101,086	64,955

* *Public Land Statistics, 1965* (Washington: Bureau of Land Management, Department of the Interior, 1965).

[1] Does not include potassium receipts of 4.7 million dollars from New Mexico and 1.2 million dollars from California. The composition of the total is oil and gas receipts of 104 million dollars, coal one million dollars, phosphate one million dollars.

economy, both the royalty provision and the leasing concept were abandoned. In 1846 Congress approved sale of lead-bearing land in Arkansas, Illinois, Iowa, and Wisconsin at a minimum of two-and-a-half dollars per acre, and the next year authorized sale of mineral lands in Michigan and Wisconsin known to have copper and other ores for five dollars per acre. Then, in 1850 Congress abandoned the sale fee in reopening Lake Superior area lands to preemption under the agricultural land disposal laws. Thus much of the present-day mineral lands, including the iron ores of Lake Superior, early passed to private ownership. Moreover, the retreat of the government from the sale, leasing, and royalty concepts left the West without any body of mineral law, and it was under such circumstances that the spectacular California gold rush of '49 developed. There the miners made their own rules, chiefly based on the right of the first finder to ownership, protection, and profit.[8] It is worthy of note that much of the basis for the western water law, or prior appropriation, evolved in these same mining camps because water was essential to the mining operation.

In 1866 the Federal government recognized most of the common law practices of the gold fields in the first mining law. This was followed in 1872 by enactment of the general Mining Laws of the United States which with revisions still regulate the use and separation of most mineral resources on public lands. In 1920 Congress enacted the second major facet of the present-day mineral regulations, the Minerals Leasing Acts, which retain ownership to and regulate disposal of coal, oil and gas, potassium, phosphate, sodium, oil shale, and sulphur (in

Louisiana and New Mexico only) on public domain land.

THE GENERAL MINING LAWS OF 1872

The Mining Laws of 1872 clearly established the legal principle that mineral wealth found on public domain lands belongs to the finder without royalty fee. (Later in 1920 the Minerals Leasing Acts excluded certain nonmetals from this right of ownership.) It should be noted that the Mining Law does not apply to private lands, but only to the remaining public domain (excepting national parks and certain other exclusions such as military reservations). Two levels of ownership regulating the transfer of public mineral wealth to private ownership were established: The claim and the patent.

THE CLAIM. Under the provisions of the Mining Law, every citizen has the right to prospect for minerals on public lands and, on finding an apparently valuable mineral, the right to locate and establish a claim, to mine the ore, and to sell it without payment of any kind or even the consent or knowledge of the Federal government.[9] In most states local laws, however, require the recording of mining claims with the county recorder. Two classes of claims are recognized: Lode and placer. The lode claim may not exceed 1,500 feet in length along a vein or 300 feet on each side, normally forming a quadrilateral of some twenty acres (Figure 14 – 6). The placer claim is similar in that it is twenty acres, but there is a provision allowing a group claim up to 160 acres in size, and a regulation that the claims must be laid out according to the rectangular land survey system (Figure 2 – 4).

On every claim the law requires that there be a mineral discovery or the claim can be invalidated. After finding a vein, marking the claim boundaries, and complying with local laws, a prospector gains possessory title. This includes the mineral wealth and the right to cut timber needed for his mining operation (not for sale), but since 1955 does not convey control over surface resources such as forest, water, or right of access.

To maintain title, the law requires that at least one hundred dollars worth of work be invested in the claim each year, but does not require mining and provides no time limit for it.[10] A prospector may stake as many claims as he wishes, but he must do the required annual assessment work for each single claim. Because of the inadequate system of recording (and in

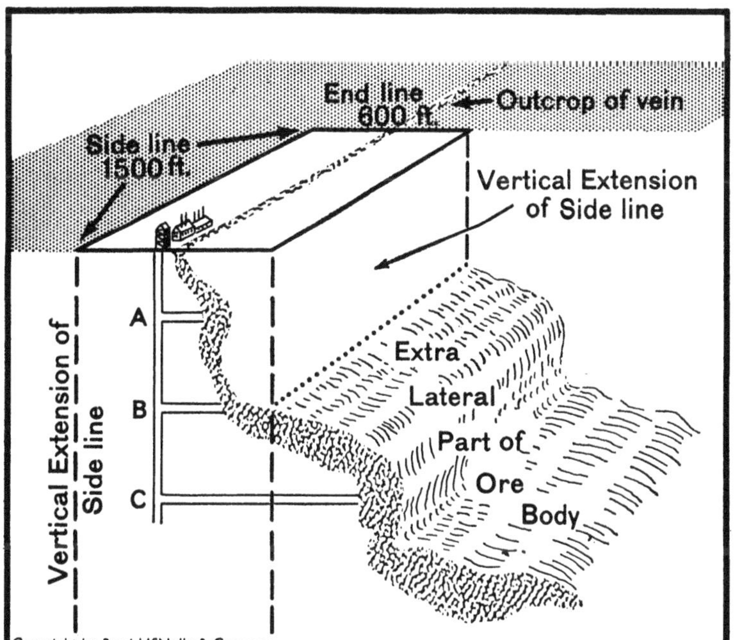

14 – 6 ***The mining claim.*** **The basic mining claim established by the law of 1872 is shown: 1,500 feet on the vein and 300 feet each side. Note that the mineral rights extend downward and also extend along the vein. This is known as extra-lateral rights and has been the basis for many legal battles. Even though the mineralized vein is under some other claim, the ore belongs to the claim containing the apex (Redrawn with permission from Robert S. Lewis, *Elements of Mining* [New York: Wiley, 1933], p. 55.)**

some cases lack of its requirement), no one knows exactly how much land is now held in possessory title under mining claims.[11]

The Mining Law thus went far to establish the right of private ownership and to profit from personal effort in connection with the public mineral wealth, but said little about the obligations of ownership or the rights of other resource values.

THE PATENT. The Mining Law established provisions for what is known as the *patent* by which a prospector, if he so desires, can gain absolute title to a valid mineral claim. To obtain a patent, the owner must have accomplished at least five hundred dollars worth of development work and must establish to the satisfaction of the Federal government that a mineral discovery has been made sufficient to justify a prudent man's investing effort and money. The prospector may then be permitted to purchase the land from the government for five dollars per acre to become absolute owner of the land. The owner of a patented mineral claim is no longer required to do annual assessment work. He owns and may mine all minerals when and if he chooses, and he also gains ownership of timber and any other surface resources and rights. By 1965 over three million acres of public domain had been transferred to private ownership through the Mining Law in some 65,000 patents.

The trend of patent issue is significant (Figure 14–7). During the last decades of the nineteenth century and early decades of the twentieth century, the first big demands for minerals to supply the needs of growing industrialization led to vigorous search for deposits. The individual (often picturesque) prospector had his heyday. In recent years fewer patents have been granted, very likely because most visible ore bodies have been found and the best patented. The future of prospecting lies beyond the range of the naked eye in the hands of the geochemists and geophysicists (prospectors with Ph.D. degrees backed by adequate capital) searching the inner earth for wealth hidden beneath the surface. For this sort of prospecting some revisions in the law will be needed.

ABUSES OF THE MINING LAW. The intent of the Mining Law of 1872 clearly was to provide major stimulus to private enterprise which would lead to exploration and development of the nation's mineral wealth. Unfortunately the generous provisions of the law and its failure to provide for regulation to protect reasonable rights of interrelated resources such as water, timber, and forage, together with the lack of definition of what constitutes a valid claim, led to various abuses. The most flagrant abuse has been the locating of low-value, if not actually fraudulent, mineral claims to gain a summer cabin site in the national forest or to gain possessory title to the land for use other than mining. To be sure, the government has always

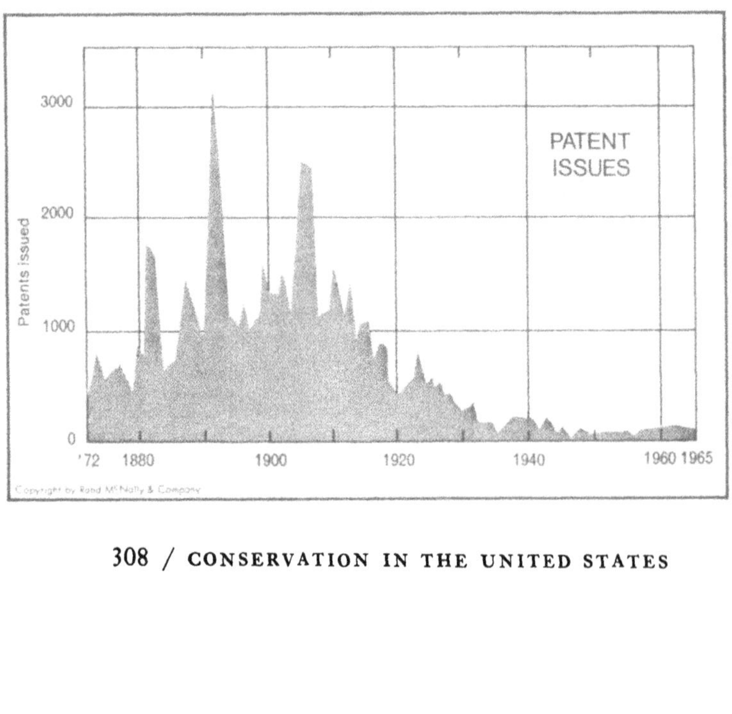

14–7 The trend in number of mineral patents issued.

had authority to invalidate such claims, but its implementation on a large scale was never very practical.

A few mining claims have abused the obligations inherent in ownership. The Mining Law reasonably and properly provides that a claim owner is permitted to cut timber for use in the mine and necessary buildings, but no provisions were made for regulation. Claim owners sometimes have cut timber illegally and often without consideration for good forest practices. Timber commonly was inadequately managed or not harvested when it should have been because the claim owner, although lacking the right to harvest and sell the timber, prevented the government from managing or harvesting the timber. More serious has been the inhibiting of other resource management resulting from mining claims having preempted access and control from other resources. In many cases a claim blocked road access to public timber or to water sources. Large mining operations sometimes caused serious pollution of streams from tailings.

In the early years these conflicts were not significant, but as mineral claims greatly increased in number, especially on forest and grazing lands, the fragmentation of ownership and abuses of the mining law came to interfere significantly with adequate management of other resources. As a result, Congress in 1955 approved the first major revision of the Mining Law in what has come to be known as the Multiple Use Act.

THE MULTIPLE SURFACE USE ACT OF 1955

The Multiple Surface Use Act of 1955, referring to nonpatented mineral claims, takes a long-needed forward step by specifically reserving the right of the Federal government to access to and management of surface resource values in the interest of the public. However, no fundamental change is made in the procedures or rights of acquiring mining claims and patents. The significant changes are summarized in the following four points:

1). Sand, gravel, pumice, pumicite cinders, and clay are specifically removed from the definition of minerals on which a mineral claim may be based. This change came about because claims based on gravels, pumice, etc., were common legal ways of abusing the law.

2). Timber cut for use on the claim must be cut with regard for good forest management practices.

3). Mining claims are restricted to normal and necessary activities of mining. This section specifically prohibits use of mining claims for summer cabin sites, cafes, etc.

4). Rights of surface use, access, and passage, as well as rights to manage and dispose of vegetative surface resources and to manage other surface resources, except mineral deposits, are reserved for the Federal government. In short, for nonpatented claims surface and subsurface rights have been separated.[12]

Although these amendments apply directly to claims made after July 23, 1955, the new law also provides specifically for the Bureau of Land Management to challenge all unpatented claims located before passage of the Act. Owners who fail to respond to the challenge or whose claims fail to meet standards of validity will lose the surface rights, thus permitting Federal management of surface resources.

The Multiple Surface Use Act corrected one of the major inadequacies of the Mining Law concerning unpatented claims, and will advance the prospects for better management of the total resource complex on public lands; but it did nothing to alleviate problems inherent in the patent provision.

Inadequacies of the Mining Laws

A few specific regulations of the mining laws governing metallic minerals seem to be inadequate and no longer in the public interest. Among these regulations the following are especially pertinent to conservation considerations. The requirements for a claim are particularly out-of-date with reference to minerals explorations seeking deposits which may exist completely hidden beneath the surface, because the provisions of the law explicitly require that there be a mineral exposure before a claim may validly be filed. It would appear to be in the public interest that some modification be made to permit exploration leases for the purpose of subsurface prospecting by modern techniques. Such leases would need to provide development rights to ore bodies that might be found as a result.

The one hundred dollar annual assessment work now required to maintain the validity of a claim would seem to be inappropriate in view of the current value of the dollar. Although legitimate mining claims probably do not abuse this provision, it is a reality that one hundred dollars of work by a man and machine is accomplished in a single day's activity—hardly enough to justify maintenance of a private mineral right on the public domain. In addition, it may well be in the public interest to consider setting a reasonable time limit during which a mineral claim would have to be developed or the claim released.

In view of the intensifying pressure and competition for land use there is need for uniform or national recording of all mining claims. It would probably be in the interest of the public as well as of the mining industry to establish a Federal registry of mining claims. At present, many claims are unrecorded and there are many cases of overlapping claims which confuse the resource rights on public lands.

At least one provision of the mineral law governing the granting of patents to mineral claims is probably no longer in the public interest. This is the statutory five dollars-per-acre purchase price. The mineral land patent, except for special cases noted under the Wilderness Act of 1964, conveys full ownership rights to all resources including those of the surface. On being patented a mineral claim may carry with it associated surface rights to public timber which could be worth one thousand dollars or more per acre. There is nothing in the mining law requiring that for a claim to be patented it must be economical to work; the law simply says that the discovery be such that it would justify a prudent man's investing time and money in its development. Thus patents have been gained for lands on which the real values finally turned out to be other than mineral. An example in recent years is the Al Sarena Case, in which a mineral land patent was granted on February 15, 1954, to 475 acres on the Rogue River National Forest in southern Oregon. The legality of the patent was questioned by many, and the case was publicized as a natural resource scandal. A committee of the Senate of the United States filed a report in which it showed that the mining company had done no mining since 1943 and that no mineral production occurred after the granting of the patent, but that more than two million board feet of former public timber was cut and sold in the year following patent approval.[13] Since then, additional timber cut has been estimated by United States Forest Service fire protection inspectors at about 6.6 million board feet, but there has been no mining.

The point here is not the question of whether the patent should have been granted; the more important observation is that the mining law is demonstrated by this case to be inadequate and outmoded in its provision of a statutory fee of five dollars per acre regardless of values on the land. In the interest of the public rather than individual personal gain, the granting of a patent to mineral lands ought to require reasonable additional payment for obvious public wealth such as timber stands. Or the public should retain surface rights on patents just as on claims under the 1955 Act.

It should also be noted that patented lands, like the claims, by fragmenting ownership within public lands, can and sometimes do interfere with adequate management of interrelated resources such as timber, grazing, water, and wildlife while being unproductive themselves.[14] In view of the growing pressures on capabilities of public land, one may question whether it is in the public interest to allow interferences by mineral patented lands to lower materially the productive level of other resources. Serious consideration ought to be given to retaining for the public most, if not all, surface rights on granting patents to mineral lands.

THE MINERAL LEASING CONCEPT FOR NONMETALS

The Mining Laws, although far from perfect, worked well for the metallic minerals, but it soon became apparent that they were inadequate for most nonmetals, especially coal and oil. Thus in the period from 1870 to 1920 there evolved the second major facet of the system for disposal and regulation of the public mineral heritage—the Mineral Leasing Act.

BACKGROUND. The United States Congress in its policies of disposal of the public domain differentiated between agricultural land and mineral land as early as 1785. But the first regulations were soon abandoned, and much of the mineral wealth of the nation passed to private ownership through entry as agricultural land. In 1873 the Coal Lands Law renewed recognition of mineral land values in requiring payment of at least ten dollars per acre for all land declared to be coal land and placed a 160-acre limit on the amount a person could acquire and a 640-acre limit on a group. In 1879 the Geological Survey was established with a mandate to investigate mining resources and to withdraw lands for mineral classification. The survey began pointing out widespread fraud in coal land acquisitions and the fundamental inadequacies of the mining law for petroleum. It urged vast withdrawals of land from entry of all sorts as well as from the operations of the mining law in the interest of the public welfare. Its reports on mineral lands gave rise to and supported growing feelings among national leaders concerning the wisdom and ethics of allowing resource wealth such as coal and oil to pass so easily from the public domain. As a result of these influences, large withdrawals were made by executive order of the President between 1900 and 1910 under provisions of the act establishing the Geological Survey. In 1900 two townships in California were withdrawn under the Mining Law and later, especially under the Roosevelt and Taft administrations, some seven million acres of oil lands were withdrawn in California and Wyoming. In 1906 President Roosevelt withdrew some seventy-five million acres of coal land in the United States and some nine million in Alaska. In 1908 all lands suspected of having phosphate deposits in Idaho, Utah, and Wyoming (totaling 4.7 million acres) were withdrawn. Later the Florida phosphate lands were added.[15] In 1909 the first of a number of oil land withdrawals was made to create naval oil reserves.

Opposition to these executive withdrawals was vigorous, and the controversy reached major proportions, but in June, 1910, Congress supported the actions by approving the Withdrawal Act. This Act granted broad authority to the Chief Executive for withdrawals of public lands from all forms of sale or entry for water power sites, irrigation, and classification.[16] In total at least one hundred million acres were thus withdrawn during the first decade of the twentieth century for classification as nonmetallic mineral lands. From then on the Bureau of Land Management has carried on a continual process of withdrawal, examination, and classification of public lands. Because it was not intended that the minerals be locked up in preservation, it then became necessary to devise some system for permitting utilization of the mineral lands and the mineral wealth while protecting the public interest. In 1910 a law had been passed which opened the coal lands to surface use by preemption or sale, but reserved the coal rights for the government. This provision was extended to all withdrawn mineral lands in 1914.[17] It was readily apparent that some form of leasing in which surface and mineral rights were considered separately was needed. The first minerals leasing act was the Alaska Coal Leasing Act of 1914. This was followed by the major milestone, the Mineral Leasing Act of 1920, which with modifications regulates a significant portion of the public mineral wealth.

THE MINERAL LEASING ACT OF 1920. The Mineral Leasing Act of 1920 established the right of the United States to hold permanent title to nonmetallic mineral resources on public lands. The main provisions of the law are summarized by the following:

1). The Act specifically and finally removed the major nonmetallic minerals (oil, gas, coal, phosphate, potassium, sodium, and sulphur) on all public lands from private claims under the general mining laws.

2). The Act provides for leasing to private enterprise of the rights to explore for and to produce these minerals from public domain lands and provides for regulation and supervision of mining in the interest of conservation, including protection of other resource values.

Of particular importance is the fact that these provisions permit and protect use of surface resource values for other purposes. Many such areas are used for grazing and farming while producing minerals of major importance.

In 1953 the United States established similar patterns of ownership and provisions for lease to millions of acres of submerged coastal lands through the Outer Continental Shelf Act.

The Mineral Leasing Act provided for rentals, lease fees, royalties, and bonuses. For most lands on which no mineral discovery has been made, the law requires leasing to the first applicant, and establishes modest rental fees. For oil and gas leases the usual fee is fifty cents per year per acre in areas outside of known oil bearing structures, or two dollars per year per acre if within areas of known producing structures. For most other leases the basic rental fee is one dollar per acre per year. Following discovery of minerals the basic fee becomes one dollar per acre until production royalties apply. Some criticism has been directed toward the low fee schedule during exploration which has led nonmining interests to secure leases and hold them as a speculation. Also, large companies, such as oil enterprises, have been criticized for maintaining leases but not producing because the fee is so small. Because no waste is involved, the problem is not one of conservation, but of immediate returns to the Treasury. Where oil or other minerals are already proven to exist, the Act provides for competitive bids through a system of bonus payments.[18]

Royalties vary. For oil and gas the usual rate is 12.5 per cent, except on Continental Shelf leases where it is 16⅔ per cent. Other commodities carry a statutory minimum such as 5 per cent for phosphate and potassium. The Act also provides for distribution of the income from exploitation of public mineral wealth: 37.5 per cent of the receipts are paid to the producing state for schools and roads, 52.5 per cent to the Reclamation Fund, and only 10 per cent to the Treasury. Revenues from the Outer Continental Shelf are not shared, but go into the Treasury general fund.

Receipts from public domain under mineral leasing are revealed in Table 14–3. A total of sixty-five million acres is shown to be under leases in twenty-two states, chiefly for oil and gas.

Conclusion

The physical characteristics of mineral resources and the economic and conservation implications of these conditions together with the evolution and present legal status form a necessary background for insight into problems of the mineral endowment. How well the nation has been endowed with mineral resources to meet the demands of an expanding economy is the subject of the following chapter.

FOOTNOTES

[1] The student interested in the study of the formation of mineral deposits should consult textbooks such as Alan M. Bateman, *Formation of Mineral Deposits* (New York: John Wiley & Sons, Inc., 1951), 371 pp., or Brian Mason, *Principles of Geochemistry* (New York: John Wiley & Sons, Inc., 1958), 310 pp.

[2] *Report of the Advisory Committee on Minerals Research to the National Science Foundation* (Washington: National Science Foundation, 1956), p. 2.

[3] *Oil Producing Industry in Your State*, 1956 Edition (Tulsa, Oklahoma: The Independent Petroleum Association of America, 1956), p. 10.

[4] For a cogent recent discussion of mineral resource terminology see F. Blondel and S. G. Lasky, "Mineral Reserves and Mineral Resources," *Economic Geology*, vol. 51, no. 7, (1956), pp. 686–97.

[5] "The Outlook for Key Commodities," *Resources for Freedom*, Report of the President's Materials Policy Commission, vol. 2 (Washington: Government Printing Office, 1952), p. 151.

[6] The contrast between this legal philosophy toward resources and that held in much of the rest of the world is worth special note. The United States law comes from the common law of England which recognized the right of surface possession to include the subsurface. In Latin America and most other areas of the world, more of the basic tenets are derived from Roman law and express the concept that natural wealth, such as water and minerals, is a heritage belonging to all of the people with the result that private exploitation is only by special permission of the government.

[7] For a review of the regulations created under these laws see *Code of Federal Regulations,* revised 1967 (Washington: Government Printing Office, 1967). Mineral resource regulations are chiefly under Title 30 and Title 43.

[8] For example see T. A. Rickard, *A History of American Mining* (New York: McGraw Hill Book Company, Inc., 1932), p. 32–35.

[9] The Bureau of Land Management indicated in 1965 that there are 770 million acres of federal lands subject to mineral development of some sort. Of this total, however, only about 600 million acres are still open to prospecting.

[10] Today one hundred dollars is a very modest requirement, but in 1872 it was a considerable investment in time or capital. Recent moves to increase required assessment work have been unsuccessful, but geochemical and geophysical prospecting is now accepted as work. Moreover, failure to do annual work does not result in loss of the claim, but only makes the claim vulnerable to preemption by another prospector.

[11] In 1963 the Bureau of Land Management estimated that six million claims had been filed in the twelve western states since 1872 and that about 160,000 claims were active.

[12] Separation of surface rights from the mineral rights is not a new idea. The resolution of the North American Conservation Conference of 1909 includes a specific recommendation that surface rights be separated from mineral or subsurface rights.

[13] "The Al Sarena Case," United States Congress, Senate Committee on Interior and Insular Affairs, *Committee Print,* 84th Cong., 2nd sess. (Washington: Government Printing Office, 1956), 58 pp.

[14] The United States Forest Service has estimated that of 915,638 acres in 34,000 patents on former USFS lands, only 15 per cent have ever been developed successfully. In 1950 it was reported that standing timber on 23,000 mineral claims in Washington, Oregon, and California was worth fifty million dollars. *Resources for Freedom, op. cit.,* vol. 1, p. 30.

[15] Based on *Annual Report of the Secretary of the Interior* (Washington: Department of the Interior, 1865–1958).

[16] The entire withdrawal concept was also legally contested in the courts in the case of *United States* vs. *Midwest Oil Company.* After the lower courts had found the withdrawals unconstitutional, the Supreme Court reversed the findings, supporting the Executive withdrawals and the Withdrawal Act. See *Resources for Freedom, op. cit.,* vol. 5, p. 4.

[17] In 1965 the Bureau of Land Management managed mineral resources on fifty-nine million acres of private land on which the government retained mineral rights. *Public Land Statistics* (Washington: Bureau of Land Management, Department of the Interior, 1965).

[18] The Outer Continental Shelf submerged oil lands are a prime example. In 1956 of 121 leases, fifty-two carried bonuses of fifteen dollars to fifty dollars per acre, while sixteen were for bonuses of over one thousand dollars per acre. See discussion in Marion Clawson and Burnell Held, *The Federal Lands, Their Use and Management* (Baltimore: The Johns Hopkins Univ. Press, 1957), pp. 101–6. A 1,850-acre potassium lease in New Mexico in 1958 carried a bonus of just over one thousand dollars per acre. *1958 Annual Report of the Secretary of the Interior* (Washington: Department of the Interior, 1959), p. 260.

15

Mineral Resources Adequacy

THE MINERAL ENDOWMENT of the United States is clearly rich in variety and in quantity. The land contains about one-third of the world's coal reserve, about 10 per cent of the petroleum, half of the natural gas, and 20 per cent of the iron reserve. It also possesses a wide range of metallic mineral deposits, especially those required in great tonnage such as copper, lead, zinc, and major deposits of nonmetals such as phosphate, potassium, salt, and sulphur. Not only are the endowments generous, but the situation for the critical deposits is generally favorable for development. A notable example is the Lake Superior iron ore, which is easily mined and favorably situated for low cost lake transport to be joined with the quality coals of the Appalachian and Interior Fields in the industrial core of the nation. This is not to imply, however, that the United States has all needed resources to support the present and future economy.

The fundamental question for consideration is not whether the nation is rich in mineral resources—that the United States is generously endowed is accepted as axiomatic—it is the question of how adequate these resources are for predictable future needs. Answers are difficult and cannot be given exactly or finally. Nevertheless, it is an obligation of each generation to strive for increasingly specific knowledge and understanding of mineral resources, to continually develop projections and conclusions concerning requirements of the future, and to consider what should be done now in preparation for meeting future needs. Toward the achievement of this kind of understanding, this chapter presents an appraisal of the adequacy of the mineral endowment.[1]

The Beginning of the Mineral Industry

Unlike the resources associated with productive land, the mineral wealth of the nation was a long time in being recognized. T. A. Rickard, in his work *A History of American Mining*, quotes numerous writers who expounded upon the apparent poverty of America as far as mineral wealth was concerned. For example, he quotes Cornelius de Pauw as saying in 1770, "In all the extent of America there are found few mines of iron, and these so inferior in quality to those of the old continent that it cannot even be used for nails."[2] Iron ore was discovered in 1585, but not until 1628 was any mined, and it was 1844 before the iron deposits of the Lake Superior region were recognized. Not until 1750 was bituminous coal mined for use in local blacksmith shops. Anthracite coal came into use about 1812 after accidental discovery of how to make it burn.[3] Coke was introduced commercially about 1850.

The large scale consumption of mineral raw materials required by the high living standard in the United States is a new phenomenon occurring only in the past few decades. Bituminous coal production in 1850, for example, was only seven million tons; by 1875 it was only fifty-two million, and by 1900 it had risen to 270 million. In 1947 coal output was 630 million tons, and since then has averaged around 500 million tons annually. Iron ore production in 1900 was only twenty-seven million tons; by 1942 it was 103 million, and since then it has averaged about one hundred million tons annually. Petroleum production in 1900 was only sixty-three million barrels, but by 1965 it had soared to 2,849 million barrels. Similar relative increases have occurred for most other minerals to satisfy requirements for fuels, machines, and consumer goods. (Figure 15 - 1).

The Fuel Resources

The fossil fuels form one of the essential elements of the resource base. In the United States the endowment is great, but present rates of depletion and projected future requirements give cause for concern. Consumption of energy has risen sixfold since 1900, while population has increased only two-and-a-half times; moreover, by 1980 it is expected that fuel requirements will be nearly twice the present consumption. Increases are expected in all areas of use. The pattern of how the nation's energy "slaves" are used at present is shown in Table 15–1.

Not only has energy use mounted at a prodigious rate, but it has been accompanied by a striking change in importance of its several sources, which has profound bearing on conservation.

PATTERN OF ENERGY SOURCES

Before 1850 wood and muscle of man and animal were the major energy sources, and the only conservation concern was for saving the forests. Between 1850 and 1900 wood and muscle lost favor rapidly to coal. As a result, in 1900 the pattern of the nation's commercial energy consumption by source was: Coal, about 70 per cent; wood, 20 per cent; oil and gas,

15 – 1 *Minerals in quantity are essential.* This illustration symbolizes the dramatic rise in importance of minerals. To supply the demands of the national economy for machines of metal and the fuels to run them plus the home conveniences and innumerable gadgets of metal, mines of the United States and of the world are being called on for ever-increasing tonnages of raw materials. (Bankers Trust Company of New York)

TABLE 15–1
ENERGY CONSUMED BY USER GROUP, 1955*

Group	Per cent
Industry	39.2
Transportation	20.2
Homes	18.6
Commercial	8.5
Government	4.6
Agriculture	1.8
Other and losses	7.1
TOTAL	100.0

* Sam H. Schurr and Bruce C. Netschert, et al., *Energy in the American Economy, 1890–1955* (Baltimore: The Johns Hopkins Press for Resources for the Future, Inc., 1960), p. 204.

about 5 per cent; and water power, most of the rest. Since 1900 evolution in consumer preference has continued as popularity of liquids and gases has risen at the expense of the solid fuels. The proportions for commercial energy (not including wood) in 1965 were approximately as follows: petroleum, 40 per cent; natural gas liquids, 3 per cent; natural gas, 30 per cent; coal, 23 per cent; and hydro, 4 per cent. The use of wood as fuel in the United States is minor and declining. Nuclear power plants in 1965 contributed 0.1 per cent, but almost certainly will increase to 10 per cent by 1980. Thus while requirements in quantity have soared, there has been a notable switch to the fossil fuels as the foundation of the economy. Of all the currently utilized energy sources only hydro is a renewable resource; all the others are depletable minerals. The rate of withdrawal from the earth is of such magnitude that the mind can hardly appreciate the quantitative significance. It is, therefore, essential that the fuel endowment be appraised in terms of adequacy for present and predictable future needs for each major source and as a composite.

COAL

The coal resource of the United States is enormous, of high quality, favorably located for mining, and widely found. The map (Figure 15 – 2) illustrates the pattern and shows comparisons of reserve and production by major field areas. The reserve, as estimated by the United States Geological Survey in 1960, totals 1.66 trillion tons. This estimate includes only coal no deeper than three thousand feet and in seams at least fourteen inches thick, but is in considerable part based on geologic inference. Lignite and sub-bituminous coals have lesser value for most uses so they should be discounted in Btu value in considering adequacy of coal. Table 15–2 shows the distribution of coal currently recoverable in terms of bituminous coal equivalent.

Study of this Table together with the map will reveal that western coal fields today are producing only a small share of the nation's coal, that they contain nearly half of the nation's reserve, and that most of the coals are sub-bituminous and lignite. In appraising the pattern from the conservation viewpoint, it is important to note the heavy demand being made on coal fields of highest quality and on those most favorably located with reference to the industrial northeast quarter of the nation. There is little reason to suppose that the major industrial center will shift notably, and it follows that the western coals would be costly in terms of transportation for supply to the eastern industrial area.

ADEQUACY OF COAL. Coal production reached its maximum in 1947 when 630 million tons of bituminous coal were produced; since then about 500 million tons have been the annual requirement. Anthracite production reached its

TABLE 15–2
RECOVERABLE COAL RESERVES* (BASIS OF 13,000 BTU COAL–BILLION TONS)

	Anthracite	Bituminous	Sub-bituminous	Lignite	Total
Eastern	6.0	218	..	..	224
Interior	0.1	205	..	..	205
Others	0.4	116	136	119	371
	6.5	539	136	119	800

* "Mineral Facts and Problems," *United States Bureau of Mines Bulletin 630* (Washington: Department of the Interior, 1965), p. 134.

peak of 99.6 million tons in 1917 and has declined steadily to fifteen million tons in 1965. For reasons presently to be made clear, it is expected that demands on coal will increase in the near future, and it is suggested that by 1980 demand will have doubled. The key to whether this estimate is high or low is how much coal will be burned in electric generating plants. Current trends in improved firing of coal, more efficient mining, and lower cost transportation indicate a good future for coal. To meet this probable demand, the coal reserve, as noted, is 1.66 trillion tons. But a realistic appraisal of adequacy must consider the percentage of actual recovery of the coal in mining and must consider the character of the reserve estimate.

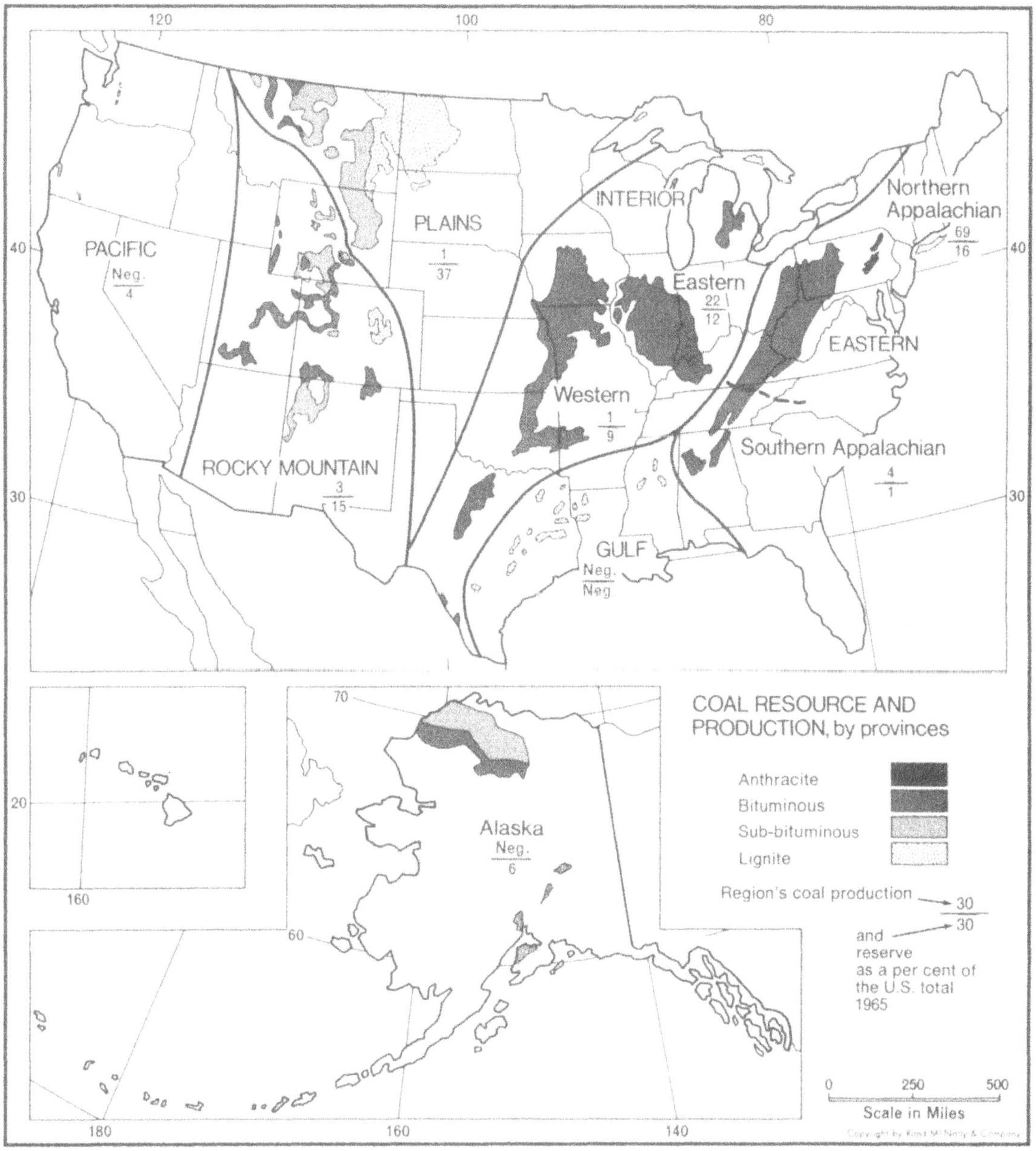

15 – 2 The major coal fields of the United States, reserves and production.

Over the years, the recovery of coal from underground mines has averaged about 50 per cent. In other words, on the average, half of the coal in any given seam is left behind in pillars, in blocks under towns and railroads to prevent the land's settling, or in low-grade sections and areas hard to mine. Surface mines recover higher percentages in actual working, but are faced with the same problems of towns and other obstacles in complete mining. There is also shrinkage during washing and sorting which amounts to 15 to 20 per cent of the coal mined. Conservational possibilities are suggested by these figures. It follows that the coal reserve estimate must be reduced by half to stay within the bounds of present-day economic reality and technology of recovery. Thus about 830 billion tons is the present estimate of recoverable coal reserve (Table 15–3).

The basis on which the reserve estimate was made will now be examined. As much as two-thirds of the coal reported is *inferred*, that is, based on relative lack of specific knowledge such as drillings or mine development. The reserve estimate includes coal in beds as thin as fourteen inches, but even with the best of modern technology, in 1965 only 1 per cent of the nation's coal was mined from seams less than two feet thick, and nearly all of this was done by surface operation. Only 11 per cent of the coal was mined from seams less than three feet thick. In total, as much as 40 per cent of the coal included in the reserve estimate is beyond consideration under present economics and technology because of thin beds (Figure 15 – 3). The estimates also include coal as deep as three thousand feet, but little coal is mined in the United States from depths over one thousand feet. Considering all these factors, many resource scholars believe that the tonnage for *proved* recoverable coal reserve should be considered as being one quarter of the total. Accepting these assumptions, the proved coal reserve is adequate for about 400 years at current rate of use, but only for about 250 years at the suggested 1980 demand rate (also see Table 15–4).[4]

RESERVES OF HIGH QUALITY MOST UTILIZED. Low-volatile bituminous coals have unusual significance because they are essential to production of coke, and when used in blending they make possible the use of reserves of high-volatile coals that would be unsatisfactory by themselves.

Reserves of low-volatile coal have been estimated as only 1 per cent of the nation's coal reserve.[5] For the most part this coal is in the northern Appalachian fields of West Virginia and Pennsylvania and in the interior fields of Oklahoma. Mining has been intensive in low-volatile coal deposits because the demand for

TABLE 15–3
RATION OF FUEL RESERVES TO PRODUCTION*

	1965 Production	1965 Reserve	Years adequate reserve 1965 rate	Suggested production 1980[1]	Years adequate reserve 1980 rate[2]
Coal (million tons)					
Bituminous & lignite	512	823,000	1607	900	907
Anthracite	14.9	8,000	537	12	648
Petroleum (billion bbls)	2.8	31.4	11	5.0	6
Natural gas liquids (billion bbls)	0.44	8.0	18	1.0	8
Natural gas (trillion cubic feet marketed production)	16.0	286.5	17	24	12

* *Minerals Yearbook, 1965* (Washington: Bureau of Mines, Department of the Interior, 1965).

[1] Author estimates.
[2] 1965 level of reserves assumed.

this quality coal for fuel and for coking usually accounts for 15 to 20 per cent of the coal mined.[6] Already the most accessible better grade coking coals have been mined, and industry has increasingly had to adjust to poorer grades and to increased blending.

The outstanding example of concentrated mining resulting from quality of coal and ease of access is the Fayette County–Connellsville area of southwestern Pennsylvania near Pittsburgh. This famous coking coal literally supported the early steel industry, providing a quarter of the nation's total output of coking coal. In 1900 the county mined fifteen million tons of coal (nine million used in coke). In the 1930's exhaustion of the high-grade coal was evident, and decline followed until by 1958 the county was producing only 2.8 million tons. The number of miners had dropped to only 2,121 in 1958.[7] During the period from 1940 to 1955, seventeen thousand persons left the county, and unemployment was high.[8]

15–3 *The continuous miner.* This machine is designed for mining in seams from five feet to twelve feet thick. Lower contour models of a machine applying the same "roto-ripper" principle will mine in seams as thin as 38 inches. Operation of the machine calls for sumping the oscillating ripper wheels, equipped with conical-shaped bits, into the coal face at the top of the seam. The cutting head is then forced down hydraulically, breaking the coal onto the mine floor where a gathering-arm loader head collects it into a chain conveyor. The conveyor then moves it back through the machine to a haulage vehicle (shuttle car). In continuous mining systems, this machine replaces cutting, drilling, blasting, and loading operations which are performed in conventional mining systems. (Joy Manufacturing Company)

TABLE 15–4

COMPARISON OF FUEL RESERVES AND PRODUCTION IN BTU EQUIVALENT*

	Reserves in quadrillion Btu				Production quadrillion Btu	
	Proved	Per cent	Potential	Per cent	1965[1]	1980[2]
Coal	5,400[3]	87.1	22,000[7]	77.2	13.8	23.4
Petroleum	215[4]	3.5	2,355[8]	8.3	15.9	29.0
Natural Gas	285[5]	4.6	1,100[9]	3.9	17.7	24.0
Bitumen from sandstone	10	0.1	60	0.1	—	—?
Oil from shale	290[6]	4.7	3,000[10]	10.5	—	—?
TOTAL	6,200	100.0	28,515	100.0	47.4	76.4

* "Mineral Facts and Problems." *Ibid.*, p. 134.
[1] *Minerals Yearbook, 1965 Ibid.*
[2] Estimates by author of probable domestic production in 1980.
[3] One-fourth of recoverable coal reserve—208 billion tons x 26 million Btu.
[4] Crude petroleum 31 billion bbls. x 5.7 million Btu plus 7.7 billion bbls. natural gas liquids x 4.6 million Btu.
[5] 276 trillion cu. feet x 1,035 Btu.
[6] Oil potential of shales containing not less than 30 gallons per ton.
[7] 830 billion tons.
[8] About 400 billion barrels.
[9] About 1,000 trillion cubic feet.
[10] Oil potential of shales containing not less than 25 gallons per ton and including inferred deposits.

PETROLEUM

Petroleum and natural gas also are found in large quantities and over considerable areas of the United States. The map (Figure 15 – 4) illustrates the pattern of petroleum and gas fields and also presents comparative data on reserves and production. Study of the map will indicate the dominance of the mid-continent fields in both production and reserves of petroleum as well as gas. The recent phenomenal rise in consumer préférence for petroleum and gas at the expense of coal has resulted in their withdrawal in enormous amounts, and both have surpassed coal in terms of percentage contributed to the national energy base.

Petroleum reserve data are notable for having steadily increased over the decades in spite of soaring use. As of December, 1965, the proved reserve of crude petroleum was estimated at 31.4 billion barrels.[9] The fact that size of reserve continues to increase is evidence of the major efforts put forth by the industry in exploration, but does not imply an endless resource. The annual demand on the reserve in 1965 reached a total of 2.9 billion barrels, indicating a ratio of production to reserve of about one to eleven, a ratio that has been more or less constant for the past two decades.

There is reason to assume that the time will come when new discoveries will no longer keep pace with production, and reserves will begin to decline, though no one can know exactly when this turning point will come. The only approach to determining when it will come lies in estimates of the total potential resource and projection of the demand curve. Thus, using geologic inference as the basis, the potential reserve of petroleum has been estimated by many scientists and by the industry. Estimates have ranged from as little as one hundred billion barrels to as much as five hundred billion barrels. In Table 15–4 potential petroleum is estimated at about four hundred billion barrels. For consideration of the adequacy of this potential resource, it is suggested that petroleum requirement will double by 1980. Some of the demand will be supplied by imports, but it is assumed that annual domestic production will rise to five billion barrels. The ratio of this 1980 production to the potential reserve is eighty years.

NATURAL GAS

Natural gas reserves are currently estimated at 286 trillion cubic feet. Gas reserve estimates, like those of petroleum, have increased annually during recent decades in spite of great increases in production. Withdrawals in 1965 totaled seventeen trillion cubic feet, indicating a ratio of production to reserve of about one to seventeen. Just as with petroleum, no one can really know when the ratio of production to reserve will start declining, although this appears to be inevitable. As an approach to the answer, it is worth noting that estimates of the gas potential reserve are about one thousand trillion cubic feet. On the basis of 1980 demand, about twenty-four trillion cubic feet annually, the indicated potential reserve would last forty-six years.

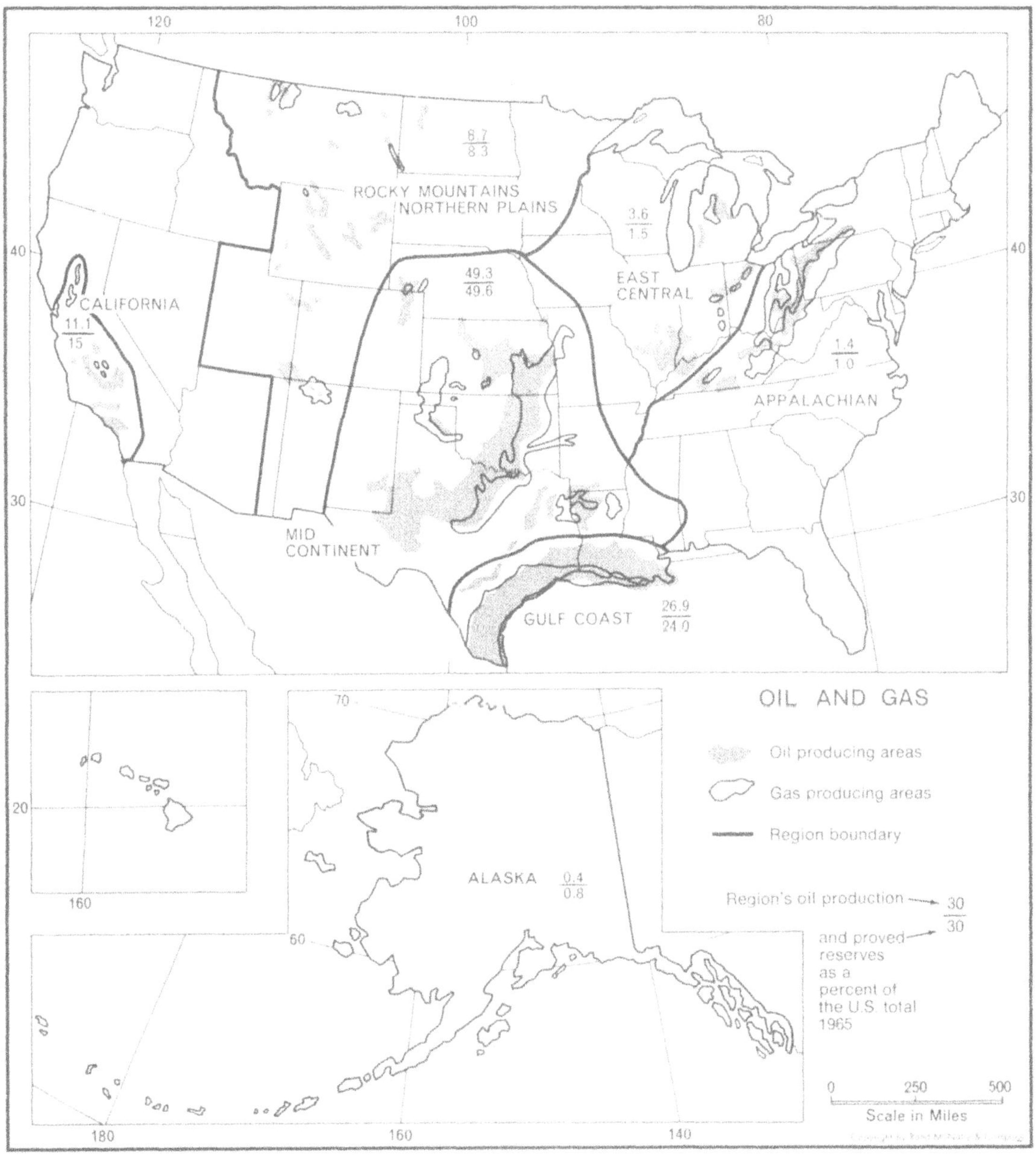

15 – 4 *The petroleum and gas fields of the United States.*

COMPOSITE OF FOSSIL FUELS

Table 15–4 assembles the data previously noted in appraising the individual fuel reserves and production in terms of the common denominator of Btu equivalents. The suggested quantifications are to be considered as indicative only, because the basis for the reserve estimate leaves much to be desired. Nevertheless there are several conclusions significant to conservation study that one can draw from the Table.

Coal is the great fossil fuel resource; it accounts for 87 per cent of the proved reserve and 77 per cent of the potential reserve. In contrast, the two energy sources that are today supplying 73 per cent of the total consumption accounts for no more than 9 per cent of the proved reserve and only 12 per cent of the potential. As gas and oil decline in the not-so-distant future, it seems certain that increased demand will be transferred to coal to supply liquid and gas fuels as well as solid ones. The evidence points strongly to a growing importance for coal.

TABLE 15–5

MINERALS WITH DOMESTIC PRODUCTION UNDER 25 PER CENT OF CONSUMPTION 1965*

	Consumption	Mine Production	Per cent sufficient	Proved reserve
Antimony (1,000 tons)	41	0.8	2	50–100
Asbestos (1,000 tons)	795	118	15	2,000
Bauxite (1,000 tons)	13,534	1,654	12	50,000 plus large low-grade resource
Beryllium (1,000 tons)	5.8	neg.	1	0.5–10?
Bismuth (1,000 pounds)	2,932	by-product	10?	10,000?
Chromite (1,000 tons)	1,582	none	0	400 plus 8,000 low grade
Cobalt (1,000 pounds)	15,408	not available	10?	7,500 low grade 140,000
Columbium-tantalum (tons)	1,762	none	0	(low grade but large) 180,000
Diamonds (1,000 carats)	14,400	5,500 manufactured	4	none
Graphite (1,000 tons)	47	?	5?	5 (large low grade resource)
Manganese (1,000 tons)	2,866	29	1	1,000 plus additional low grade
Mica (sheet) (1,000 tons)	4.5	.01	0.2	2
Nickel (1,000 tons)	172	16	1	243
Platinum Group (1,000 troy ounces)	1,186	35	3?	50?
Quartz Crystal (1,000 pounds)	315	none	0	none
Rutile (titanium) (1,000 tons concentrate)	117	8 (1964)	7	1,000
Tin (Primary) (1,000 tons)	58	negligible	1	3 low grade

* Based on Bureau of Mines data.

The second conclusion concerns the relationship between energy requirements and indications of reserve. In 1965 the United States economy consumed about fifty quadrillion Btu equivalents in the fuels noted (not including hydro). Even if no allowance is made for increasing energy requirement to match growing population and increasing per capita use, the proved fuel reserves would last only 125 years. The relationship between the 1980 demand estimate and total potential reserves is a little better, about 370 years.

The inescapable conclusion from these data is that the fuel resources are adequate for the present, but that fossil fuels are more limited than generally realized.

The indicated short life of the fossil fuels is not considered by the authors as dooming the nation to decline; rather it is advance warning that there is need to support research leading to discovery and perfection of the means of making the transitions to other sources of fuel such as nuclear, geothermal, and solar energy.

Nonfuel Minerals

The United States has a generous endowment in nonfuel minerals as a result of its large size and wide variety of geologic environments. Adequacy of known reserves, however, varies greatly from mineral to mineral, and it is difficult to give a definitive appraisal due to inadequacies of exploration and the variables of price. It is also noteworthy that international trade figures more strongly in nonfuel mineral supply.

Tables 15–5, 15–6, and 15–8 present basic data on the major minerals arranged into groups according to degree of current domestic sufficiency. In using the tables the characteristics of mineral resources must be kept in mind lest erroneous conclusions be drawn. The per cent of self-sufficiency reveals the nation's position in mineral commodities from an economic point of view specifically for only one year. The fact that a mineral is imported does not always imply inability of the domestic mining industry to meet demands. Moreover, importing is not necessarily a national weakness. In many cases imports are an indication of low-grade ores and resultant high cost in comparison with imports. There is also the fact that imports aid economic and political foreign relationships.

MINERALS IN SHORT SUPPLY

Domestic production is less than 25 per cent of the nation's consumption for at least twenty minerals. Here is an outstanding example of the principle that no nation has all needed minerals. Questions of mineral resource adequacy and conservation inevitably take on world-wide proportions, and shortages are rarely world wide. Most, if not all of these minerals are in some quantity in the United States, but the deposits are small or of such low grade that exploitation is not economical in a competitive world economy. Manganese and bauxite are especially good examples of imported minerals in which the nation has large low-grade possibilities. The adequacy question for these is not a matter of quantity but of quality, and greater use of domestic sources is inhibited by a high cost in comparison with imports. For a few minerals such as tin, platinum, diamonds, and crucible graphite the known deposits in the United States are insignificant. Study of consumption and reserve data in Table 15–5 will reveal that, were it suddenly necessary for the United States to meet requirements for these minerals by domestic mining, the proved reserves of most of them would be exhausted in one to several years.

It is not surprising that these mining industries have received some form of Federal government aid to stimulate exploration and experimentation on new processing techniques. Such aid is highly desirable and is a good example of the proper role of government in supplying leadership and financial aid in basic research to extend or expand mineral resources.

It is not the belief of the authors that the nation should attempt to achieve an uneconomical degree of self-sufficiency in these minerals. To do so would be wasteful of other resources, especially energy and capital. The major share of these minerals should generally

TABLE 15–6

MINERALS WITH DOMESTIC PRODUCTION OF 25–65 PER CENT OF CONSUMPTION 1965*

	Consumption	Mine production	Per cent sufficient	Proved reserve
Barite (1,000 tons)	1,388	846	61	46,000
Cadmium (1,000 pounds)	10,431	by-product	40?	75,000?
Fluorspar (1,000 tons)	931	241	26	17,960
Lead (1,000 tons)	1,241	301	25	6,000
Mercury (1,000 flasks)	76	19	25	75
Silver (1,000 troy ounces)	137,000 (industrial)	39,800	29	590,000
Tungsten (1,000 pounds)	13,868	8,280 (1962)	60?	53,000
Zinc (1,000 tons)	1,354	611	45	12,000

* Based on Bureau of Mines data.

be imported, providing the raw material at lowest cost to consumers, adding to the inservice and stockpile reserve, and at the same time assisting the economy of other nations from which the United States imports.

MINERALS IN INTERMEDIATE SUPPLY

Eight of the more common minerals required by the United States economy come within the range of 25 to 75 per cent self-sufficiency and are shown on Table 15–6. In all of these there is a major domestic mining industry. The significant importation to supply annual requirements reveals the ability of foreign producers, in many cases with richer ore and usually lower labor cost, to compete in the American market. Mercury is a notable example of the influence of price and government aid on production (Figure 15 – 5). During 1942 and 1943, when the price reached $196 a flask, domestic mine production surpassed consumption; when the price fell to $73 a flask mining all but ceased, while imports continued at a high level. See Table 15–7 for an illuminating series of data on the concept of reserve as a function of price. Depressed price is a significant problem facing the lead and zinc mining industry of the nation. One of the chief objectives of government regulation in the interest of conservation of these and other minerals must be to provide reasonable stability of price, for only in a stable business climate can there be adequate conservation practices. Study of the ratio of consumption and production to reserves will reveal that these minerals also are limited and

TABLE 15–7

UNITED STATES MERCURY RESOURCE AS A FUNCTION OF MARKET PRICE

Market price dollars	Resource[1] flasks	Projected potential production level
100	40,000	
200	140,000	19,000
300	379,000	30,000
500	827,000	53,000
1,000	1,287,000	81,000
1,500	1,465,500[2]	90,000

[1] Based on 1961 costs and technology.

[2] Rescue estimates are accumulative. "Mercury Potential of the United States," *Bureau of Mines Information Circular 8252* (Washington: Department of the Interior, 1965).

15–5 *Trends of United States mercury production and imports.* Low-grade ore, scattered deposits, and high cost of labor resulting in high-cost production compared with mercury from Spain and Italy causes domestic output to fluctuate with price and stimulative actions of government. Thus in the war years of the 1940's with higher prices, domestic producers were able to meet all consumption needs, but by 1950 the depressed price had nearly closed down domestic output, while imports remained high.

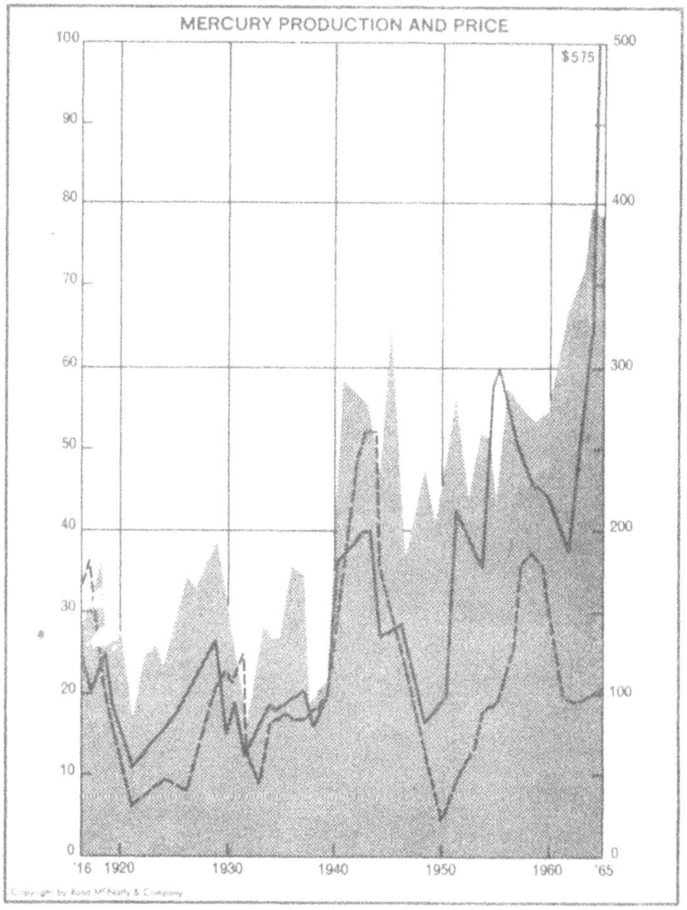

could supply total needs for no more than ten to thirty years at the most.

MINERALS IN GENEROUS SUPPLY

Minerals which are produced in surplus or near self-sufficient amounts include eleven of the more common minerals (Table 15–8). Nonmetallics such as phosphate dominate the list. For the most part, these minerals are found in great deposits in the nation, and they are used in major quantities. Greatly increased imports of iron ore and copper reflect the economics of the market and the passing of the "cream skimming" era of the early mining years in the United States, but in no case is there any reason to suggest that the country is becoming a "have-not nation" as once was feared.

The reserves of these minerals are shown to be large, but study of the relationship between demands now being put on them and probable future requirements raise the question of how long such demands can be met. For example, note that proved high-grade iron ore reserve is indicated to have a life of only thirty-five years at the current rate of production. The obvious conclusion is that industry must turn to lower grade ores and to other sources such as imports.

REQUIREMENTS INCREASING

Just as in the case of the fuels, the need for nonfuel mineral commodities in the next decades is certain to be larger. The President's Materials Policy Commission in 1952 concluded that major increases shown in Table 15–10

TABLE 15–8

MINERALS WITH DOMESTIC PRODUCTION OVER 65 PER CENT OF CONSUMPTION 1965*

	Consumption	Mine production	Per cent sufficient	Proved reserve
Boron (gross (1,000 tons)	not available	807	100	very large (120,000)
Copper (1,000 tons)	2,005	1,352	70	33,000
Diatomite (1,000 tons)	not available	580	100	large
Iron ore (1,000 tons)	131,888	87,842	66	15,000,000 (over 25 per cent F)
Magnesium (1,000 tons)	69	81	100	sea water–enormous
Molybdenum (1,000 pounds)	68,112	77,372	100	3,000,000
Phosphate rock (1,000 tons)	19,500	26,400	100	7,100,000
Potassium (1,000 tons)	3,390	3,140	92	450,000
Salt (1,000 tons)	36,409	34,687	95	60,000,000,000 (excluding sea water)
Sulphur–native (1,000 tons)	5,775	6,116	100	50,000–200,000
all forms (1,000 tons)	7,959	8,212	100	100,000
Vanadium (1,000 pounds)	9,416	10,452	100	by-product of uranium and phosphate

* Based on Bureau of Mines data.

TABLE 15–9

CONSUMPTION OF SELECTED MINERAL RESOURCES 1960 AND PROJECTED 1975 CONSUMPTION*

Energy resources	1960	1975
Bituminous coal mined (thousand tons)	380,429	800,000
Anthracite coal (thousand tons)	17,600	10,000
Natural gas (billion cubic feet)	12,509	25,619
Petroleum (million barrels)	2,963	6,530
Natural gas liquids (million barrels)	340	6,530
Iron ore (thousand tons usable)	108,050	164,000
Manganese ore (thousand tons)	1,946	3,090
Molybdenum (thousand pounds)	31,837	87,800
Tungsten ore (thousand pounds W content)	11,605	16,870
Chromite ore (thousand tons)	1,220	2,555
Bauxite (thousand tons)	8,883	27,550
Copper (thousand tons) refined	1,350	2,080
Lead (thousand tons)	1,021	1,610
Zinc (thousand tons)		2,150
Titanium (contained T_1O_2) (thousand tons)	868,080	1,220,500
Mercury (thousand flasks)	51,167	104,000
Potassium (thousand tons)	3,997	6,250

TABLE 15–9 (Continued)

CONSUMPTION OF SELECTED MINERAL RESOURCES 1960 AND PROJECTED 1975 CONSUMPTION*

Energy resources	1960	1975
Phosphate rock (P_2O_3 content) (thousand tons)	13,337	27,200
Salt (thousand tons)	26,114	53,950
Sulphur (thousand tons)	5,859	9,700
Cement (thousand barrels)	325,557	660,000

* *Minerals Yearbook, 1960,* and *1965. Op. cit.*

TABLE 15–10

PROBABLE INCREASE IN MINERALS CONSUMPTION BY 1975*

Mineral	Per cent of increase
Magnesium	1,845
Titanium and cadmium	324
Cobalt	344
Bauxite	291
Fluorspar	187
Molybdenum	170
Tungsten	150
Phosphate rock	150
Potash	150
Sulphur	110
Chromium	100
Nickel	100
Antimony	81
Iron	54
Lead	53
Manganese	50
Copper	43
Zinc	39
Platinum	30
Mercury	25
Tin	18

* *Resources for Freedom,* the President's Materials Policy Commission, vol. 1 (Washington: Government Printing Office, June, 1952), p. 24.

should be expected by 1975. The enormous increase in demand for magnesium is provocative, but the resource, especially that in sea water, is fortunately superabundant. The bauxite requirement is expected to be nearly three times as large. Other increases are less spectacular, but in all cases a large rate of consumption was predicted. In the 1960's the Bureau of Mines began making projections to 1975. Table 15–9 indicates 1960 consumption and projections.

Conclusion

One may properly conclude that the United States is richly endowed with minerals. Comparison of consumption with proved reserve data, however, makes it clear that to meet future needs—not in a few hundred years, but in just a few decades—intensified exploration and new technology will be required. Clearly the all-too-common attitude of complacency engendered by the sense of abundance must soon be replaced by realization of the need for expanding knowledge and of the obligation of the present to future generations. In the following chapter conservation needs and some of the more hopeful practices and prospects are discussed.

FOOTNOTES

[1] An outstanding reference is "Minerals Facts and Problems," 1965 edition, *Bulletin 630* (Washington: Bureau of Mines, Department of the Interior, 1965), 118 pp.

[2] T. A. Rickard, *A History of American Mining* (New York: McGraw-Hill Book Company, Inc., 1932), p. 2.

[3] *Ibid.,* p. 15.

[4] For illuminating summaries of the fuel reserve and adequacy problem see "Report of the National Fuels and Energy Study Group on an Assessment of Available Information on Energy in the United States," *Senate Document No. 159,* 87th Cong. 2nd Sess. (Washington: Government Printing Office, 1962), 501 pp; and

W. A. Vogely and W. F. Morrison, "Patterns of Energy Consumption in the United States, 1947 to 1965 and 1980 Projected," paper 83, 1A (Tokyo: *World Power Conference, 1966)*, 24 pp.

[5] *Ibid.*, p. 17.

[6] *Resources for Freedom*, Report of the President's Materials Policy Commission, vol. 3 (Washington: Government Printing Office, June, 1952), p. 27.

[7] Data from *Mineral Resources of the United States, 1901* (Washington: Government Printing Office, 1901), and *Minerals Yearbook, 1920* and *1958* (Washington: Bureau of Mines, Department of the Interior, 1920 and 1958).

[8] *Monthly Business Review* (Cleveland: Federal Reserve Bank of Cleveland, July, August, 1956), p. 6.

[9] *Minerals Yearbook, 1965* (Washington: Government Printing Office, 1965). Of the total, Texas accounted for 45 per cent, Louisiana 17, California 15, Oklahoma 5, and Wyoming 4 per cent, making a total of 86 per cent in the five states.

16

Mineral Conservation: Practices and Prospects

THE NEED FOR MINERAL RESOURCE CONSERVATION is indicated by the evidence presented in preceding chapters which points to greater demands for minerals to supply the growing population and rising material standard of living. Although mineral resources cannot be managed for sustained yield, the objectives of conservation are much the same as those for other resources: To assure society, present and future, of an uninterrupted supply of needed resource commodities and of reasonable maintenance of environmental quality. Conservation with reference to mineral resources, therefore, is concerned with the means of assuring a continual supply of fuels and other needed mineral raw materials, and with avoiding unnecessary damage to associated resource values. The specific concern is to assure that mineral deposits be efficiently developed and mined and to minimize waste in utilization.

In broad terms, there are five main avenues to conservation in the field of mineral resources, each with numerous specific practices. Some minerals are more amenable to one practice, few are subject to all. The following outline presents the major possibilities and notes a few specific cases, most of which will be discussed briefly.

I. Practices and possibilities for increasing the resource include
  - A. Increasing recovery in mining
    1. Improved petroleum recovery
    2. Improved coal mining practices
    3. Reduced "high grading" of mines
  - B. Utilizing lower grade materials such as in
    1. Iron ore
    2. Coking coal
    3. Copper, bauxite, and other minerals
  - C. Recovering material mined more completely by
    1. Improved processing methods
    2. Added by-product recovery
    3. Stockpiled tailings and low-grade material

II. Importation as conservation

III. Life of minerals in use extended
  - A. Improved efficiency in fuel burning
  - B. Allocation to highest beneficial use
  - C. Increased use of scrap
  - D. Controlled corrosion
  - E. Substitutions for scarce minerals

IV. Damage to related resource values minimized
   A. Regulation of surface mining and rehabilitation
   B. Control of stream pollution
   C. Control of air pollution
V. Intensification of exploration with modern technology

In nearly all of these conservation practices, some segments of the industry and government have long been pressing research toward improvements. Nevertheless, in too many cases the competitive national and international economy is operating to inhibit conservation. The goal of government regulation, taxation, and aid should be to foster the best conservation practices.

Increasing Recovery in Mining

The most immediate means of extending mineral resources is to achieve a higher percentage of recovery in mining. Coal mining on the average recovers only half of the available coal; petroleum recovery is still as little as 35 per cent; and for metallic minerals in general, unknown tonnages of low-grade resources have been irretrievably lost in abandoned mine workings where only the high grade was produced. The inhibiting factors are cost, competition, and instability of the market. Nevertheless, the mineral industries have an obligation to exploit mineral resources in such a way that waste in mining is minimized. Government, however, has an obligation to the industry to regulate and aid in such a way that conservation practices are economically feasible and rewarded. In many cases government assistance is probably essential to provide long-term stability, and in some cases it may be desirable to develop a system of premium price or compliance payments for conservation practices which can be shown to be desirable in the public interest yet noneconomic to the enterprise.[1]

IMPROVING PETROLEUM RECOVERY

The petroleum industry has come a long way in increasing percentage recovery, but there is still more that can be accomplished (Figure 16–1). In the early decades of the petroleum industry until about 1925, the recovery was as little as 10 to 20 per cent of the resource.[2] Worse yet, trillions of cubic feet of gas were flared and lost without having been utilized. This deplorable waste resulted from two factors: (1) lack of knowledge concerning the behavior of petroleum in the ground; and (2) lack of adequate regulations to curb "rule of capture" and require conservation practices.

CHARACTER OF PETROLEUM PRODUCTION. Petroleum occurs in the pore spaces of the reservoir rock. Wherever a well is drilled a lower pressure area is created and petroleum under pressure from gas and water slowly migrates toward the well bore. It was some years before oil scientists and producers recognized that unregulated production caused rapid decline and that irregular pressure developments left large quantities of oil unable to migrate to a well. Moreover, the importance of the gas and water drive in creating the energy to produce oil was not fully understood at first.[3]

Even more inhibiting to conservation was the legal interpretation of ownership known as the "rule of capture" based on the English common law and established by early Pennsylvania Supreme Court opinions.[4] This legal interpretation concluded that the oil belongs to the first person bringing it to the surface regardless of where it came from in its underground movement. As a result there were frantic competitions to drill and to produce as fast as possible before the owner of adjoining property gained the oil. Under such conditions there was frequent overproduction, waste of capital and gas and oil, and low total percentage recovery from the fields. Fortunately, such practices have now been largely stopped by state regulations designed to prevent unnecessary waste, through prorating production and requiring well spacing and unitized fields.

As a result of the "rule of capture," far too many wells were drilled. Thus the United States at the end of 1965 had about 590,000 wells in operation with an average daily production of only 13.3 barrels.[5] Very likely this is more wells than are needed for efficient pro-

16 – 1 *The petroleum industry yesterday and today.* The United States oil industry began a century ago with the Drake Well near Titusville, Pennsylvania, shown at the left. In the early years lack of regulation and lack of knowledge led to drilling of oil well "jungles" like Signal Hill, shown below left. Such competitive production recovered as little as 10 to 20 per cent of the oil and wasted most of the gas. Today wells are spaced, as in the West Edmond field of Oklahoma above. Such managed production results in recovery of as much as 70 to 80 per cent of the resource. (Drake Well Park, Shell Oil Company, and Standard Oil Company)

duction from a conservation viewpoint. For example, in the Oklahoma City field, three hundred wells were estimated to be plenty, but seven hundred were actually drilled, expending thirty-one million dollars more than needed, as well as resulting in inefficient pressure utilization.[6]

An oil field in past decades typically began production with a short period of flush flow followed by rapid decline. For example, the Spindletop field of Texas reached annual production of 17.4 million barrels in its second year (1902), declined to 8.6 million in the third year, then to 3.4 million in the fourth, and production was less than one million barrels by 1910. In 1926 a deeper formation was tapped which followed the same history. Conservational practices have resulted in extending high-level flow through a longer time as well as improving total recovery. The improved pattern of production under reservoir management is illustrated by the Webster field (Figure 16 – 2).

The production pattern illustrated by the Spindletop and Webster Fields indicates the reason why the petroleum industry must be continuously exploring for and developing new wells to maintain production. Increasing difficulty in finding new fields, and increasing production cost as old wells decline in output and must be pumped, have been partly offset by advances in technology. Depth of drilling has greatly increased. Drake's well was only 69 feet deep; by 1890 the deepest bore was 6,000 feet, in the 1930's depths over 10,000 feet were reached, and in 1958 the deepest well went down 25,340 feet.[7] In recent years forty to fifty thousand new wells have been drilled annually, with 20 per cent in new areas (wildcat drilling) which average four thousand feet per well. The ratio of success to failure for wildcat wells is about one to nine. This high risk combined with the rapid decline in individual well production is the reason the government allows the industry a 27.5 per cent depletion allowance for income tax purposes.

RISE OF CONSERVATION REGULATION IN PETROLEUM. The need for petroleum conservation practices, including regulation in the public interest, was recognized early.[8] The Geological Survey in *Bulletin 394* (1909) called attention to the need. In 1913 the American Institute of Mining and Metallurgical Engineers established a committee to study the problem. In 1915 the State of Oklahoma approved the first proration act to control production. It should be noted that proration laws were aimed at economic stabilization and only indirectly resulted in conservation. By the 1920's petroleum scientists realized the role of gas in oil production, and about the same time water flooding as a secondary recovery process was established. The Federal government made a major advance in the Minerals Leasing Act of 1920 which allowed leases for petroleum on public lands and required conservation measures. In 1924 President Coolidge took a major step directed specifically toward the oil industry in establishing the Federal Oil Conservation Board. In doing so he said, "It is evident that the present methods of capturing our oil deposits is wasteful to an alarming degree in that it becomes impossible to conserve oil in the ground under present leasing and royalty practices if a neighboring owner or lessee desires to gain possession of his deposits."[9] Although the commission did accomplish a great deal, especially in legalizing unitization, it was inadequate for the national problem. During this period most states developed conservation regulations which were usually for proration of production with provisions for well spacing and minimizing waste. These state acts were then supported by the Connally Act of 1935 which made it illegal to transport petroleum produced in violation of state regulations. In the same year the Interstate Oil Compact Commission was established to advance the cause of conservation, and compulsory unitization was authorized on Federal lands. In 1938

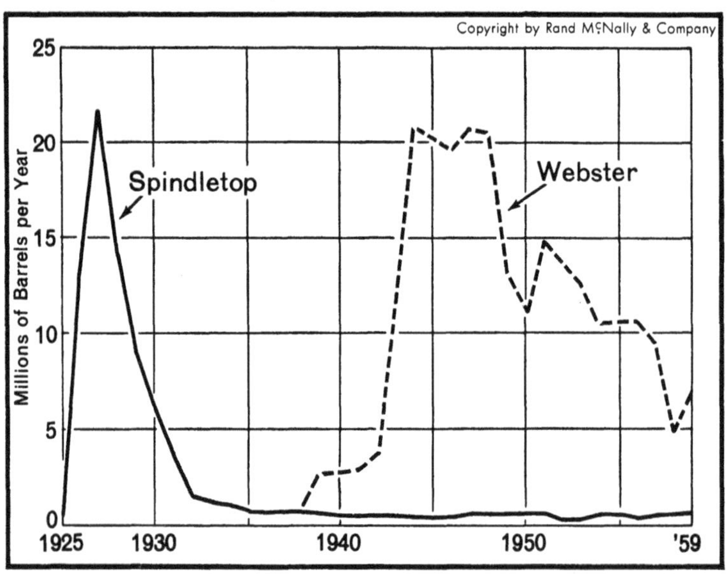

16-2 The chronological pattern of production, Spindletop and Webster, Texas.

interstate flow of natural gas came under regulation by the Natural Gas Act.

The states are primarily responsible for regulation of petroleum and gas production, conservation, and industry stabilization. The Federal government exercises full control over oil and gas production on public lands and regulates interstate and international petroleum and gas movement. Some students of the petroleum conservation problem believe that there should be some nationwide regulation to assure conservation practices in production. One reason is that it is unlikely that a single state can adopt strong conservation regulation to maximize recovery unless all do so. Regulation that is chiefly tied to economic stabilization and concerned mainly with the present profit motive is likely to be inadequate for conservation.

PRACTICES OF PETROLEUM CONSERVATION

Petroleum scientists and engineers have identified a number of production practices that foster maximum recovery from the oil bearing structures. Six major practices which foster such conservation are:

1. Operation of fields as single units.
2. Limiting the number of wells to the minimum number necessary for efficiency.
3. Producing only from points where reservoir energy is best used and closing off other wells.
4. Adjusting the rate of production to the most efficient rate for greatest production in the long run.
5. Returning gas to the oil formation.
6. Using secondary recovery techniques.

Even with best practices, probably no more than 50 per cent recovery can be achieved in the primary operation, but use of secondary techniques can increase recovery to as much as 70 to 80 per cent. As normal production continues, pressure declines, artificial lift is required, and in time even pumping no longer economically produces oil. At this time secondary methods, notably gas injection or water flooding, must be initiated or the well abandoned. In the future it is thought that mining depleted fields may become practical.[10] The following two examples are given as illustrations of forward-looking industry developments.

THE ELK CITY FIELD. The Elk City field of western Oklahoma with 304 wells on 7,700 acres is a good example of a unitized field and the returning of gas to the formation. In this field the gas to oil ratio is especially large, and total recovery of oil would be small if unregulated. Petroleum men studied the field behavior and decided that repressuring would add twenty-five million barrels to the recovery expected by normal means. With the cooperation of nine operating concerns and scores of farm owners, the field was unitized. Facilities were then constructed, including a processing plant, compressor plant, storage facilities, and injection pumps and wells. In the operation, well output was moved by gathering lines to the central processing plant where butane, propane, and natural gasoline were separated from the wet gas. The remaining dry gas was compressed to 4,500 pounds per square inch and injected nine to ten thousand feet below the surface where it was reused in driving more oil to the wells. In 1960 oil depletion reached a level which made return of gas no longer feasible. Produced gas was contracted to a pipeline firm for sale. The technique is illustrated in Figure 16 – 3.

THE BENTON FIELD. The Benton Field in southern Illinois is an example of secondary recovery by water flooding. The field was discovered in 1941 and in eight years produced twenty million barrels, but by 1949 was nearing exhaustion with recovery of only seven barrels daily per well. A major oil company proposed a water flooding project.

Unitization was accomplished with cooperation of 275 owners, involving 3,500 acres and 237 operating wells. Facilities for water pumping, including a five million gallon water reservoir, a water treatment plant, injection pumps, and an oil treating plant for separating oil and water, were constructed at an initial cost of one and a half million dollars.[11] The technique is illustrated in Figure 16 – 3. Water similar to the natural water of the formation is pumped under pressure around the margins of the field to

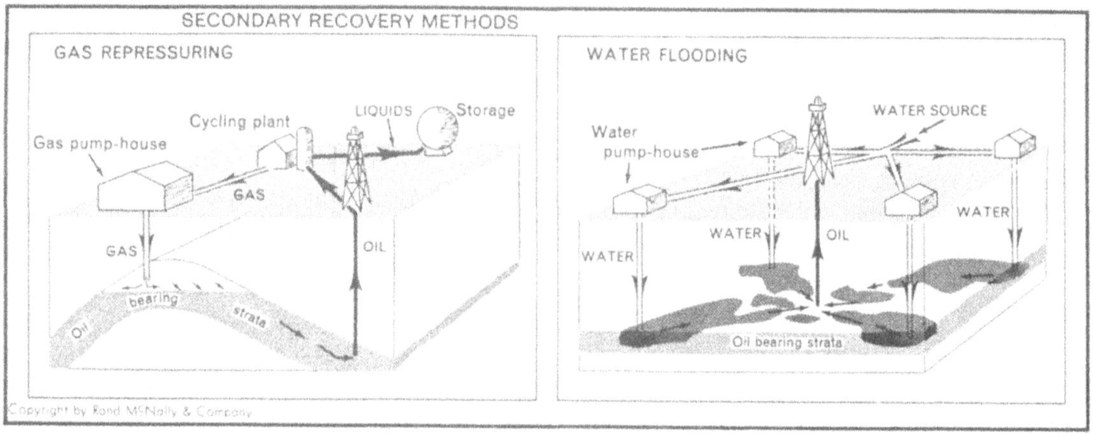

16 – 3 ***The technique of gas repressuring and water flooding.***

re-establish the pressure gradient and drive oil to the well in the center. In 1949 the operation started, and the rate of petroleum flow increased tenfold.[12] Reserve estimates in 1949 were six million; ten years later the field had produced fourteen million barrels and had a reserve of 7.4 million barrels.[13] The flooding project resulted in a gain of about sixteen million barrels for the field, or ten to fifteen years of profitable production.

IMPROVING COAL MINE RECOVERY

Coal mining has passed through an economic and technical revolution in the past half-century. Nevertheless, underground recovery percentage remains low, averaging about 50 per cent. Widespread application of conservational mining practices is one approach to achieving expansion of recoverable coal reserve at modest cost. Surface coal mining recovers about 90 per cent of the coal uncovered.

CHANGES IN THE COAL INDUSTRY. The technical revolution in coal mining is summarized in the evidence of comparative data. In 1910 output per man was three and a half tons per day, and 1965 it was fourteen tons per day for underground mines and thirty-two tons per day for surface coal mines. In 1910 only 4 per cent of coal produced was cleaned, but in 1965, 65 per cent was mechanically cleaned. The evolution toward complete mechanization is illustrated by Figure 16 – 4.

Mammoth machines have been introduced for surface stripping. Shovels to dig overbur-

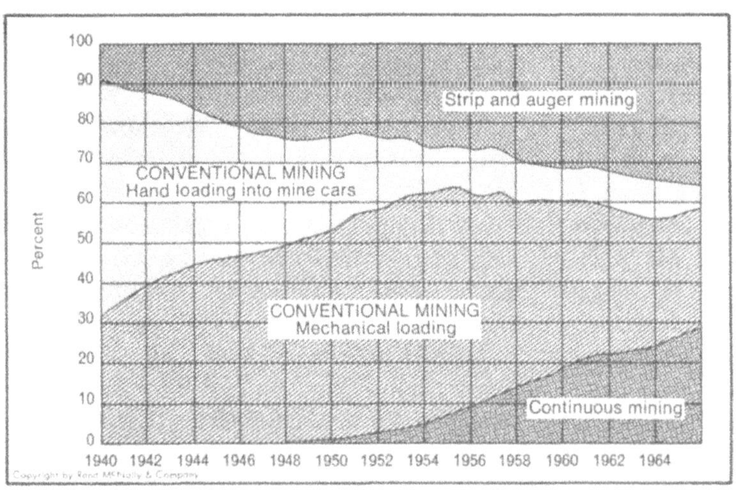

16 – 4 ***The continuing trend to mechanize coal mining.*** **Increasingly the miner is a director of machines which greatly increase per-man-day output.**

den and coal are electric or diesel-electric and the largest can handle as much as 180 cubic yards at a bite. Wheel excavators are even bigger (Figure 16 – 5). As a result, up to one hundred feet of overburden are being removed to reach coal seams. Underground, several types of electrically driven continuous coal miners are now in use, as well as numerous loading machines and conveyor belts. Thus the coal miner has become a highly skilled, productive, and well-paid operator of expensive machinery.

Mines have become large. In 1965 there were 7,865 coal mines of which 5,989 were underground. Those with capacities of less than ten thousand annual tons (49 per cent of the total mines) accounted for only 3 per cent of the coal (Figure 16 – 6). A major problem for conservation is this very large number of small mines, for they are less likely to practice adequate conservation mining than the large well-capitalized organizations.

CAUSES OF COAL LOSS IN MINING. The losses in mining coal result from numerous practices, some necessary and others possibly avoidable. Major loss occurs in various pillars and barrier zones left as supports after mining, in boundary zones separating different ownerships, in pillars left around oil and gas wells, and in areas under rivers, railroads, and towns left to prevent surface shifting and subsidence. A layer of coal sometimes remains on the floor and roof

16 – 5 *The wheel excavator, newest advance in surface mining.* Contrast the size with the cars and the big shovel in the background. The wheel excavates the top layer of overburden, placing it far back on top of the spoil piles; the normal shovel handles the lower layers. Thus the depth of economical operation has been much increased. This wheel is powered by electricity. The cutting wheel is twenty-four feet in diameter and has nine one-yard buckets. It can cut from seventy-five feet down to twenty-six feet above the coal layer and place the stripped material back 225 feet to a height of 110 feet. (Bucyrus-Erie Company)

of mine workings. Major losses also occur through failure to mine low-grade areas or hard-to-reach corners. Thin seams are generally not mined, especially if they occur close to a thick seam. In any case, seams under two feet thick are seldom worked.[14] The significant point for conservation is that coal left in a mine is almost always lost forever because roof collapse and invasion by water make later recovery a perilous as well as an uneconomic venture.

PROSPECTS OF HIGHER RECOVERY. Many studies have shown that, on the average, about half of the loss sustained in mining can technically be avoided.[15] The problem is how to provide adequate incentive. Support of the roof is a major problem, and it is for this purpose that most pillars are left. The most common system of underground mining is the *room and pillar* in which advancing entry is made into the seam, and the coal is recovered in alternate areas (called "rooms") separated by pillars of coal which support the roof as mining proceeds to the limits of the property. On the average, about half of the coal is left in pillars, the actual amount varying with roof conditions. The most immediate increase in percentage coal recovery would come from removal of pillars where consistent with safety. These pillars can be removed as a final mining operation, usually starting at the outer limits and retreating toward the entry, allowing the roof to fall as the retreat progresses.

A second mining method, the *longwall* system, is generally regarded as a conservational mining practice because it requires less pillar area. In this system, coal is mined from faces of the seam which are usually several hundred feet long. Steel supports are used for temporary support of the roof. As mining progresses, supports are removed and the roof is allowed to fall behind the operation. Longwall mining can be accomplished with usual mining machinery. However, the coal planer (or plow) is one of the newer machines designed especially for use in the system (Figure 16 – 7). Typical recovery with longwall methods is estimated to be 80 per cent compared to 50 per cent with the more usual room and pillar method. In addition, more complete pick-up of coal and more complete mining of the roof, floor, and low-grade areas are possible.

Improved machinery for thin seams appears likely to encourage recovery of much of the coal that is now seldom considered. Remote control mining machines are coming into use. The possibility of extracting the energy and chemical values from coal by underground combustion is also under intense study. The idea apparently originated in 1868 with Sir William Siemen, but so far research has not shown the process to be economical in the United States.[16] Coal research scientists believe that in the future it will be practical to utilize many seams, especially steeply dipping and broken seams, without mining. Possibilities for surface working of coal seams, providing 85 to 95 per cent recovery, are being increased by new machinery permitting deeper stripping.

ECONOMIC DILEMMA. Coal mining is placed in a conflict between maximizing recovery at in-

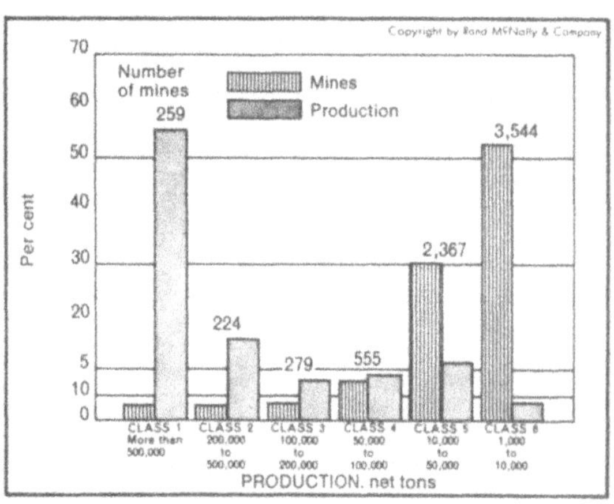

16 – 6 *Number of coal mines and production in percentage by size classes in 1965.* Note that 70 per cent of the nation's coal is supplied by 7 per cent of the total mines. (Bureau of Mines)

creased cost for long-range benefits in the interest of society, and the competitive economy which encourages—even forces—minimizing cost by concentrating on high grade and easy-to-get coal for short-range economic advantage. Conservational mining is certain to cost a little more. The following quotation from a 1936 report of the Bureau of Mines gives the essence of the conflict and the reason why coal recovery is still less than it could be. "The individual operator, acting alone, can hardly avoid the use of wasteful methods. To assume additional expense, unless it is also borne by his competitors, would soon put him out of business. At best, his own direct incentive to increase recovery goes no further than the point that would yield maximum profit in the long run. For much the same reasons, it is impossible for the individual state to act alone. In the division of labor between the state and Federal government, the states have the constitutional powers to regulate waste, but under the present conditions no single state can well take action. . . ."[17]

Unfortunately there still is little conservational regulation in the coal industry. National regulation is aimed at economic stabilization under the Bituminous Coal Act of 1937, and mine safety under the Coal Mine Safety Act. Coal exists under about one-ninth of the nation, is mined in twenty-six states, and the industry employs 133,000 (1965) workers, making it a major industry of national concern. It therefore may well be that leadership in conservation can best come from the Federal government. Serious consideration by Congress of how to provide incentive or regulation for the best conservational mining is needed.

REDUCING "HIGH GRADING" OF MINES

One of the most difficult problems is how to assure maximum mining of developed ore bodies. For every mine, the operators must decide on the lowest grade to mine and how completely to mine the deposit. It is clear from study of the characteristics of minerals that every mine will have a "cut-off" level determined by the cost of mining compared to selling price. Through time the level may vary, and it may also vary with different owners. During unstable market periods, when the price is low or there is concern that prices will soon drop, there is strong motivation to work rapidly and selectively only the most easily mined and the highest grade ore to maximize economic profits and reduce risks. Unfortunately, this "high grading" often results in permanent loss

16–7 Electrically powered coal planer operating on a longwall. The planer mines the bottom layer of coal, the top portion falls or is broken down. The planer head, a plow-like cutter with V-shaped knives, is towed 250 to 500 feet in distance at about seventy-five feet per minute back and forth along the seam, mining about six inches of coal each time and loading it onto a conveyor. At the right of the picture note the temporary supports which are recovered as the planer advances toward the left. Pneumatic shifters every twenty feet continually force the cutting head against the coal face. (Bureau of Mines)

of large tonnages of low-grade ore because a closed mine deteriorates rapidly and it is usually uneconomic to reopen it for the remaining low grade. In most cases low-grade ores and hard-to-get sections can be worked economically only in association with high-grade mining over a long period.[18]

Small mines are the most likely to be pressed into high grading which is not in the interest of conservation of mineral resources. The large, well-capitalized companies, with expectations of long-term operations through several generations of stockholders, carry out mining practices that are generally in the interests of conservation. They are most likely to maximize production of the resource over a long period and to be less concerned with immediate dollar profits.

Governmental leadership in stabilizing market conditions over a long period, in taxation programs that reward conservational mining, in establishing incentive prices for producing low-grade ores, and in financial aid for exploratory development work would help foster maximum extraction of the resources from operating mines.

Utilizing Lower Grade Deposits

The strongest possibility for expanding mineral reserves is through improvements in technology (or increased price) which make economical the use of lower grade material. There are numerous examples, of which iron ore is the most outstanding.

UTILIZING LOWER GRADE IRON MATERIAL

Iron ore is the major metal foundation of the national economy, being mined in volumes many times greater than any other metal and accounting for 35 per cent of the value of all metals. It, therefore, merits special consideration; moreover, the transition to lower grade iron ore is representative of developments in other metal resources.

DEPLETION OF HIGH-GRADE IRON ORE. The nation is rich in iron resources in quantity, quality, and favorable location. Nevertheless, in the 1940's and 1950's serious concern was expressed in Congress and newspapers that the United States was on the verge of running out of iron ore. An examination of the evidence makes clear that the concern stemmed from rapid depletion of the most accessible, easy to mine, high-grade ore deposits of the Lake Superior area, notably the Mesabi Range. During the twentieth century, the Lake Superior deposits have supplied 75 to 80 per cent of the nation's iron ore. Even before the turn of the century these five iron ranges were renowned: Marquette, developed in 1856; Menominee, in 1877; Gogebic, in 1884; Vermilion, in 1894; and Mesabi, in 1894, the youngest but destined to be the greatest.

Tables 16–1 and 16–2 give an abstract view of the pertinent data. Study of these Tables will reveal that the Mesabi Range alone has supplied half of the nation's all-time iron ore output. The high-grade reserves under such demand had only ten to fifteen years remaining life (Table 16–2). Then, under the pressure of impending need, science and technology were applied to the problem of how to use economically the large quantity of low-grade material known to exist in the area. The result was the perfection of magnetic taconite processing. Immediately about six billion tons of low-grade ore equal to about two billion tons of high-grade ore were made available. In short, the perfection of taconite processing had an impact about equal in importance to the all-time output of the Mesabi Range. The increased use of lower grade ores, however, requires increased resource-converting techniques and inputs of energy (Figure 16 – 8).

TACONITE PROCESSING. Taconite is a low-grade siliceous iron-bearing material containing 20 to 35 per cent iron. For use it is processed to hard pellets containing 60 to 64 per cent iron, thus requiring two to three tons to produce one ton of pellets (Figure 16 – 9). Much of the added cost is offset by the high quality and more uniform material sent to the blast furnace and some reduction in bulk shipping. The important concept is not that it costs a little more, but that

TABLE 16–1

THE IRON ORE SITUATION

Area	Resource estimates[1]		Production[2]		
	Reserve	Potential	1957	1964	All time[3]
	Million tons	Million tons		Million tons	
Lake Superior	4,000[4]	50,000[5]	83.5	63.1	3,569
Southeast	610	11,220	6.8	2.5	386
Birmingham, Alabama area 80 per cent, Georgia					
Northeast	300	2,850	5.0	5.2	240
New York, New Jersey, Pennsylvania					
Western Areas	490	500	8.0	5.1	119
California, Utah, Wyoming					
Central and Gulf	50	116	2.0	7.8	65
Missouri and Texas					
Alaska	5	large	..	..	....
TOTAL	5,500	65,000	105.3	83.8	4,379

[1] Martha S. Carr and Carl E. Dutton, "Iron Ore Resources of the United States Including Alaska and Puerto Rico, 1955" (Washington: Geological Survey Bulletin, 1082-c. 1959), p. 87.

[2] *Minerals Yearbooks 1957 and 1964* (Washington: Bureau of Mines, 1958, 1966).

[3] 1955 estimates by Carr and Dutton extended by author.

[4] Tonnage of shipping grade ore plus concentrates from taconite, measured, indicated, and inferred.

[5] Crude ore iron content average 22 per cent minimum.

TABLE 16–2

IRON RESOURCES AND PRODUCTION, THE LAKE SUPERIOR RANGES

Range	Crude iron resource[1]				Production[2]		
	Iron content per cent	Measured	Indicated Inferred	Potential	1957	1964	All time
		million tons				million tons	
Mesabi	50	855	500		65.9	47.3	2,466
	22		5,000–6,000	15,000			
Gogebic	52	37	100		4.4	1.6	319
	25–45			7,750			
Marquette	51	65	185		6.5	7.9	330
	25–45			17,500			
Menominee	50	62	75		4.2	4.5	284
	25–45			4,320			
Vermilion	56	13	25	330		1.3	101
	35			large			
Cuyuna	43	46	50	4,400	2.4	0.5	68
TOTAL		1,068	6,435	49,300	83.4	63.1	3,569

[1] Martha S. Carr and Carl E. Dutton, "Iron Ore Resources of the United States Including Alaska and Puerto Rico, 1955" (Washington: Geological Survey Bulletin, 1082-c. 1959), pp. 94–95.

[2] *Minerals Yearbook, 1965*, vol. 1 (Washington: Bureau of Mines), p. 534.

the nation is again assured of a long-range iron resource adequacy. The Lake Superior region is still the dominant iron ore supplier to the nation's all-important iron and steel industry (Figure 16 – 10).[19]

USING LOWER GRADE MATERIALS IN OTHER MINERALS

Transitions can be expected to follow a similar pattern in most minerals. Coking coal, for example, is now being utilized with sulphur content that would have precluded its use twenty years ago. Blending of best low-volatile coal with lower grades has been growing in importance. Copper ores mined today average under 1 per cent copper content, where in 1900 it was 2 per cent, and in 1890 5 per cent was about average. Bauxite of only 30 to 35 per cent recoverable alumina is now being used where twenty years ago the lower limit was 50 per cent. Research in processes that will open the way to use of vast tonnages of alumina-bearing material found in the nation is continuing.

More Complete Recovery from Material Mined

Increasing recovery from the ore mined is another approach to expanding mineral re-

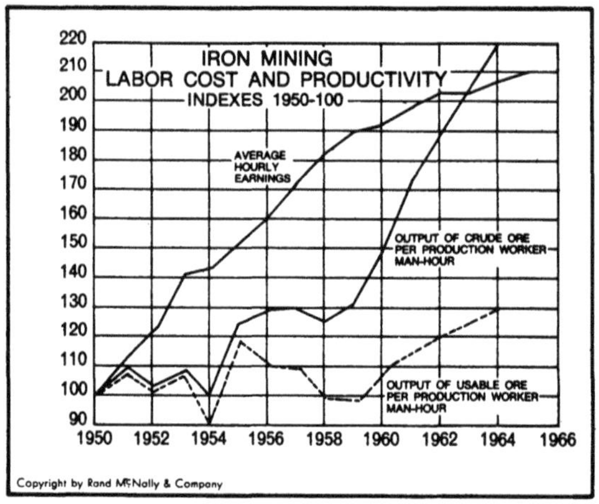

16 – 8 ***Relationship of labor cost and productivity in iron ore mining.*** **Note especially that the gains from increased use of improved resource converting technology is largely offset by the need to produce greater tonnage of crude ore to provide the needed usable ore. (Bureau of Mines)**

16 – 9 ***The E. W. Davis taconite processing plant at Silver Bay, Minnesota.*** **This industrial complex costing about $350,000,000 can produce about 10,800,000 tons of iron ore pellets annually. The plant is forty-seven miles by rail from the mine on the eastern end of the Mesabi Range. Water supply, deep water transportation, and disposal areas in the depths of the lake favored selection of the site. Note the pellet storage areas in the foreground and shipping facilities. In the background is the community, housing about 5,000 persons in 1960. The 125,000-kilowatt power plant is to the right. The taconite ore entering in the center background is about 25 per cent iron and 35 per cent silica, the processed pellets as they leave the plant contain 62.5 per cent iron and only about 8 per cent silica. (Courtesy, Reserve Mining Company)**

sources. At least three specific concepts are worthy of note: (1) Improved processing, (2) additional by-product recovery, and (3) stockpiling of tailings and other low-grade material for future processing.

IMPROVED PROCESSING

In the advancement of recovery by better processing, a conflict between technology, economics, and conservation is apparent. In a particular case, processing to recover 70 or 80 per cent may be technically possible and conservationally desirable, but it may not be profitable. Nevertheless, improvements in machinery and in processing do result in increased recovery and thus are conservational (Figure 16 – 11).[20]

Development of *flotation* using selective chemical additives, for example, was a major advance because it made possible use of lower grade material reserves, especially of sulphide ores such as lead, zinc, and copper. Since its introduction in the 1920's notable advances have been made in reagents, automation, and increased size of equipment.

PETROLEUM REFINING. Many improvements in handling and refining petroleum and gas have made important contributions to conservation. Only thirty years ago an 8 per cent loss by evaporation was common. Today crude and refinery products are carefully handled to reduce spillage and evaporation. Pipelines are better, controls are more adequate, tank design and construction are much improved, and usu-

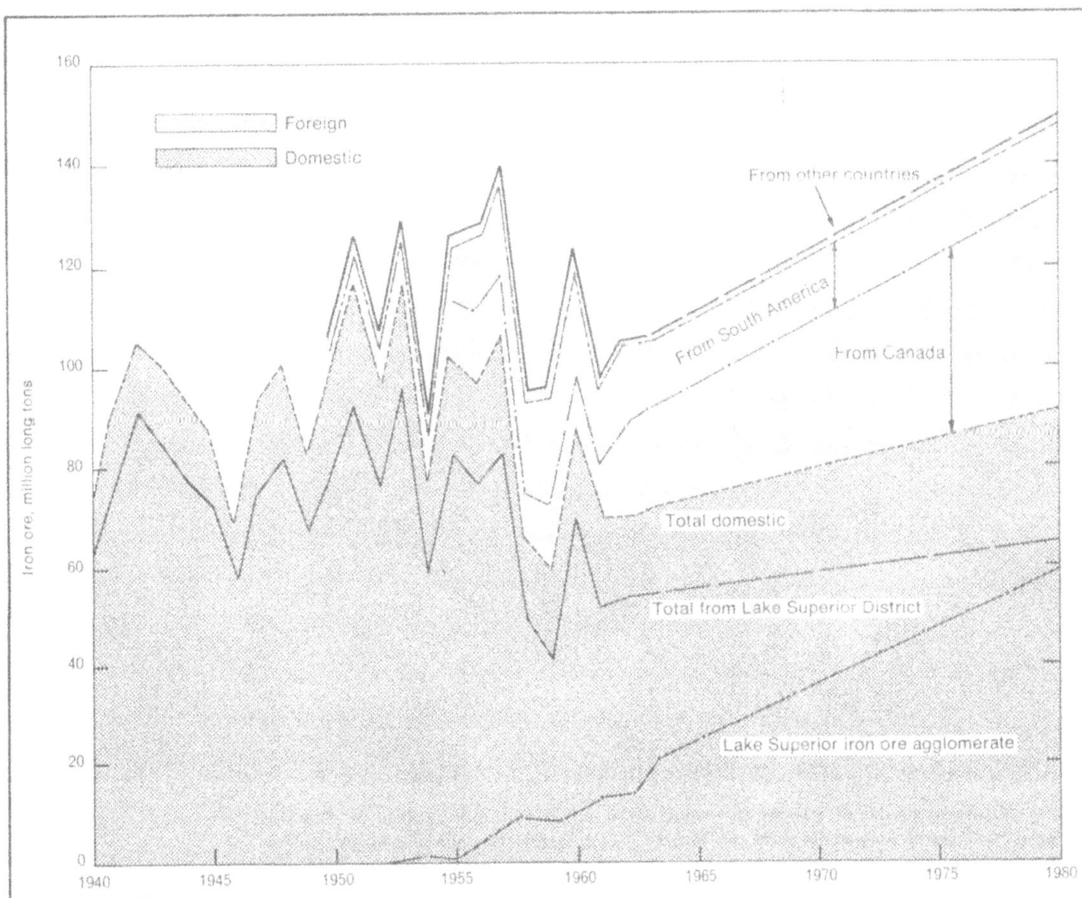

16 – 10 ***United States iron ore supply resources, 1940–1980.*** **Note the real and projected increases from foreign sources. (Bureau of Mines)**

ally tanks are filled from the bottom to reduce evaporation loss as well as to limit fire danger.

Major improvements in refining have done much to extend the petroleum resource. Development of thermal and catalytic cracking processes and catalytic reforming which resulted in raising gasoline yield from 25 to 50 per cent are estimated to save about one billion barrels of crude oil annually by reducing the number of barrels needed to supply the nation's gasoline requirement. By-product recovery also has been greatly expanded, creating the base for a vast new petro-chemical industry.

OTHER EXAMPLES. There are, of course, numerous other cases of improved processing. Application of magnetic separation to taconite ore is a major example. The combination process for refining bauxite developed in the 1940's made possible economic utilization of ores having 8 to 25 per cent silica content. As a result, large expansion in reserves was achieved. Coal is now generally processed to remove impurities and also to recover *fines* once wasted. Humphrey spirals once mainly used for gold concentration are now widely used in beneficiation. Numerous improvements have been made

16–11 ***The Homestake Mine.*** **This world-famous United States gold mine located in 1876 has produced continuously except during World War II. It is a notable example of industry continuously improving technology for conservation. The mine has 34 working levels between 1,700 and 6,800 feet. The sand fraction of the ore goes back into the mine for backfill after the gold is extracted. Only 0.31 ounce of gold is recovered from each ton of ore. Amalgamation recovers 63.3 per cent of the gold, and 31.9 per cent is extracted by cyanidation of the sand and clay (slime) mill tailings; thus 95.2 per cent of the total gold value is recovered. (Courtesy Homestake Mining Company)**

in leaching, and electrostatic concentration is becoming important.

EXPANDING BY-PRODUCT RECOVERY

Many by-products such as cadmium from zinc ores and cobalt from copper ores have long been recovered. The full possibilities, however, are seldom achieved. From the viewpoint of conservation it is desirable that more complete by-product recovery be carried out, or that tailings be stockpiled for processing in future periods of higher price. Germanium is now recovered from coal and zinc ores, whereas twenty years ago it had no known industrial use. Gallium is recovered from bauxite and zinc. Sulphur is a notable example of by-product recovery. In the 1960's over a million tons of sulphur are recovered annually from oil refineries and natural gas plants and nearly half a million tons from stack gases of copper, lead, and zinc smelters. Not only does this recovery add materially to conservation of sulphur, but it has done much to reduce air pollution and damage to resource values surrounding smelters. Vanadium today is in surplus because of its recovery as a by-product of uranium in the Colorado Plateau. The coke oven offers a notable example of by-product recovery, including gas for fuel and a host of coal hydrocarbons which support a great family of industries.

STOCKPILING TAILINGS AND LOW-GRADE MATERIAL

One of the conservation practices in industry is to stockpile tailings and low-grade material during low-level market conditions. An interesting example of a waste becoming a valuable resource is the anthracite tailings mined near Pottsville, Pennsylvania. In 1960 a one hundred million dollar plant was established to recover chemicals from anthracite coal silts and other waste discarded during past decades, fifty million tons to be used for electric generation and 350 million tons to be turned into valuable gas and chemicals.[21]

Importation as Conservation

Importation of minerals serves conservation in two ways: First, by lessening the demand on domestic deposits; and second, by adding materially to in-service and stockpile reserves. In recent years there has been a notable trend toward increased dependence on foreign sources for minerals. There is little evidence, however, that the motivation has been conservation; rather the trend reflects the increasing gap between high-cost domestic production and low-cost foreign mineral production (Figure 16 – 12). Nevertheless, conservation objectives are served by saving domestic resources while using the imported mineral raw materials in the United States.

Dependence on importation raises two profound questions: (1) To what extent can the nation afford to be dependent upon overseas sources in terms of national security? (2) How can competing domestic mining be adequately supported in order that a healthy domestic mining industry be perpetuated? There is no simple answer to the problems of the domestic mining industry. But it should be noted that for at least fifteen minerals (Table 15–5) domestic resources are so limited that self-sufficiency is out of the question. The approaches to protection of the domestic industry have been through the tariff, quota systems, tax benefits, and premium prices for domestic producers. The chromium industry is a prime example. Chromium is essential, but domestic resources are economically marginal due to the small size and low grade of deposits. Normally, domestic mining supplies only about 2 per cent of the need. The Defense Production Act of 1950 established an incentive price of 115 dollars per ton for standard grade and a guaranteed purchase by Federal depots. In contrast, imports of equal grade sold at from forty-five dollars to fifty-five dollars per ton at United States ports. Also, tax benefits of 23 per cent were allowed, in addition to the usual depletion allowance for ore. When the incentives were ended the number of active mines declined from ninety to only one. Between 1961 and 1966 no chromite was mined in the United States, even though consumption continued at a high level.

It is evident that the United States imports minerals in a number of cases because of economic advantages rather than because of physical lack of resource. There is considerable difference of opinion between those who believe it most desirable to import all we can as a conservational measure and those who favor a policy of near self-sufficiency. It would appear that a mineral policy approaching self-sufficiency would have undesirable repercussions on the economies of other nations as well as undoubtedly resulting in higher costs for some mineral supplies.[22]

The problem of how to maintain national security in spite of dependence on foreign minerals is equally difficult. The solution has been to maintain stockpiles of minerals deemed critical. This solution is adequate, but there are potential problems. Other countries are rapidly industrializing and moving to limit and regulate exports of raw materials in efforts to maximize benefits for their own people. In decades ahead there may be increasing difficulties in securing adequate tonnages of low-cost minerals by import. Conservation of minerals and provision of future supplies for the United States are concerns of international scope, but it would seem to be unwise for the United States to be complacent in the belief that imports of mineral wealth from other lands will solve mineral requirement problems.

Extending Life of Minerals in Service

Another important aspect of conservation is prudent in-service use of mineral commodities to minimize waste and so extend their useful life. The five most likely prospects are: (1) Improved efficiency in fuel use, (2) allocation of scarce resources to highest utilization, (3) increased use of scrap, (4) corrosion control, and (5) substitutions for scarce minerals.

IMPROVED EFFICIENCY OF FUEL USE

There have been notable advances in the technology of fuel burning for space heating, propulsion, and generation which have major significance to conservation. The best example is the reduction in pounds of coal required to produce a kilowatt-hour of electricity by electric utilities. In 1900 the average consumption of coal to produce a kilowatt-hour of electricity was eight pounds. By 1920 improve-

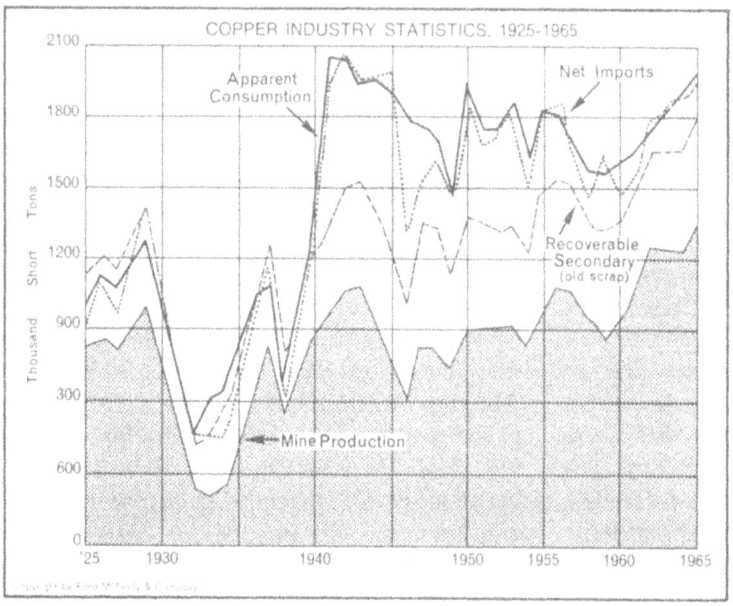

16 – 12 *The pattern of copper supply.*

ments had reduced consumption to three pounds, and by 1965 to only nine-tenths of a pound (Figure 16 – 13). Because the electric utilities burned 155 million tons of coal in 1965, this improvement of technology over 1920 resulted in the equivalent of saving three hundred million tons of coal by the electric utilities alone. The automobile engine has been made at least 50 per cent more efficient, but the conservational gain has been nullified by deference to comfort, bigness, and speed. The comfort is certainly desirable, but it would seem that automobiles are far overpowered. There is little need for such speed potentials and pickup, and these could, and should, be reduced in the interest of operating economy and resultant conservation of gasoline. An additional major saving could easily be achieved by regular engine tune-up which would give better engine performance, save gas, and reduce air pollution.

ALLOCATION TO HIGHEST BENEFICIAL USE

A conservation possibility that is little used (except indirectly by the effect of price level) is the allocation of mineral commodities to their highest use. An example will illustrate. There is a critical shortage of high quality coking coal, yet only about one-fourth of the low-volatile coal mined today is actually used for coking—the rest is burned for space heating or other fuel uses. It would be prudent to find some system whereby high-grade coking coals could be reserved for use only as coking material. The chief problem in approaching any such allocation, whether it be for petroleum, coal, or other minerals, is that of determining what the highest beneficial use is.

INCREASED USE OF SCRAP

For many metals a major conservation factor is the large scrap potential. Not all minerals offer re-use possibilities. Fuels, for example, are lost completely; most pigments in paints and, in general, the nonmetals are consumed when put into use. The percentage of the return to re-use for the major metals depends on the type of use and is estimated in various ways, but there is general agreement on the following potential average levels: Aluminum, 40 per cent; copper, 65 to 75 per cent; iron and steel, 65 per cent; lead, 75 to 80 per cent; nickel, 20 per cent; tin, 20 per cent; and zinc, 25 per cent.[23] The length

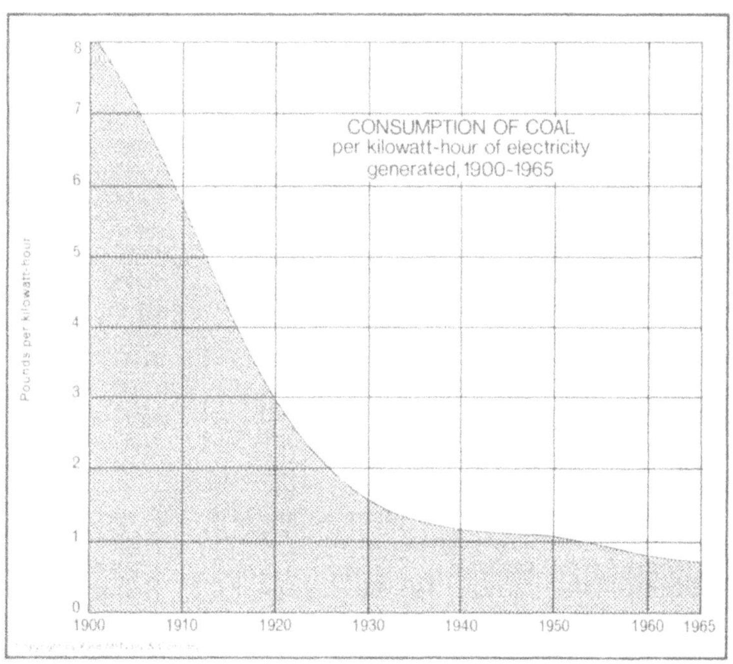

16 – 13 The increasing efficiency in coal burning by utilities.

of time before metal once put into use is returned through scrap channels varies with the commodity, but in general a period of twenty years is involved. The growing significance of scrap resources in an industrial economy is illustrated by copper (Table 16–3). In recent years a quarter of the nation's copper needs have been supplied by recovery of old scrap. It is estimated that 65 to 75 per cent of the copper put into use will in time be reclaimed as scrap, but because return is actually at a lower rate, there is a continual increase in the in-service resource. In 1908 there was a per capita in-service copper reserve of 68 pounds, in 1948 of 304 pounds, and in 1965 of 372 pounds. A similar analysis can be made for lead, zinc, iron and steel, and other minerals having a high ratio of re-use. Thus it is estimated that in 1900 there were about 3,300 pounds per capita of steel in service in the nation; in 1965 this reserve had increased to at least eight tons per capita. The scrap business is a major enterprise, with at least forty million tons of ferrous scrap alone being collected and marketed by dealers annually (Figure 16 – 14).[24]

The potential recovery from steel in use varies greatly with use categories, but there can be no doubt of the importance of this type of conservation. In recent years about half the iron raw material charged into steel furnaces has been scrap—about seventy-five million tons annually. Of this amount an estimated four million tons comes from the nearly five million automobiles and trucks that are scrapped each year. According to the Institute of Scrap Iron and Steel Inc., recovery rates range from a high percentage for automobiles to a low of 13 per cent for containers (tin cans). One of the biggest single wastes in tonnage of steel today is the loss of the steel put in tin cans which is not

TABLE 16–3

IN-USE COPPER RESERVE*

(Thousands of tons)

	Consumption			Reserve	
Year	New copper	Old scrap	Total	Increase 75 per cent of total minus old scrap	Total
Accumulation to 1947					21,713[1]
1948	1,214	505	1,719	783	22,496
1949	1,072	383	1,455	708	23,204
1950	1,447	485	1,932	964	24,168
1951	1,304	458	1,762	863	25,031
1952	1,360	415	1,775	916	25,947
1953	1,435	429	1,864	969	26,916
1954	1,235	407	1,642	824	27,740
1955	1,336	515	1,851	874	28,614
1956	1,367	468	1,835	908	29,522
1957	1,239	444	1,683	818	30,340
1958	1,157	411	1,568	765	31,105
1959	1,183	471	1,654	769	31,874
1960	1,148	429	1,577	754	32,628
1961	1,237	411	1,648	825	33,453
1962	1,352	416	1,768	910	34,363
1963	1,423	422	1,845	962	35,325
1964	1,495	474	1,969	903	36,228
1965	1,526	513	2,039	1,016	37,244

* Extended by author. Data: *Minerals Yearbook* 1949 through 1965 (Washington: Bureau of Mines, Department of the Interior).

[1] Charles White Merrill, "The Accumulation and Conservation of Metals-In-Use," Proceedings of the United Nations Conference on the Conservation and Utilization of Resources, vol. 11 (New York: United Nations, 1951), p. 34.

commonly recovered because of the costs in collecting the cans and in removing the tin coating from the steel. A significant beginning is being made by "garbage" disposal agencies in sorting out tin cans for recovery of the steel scrap and tin.

CORROSION CONTROL

One of the major causes of waste of mineral resources and an enemy of conservation is corrosion. The annual cost of corrosion control and metal replacement due to corrosion is estimated to be five to ten billion dollars.[25] Pipeline replacement alone costs about 600 million dollion dollars.[26] Not all of this loss can be prevented, but intensive research is being applied to the problem by industry to discover better protection methods. Even a 20 per cent improvement would save at least a billion dollars and thousands of tons of steel and other metals annually.

SUBSTITUTION FOR SCARCE MINERALS

Substituting more generously available minerals for scarce minerals is another conservational measure related to the concept of allocating minerals for their highest use. A few examples will indicate the possibilities. Noncoking coals can certainly be substituted for low-volatile coking coals in space heating with little problem. Aluminum, which is abundantly available in the earth's crust, has come into wide usage in building construction and in automobiles. It is being substituted more and more for tin in foil, tubes, and cans, thus tending to reduce national requirements for a metal that is relatively scarce. Magnesium and titanium are other structural metals for which potential earth resources are enormous. Plastics and other nonmetal materials have been developed for a wide range of uses, many of which have replaced metals, such as the recently increased use of plastic pipes. When and if increased cost of metals provides incentive, there will be considerable replacement of metals by plastics.

INTERNATIONAL STABILITY ESSENTIAL FOR CONSERVATION

One of the most significant and most deplorable factors inhibiting conservation is international tension and war. During periods of war concern for resource stewardship is necessarily put aside. Moreover, there is terrific actual waste of all kinds, especially of minerals. There is no scrap recovery from a battleship at the bottom of the ocean. Unrecorded tonnages of mineral materials are lost, and fuel waste is enormous. Even in peacetime, military defense is one of the nation's largest fuel users. Surely war is the greatest waster of resources and a prime enemy of conservation.

Minimizing Damage to Related Resource Values

The great expansion in mining and mineral processing during this century has brought

16 – 14 *The ubiquitous scrap yard.* **The scrap metal dealer is one of the most important resource producers. For every ton of scrap melted the steel industry saves about three tons of natural resources, and the scrap industry handles about forty million tons of scrap annually for a major contribution to conservation. (Institute of Scrap Iron and Steel)**

with it conflict and deterioration of related resource values that are more widespread than is generally realized. A prudent society can ill afford to permit benefit from one resource needlessly to impair related values. The most significant types of damage are destruction of land values by surface mining, stream pollution, and air pollution. Prospects for greatly increased pressure on all resource values make it imperative that such deterioration be recognized and steps be taken to minimize impairment.

LAND DETERIORATION BY SURFACE MINING

By far the most noteworthy destruction of land values follows in the wake of surface mining, notably for gold, iron, copper, and coal. The problem is especially acute with coal mining because much coal is under the surface in areas climatically suited to agriculture and forest land-uses.[27]

SURFACE COAL MINING

The ability to work coal seams from the surface economically has made great forward strides as modern machinery has raised efficiency in excavation and earth moving. In 1920 there were only 174 strip coal mines in the United States producing 1.5 per cent of the nation's coal. By contrast, in 1965 there were 1,541 strip mines removing an average of forty-five feet of overburden, with a few removing seventy to one hundred feet.[28] Strip coal mining has worked over half a million acres, and about twenty-five thousand acres are now being stripped each year to produce 165 million tons, or about a third of all coal mined.

After the big shovels are through, not only is the coal exhausted, but the surface has been overturned, leaving only barren, steep-sided spoil banks tens of feet high where once there were productive farms and woodlands. The coal should be strip mined because surface mining provides the most efficient technique, but much more can be done to rehabilitate the surface after mining. The problem is complicated. Rehabilitation of spoil banks by reforestation or establishment of pasture encounters unique difficulties resulting from the unnatural terrain and physical character of the surface material in texture, structure, and resultant changes in drainage and aeration. The pH values are frequently low as a result of sulphuric acid formed from sulphur-bearing material in the bank. In general, values lower than pH4 are toxic to vegetation. Moreover, the type of grading significantly influences micro-climate and drainage.[29] The case of Harrison County in southeastern Ohio is a good example of what can be done.

HARRISON COUNTY, OHIO. The southeastern half of Harrison County, centering around Cadiz, is rolling land generally devoted to dairy and livestock production. Under the surface at a depth of thirty-five to sixty-five feet there is a four-and-a-half-foot coal seam. With the use of giant shovels, Harrison county coal production rose to 6.8 million tons in 1957 (about 70 per cent of this was done by surface mining), making it the leading coal-producing county in the state.[30] Spoil banks rapidly became the major features of the landscape. Much of this stripping took place on the county's best agricultural land, and the impact on agriculture is revealed in decreasing farm numbers. In 1940 farms numbered 1,818; by 1950 there had been a decline to 1,301; and in 1958 there were only 977.[31] Gradually, public concern resulted in regulation to assure reasonable rehabilitation of spoil banks. In 1949 the Ohio Strip Mine Law requiring specified minimum rehabilitation was approved. As amended, the law requires posting bond of 220 dollars per acre of coal land before mining, and rehabilitation of all spoil banks by grading to gently rolling topography, construction of an earth dam in the last cut, and planting of trees, grasses, or legumes wherever revegetation is possible. Failure to rehabilitate can result in loss of license for stripping operations as well as loss of the surety bond. Prior to the 1949 law, some forward-looking companies had recognized the obligation to restore surface values and were reforesting. But the adoption of the state law had beneficial impact in upgrading formerly destroyed land values after stripping and may well be taken as a model for other states.[32]

The law requires only a minimum of rehabilitation, but many companies carry out a higher degree of restoration, as sound economy and in recognition of their obligation to restore surface values as nearly as possible. The Harrison County Extension Agent believes that a forage and pasture program can be set up on rehabilitated land to return it to its original production. Cultivation of the land is considered impractical because of the large number of rocks. "A few farms in the county have been stripped and the soil restored to almost original topography, when the owners did not want to leave the farm. These are in good production and it is hard to tell that anything has been disturbed."[33]

The largest company in the area for the past ten years has devoted major effort to finding the most effective solution to rehabilitation. Field studies have been made to determine the best mixture of grasses, legumes, and fertilizer. Since 1950 about 65 per cent of their spoil banks have been leveled and seeded to pasture for support of Hereford cattle (Figure 16 – 15). This reclamation is costing 150 to 180 dollars per acre. Although this is generally more than the market value of the land, it is economically sound for several reasons:(1) The rehabilitated land will again produce commodities and profits from farming as well as provide a higher tax base,(2)the cost of restoration is only about 1 per cent of the gross return from the coal recovered, or about three cents per ton of coal, (3)there are important intangible values in good public relations. It is expected that one-third of the county will be stripped, making rehabilitation essential.

STREAM AND AIR POLLUTION

Stream and air pollution resulting from mining and mineral processing is less obvious and probably less significant than the conflict between agriculture and surface mining, but it needs control. There was a time when no vegetation could grow within miles of smelters handling sulphide copper, lead, or zinc ores, so heavy was the air pollution by sulphur fumes. In general, this waste and pollution has been curtailed, and in many cases recovery of valuable by-products, notably sulphur, has more than paid the cost. The cement industry is another notable cause of air pollution. Fine dust lost to the air by a major cement plant may cover the landscape for miles (Figure 16 – 16). It is, of course, unrealistic to attempt to curtail air pollution completely, but controls are possible, and certainly regulation to minimize air pollution is desirable everywhere. Here, too, local regulation is usually inadequate, for in a

16 – 15 *Rehabilitation of spoil banks in Ohio.* **Spoil banks such as at the top can be rehabilitated for forest or pasture. Regulation and development of knowledge of best techniques is essential to the future of such areas. (Hanna Coal Company)**

competitive industrial economy a single city, county, or even state is not likely to require costly pollution control if neighboring areas with equally good industrial sites do not also require equal standards. In the long run the welfare of a clean city may be better served by having controlled industry. (Also see Chapter 17 on air pollution.)

Stream pollution by sediments from placer mining or by toxic material from mine waters also needs control. Most states now have regulations designed to protect fish in streams from such pollution. Toxic runoff from mining areas can be impounded during periods of low stream flow. Sediment from placer and dredge mining is often controlled in much the same way. Idaho, for example, passed regulations in 1954 under the Dredge Mining Protective Act requiring operators to construct settling ponds to reasonably clarify water before discharging it into rivers.

Intensification of Exploration

Major advances are being made in the technology of mineral exploration that may have an important impact on the nation's future mineral position. Mineral deposits in the past were commonly found by discovery of a surface outcrop or other physical indication, but now exploration (prospecting) has become a sophisticated science through which large areas are studied with expensive instruments for recording and analyzing subsurface structures and earth materials. Notable advances are being made in the capability of remote sensing equipment which are providing the means of expanding knowledge of mineral environments. Geochemical and geophysical scientists are applying their special skills and knowledge to the problems of better understanding the formation of mineral deposits, as well as to exploration. Among new techniques the following are a few examples. Gravity meters for measuring rock density have been developed and are in use. Seismic reflection techniques permit analysis of structure to great depths. Airborne magnetic and electromagnetic surveys are being used to map structure in large areas and locate magnetic ore bodies. Geochemical analysis is being studied as a means of refining surface prospecting techniques.

Scientists are intensifying studies toward better understanding of how mineral deposits form, and of the characteristics of mineral environments. Out of these studies will come the

16–16 ***Air pollution control at a cement plant.*** **This cement plant at Tijeras, New Mexico, is an outstanding example of pollution control to maintain a clean operation. Kiln exhaust is passed through the usual cyclonic collectors and then forced into glass-bag dust collectors which filter the dust from the exhaust. Dust recovery amounts to about 5 per cent of the kiln output. (Ideal Cement Company)**

diagnostic skills needed to interpret the geophysical and geochemical data provided by the sophisticated sensing equipment.

POSSIBILITIES OF UNDERSEA MINING

In recent years major research attention has been directed to the possibilities of extending exploration and development to mineral deposits under the seas, and to recovery of minerals from seawater. So far, minerals production from the seas has been limited to recovery of various salts such as magnesium and sodium chloride from seawater, petroleum and sulphur from the continental shelf, and a few minerals such as titanium ores from dredging beach sands. In addition, a few conventional mines have extended workings to areas under adjacent seas. The new exploration proposes a much greater and broader scale involving near-shore areas and the ocean bottom.

At present, marine beaches of the near-shore areas would seem to offer the most likely prospect (Figure 16 – 17). Shallow water phosphoric beds and deep water (10,000–20,000 feet) sea bottom beds of manganese nodules are considered by research scientists to have major possibilities. Offshore areas are also known to contain beach deposits formed during Pleistocene times when the sea level was lower and streams discharged in areas now submerged.

The problems are great, but researchers believe that economical solutions can be developed. A basic problem is that of developing methods and equipment for large-scale exploration and mining in the undersea environment. Some research currently is directed toward devising means of placing men and mining devices on the sea bottom (Figure 16 – 18). Other researchers are oriented to perfection of dredging equipment that would operate from the surface

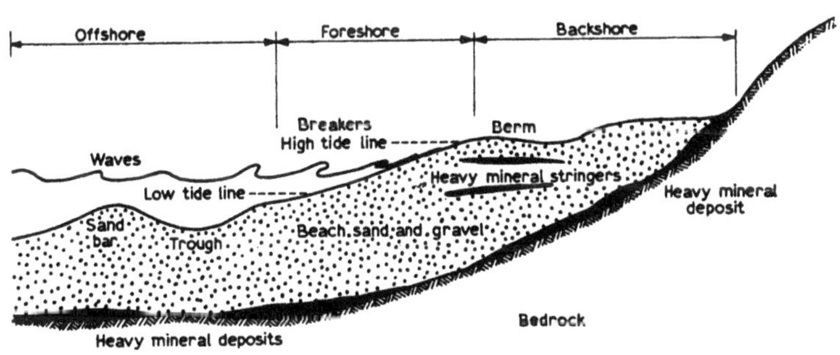

16 – 17 *Cross section of a marine beach indicating zones of minerals concentration.* (After J. L. Mero in *The Mineral Resources of the Sea*, [New York: Elsevier, 1965.] With permission)

16 – 18 *A concept of a mining complex for operation on the sea bottom.* Here the artist has embodied concepts of researchers who are attempting to devise means for undersea mining with men and machines on the sea bottom. This dredge is conceived to have habitation for men and primary concentration machinery inside, and to be capable of moving by the tractor elements. (Courtesy of Frank Pennock, *Oakland Tribune*)

of the sea (Figure 16 – 19). In addition to problems of mining, the deposits now under consideration are low grade in comparison with deposits that are now being mined. Economical development of undersea deposits, therefore, almost certainly would involve very large-scale operations that would require the investment of millions, perhaps tens of millions, of dollars to initiate production. The implications for the mineral markets, as well as for conservation, are provocative. In summary, although many minerals are known to exist in and under the seas, mass mining of undersea deposits would seem to be a possibility of the future—nevertheless there is the possibility.

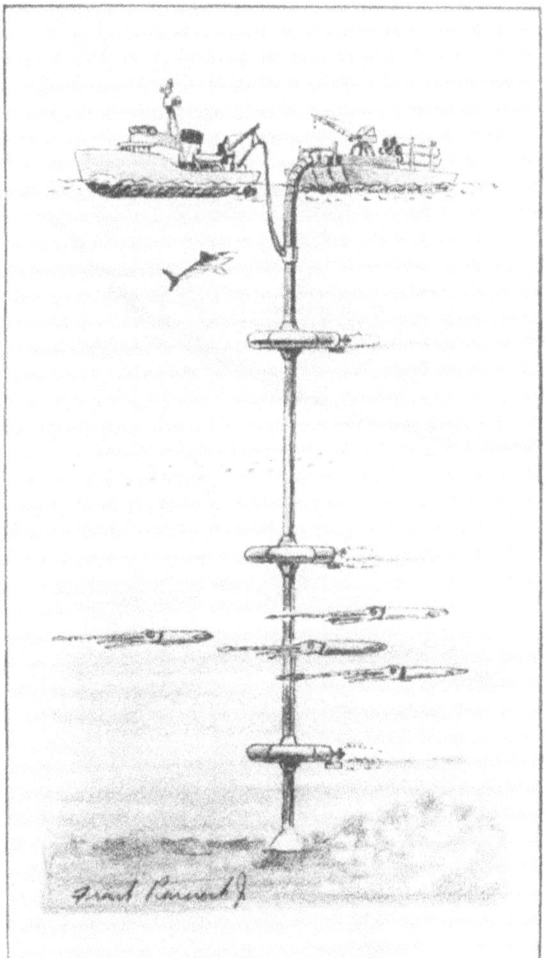

16 – 19 *An artist's conception of proposed airlift-type dredge to operate from the surface of the sea.* **Note that a series of floats support the long "tube" of the dredge and that each float is propelled so that the suction tube can be swept over the sea bottom. (Courtesy of Frank Pennock, *Oakland Tribune*)**

Relationships of Government and Industry

Interrelationships and responsibilities of government and industry have been noted throughout this discussion. In an appraisal of conservation it is important to note that in a broad sense democracy provides for a system of checks and balances between government, industry, and individual citizens from which regulations are derived at the local, state, and Federal levels. Only a few highlights significant to mineral conservation are noted in the following.

ROLE OF GOVERNMENT

Government responsibilities and activities at all levels play important roles in fostering or inhibiting mineral conservation practices. The more important activities include regulation to assure the long-range public interest and safety, economic aid when needed to provide industry stability and stimulation, leadership in long-range scientific and technological research, and taxation.

Taxation practices have direct influence on conservation. States often utilize ad valorem systems which, by taxing reserves, in some cases may inhibit conservation by penalizing companies for carrying out long-range mineral developments. Federal taxation is mainly through the income tax laws, and in general has taken account of conservation through the depletion allowance granted as a means of amortizing the depletable resource and by deductions allowed for exploration and development. The important point is that at all levels of government taxation of resources ought to be such that desirable conservational practices are fostered.

State governments, in general, exercise authority to regulate production for conservation and have developed many desirable laws. All the states maintain a department to assist the development of mineral industries and carry out provisions of mining laws. Their size, importance, and type of activities vary in keeping with the mineral industry of the state.

The Federal government, through the executive branch, the Congress, and various agencies, has long been active in research, economic aid, and regulation of the mineral industries. Regulations for disposal or leasing of public domain minerals are the prerogative of Federal agencies. Interstate and international commerce is regulated in the public interest. A vast program of basic research in science and technology is maintained, providing leadership and supplementing industrial activities. Aerial photography and mapping of topography and geology is a prime responsibility of Federal agencies. Many agencies are involved in activities which influence the mineral resources and industries, but the chief responsibility has been delegated to the Department of the Interior which has a cabinet rank secretary. The following notes indicate the most important agencies of this Department and their responsibilities.

Assistant Secretary for Mineral Resources

1. Discharges responsibility of the Secretary of the Interior with regard to mineral resources.
2. Supervises the following except the Bureau of Land Management.

Bureau of Mines

1. Conducts programs in minerals and metals technology.
2. Conducts research in mining technology.
3. Conducts research in fuels technology.
4. Bears responsibility for helium production, distribution, and research.
5. Regulates mine safety.
6. Conducts economic and statistical studies.

Geological Survey

1. Bears responsibility for topographic and geologic mapping.
2. Conducts field research on appraisal of mineral deposits.
3. Classifies public lands as to mineral value.
4. Supervises mining under public lease and permits.
5. Administers Connally Oil Act.

Office of Coal Research

1. Promotes and assists research to develop improved methods of mining, preparation, and utilization of coal.
2. Cooperates with and assists other agencies and departments with coal programs.

Office of Minerals and Solid Fuels

1. Informs and advises the Secretary of the Interior on matters of domestic mining industry, stockpiling and disposition of mineral surpluses.
2. Responsibility for planning and programming to assure adequate supply of minerals for civilian and military needs in the event of mobilization.

Office of Oil and Gas

1. Studies conservation of oil and gas for programming and policy at Federal level.
2. Prepares plans for assuring adequate supplies of oil and gas during national emergencies.

Bureau of Land Management

1. Administers general mining laws.
2. Administers Mineral Leasing Act and Outer Continental Shelf Lands Act.

ROLE OF INDUSTRY

Individuals, industry, and various private organizations play important roles in fostering desirable regulations by advising state and Federal lawmakers and the public of needs, and by providing fundamental information. Nearly every specific mineral resource has an industry organization dedicated to the welfare of its constituents. In addition, there are national organizations which represent all or large segments of the mineral industry. For example, the American Mining Congress, incorporated in 1898, has long been a major influence in promoting the welfare of mining and was instrumental in the establishment of the United States Bureau of Mines.[34] Other nationwide industry associations include the American Iron and Steel Institute, the Institute of Scrap Iron and Steel, the National Coal Association, and the American Petroleum Institute.

These organizations maintain effective staffs and carry on public information programs in the interest of their industry, appear as witnesses at Congressional hearings, and provide necessary data to state and Federal legislators. Cooperation, communication, and understanding between these industry representatives and government agents at all levels are needed. When there is full and open debate on resource issues, the country will not go far wrong in regulations and aid for conservation.

Conclusion

It appears certain that the decades ahead will be marked by increased demand for mineral resources. Per capita consumption is expected to rise, and by the end of the century requirements for many minerals are projected to be double the present use. Moreover, world-wide industrialization, growing population, and rising material standards of living portend major acceleration of demand outside of the United States.

There have been many changes and developments significant to mineral conservation. Exploitation of mineral deposits is changing in character from selective mining of richer outcroppings and placer deposits to emphasis on mass mining of huge low-grade ore bodies requiring great capital investments for mining and processing. Conservation practices and regulations have been developed and implemented. The percentage of loss in mining has been reduced, and oil well spacing is generally required. Natural gas wastage has been much reduced through recycling and repressuring. There is growing recognition of the need for control of pollution originating from mining, and the desirability of multiple use of the land surface has been recognized in the mining law. Recovery of by-products in smelting and refining now produces important products that once were discarded. Revolutionary techniques of exploration have been introduced which may have major implications. Research on means of economically developing low-grade deposits is being intensified, and major research is being directed to potentials of the sea and the sea bottom.

The implementation of rational conservation of mineral resources clearly is a problem that ought to be of fundamental concern. For the present, programs and regulations ought to encourage maximizing recovery in mining from mineral deposits, and foster preventative practices that minimize damage or interference with related resource values and minimize waste in utilization. In addition, there is need to reappraise potential demands continually, and to consider how these foreseeable requirements can be met. The concept of conservation includes a fundamental obligation of the present generation to point the way forward by searching for solutions to problems before they are fully upon the nation.

FOOTNOTES

[1] Proposals presented to the House and Senate of the 85th Congress on behalf of the Secretary of the Interior under the title *Domestic Minerals Act of 1958* embodied these ideas, but were defeated in the House after being passed in the Senate.

[2] *Energy Resources and National Policy* (Washington: National Resources Committee, 1939), p. 196.

[3] Early studies of the United States Bureau of Mines indicated possibilities. A major experiment in which underground conditions were simulated showed 51 per cent recovery when gas was "shut-in" and oil was produced slowly, only 22 per cent when gas was conserved but oil was produced rapidly, and only 11 per cent recovery when gas was wasted. *Petroleum Investigation 1934*, Hearings on House Resolution 441, 73rd Cong. (Washington: Bureau of Mines, Department of the Interior, 1934), pp. 1220–21.

[4] George W. Hazlett, "Property Rights and Oil Production," *Oil For Today and For Tomorrow* (Oklahoma City: Interstate Oil Compact Commission, 1953), pp. 32–45.

[5] *Minerals Yearbook, 1965*, vol. 2 (Washington: Bureau of Mines, Department of the Interior, 1967).

[6] *Energy Resources and National Policy, op, cit.*, p. 203. For a recent discussion of well spacing see Harold Decker, "Proration and Well Spacing in Texas," *Journal of Petroleum Technology*, vol. 12, no. 3 (1960), pp. 18–21.

[7] *Oil and Gas Journal*, vol. 57, no. 4 (January 26, 1959), p. 130; and *Energy Resources and National Policy* (Washington: National Resources Committee, 1939), pp. 148–51.

[8] See for detailed appraisals *Energy Resources and National Policy, op. cit.*, pp. 186–236, 371–401; Erich W. Zimmerman, *Conservation in the Production of Petroleum* (New Haven: Yale Univ. Press, 1957); an interesting, pictorial review of development in "Petroleum Panorama," *Oil and Gas Journal*, vol. 57, no. 5 (January 28, 1959), 635 pp.

[9] *Energy Resources and National Policy, op. cit.*, p. 198.

[10] For a nontechnical review of conservation methods see Stuart E. Buckley, ed., *Petroleum Conservation* (New York: American Institute of Mining and Metallurgical Engineers, 1951), 304 pp.

[11] A personal communication to Professor Jensen on September 6, 1960, from Shell Oil Company. Investment to January 1, 1960, has amounted to two and one-half million dollars.

[12] *Waste Not Want Not*, revised booklet (New York: Shell Oil Company, 1959), p. 9.

[13] *Oil and Gas Journal*, vol. 57, no. 4 (January 26, 1959), p. 142.

[14] The 1960 Bureau of Mines survey of seam thickness shows that only 1.4 per cent of underground production came from seams under two feet, and only 11.5 per cent from seams under three feet. *Minerals Yearbook, 1965*, vol. 2, *op. cit.*

[15] *Energy Resources and National Policy, op cit.*, p. 95. R. W. Stahl, "Extracting Final Stump and Pillars and Pillar Lifts with Continuous Miners," *Report of Investigations 5631* (Washington: Bureau of Mines, Department of the Interior, 1960), 13 pp.

[16] "Basic Data Relating to Energy Resources, 1951," 82nd Cong. 1st sess. *Senate Document No. 8* (Washington: Government Printing Office, 1951), p. 220.

[17] *Energy Resources and National Policy, op. cit.*, p. 384.

[18] For a good essay on this subject see Donald Carlisle, "Maximum Total Recovery Through Mining High Grade and Low Grade Together is Economically Sound," *Canadian Mining and Metallurgical Bulletin*, vol. 46, no. 489 (1953), pp. 21–27; and "Long-Range Program for Minerals Industry," Hearings before subcommittee on Minerals Materials and Fuels, 1958, United States Senate, 85th Cong., 2nd sess., part 2 (Washington: Government Printing Office, 1958), pp. 662–63.

[19] For an account of taconite see Clyde F. Kohn and Raymond E. Specht, "The Mining of Taconite, Lake Superior Iron Mining District," *Geographical Review*, vol. 48, no. 4 (1958), pp. 528–39.

[20] For an interesting review see Nathaniel Arbiter, "Minerals Beneficiation 1908–1958," *Mining Engineering*, vol. 11, no. 2 (1959), pp. 169–76.

[21] *Coal Age*, vol. 65, no. 4 (April, 1960), p. 26.

[22] For observation on this question see "Depressed Domestic Mining and Mineral Industries," 86th Cong., 1st sess., *House Report 708* (Washington: Government Printing Office, July, 1959), pp. 79–88. See also "Long Range Program for Mineral Industry," *op. cit.*

[23] For example see H. J. Miller, "Mineral Supplies and Their Management," *Proceedings of the United Nations Conference on the Conservation and Utilization of Resources*, vol. 2 (New York: United Nations, 1951), pp. 24–32.

[24] *Minerals Yearbook, 1965*, vol. 1, *op. cit.*

[25] For example see Herbert H. Uhlig, "Cost of Corrosion to the United States." *Proceedings of the United Nations Scientific Conference on the Conservation and Utilization of Resources*, vol. 2 *Mineral Resources* (New York: United Nations 1951), p. 213.

[26] *Machine Design*, vol. 31, no. 24 (November 26, 1959), p. 41.

[27] For example see Lee Guernsey, "The Reclamation of Strip Mined Lands," *Journal of Geography*, vol. 59, no. 1 (January, 1960), pp. 5–11 and Arthur H. Doerr and Lee Guernsey, "Man as a Geomorphological Agent: The Example of Coal Mining," *Annals of the Association of American Geographers*, vol. 46, no. 2 (June, 1956), pp. 197–210.

[28] *Minerals Yearbook, 1965*, vol. 2, *op. cit.*

[29] For a good analysis see G. A. Lindstrom, "Forestration of Strip-Mined Land in the Central States," *Agricultural Handbook*, no. 166 (Washington: Department of Agriculture, 1960), 75 pp.

[30] *Minerals Yearbook, 1958*, vol. 2, *op. cit.*, p. 115.

[31] Communication from Harry G. Smith, Harrison County, Ohio, Extension Agent, 1960.

[32] For a report on Harrison County see *Monthly Business Review* (Cleveland: Federal Reserve Bank of Cleveland, July-August, 1956), pp. 10–13. In 1964 Pennsylvania enacted a stronger law which requires restoration as nearly as possible to the original surface and provides major penalties for noncompliance.

[33] Communication from Harry G. Smith, Harrison County, Ohio, Extension Agent, 1960.

[34] For illumination on this point see J. H. Richards, "Presidential Address 1905," *Papers and Addresses of the Eighth Annual Session of the American Mining Congress* (Denver: American Mining Congress, 1906), pp. 7–19.

17

Air Pollution

THE AMERICAN WAY OF LIFE is held up for all the world to see as a shining example of man's technical capabilities, but the image is becoming difficult to see through the pall of atmospheric pollutants that occurs over more and more of our cities with ever increasing frequency. Ironically, it is those very same devices and services which help make our way of life what it is today that are responsible in large measure for what has come to be recognized as a serious threat to our health and welfare. The manufacture of countless labor-saving and labor-magnifying devices, the generation of power to run them, our much-vaunted and highly diverse transportation facilities, and even the heating of the buildings in which we work and live, all involve the use of fuel in quantities and spatial concentrations heretofore unimagined. The seemingly limitless atmosphere which, with infrequent exceptions, has absorbed our waste readily enough in the past has been found to have finite limits, and locally to be increasingly unequal to the tasks set before it.

Air pollution problems are not new; with diligent search, references to their occurrence may be found scattered back through time a surprisingly long way. Even legal reference to air pollution had its beginning centuries ago; the use of "sea coal" for heating was forbidden in England on penalty of death in 1306.[1] A feature of urban settlement, air pollution problems have increased in intensity and frequency as the number of people in cities has grown. Today over half the world's people are concentrated on about 1 per cent of the land area. In the United States, about 70 per cent of the population is urban,[2] and the percentage is increasing. To one degree or another, every major city in the world today has an air pollution problem, and for many the tolerance level is marginally acceptable (Figure 17 – 1). Los Angeles, the classic American example, is believed by some to have begun its control program only just in time to avoid a major disaster or a mass exodus. The increasing sophistication of industrial technology, the extension of urban settlement with its concentration effect over more and more area, and sheer numbers of people whose daily needs must be served have caused air pollution to assume staggering proportions. It is estimated that in the United States over 130 million tons of pollutants are put into the atmosphere annually, a figure which approximates the nation's annual steel production.[3]

Until recently the public reaction has been typified by a general apathy, interrupted only occasionally by indignation. Seventeen deaths attributed to air pollution at Donora, Pennsylvania[4] in 1948, for example, caused relatively few people to take a hard look at the implications. As lesser incidents have increasingly involved greater areas and more people, however, committees have been formed and investigations carried out which reveal the complexity of the problem. Air pollution has emerged as a problem for which the blame cannot be laid on the doorstep of any one particular group of people or activities. Its causes are both man-made and natural, and its solution will involve concerted organized effort, time, and capital expenditure both by groups and by the individual citizen himself.

The Causes of Air Pollution

Before beginning a consideration of the situation in which the atmosphere has such a concentration of contaminants (principally resulting from man's activities) that it is classified as polluted, it is well to remember that in its natural state the atmosphere contains many substances other than the gases which constitute "pure air." Salt particles, meteoric and terrestrial dust, forest fire smoke, bacteria, spores, pollen, gases such as NO_2 produced during lightning discharges, and SO_2, HF, and H_2S resulting from volcanism were present in the atmosphere before man existed. Indeed, these may be present on occasion today in sufficient concentrations to cause problems analogous to those which result from man's activities. In this treatment, however, natural pollutants will largely be ignored, and attention will focus on those substances put into the air through man's actions which cause problems.[5]

NATURAL FACTORS AFFECTING DISPERSION

The mass of the earth's atmosphere is enormous, and it is capable of holding unbelievable quantities of substances. In the 6×10^{15} tons of air which surround the earth there are believed to be some ten million tons of solid pollutants.[6] Most of the time we are not aware of the presence of this material, even though it is not evenly distributed, so vast is the volume throughout which it is spread. Still, it is chiefly the lower atmosphere which contains most of the pollutants, dispersal into the upper layers

17 – 1 ***Air Pollution, New York City.*** **The "clear" view on the left was taken on Saturday afternoon, November 26, 1966. The "smog" view was taken on Friday, November 25, 1966 at 10:30 a.m. (Courtesy of *The News,* "New York's Picture Newspaper")**

being limited by the structure of the atmosphere and the physical laws which govern motion within it. So long as the air moves freely, both vertically as currents and horizontally as winds, the atmosphere is capable of dispersing the quantities of pollutants presently being exhausted in most urban areas most of the time. As air motion slows or is inhibited, however, the danger of a serious problem increases rapidly. Air motion is governed notably by atmospheric circulation and stability and by landforms. The following paragraphs examine each briefly.

ATMOSPHERIC CIRCULATION. Basically, air moves in response to pressure differences, but the air in motion is affected by other factors giving rise to hierarchies of circulation systems. Among these systems in the mid latitudes are the familiar anticyclone and cyclone. Air movement associated with a cyclone typically favors clearing the air because winds may be brisk, air rises in the system, and accompanying precipitation "washes" the air. With an anticyclone, however, winds are light or even calm, and air subsides from aloft, inhibiting the vertical mixing process. Moreover, the subsiding air aloft is warmed adiabatically, often giving rise to a temperature inversion which acts as a "lid," preventing vertical motion above it and dispersal of entrained pollutants. Although pollution may occur in the absence of anticyclones, areas frequented by anticyclones are particularly vulnerable to serious air pollution problems, especially on those occasions when the systems stagnate. A persistent anticyclone with a subsidence inversion was associated with each of the major air pollution tragedies in recent years[7] (Donora, Pennsylvania in 1948 and London, England in 1952 and 1962). Interestingly, cities are endowed with a limited natural resistance to surface inversions by virtue of the heat they generate, thus offsetting to some extent the very problem they help create. Perhaps there would be more air pollution tragedies on record were it not for this factor.

STABILITY. The stability of air is indicated by its reaction to vertical displacement. Unstable air which once disturbed will rise of its own accord usually facilitates vertical mixing and the dispersion of entrained pollutants. Stable air resists rising; in fact, on occasion it will sink, thus containing pollutants in a fixed or diminishing volume of air. Stability may be induced in various ways; cooling the lower layers of the air is a common way. Autumn in the mid latitudes, with its longer nights for radiational cooling and fewer cyclones, is the time of the most frequent and persistent inversions.[8] Radiational cooling above a snow cover and under the clear skies of an anticyclone often results in a surface temperature inversion with unusually stable air at or near the surface.

LANDFORMS. The earth's surface form, where irregular, may aid in the dispersal of pollutants by inducing turbulent flow in moving air, and thus vertical mixing of the air. Surface form also may serve to enclose or shelter areas, however, reducing the ventilating effect of free access by winds. In the case of a valley or a basin, cool air drainage down the surrounding slopes may cause development or persistence of an inversion that otherwise would dissipate. The role of terrain is often a complementary rather than primary one, but it is no less important, as the situation in the Los Angeles basin illustrates. Increasingly, industries are being urged to assess the "pollution environment" among the site selection factors they consider,[9] and terrain constitutes a significant aspect of such environment.

There are other aspects of the natural environment which affect air pollution directly or indirectly, but most are beyond the scope of this treatment. An effort has been made here to describe the principal natural factors affecting dispersion of pollutants, and to illustrate why some understanding of atmospheric processes is essential to the solution of the problem. As suggested earlier, consideration of the pollutants themselves will focus on those resulting from man's activities.

SOURCES OF POLLUTANTS

As the results of air pollution studies carried out in the larger urban areas of the nation began to come in, the emerging picture surprised many people. Finally aroused and prepared to fight the battle with the alleged culprit, industry, the average American was not prepared to learn that he was at least as guilty, if not more so. The identification of the family auto, the home furnace, and even the backyard incinerator as major contributory agents, however, is perhaps not such a surprise, considering their numbers and the inefficient way in which most are operated.

If we will expand the auto category to include all forms of transportation, the furnace category to include space heating in general, the incinerator category to cover all refuse burning, and add industry and utilities (especially thermal electric power generation), we have the five principal man-generated sources of air pollution. Of these, transportation ranks first, and the private auto is the worst offender. "... As cities have grown more crowded and commuting more difficult the rush-hour traffic jam has become civilization's most effective method of multiplying atmospheric poisons."[10] Some efforts to solve traffic problems have inadvertently intensified the air pollution problem, because the more efficiently autos are able to move into and out of urban areas, the greater the number of autos which do so. The result is occasional rush-hour concentrations of fifty parts per million of CO in the air at busy traffic points in New York,[11] and over one hundred parts per million in Los Angeles;[12] both figures are above the toxic threshold for continued exposure. Additional pollutants contributed by autos result from the pounding, skidding, and rubbing of millions of tires against the highways; the exact nature of these products can be argued,[13] but they must exist in some form. Hydrocarbons, NO, and SO_2 are other byproducts of the internal combustion engine which are being poured into the atmosphere daily by the approximately seventy million autos in the United States. "In the course of using up a thousand gallons of gasoline, motor vehicles typically discharge 17 pounds of SO_2, 18 pounds of aldehydes, 25 to 75 pounds of oxides of nitrogen, and more than 3,000 pounds of CO."[14] With concentrations of pollutants already approaching the lethal point, and with an eye to the increased numbers of autos the burgeoning urban populations will demand, some planners are considering completely banning autos from cities or encouraging development of electrically powered autos.[15]

Before there were so many autos and before any effort had been made to control gaseous and particulate waste emission from factories, the billowing industrial smoke plume pointed clearly to a major source of the atmospheric pollution of that day. Much has changed since then, notably the efforts by many industries to become good neighbors and the emergence of the citizenry as comparable culprits. Still, industry must shoulder its share of the blame today, because its expansion has offset much of its conservation effort. The fact that in a few short generations we have consumed a substantial proportion of the fossil fuels that were accumulated over the past five hundred million years is a phenomenon unique in history,[16] but this high use-rate will increase in the foreseeable future. Combustion is essential to many industries and, although inefficient combustion continues to make its contribution to air pollution, many of the more irritating pollutants are created in quite efficient combustion.[17] Additionally, the mere collection and exposure of some industrial materials results in the concentration of escaped substances in the air above our cities. In Los Angeles County where air pollution restrictions today are among the most severe in existence, many tons of hydrocarbons daily evaporate from crude oil production centers. Despite the considerable progress made in recent decades, industry faces a challenging task as its volume grows and the diversity of its effluents multiplies.

About one half of the energy consumed in the generation of the nation's electrical energy comes from coal.[18] Thermal electric plants produce large amounts of waste material, especially the particulate fly ash and the gas SO_2. Installation of precipitators in many plants has substantially reduced the escape of particulants but economical methods for trapping SO_2 are still wanting. Several devices have been developed which will do the job, however, and with the probable continued significant use of coal in electricity generation, attention must be focused on making them more economical.

While gas and oil have made strong inroads in the space heating field, coal continues to serve here also, and the domestic furnace adds its substantial contribution of waste to those already mentioned. It is estimated that the chimney exhaust from 100,000 homes burning an average grade of bituminous coal would include over 300,000 pounds of pollutants per day, about two-thirds of which would be solid material.[19] If fuel oil were used instead of coal, the same 100,000 homes should produce 42,000 pounds daily, and with gas only ten pounds. Such figures seem to indicate the obvious direction in which the solution for many problems lies. The distribution, availability, and economics of coal and natural gas supplies, however, demand equal attention in any considerations for change.

One of the more disagreeable facts to emerge from the investigations of increasing metropolitan air pollution is that garbage dumps and domestic incinerators constitute one of the five major sources of pollutants. While the literal accuracy of the statement that some urbanites "inhale a portion of their own garbage"[20] may be questioned, it is close enough to the facts to give rise to terms such as "aerial sewers"[21] and the like in the literature on air pollution. Americans generate between a quarter and a half million tons of trash and garbage per day, much of which is disposed of in municipal or home incinerators. Too often neither the job done at the city dump nor in the backyard is accomplished very efficiently; so the effluent contains particle sizes and substances (not to mention odors) that could be eliminated with more efficient incineration. In some cities where large modern disposal plants have been installed, the cost of the service to the citizen does not encourage him to make use of the facility. The planner who looks ahead to a doubled population with more than doubled waste to be disposed of is confronted with the possibility of a citizenry partially buried in its own waste unless more adequate measures are developed.

Los Angeles: A Case Study

Now that the basic causes of increasing air pollution in American metropolitan areas have been outlined, it may be instructive to look at the case of the city which currently has the most publicized air pollution problem. Although most of the causal factors discussed influence air pollution in any major metropolitan area, they vary in degree of significance from place to place. At Los Angeles a majority of these factors act in combination and with unusual effectiveness to create a truly challenging problem. Indeed, despite recently adopted measures that seem strong in comparison to those adopted by most cities, the problem there remains a long way from a satisfactory solution.

One of the principal attractions the Los Angeles area offered the millions who migrated there was a mild sunny climate. Cyclonic storms and strong winds are infrequent, and during the warmest part of the year day after day of sunny skies result from the dominance of the large semipermanent anticyclone in the eastern North Pacific. Unfortunately from the air pollution standpoint, this means subsidence and often an inversion to cap any rising air currents that might disperse pollutants; this combination is pronounced for protracted periods in the Los Angeles area. Los Angeles also is situated in a topographic basin which opens to the sea but is surrounded landward by mountains and hills which restrict ventilation by horizontal air movement. This set of natural circumstances, tailor-made for an air pollution problem, needed only one ingredient: man.[22] Man came in ever increasing numbers, attracted in part by the rapidly expanding industry that owed its development in some measure to the petroleum resource the basin possessed. The petroleum-based industries, expanding rapidly from a substantial reserve, added their own peculiar contribution to the contaminants being poured into the sunny California air by a rapidly diversifying industrial complex. As though these factors were not enough, Southern Californians elected to remain largely independent of public transportation facilities, and the number of private autos soared with population increases. Even the California sunshine now appears to be a mixed blessing, because it has been demonstrated that new and more dangerous contaminants are produced by photochemical action on hydrocarbons from auto exhaust gases.

Developments at Los Angeles will be of prime interest to students of the air pollution problem for several reasons. First, whatever success attends the efforts there will be achieved through limiting man's activities; the physical controls are not amenable to modification.[23] Secondly, it seems increasingly clear that the actions of the individual citizen are basic to any real solution in Los Angeles; the family auto looms as a greater threat than industry or any other cause of pollution. Finally, the amount of lead time before conditions become intolerable is reduced with each new immigrant—and it is small enough to start with. In one way or another, these same three factors, especially the last one, require consideration in any area with an air pollution problem. Hence, any success in the struggle against pollution at Los Angeles will come none too soon for the other cities of the nation. Atmospheric conditions capable of creating a problem may occur almost anywhere in the country, and as quantities of pollutants increase, so do the possibilities for tragedy. While landforms may make pollution a greater problem, their contribution is not essential; London, England, where four thousand persons died during severe air pollution conditions in 1952, is not surrounded by terrain barriers.

The Effects of Air Pollution

Recently, James Quigley, Assistant Secretary of Health, Education and Welfare, estimated the cost of deterioration and repair of public buildings resulting from air pollution to be eleven billion dollars per year. It is fairly well established that in many large metropolitan areas today, the soot and dustfall is on the order of fifty to one hundred tons per square mile per month, contributing substantially to such deterioration. Particulate pollution, however, is by no means the major offender; even greater tonnages of gaseous pollutants are at work corroding, dissolving, and defacing property at rates more difficult to assess. While many of these substances are familiar and are being measured, newcomers are constantly showing up as analysis becomes more sophisticated.

There are so many ways that pollution affects the pocketbook of the citizen that an exhaustive summary is impractical here, if indeed it is possible at all. Cleaning bills, higher food costs,[24] repainting and repair, depressed real estate values, costs added by manufacturers and taxes to defray the expense of pollution abatement equipment, and lost labor time due to illness are only a few of the more obvious expenses borne by the urban dweller. Some of these will increase as necessary additional efforts to control pollution are made, but others will decrease as air conservation measures take effect. Another factor which should offset pollution control costs is implicit in the recognition of the wastefulness of air pollution. Hundreds of tons of substances (such as sulphur) are lost annually, some of which could be trapped and sold as byproducts, thus helping to defray the cost of the entrapment equipment. The processing of garbage to recover or produce salable materials is in itself not new, but procedures developed in some European cities indicate far greater possibilities for economic return from such wastes.

It little matters, however, whether pollution control costs will be defrayed entirely, or partly, or at all; control is rapidly becoming essential. Several times in recent years the lethal qualities of man-caused air pollution have been convincingly demonstrated. In 1930 in the heavily industrialized Meuse Valley of Belgium sixty persons died and uncounted thousands became ill during a persistent anticyclonic condition with an accompanying subsidence inversion and fog. Subsequent investigation held industry to blame for the pollutants concentrated by the weather conditions, although the factories were proved to have been carrying on no more than their normal operations.[25] In 1948 at Donora, Pennsylvania again the combination of an industrialized valley and a persistent anticyclonic system with subsidence and fog caused disaster. As mentioned earlier, seventeen persons died and nearly six thousand (half the town's population) became ill in the few days the pollution condition existed. At London, England, in 1952 the same general type of weather condition proved terrifyingly effective over a broad shallow basin as some four thousand deaths believed attributable to the air pollution were recorded. In this case, although there were other contributory sources, the coal-burning home furnace was considered to be the chief offender.[26] "In New York City, some 200 excessive deaths were noted retrospectively during a recent period of excessive (air) pollution."[27]

It should be pointed out that some do not agree with the studies of these disasters, and feel that the whole question of the effect of air pollution on human health is quite controversial It is suggested, for example, that four to six parts per million of SO_2 (a level rarely reached) could be harmless to a healthy person while one part per million could be too high for someone already ailing.[28] It is true that the relationships between pollutants and human health have not been subjected to exhaustive study; indeed, the interaction of some extant pollutants is not understood, and newly-discovered pollutants are still emerging. Nonetheless, episodes of smarting eyes, coughing, shortness of breath and the like have become all too familiar during periods of obvious air pollution and in themselves provide adequate bases for action. Few people who have experienced a severe air pollution situation are willing to accept the prospect of more of the same, let alone with increasing frequency or intensity. The prospects for both the latter, unfortunately, are very real as industry and services increase in complexity, as urban populations swell, and as the numbers of autos multiply.

Though of different dimension and temporal character, two other aspects of air pollution perhaps deserve mention in passing. The topic of radioactive wastes in the air has not been raised in this treatment because it is obviously a special case. Additionally, there is the concern on the part of some earth scientists over the increase in CO_2 content of the atmosphere and its possible effect on the earth's radiation balance and climatic change.[29] While neither of these topics is treated here, mention of them serves to suggest the scope of the air pollution problem.

Air Conservation

A reasonable use of the air for community and industrial waste disposal, as with the use of water bodies, is not only justified, it is necessary. With present technology, attempts to avoid completely the release or escape of substances into the free air either would result in exhorbitant costs or simply be impossible. With the acceptance of the above dictum, there is a corollary fact to be faced: The best efforts toward air conservation are not likely to return the atmosphere to its original pristine state. Lest this unduly concern some persons, let us remember that, in the main, mankind has gotten along without a pollution-free atmosphere for quite some time. There is a veritable spectrum of stages of contamination between pristine purity and intolerable or lethal pollution. In the immediate future a sufficient task for many metropolitan areas will be to reduce their proximity to the intolerable end of the spectrum. As success attends these efforts, methods may evolve by which to move substantially toward the other end of the spectrum, if not all the way to it.

The fact that air pollution control efforts have been underway in a modest fashion for decades is often overlooked. Some devices (modified versions are currently in use) were in use a half century or more ago in this country. Most of the larger industrial firms have spent many millions of dollars in control research and equipment.[30] Precipitators, centrifugal collectors, cloth bag filters, scrubbers, higher smoke stacks, absorbants in baffles—these and countless other devices are being employed by many firms, some of which have adopted more stringent pollution control than the state or local regulations in their locality demand. Some communities enforce trash-burning restrictions, and in California modification of auto exhausts and limitations on the use of certain industrial and heating fuels have legal backing. These scattered efforts are insufficient, however, and it is gradually being recognized that a concerted, unified effort will be required to achieve significant progress. This fact has been made clear as communities and states have adopted pollution control legislation which they later found difficult, if not impossible to enforce. With the entry of the Federal government into the field in 1955, the beginning of an organized effort emerged and was further strengthened with the passage of the Federal Clean Air Act of 1963.

THE LAW AND AIR POLLUTION[31]

As of 1960, some sixty-five million Americans lived in urban centers of fifty thousand or greater population with an air pollution problem of some degree. About half the population of such areas is covered by a local control program. In the main, it has been at the local level that control efforts have been made to date; this probably will continue to be the case, as it seems to be the most desirable arrangement. A chief shortcoming has been the problem of intercommunity or interstate disagreements, but the Federal Clean Air Act of 1963 sought to correct that. An outstanding example of the kind of work a local group can carry out is afforded by Los Angeles County. In the first thirteen years following its inception in 1947, regulation was responsible for cutting in half the volume of contaminants produced by stationary devices, despite the fact that during this time the population and number of industries in Los Angeles both doubled. Most programs are considerably more modest; in 1961 Los Angeles' expenditure on air pollution control represented 40 per cent of the total local effort in the nation.

California (1947) and Oregon (1948) were the first states to enter the air conservation field. Only five states had done so by 1955, the year of the first Federal air pollution statute. Between 1955 and 1963, following the entry of the Federal government into the field, twenty-two more states followed suit and the number continues to grow. The extent of state statutory authority varies[32] (see Table 17–1), and implementation has lagged, but major conservation laws have been passed in some states and more are considering them.

TABLE 17–1

AIR POLLUTION: EXTENT OF STATE STATUTORY AUTHORITY

State	Control by state officials authorized	Local option specifically authorized by state	Research and technical assistance program authorized	Any other authority
Alabama				
Alaska	*		*	*
Arizona		*	*	
Arkansas				
California	* (P)[1]	*	*	*
Colorado			*	
Connecticut			*	
Delaware	*		*	
District of Columbia	* (L)[2]			
Florida	*	* (L)	*	
Georgia				
Hawaii	*			
Idaho	*		*	
Illinois	*		*	
Indiana	*			
Iowa		* (L)		
Kansas				
Kentucky		*		
Louisiana				
Maine				
Maryland	*		*	
Massachusetts	* (L)	* (L)		
Michigan				
Minnesota				
Mississippi				
Missouri		* (P)		*

TABLE 17–1 (Continued)

AIR POLLUTION: EXTENT OF STATE STATUTORY AUTHORITY

State	Control by state officials author-ized	Local option specifically authorized by state	Research and technical assistance program authorized	Any other authority
Montana				
Nebraska		* (P)		
Nevada		*		
New Hampshire				*
New Jersey	*		*	*
New Mexico				
New York	*		*	*
North Carolina			*	
North Dakota				*
Ohio			*	
Oklahoma			*	
Oregon	*	*	*	*
Pennsylvania	*		*	
Rhode Island		* (P and L)		
South Carolina		* (P and L)		
South Dakota				
Tennessee			*	
Texas				
Utah				
Vermont				
Virginia				
Washington		*	*	
West Virginia	*		*	
Wisconsin		*		*
Wyoming				
TOTALS	13, and 3 partial or limited	6, and 7 partial and/or limited	19	8

* *Air Conservation*, Publication No. 80 (Washington: American Association for the Advancement of Science, 1965), p. 224.

[1] P, partial, not all aspects of air pollution covered.
[2] L, limited, not all jurisdictions in state are affected.

Federal action specifically aimed at air pollution control began with passage of Public Law 84–159 in July, 1955.[33] Its stated purpose was to provide research and technical assistance to aid air pollution control through the office of the Secretary of Health, Education and Welfare. In 1960, the Division of Air Pollution was established in the Public Health Service, concentrating Federal air conservation activity in a specific administrative unit. From the beginning Federal legislation has emphasized that the chief objective of its program was and would continue to be to provide leadership and assistance to developing control programs. Federal expenditures in air conservation rose from 1.7 million dollars in 1955 to 13 million dollars in 1964, and 35 million dollars is authorized for 1967.[34] The vast majority of the expenditures to date have gone to support research, although some have been for technical assistance and training. The primary responsibility for actual control remains with state and local governments.

The passage of The Clean Air Act of 1963 strengthened the Federal government's ability to assist state and local efforts in several ways.

1. On a matching basis (two or three dollars for one) Federal funds may be

awarded directly to air pollution control agencies to aid in the initiation, development, or improvement of air conservation.

2. The Federal agency will seek the development and promulgation of air quality criteria for use by state and local authorities in establishing standards to fit their particular areas.
3. Limited legal regulatory authority is afforded by the Act, giving the Secretary the power to stop airborne contamination which local or state agencies are unable to cope with. The word "limited" stresses the implicit supplementary character of this authority; it is supposed to be invoked only after local or state efforts have been unsuccessful. A recognition of the problem posed by activities in one state affecting air pollution in another state or states is behind this part of the statute. The Federal government may intervene in an intrastate problem also, but in this case only by invitation by the governor.

The nature of the change in the role of the Federal government between 1955 and 1963 implies recognition of the need to move forward promptly in air conservation. Also, while the Federal government continues to prefer that control rest with state and local agencies, it is now in a position to back up those agencies. It seems reasonable to assume that if progress in pollution abatement lags, the Federal role will be changed further to fit the situation.

FUTURE CONSERVATION MEASURES

Any consideration of methods for solving the growing air pollution problem quickly leads to the recognition of need for more information. Before major legislative changes can be considered or plans to implement legislation formulated, we must know the problem more thoroughly to avoid contradiction or courses of action which do more harm than good. Research on causes and effects of air pollution thus needs to be expanded and moved forward as rapidly as possible.

Time is short, however, and the nation cannot afford to mark time until all the results of research are in. While there is much to learn, the use of proven methods and devices should be expanded as rapidly as is reasonable. The universal adoption of general regulatory measures that have proved viable will be acceptable if the measures are modified to fit individual area conditions. Federal and state tax incentives also can hasten the adoption of appropriate devices and measures. Few governmental units—local, state, or Federal—are without planning staffs today. These groups should give greater attention to the air pollution problem in their future efforts.

Finally, there is a selling job that needs to be done. People are the key to the air pollution problem: they cause it, they complain about it, they support or resist efforts to do something about it.[35] "... [I]f man is willing to recognize that the problem exists, if he is prepared to bring to it his political wisdom, scientific knowledge, and technological skills, and if he is willing to work with nature instead of against it, then he can leave to his children... something more valuable and more necessary to human life than any of the manufactured products of his civilization. He can bequeath to them the blessing of clean air."[36]

FOOTNOTES

[1] *Air Conservation*, Publication No. 80 (Washington: American Association for the Advancement of Science, 1965), p. 212; see also Edward Edelson and Fred Warshofsky, *Poisons in the Air* (New York: Pocket Books Inc., 1966), pp. 82–86.

[2] *Statistical Abstract of the United States* (Washington: Bureau of the Census, 1966), p. 26.

[3] Edmund K. Faltermayer, "We Can Afford Clean Air," *Fortune* (November, 1965), p. 159.

[4] See Berton Roueche, *Eleven Blue Men and Other Narratives of Medical Detection*, (Boston: Little, Brown & Company, 1953), pp. 194–215.

[5] There are numerous and varying definitions of air pollution which emphasize different aspects of the problem as the uses of the definitions differ. For example see Louis C. McCabe, "The Identification of the Air Pollution Problem," *Air Pollution* (New York: Columbia Univ. Press, 1961), p. 40.

[6] Louis J. Battan, *The Unclean Sky* (Garden City: Doubleday & Company, Inc., 1966), pp. 15, 16.

[7] For a discussion of this point see Donald H. Pack, "The Meteorology of Air Pollution," *Science*, vol. 146, no. 3648 (1964), pp. 1119–27.

[8] *Ibid.*

[9] "Air Pollution: Special Report," *Power*, vol. 109 (August 1965), pp. S1–S48.

[10] "Pollution of the Environment," *American Scientist*, vol. 54 (June, 1966), p. 169A.

[11] *Power, op. cit.*, p. S14.

[12] Faltermayer, *op. cit.*, p. 163.

[13] *Air Conservation, op. cit.*, p. 249.

[14] Lewis Herber, *Crisis in Our Cities* (Englewood Cliffs: Prentice-Hall, Inc., 1965), p. 14.

[15] For an elaboration of the culpability of the auto see Donald E. Carr, *The Breath of Life* (New York: W. W. Norton & Company, Inc., 1965), 175 pp.

[16] *American Scientist, op. cit.*, p. 169A.

[17] *Power, op. cit.*, p. S17.

[18] *1964 Minerals Yearbook*, vol. 2, *The Mineral Fuels* (Washington: Bureau of Mines, 1965), p. 8.

[19] *Power, op. cit.*, p. S14.

[20] Faltermayer, *op. cit.*, p. 159.

[21] "The Aerial Sewer," *Progressive Architecture*, vol. 46 (September, 1965), p. 47.

[22] It is suggested that the first white man to see the Los Angeles basin saw it under a layer of smoke from Indian campfires spread horizontally under a subsidence inversion. Prophetically, he named what later became San Pedro Bay the "Bay of Smoke." See Herber, *op. cit.*, pp. 48, 49.

[23] For an interesting commentary on suggestions for modification, see Morris Neiburger, "Weather Modification and Smog," *Science*, vol. 126 (October 4, 1957), pp. 637–45.

[24] Visible damage to certain crops from air pollution in parts of California during one year has been estimated at over $6,000,000. See Moyer D. Thomas "Effects of Air Pollution on Plants," *Air Pollution, op. cit.*, p. 274.

[25] Harry Heimann, "Effects of Air Pollution on Human Health," *Air Pollution, op. cit.*, p. 164.

[26] *Ibid.*, p. 181.

[27] William S. Spicer, Jr., "Relation of Air Pollution to Disease," *Archives of Environmental Health*, vol. 9 (November, 1964), pp. 600–05.

[28] R. J. Bender, "The Struggle For Clean Air," *Power*, vol. 109 (August, 1965), pp. 195–200.

[29] Pack, *op. cit.*, p. 1127. Extrapolation of present measurements of CO_2 present in the atmosphere indicates that there will be a global increase in this constituent of about 40 per cent by the turn of the century.

[30] Faltermayer, *op. cit.*, p. 161. The member companies of the Manufacturing Chemists Association had invested over $200,000,000 on facilities up to 1965.

[31] *Air Conservation, op. cit.* See Chapter 4 on this topic.

[32] *Ibid.*, p. 224.

[33] See Vernon G. MacKenzie, "National Policy on Air Pollution Control," *Journal of the Sanitary Engineering Division, Proceedings of the American Society of Civil Engineers*, vol. 90, no. SA6, part 1 (December, 1964), pp. 51–58.

[34] *Ibid.*, p. 229.

[35] *Power, op. cit.*, p. S34.

[36] *Air Conservation, op. cit.*, p. 19.

18

Cultural Erosion of the Land

DURING THE FOUR CENTURIES since the white man first settled on United States territory the landscape has been markedly altered.* The decisions and efforts of fifteen generations, totaling several hundreds of millions of Americans, have had a role in the processes of alteration—perceiving environmental utility, establishing the appurtenances for living and for bringing resources into use, and establishing the resource use, management, and service systems. In these processes forests have been cleared; grasslands have been plowed; crops, grass, and trees have been planted; water has been drained; reservoirs have been constructed; irrigation projects have been built; livestock has been substituted for wildlife; transportation networks and power grids have been laid out; and industrial plants, commercial structures, residences, towns, cities, and billboards have been scattered across the land. Indeed, essentially every square mile of the nation has felt the impact and modifications of man in some way.

These processes of alteration, resulting in the present occupancy and economy, have produced an affluent society—affluent in terms of such measures as per capita consumption of goods, property ownership, education, social mobility, and the like. But they have also produced a physical and social environment that is beset with many ills and is nurturing many more to plague future generations. These ills are evident in the current slum areas in our large cities; in pollution of varied sorts; in clogged streets, cluttered roadsides, junkyards, urban sprawl, and mixtures of graceless buildings; in noise and vandalism; in park deterioration; in a general decline in the beauty of the countryside; and in the significant consumption of high quality land for secondary and tertiary activities. They are also evident in the increasing amount of public budgets required for corrective measures.

In view of such evidence, one can wonder if some aspects of the philosophy that has underlain the quest for and accomplishment of general affluence are compatible with the rational progress of a mature society. It would seem that in fostering and protecting their rights to the profits of individual enterprise, Americans have largely fulfilled their dream of affluence, but in so doing "have largely ignored the other half of civilized life: a whole spectrum of human needs that can be met only through communal action."[1] Not the least of these is the need to provide adequate resources for the future.

This discussion is not intended to lay the blame for environmental deterioration wholly upon the American system of enterprise beliefs and enterprise capitalism, but rather to suggest that some of the institutionalized features of the society—those that shape citizen attitude and responsibility toward the total society and have also to do with the laying down and carrying out of communal responsibilities—have not kept pace with the needs of today or the future.

A significant feature of this lag is the failure to clearly reckon with the mathematics of population growth and finite space.[2] On the contrary, we view progress largely in terms of economic growth, and all concerned with economic activities (individuals, companies, corporations; or local, state, or national governments), are promoting increased consumption. Thus population growth is viewed as essential to progress—more consumers mean more consumption, and more consumption means more production. This view, long fostered by a rich resource base and a highly favorable resource/population ratio, is now supported by a rapid advance and application of resource-converting and space-adjusting techniques.

The appraisals of the separate categories of resources in the preceding chapters indicate that the nation remains in a fortunate position in terms of quantities in most categories. In fact, as noted in agriculture, capabilities for overproduction have been significant economic and political problems for some time. Inherent in the attitudes toward and the means of attaining such capabilities, however, are more serious and basic problems for the future.

The advance of construction technology has given Americans tremendous capabilities for altering the landscape—and such alteration is moving forward at an alarming rate. These capabilities, of course, can provide the nation with important tools for improving the quality of life and meeting the resource needs of the future. As implied earlier, however, they have for some time also been a feature of our capability to foul our environment—especially when fostered by short-range profit motives and limited concern for the interaction of new developments with the features of the existing occupancy or new features to be attracted.

In many ways the deterioration of the quality of the environment—by the cultural erosion of the land—is the most serious conservation problem of the nation. The majority of Americans live daily with the symptoms of environmental deterioration noted in the second paragraph of this chapter. Let us turn to a brief examination of some of the major elements of this problem.

GROWTH OF URBANISM

Since the turn of the century the United States has been transformed from a predominantly rural nation into a strongly urbanized nation. In 1900 40 per cent of the 76,094,000 people lived in the city; the rest were scattered through the countryside where they lived and worked. By 1950 the population had grown to 150,697,000, and the urban dwellers had increased to 64 per cent of the total. During the following decade this trend continued—in 1960 out of a population of 179,323,000, around 70 per cent lived in urban areas, and a considerable portion of the remainder lived in communities along the highways and the crossroads of the nation and were engaged in nonrural work. This trend continued during the first seven years of the 1960's, with most of the twenty million population increase occurring in the urban centers. Estimates suggest that within fifty years as much as 90 per cent of the population will be living in urbanized areas.

This rapid evolution from rural to urban living has resulted from multiple causes. Six, however, may be isolated as being particularly significant.

(1) The commercial orientation of the economy which requires the concentration of business activity in central places of efficiency.

(2) The development of mass-production manufacturing requiring the concentration of labor forces in central places. This trend has had snowballing effects; the establishment of one plant often has attracted others, multiplying the need for labor. This in turn has attracted supply firms.

(3) The increase in efficiency, speed, and capacity of transportation facilities based on mechanical power. These improvements have extended the sphere of influence of urban centers, lengthening the radii for the supply of raw materials and foodstuffs and the market for goods and services. At the same time the improvement of the arteries of transportation within the centers and their immediate environs has encouraged urban growth by permitting

relatively easy movement to and from work, school, shopping centers, etc.

(4) The increasing efficiency of labor made possible by machines and other scientific and technical innovations in all aspects of primary production based in the rural areas. This trend continually has reduced the labor needs and the employment opportunities in agriculture, forestry, and mining.

(5) The progress in the ability of cities to provide sanitary and related facilities. Modern sewage and water supply systems and such services as garbage collection and fire protection have made it possible for large numbers of people to live close together without danger to their health.

(6) The urban pull. Under this term may be noted a composite of factors which tend to draw people to cities. Some are tangible, such as the opportunities provided for employment; others are less tangible, including cultural attractions and recreational advantages and the somewhat gregarious nature of people.

Although these six causes are noted as primary, it bears repeating that many factors have, and are causing city growth. A longer listing could well include the growth in governmental functions which are best performed in cities with central positions within regions, and the increase in the number of retired persons who favor urban living.

With the modern increase in business and manufacturing acivities and the growth in population numbers, the appearance of urban centers has been undergoing rapid change; widespread sprawl is now a common characteristic, with no clear demarcation between town and country. The desire for home ownership (at present a possibility for most people because of easily obtained home loans) and the resulting need for convenient shopping centers, schools, transportation arteries, and other service facilities, are pushing urban boundaries ever outward. Population density within the urban centers (especially within the suburbs) has dropped sharply in recent years as the result of a change in house styles away from multi-story dwellings to single story ramblers, demanding sixty- to one hundred-foot lots (or more), in contrast to the common forty- to sixty-foot lots of a few decades ago. Likewise, modern shopping centers and other facilities are now consuming more space. Cheaper land, more space, lenient taxes, and fewer restrictions on such features as noise, unpleasant odors, and parking facilities also draw many industrial plants and commercial firms to the urban margins. Following World War II the Federal government encouraged decentralization of industry, giving additional impetus to this trend. In many instances neighboring cities have merged, giving rise to numerous multicenter complexes, a few gigantic super-urban areas, and many areas of disorderly, unrelated land uses, congestion, and ugliness.

SPACE AND LOCATION REQUIREMENTS

As suggested, urban centers exist to provide those useful and economic functions of areas that need to be carried out in central places. As features of an occupancy or economy system, cities overcome the handicaps of distance for business and manufacturing activity. They are primarily the products of fruitful land or advantageous location for concentration, storing, processing, packaging, and distributing commodities, and for business activities of all kinds; the richer the area served the greater the growth opportunity for the urban center or centers. It is not surprising, therefore, that much of the lands of highest capability are within the immediate environs of the growing metropolitan areas (Figure 18 – 1).

Transportation facilities are established to provide the necessary circulation system within urban centers and to connect population and production areas. They are absolutely essential to the function and growth of the national economy. Streets occupy from 20 to 25 per cent of the built-up area in most cities.[3] In the interest of speed, efficiency, and low cost construction, the prevailing practice in highway location is to connect desired points in the shortest distance possible over the smoothest land possible. Thus broad, modern highways now pass through the nation in all directions, with the most dense nets occurring in the rich and populated areas. Improvements are continuing; construction is now progressing rapidly on the National System of Interstate and Defense Highways. In 1966 more than twenty thousand miles of the four-lane, modern design, limited access system were open to traffic. When completed in 1972 forty-six billion dollars of Federal funds will have been expended for this forty-one thousand-mile system. Airports demand lands of smoother surface than highways, and have become notable space users on urban fringes.

It has been estimated by Department of Agriculture personnel that, by the end of 1959, urban areas occupied 27.2 million acres.[4] Data are not available for the smaller communities, but it is reasonable to assume that these, plus other nonfarm rural residences, industrial sites, commercial establishments, and government-owned institutional sites occupied another six to eight million acres.

Rural transportation facilities (exclusive of those within built-up areas and on farms) were estimated to have occupied 25.3 million acres in 1959. This included 20.5 million acres in highway and road right-of-ways, 3.4 million acres in railroad right-of-ways, and 1.4 million acres in airports.[5]

Taking the nation as a whole, the amount of land (about 2.5 per cent of the total) occupied by urban and transportation facilities is neither significant nor alarming. From the points of view of high quality land and local areas, however, the amounts so occupied are significant. Furthermore, more impressive than the data cited above are those relative to the recent growth rate of urban and service expansion. The Soil Conservation Service, on the basis of a nationwide survey, estimated that seventeen million acres of cultivatable land in capability classes I, II, III, and IV were diverted to nonfarm uses between 1942 and 1956. The study indicated that during the decade of the 1950's approximately one million acres of rural land were diverted to urban and other nonfarm uses, which included urban and suburban homesites, commercial and industrial uses, highways, airports, defense establishments, and the like each year. This includes an annual diversion of an estimated 500,000 acres to urban, suburban, and scattered homesites. Land has been shifted to urban and special uses at a slightly higher rate during the 1960's.

It is significant to notice that a major share of the recent diversion is at the expense of agricultural land (an estimated 40 per cent from cropland and grassland pasture, and another 40 per cent from forest land).[6] The decline in cropland and the increase in special uses are by no means uniform. An estimated 10 per cent of the agricultural counties of the forty-eight conterminous states had increases in cropland acreage between 1950 and 1960.[7] The increases,

18 – 1 *The sprawl of Los Angeles.* This aerial photograph shows part of the freeway system with the civic center in the background. The built-up area extends in all directions from this central view. (Los Angeles Chamber of Commerce)

however, were usually more than offset by decreases in other counties, often nearby. These changes in uses in many instances have meant improvement and development of quality land for crops, and shifts of poor land from crop use. The exceptions to the latter, however, are of major proportions in the case of diversions to urban use.

PROBLEMS ATTENDING GROWTH IN URBAN AND TRANSPORTATION LAND-USE

Several categories of problems have become increasingly apparent during the current explosive growth in the use of land for urban and service purposes. Some categories adversely affect these uses, others adversely affect other resources and the use of land. All need attention in the interest of the national conservation program.

As suggested above, the sprawl of urban areas is, in a large measure, at the expense of high-quality land. Suburbia and the urban fringes are expanding, in many cases, over the smoothest land of their locales. Thus, every year many acres of high agricultural potential are being permanently lost.[8] The situation is not limited to large cities—it is a common characteristic of essentially all urban communities which are experiencing growth. It is especially important that notice be taken of the qualities of adaptability of the land being diverted. For example, significant acreages of former citrus and winter vegetable land (of which the national totals are limited) are now in urban developments. There are those who will argue that this is not necessarily alarming and offer the following major points in support of such opinions: (1) It is a normal and usually desirable adjustment to changing economic conditions; (2) urbanization of such land is justifiable on the basis of competitive economic grounds; (3) technology can be applied to lands of lower quality to meet agricultural and other biotic needs; and (4) in the long run, competition will price agricultural use back into a favorable position. The authors of this book would note that a conservation program cannot be built entirely upon current economic considerations which by their nature can cover only the short-run and deal with cost-profits, and not quality of environment; that the welfare of future Americans who presently have no voice in the matter must be considered; and furthermore, that it is probably unrealistic to believe that agriculture and other rural uses will be able to compete favorably with the high economic values that urban and industrial uses can force upon quality land soon enough to meet future needs at reasonable costs.

Closely related to the problem of sprawl upon quality lands is the common problem of scattering of the built-up areas and resulting incomplete land use within urban complexes. There is a large amount of empty land (a lot, an acre, or a few acres here and there) even in the large cities. Even though vacant land averages about 20 per cent of the areas occupied by urban centers, population growth exerts a centrifugal effect which tends to push city boundaries ever outward. In suburbia and on the urban fringes the incomplete use is greater; empty spaces of several acres surrounded by built-up areas are common. This leapfrog nature of urban growth results from several factors, but the land ownership pattern is often a major cause. In some cases farmers withstand the pressures to sell; in other instances the land may be held by speculators or in estates, or the land may be less desirable for buildings; in still other cases, the trend of development is such to encourage the speculative builder to jump beyond the built-up margin for his new subdivision. Thus land is taken out of production before the need for urban use has arisen. The bypassed areas, generally too small for agricultural uses, are unused or left to low quality and often undesirable uses which lower residential values. In this manner unguided development tends to degrade the aesthetic quality of the urban landscape.

This sort of sprawl and scatter results in many economic problems for the urban areas. It requires large expenditures and high per-unit cost for the extension of transportation facilities and water and sewage systems. It brings about traffic congestions, requiring more time for residents to move to and from places of employment and shopping centers. During the morning going-to-work period and evening returning-to-home period, an important share of the nation's ninety-three million motor vehicles converge upon the streets of the urban centers; and the growth in the numbers of automobiles is keeping pace with the population increase.[9] The development of local shopping centers in suburbia places a serious competitive strain on the establishments in the central cities as well as in neighboring shopping centers.[10] In many urbanizing areas, subdivisions with their shopping and service facilities adjacent to the central cities have incorporated. In such cases, they have developed municipal government, local pride, and vested interest in the perpetuation of identity. This often complicates the problems of unifying goals and solving mutual problems, and tends to intensify conflicts. Also, in the process of sprawl, urban centers lose their advantages of compactness.

Perhaps the most significant problem of all is the general decline in the quality of the urban environment. Seemingly, under the pressures of urban economics, aesthetics receive little consideration. Cathedrals may be surrounded by expansive parking lots, gasoline service stations may border parks, and slums may provide the foreground view of a high-class apartment. People from the suburbs may drive to work down a billboard alley, in the midst of slow-moving traffic and the smell of exhaust. Indeed, the automobile has become both an essential and a culprit for the American Society—and one of our greatest needs is to learn to live with it more rationally (Figure 18 – 2).

Urban concentrations, particularly those involving large populations and growth without plan, commonly have some adverse effects on resources other than land. Securing and maintaining quality water supplies are major concerns of essentially all cities. This is an understandable result of the crowding of people and industry into small areas. Local supplies are overtaxed, and it becomes necessary to tap

18 – 2 ***A modern shopping center.*** **Notice the proportions of the land required for automobile parking spaces. (Soil Conservation Service)**

sources farther afield at increased costs to the users. Because of the multiple uses of water, some of which are incompatible with others, pollution is a common and serious problem. Few cities can draw on adequate supplies which can be used without expensive treatment. In areas of high-intensity precipitation, the urban centers, with a large portion of the land they occupy covered with roofs, streets, and parking areas, have created a severe runoff problem which requires costly drainage facilities.

Large urban concentrations also place great demand on local recreational facilities and attractions. Outdoor theaters and other amusement centers have become noteworthy users of land. The pressure on local natural attractions has reached a stage, in many cases, in which deterioration is imminent.

Many cities are now faced with serious problems of air pollution. Smoke, fumes, dust, and dirt are common. Those centers located in basins are plagued with smog or smaze. These pollutants result in undesirable and unpleasant conditions for urban dwellers and their activities, and in some instances adversely influence the use of neighboring rural land. For example, yearly crop losses due to air pollution are estimated to amount to one hundred million dollars in California alone. Industry and cities presently are spending large sums in research on air pollution control.

The development of the modern highway system of the nation has brought in its wake a number of land-use problems. Highways occupy large quantities of land, some of which have been highly productive. A modern freeway has a 250 to 300 foot-right-of-way (Figure 18 – 3). It cannot, of course, be argued that highways can be constructed entirely upon low-quality land; they are essential to the functioning of the economy for the movement of

18 – 3 *A segment of the Calumet Expressway, a freeway which is an extension of the Calumet Skyway, Chicago.* **This system serves the southern suburbs and industrial districts. (Portland Cement Association)**

goods and people. Nevertheless, it is probable that with planning and the exercise of choice, the use of fertile land in the future could be minimized without reducing the desired qualities of speed and safety or seriously adding to construction costs (Figure 18 – 4).

The highway right-of-way, however, is not the most serious offender; rather, it is the economic impact that highway development generates. Highway margins have become a mecca for urban homes, industries, billboards, and commercial enterprises indiscriminately mixed and commonly presenting an unpleasant landscape. This is particularly true near metropolitan centers; but the highways also have become a major force in shaping the pattern of settlement in the rural countryside, fostering road towns, crossroad towns, and part-time farms (Figure 18 – 5).[11] The result is often a degeneration of the primary function of the highway—to speed movement from one center to another. As the highway border development continues, it becomes necessary to reduce speed limits to decrease accidents and protect lives. The ultimate result of this is the necessity to establish new thoroughfares at the expense of more land and higher construction costs.[12]

Anyone who has traveled on the new Interstate Highway System recognizes the tremendous new force it is exerting on the occupancy and land-use pattern. As the results of bypassing towns and cities, concentrating the flow of traffic (and customers), and making the countryside accessible, it is spreading the urbanizing influence along its routes across the breadth and width of the nation. It creates the need for

18 – 4 ***Highway development through productive land.*** **This development divides the farm in the foreground and will require a redesign of fields and strip cropping pattern. (Department of Agriculture)**

roadside services; it allows urban workers to have their homes several tens of miles into the countryside; and it attracts large space-requiring industrial and commercial plants and businesses to lower cost lands where lenient zoning restrictions are also attractions. The focal points are the interchanges, many of which, in less than a decade, have become the centers of communities of several thousand residents. Unfortunately most communities, in the absence of land-use controls, have grown without rational pattern and with indiscriminate mixing of structures.

It is clear that insufficient consideration has been given to the stimulating forces and interactions set in motion by the Interstate System. Land values at interchange sites usually increase several times. For example, some sites along Interstate 5 between Salem and Portland, Oregon, that were valued at 300 to 500 dollars per acre as agricultural land less than a decade ago are now valued at 15,000 to 20,000 dollars or more per acre.[13] Such values, of course, encourage farmers to sell; they cannot afford the increased land taxes, and often the encroaching urbanization is not compatible with their established land uses. Parcelling of farms and disrupted farm drainage are other problems associated with the freeway. It is not unusual for a farmstead to be separated from the farm land and for an interchange providing access to be several miles away. The most serious problem, however, relates to the general lack of land-use controls. More often than not, developments are disorganized and unsightly—not providing efficient occupancy or maximum convenience, maximum land values, or maximum tax revenues.

The improvements in transportation, along with limited controls, have contributed to the deterioration of the landscape in a number of other ways. Few parts of the nation are now beyond the reach of motorists—likewise the motorist is seldom out of sight of signs, billboards, and junkyards, and it is easy for him to make his contribution to the latter (see Figure 18 – 6). It is possible that the American economy and society create more junk than the remainder of the world combined—waste paper, cartons, cans, bottles, and the like. About six

18 – 5 *Machias, New York, looking toward Franklinville.* This illustrates the influence of a crossroads location, as well as the process by which the full value of a highway is destroyed when it serves as a congested street. (Department of Agriculture)

million automobiles are junked each year, and the economics of the steel industry currently favors use of only a small part of them. What to do with junk has become a national problem of large proportions.[14]

PLANNING, ZONING, AND RENEWAL

Forecasts of population and economic trends make several points clear: (1) The population of the nation will continue to grow at its present pace for some years to come; (2) a major portion of the increase will be employed in nonbasic activities; and (3) the number of people required for rural production will, in all likelihood, decrease below the present level. The majority of the new Americans of the future will live in or near urban centers (Figure 18 – 7). Therefore, urban areas will grow with more space being required for homesites, manufacturing, commercial plants, and all kinds of service facilities, parking, streets, etc. There will be continued demand for new highways and increased demand upon the open country for rural residences, part-time farms, and industrial sites.

The serious question relative to conservation and the future is: Can the nation permit the continued aimless sprawl of suburbia, the alienation of quality rural land to nonproductive uses, and the continued decline in the quality of both the urban and rural environment? The answer is clearly no. Quality land is an asset too precious to squander. Once covered with macadam, concrete, brick, and wood it must be stricken from the registery of category one

18 – 6 ***Trash in the countryside.*** **Here the natural beauty has been degraded. Such scenes are becoming too common in America. (Soil Conservation Service)**

lands.[15] Once reduced to small disconnected plots surrounded by urban developments it loses value for modern, large-scale machine agriculture and most other biotic uses. Even though most people dwell in the city, they must not lose sight of the basic fact that it is the natural resources of the country that, in reality, support them. Nor should the people allow their present ability to overproduce and their great faith in the prospects of science and technology blind them to their sense of responsibility to themselves and posterity. There is another reason why hodgepodge scattering should not be allowed: It is bad for everyone, especially those most directly concerned, the developer, the investor, the businessman, and the homeowner. Loss of compactness adds to the costs of all kinds of services, and it creates difficulties in traffic circulation. Without controls on development, indiscriminate mixing lowers property values.

Recognizing that the pressures on land as space for living, industrial plants, service facilities, and the like will continue to grow, Americans must direct their development along orderly and desirable lines. It is possible to minimize the amount of quality land for urban and service facilities and to protect such land for sustainable, improvable production. At the same time it is possible to provide for the growth of more wholesome, attractive, and workable communities, towns, cities, highways and other land-use requirements than currently are common. Success will require a combination of full understanding of the problems and consequences by the total populace, more willingness to compromise conflicting interests, and sincere cooperation for the good of all.

18 – 7 *Housing development in Montgomery county, Maryland.* This kind of pressure on quality lands can be expected to continue. Planning can, however, reduce the amount alienated to nonproductive uses. (Department of Agriculture)

In this regard the nation will not be starting from scratch. Noteworthy progress has been made during the past three decades, but intensified effort is required if irreparable damage is to be avoided and problems now on the horizon minimized. It is increasingly clear that ordered development can be achieved only by cooperative and sound planning. The countries of western Europe are living examples of the fact that growth and development do not necessarily spell the end of environmental quality and town and country beauty.

The first step in this process is for the community, city, or area to take a critical look at itself, appraising and mapping in detail its resources and development to date, and isolating the problems that need particular attention. This should be followed by an evaluation of the prospects for economic development and population growth. If this information is to serve as the base for a viable plan, it must be broad in view and firmly grounded on regional as well as national relationships. Knowing these facts, it is then possible to blueprint the future—to lay out a master plan which, considering all aspects of activity in accordance with community goals, assigns location, space, and standards for all types of developments to come. The plan must be constantly evolving as the community or region grows and experience accumulates. Many aspects of such a plan will be carried out voluntarily—their benefits are readily apparent to the developer. Total public compliance, however, requires education to the values of a plan and majority backing to provide the legal tools for enforcement.

One such tool, which is not a new concept, is zoning. Zoning had its genesis in the colonial period. In 1692 several towns in Massachusetts were granted power to influence the location of "offensive" industries in the interest of public health and safety.[16] Since that time many communities have used zoning ordinances to attain other purposes. In general, all zoning must serve a public purpose, such as protection of health, safety, morals, or the general welfare. Four basic types are used today: (1)Those designed to keep apart land uses and activities that conflict; (2) those that set the lower limits on the size of building tracts; (3) those that limit the height and size of buildings and structures; and (4) those that regulate the density of population.[17] The authority to grant zoning powers to counties, towns, and townships rests with the state legislatures. The passage of enabling acts does not zone communities, but rather indicates the scope—the areas that can be zoned, the zoning tools that may be used, and the way in which they may be used. Thus within this granted authority, zoning ordinances are local laws that are adopted by the local people, either directly through zoning elections or indirectly by the legislative body of the community. Many towns and cities have been granted zoning powers, but the legislatures have been slow in extending such powers to rural governments. At present about one-half of the counties in the fifty states have been granted zoning powers. In some states all counties may zone; in others these privileges have been given only to the more populous counties. About 80 per cent of the present rural zoning enabling acts grant all four classes of zoning powers noted above.[18] This does not mean, however, that they are being applied effectively in this portion of the nation, particularly regarding rural land and the highway fringe. Moreover, zoning cannot eliminate past mistakes; established land uses have "vested" rights to remain. Nevertheless, it is important to note that zoning ordinances are now available and being used in many communities, cities, and counties. Others will gain them, and improvements in their authority will be made as public pressures demand. Zoning can become the major tool in the directing of desirable land use and orderly growth within the nation.

To provide for the constantly evolving demands for urban and service facilities, planning and zoning must be accompanied by a continuous program of upkeep and renewal (replacing or remodeling antiquated buildings and other facilities on the lands they occupied) (Figure 18 – 8). This will help restrict sprawl by eliminating the common process which results in blight, lower quality use, and migration to new land. The possibilities for ordering developments are now within the reach of every urban, rurban, and rural area.[19] Favorable action awaits the demand of the citizenry.

Two kinds of problems, however, in many instances will have to be solved before major

18–8 *Downtown Pittsburgh.* In the years since World War II Pittsburgh has had an active redevelopment program that is uplifting the face of the city. This includes removal of old buildings and replacement by modern buildings as well as a large program of smoke control. (Pittsburgh Chamber of Commerce)

progress can be made. One has to do with the proliferation of governmental units, and the other with public money. In the first case, a single metropolitan area may be comprised of several score of governments—city, county, fire districts, water districts, sanitary districts, and so on. With the vested interests involved there are, as a result, many difficulties in agreement on development and rehabilitation plans and on the zoning and taxing base to carry them out. In the second case essentially every local government is short of money to carry out public projects, and in many instances, even to finance planning. Although urban dwellers pay a major share of the state and Federal taxes, cities themselves are poorly financed. Yet, by tradition, they provide most of the costs of education, transportation, water and sewage facilities, parks and cultural and recreational developments. Many of the nation's social problems have also been thrust upon them with the concentration of low income groups. Most cities rely upon property taxes for as much as 90 per cent of all local tax revenues. They need more sources of income. In part this income might come from higher charges for services provided; it seems, however, that much of the additional support must come from the state and Federal governments.

The property taxing procedure, common in most cities, is itself a problem in the way of orderly development and renewal. In most instances the procedure tends to place the highest assessment on the structures on the land which in turn tends to encourage owners to neglect the appearance of buildings to reduce appraised values. Moreover, assessment standards vary greatly from place to place.[20]

FOOTNOTES

[1] Edmund K. Faltermayer, "The Half-Finished Society," *Fortune*, vol. LXXI, no. 3 (March, 1965), p. 96.

[2] See George Macinko, "Saturation: A Problem Evaded in Planning Land Use," *Science*, vol. 149, no. 3683 (July 30, 1965), pp. 516–21.

[3] Harold M. Mayer, "Cities, Transportation, and Technology," *Land, the Yearbook of Agriculture, 1958* (Washington: Department of Agriculture, 1958), p. 494.

[4] Hugh H. Wooten, et al. "Major Uses of Land and Water in the United States, Summary for 1959," *Agricultural Economics Report No. 13* (Washington: Research Service, Department of Agriculture, July 1962), Table 5.

[5] *Ibid.*

[6] *The Farm Index* (October, 1962), p. 7.

[7] Hugh H. Wooten, et al., *op. cit.*, p. 13.

[8] For an excellent case study of this problem see Howard F. Gregor, "Urban Pressures on California Land," *Journal of Land Economics*, vol. 33, no. 4 (1958), pp. 311–25.

[9] For an enlightening coverage of problems of the immediate future see "By 1975 What City Pattern?" *Architectural Forum* (December, 1956), pp. 103–37.

[10] For a brief and very readable commentary on these and other urban problems see editors of *Fortune*, "The Exploding Metropolis" (Garden City, N.J.: Doubleday & Company, Inc., 1958), 177 pp. See also Edward Higby, *The Squeeze, Cities Without Space* (New York: William Morrow and Company, 1960), 348 pp.; and Jean Gottman (ed), *Metropolis on the Move* (New York: John Wiley & Sons, Inc., 1966), 216 pp.

[11] See Keneth C. Nobe, "The Urbanization of the Open Country," *Journal of Land Economics*, vol. 60, no. 2 (1958), pp. 352–60.

[12] This process is explained concisely in reference to Michigan in Louis A. Wolfanger, "Our Biggest Rural Program," *Prairie Farmer* (January 15, 1955).

[13] John Phillip Preston, "The Impact of National Interstate Route 5 on Adjacent Non-agricultural Land Uses in the Willamette Valley, from Salem to Portland," Unpublished Masters Degree thesis in Geography (Oregon State University Library, 1967), pp. 72–77.

[14] See Edmund K. Faltermayer, "How to Wage the War on Ugliness," *Fortune*, vol. LXXIII, no. 5 (May 1966), pp. 130–134, 250, 255, 256.

[15] For definition see Chapter 3.

[16] Erling D. Solberg, "The Why and How of Rural Zoning," *Agriculture Information Bulletin No. 196* (Washington: Agricultural Research Service, Department of Agriculture, December, 1958), p. 1.

[17] *Ibid.* See also Erling D. Solberg, "Rural Zoning in the United States," *Agriculture Information Bulletin No. 59* (Washington: Bureau of Agricultural Economics, Department of Agriculture, January, 1952), 68 pp.; and "How to Keep Land Open: Some Useful Precedents," accompanying the "City's Threat to Open Land," *Architectural Forum* (January, 1958), pp. 87–90, 164, 166, 168.

[18] Solberg, "The Why and How of Rural Zoning," *op. cit.*, p. 2.

[19] See also Erling D. Solberg, "Talks on Rural Zoning" (Washington: Agricultural Research Service, Department of Agriculture, January, 1960), 96 pp.

[20] See "The Great Urban Tax Triangle" (a report of the proceedings of a round table discussion at Claremont Men's College, Claremont, California, attended by many eminent tax and municipal experts), *Fortune*, vol. LXXI, no. 3, (March 1965), pp. 106–07, 188, 190, 195–96, 198.

19

Conservation and the World Community

WORLD INTERDEPENDENCE HAS INCREASED at an accelerated rate during the first half of the twentieth century, and it appears reasonable to expect that international interconnectivity, exchange of resources and products, and mobility of people will continue to increase in degree and in complexity, leading to intensified world-scale problems of resource adequacy and of maintaining quality of environment. In this context most if not all of the questions and problems of conservation in the United States have international implications and need to be considered in terms of their international and world-scale ramifications.

The United States is a major importer as well as major exporter of natural resource commodities and products and thus questions of national resource adequacy must be considered in the context of world supply and demand. Development and management of a significant number of resources inevitably have international ramifications, for example, international rivers, fisheries, and the atmosphere. The increasing magnitude of world population and probable betterment of living standard on a world scale almost certainly will result in major reappraisals of United States relationships to world resources, and changes in the outlook of the nation toward conservation.

There can be no doubt that the world society of man needs to strive for rational and continuously harmonious rapport between occupancy and the resource environments of the earth. It is not so clear, however, what the relationship and role of the United States ought to be. What is the potential impact of world-scale realities of population and resource requirements on United States resource adequacy? What responsibilities does the nation have—or ought it to have—in world-scale conservation? If there is need for awareness of the problems and responsibility for conservation at the national level, is there also need for awareness and responsibility on the world scale?

This chapter does not presume to answer the questions, but rather to provide some of the background and to summarize major involvements of the United States. The dynamics of world population and potential impact are reviewed first, followed by a summary of major world conservation programs supported by the United States.

World Demographic Prospects and Implications

One of the world-scale realities of the mid-twentieth century is the existence of major differences in social and economic living standards between countries, expressed by the cliche "developed" and "underdeveloped" nations. The United States, for example, has achieved an economy of mass abundance characterized by a high rate of exchange and consumption of material goods, while a large share of the world's people have only just begin the struggle for economic development. Although a number of world nations are still facing the critical problem of balancing food supply and population, it is evident that a "consumer revolution" is underway in the world and that large numbers of people, especially in the so-called "underdeveloped countries," who now consume little will gradually increase their material consumption, not only of foods but of other basic raw materials and products in decades ahead. Few scholars expect these increases to bring consumption levels up to equal those of the United States; nevertheless, consumption levels of the United States may be looked upon as measures of potentials. The probability of increased per capita commodity consumption in the world outside of the United States probably would raise few problems were it not that world scale populations are expected to increase in staggering magnitude during the next few decades.

WORLD POPULATION PROJECTED GROWTH

Undoubtedly the most significant trend that will condition United States involvement and attitude toward world scale conservation problems is the rate and magnitude of population growth that is being projected for large portions of the world community (see Figure 19 – 1). As is well known, population in many areas of the world is increasing at a rate and in magnitude without precedent in the history of mankind. The diffusion of technology and science of public health and medicine to nearly all areas of the world is resulting in significant de-

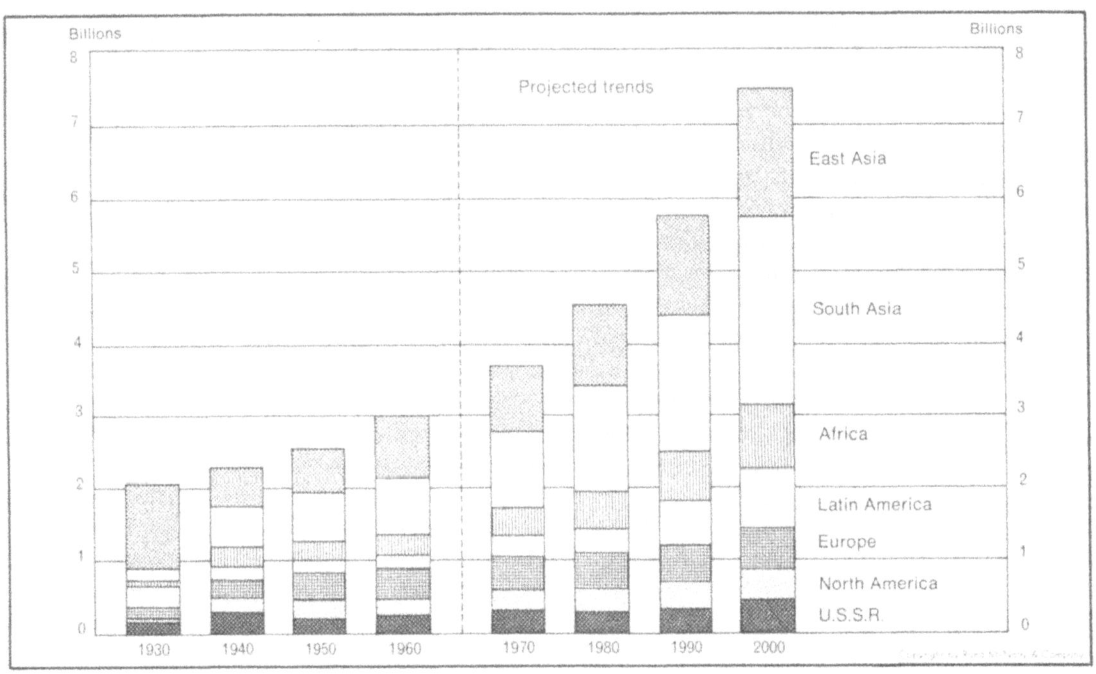

19 – 1 ***World population, 1930–2000.*** **Assuming continuation of present trends, from a base of three billion in 1960, with birth rate continuing high and death rate reductions resulting in the world average annual net increase rising from 2 per cent to 3 per cent, world population for the year 2000 is projected to 7.4 billion. Were the birth rate to decline to result in a world average net increase of only 1.8 per cent, the projection is for six billion by 2000. The United States' population would decline from about 6 per cent to 4½ per cent of world population. (United Nations Provisional Report on World Population Prospects as assessed in 1963)**

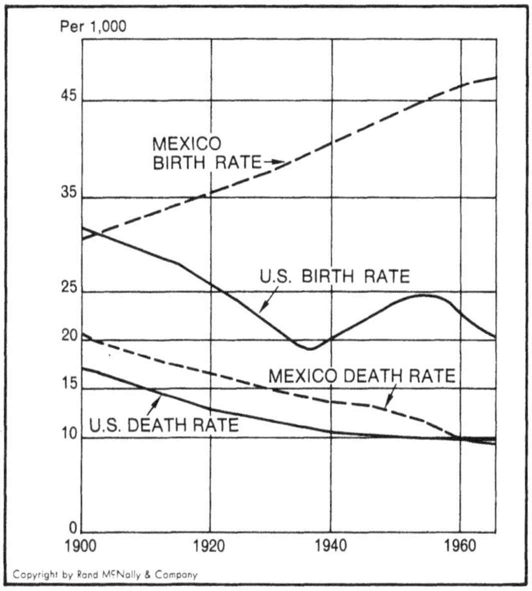

19 – 2 ***Mexico and United States birth rate and death rate, 1900–1965.*** **Mexico is an example of a "developing country" in which birth rate has remained high and even increased, while death rate has declined. Thus Mexico is recording one of the higher annual net increase rates (3.5 per cent in 1965). As a result, Mexico's population increased from sixteen million in 1930 to forty-four million in 1965, and is projected to reach 70.6 million by 1980 and 125 million by 2000.**

cline in death rate and lengthening of the life span. Although there is much yet to be accomplished, there is reason to believe that the world-scale battle for health is being won. In the United States, and in some other industrialized nations of the world, birth rates along with death rate also tended to decline, but in many countries birth rates have continued to be high and have even increased, perhaps in response to better economic conditions. This reality of world population dynamics is illustrated by the comparison of birth and death rates recorded by Mexico and by the United States (Figure 19 – 2). Thus, from the basis of about three billion population in 1960, United Nations demographers have projected a world population of 4.5 billion by the year 1980 and 7.4 billion at year 2000.[1] This enormous population projection is based on the assumption that a declining death rate in countries of Africa and Asia will not be balanced by significant decline in birth rate with the result that net increase will rise to about 3 per cent annually. Should there be major declines in birth rate, for example to 1.8 per cent, the population would only increase to 4.3 billion by 1980 and to 6 billion by 2,000. It is perhaps significant that the United States' portion of world population would decline from 6 per cent to about 4½ per cent.

In the long run the current rate of increase probably will not be sustained. The reality for the next few decades, however, appears to be that nations will plan *for* the projected population increases, rather than effectively planning *the population.* In other words, the major effort undoubtedly will be to increase resource capabilities to match the population increases. The pressures that population magnitude of the future will place on world capabilities to produce food, fiber, minerals, and to maintain quality of living environment, are not easy to comprehend. Can man's genius in technology really develop the earth environment to provide the projected needs? What will be needed to conserve acceptable quality of total environment? What will be the impact of greatly increased demand and competition on United States resources?

THE IMPACT OF POPULATION PROJECTION

Many scholars are concerned by the potential impact of population increases that are being projected, and some have gone so far as to state that population growth and war are the two greatest problems which society has yet to solve.[2] From the point of view of resource adequacy and environmental quality, it seems clear that the projected magnitude of population will result in (1) increased competition for material resources and for utilization of land space, (2) increased international resource problems, (3) increased competition for resource supplies which now are provided to the United States, and (4) continued difficulties in balancing food and material needs of "developing" countries.

BALANCING LAND CAPABILITIES AND POPULATION. There is little doubt that the major increases in population now being recorded in most of the "less developed" nations are nullifying a large share of their social and economic advance. Figure 19 – 3, for example, illustrates the trend of world food production. Total world food production has steadily increased, in large part as a result of improvements in agricultural science and technology, but population increases have tended to nullify the increases. It may be argued that so long as the rate of economic advance is greater than population growth, there is really no problem. Thus, Mexico in the 1960–65 period recorded economic growth of 6 to 7 per cent annually, which was superior to the 3.5 per cent population increase. One wonders, however, whether the high rate of population growth has in fact "robbed" Mexico of nearly half the economic advance. Can the developing nations of the world meet the challenge of balancing population and food supply themselves? It is not appropriate to attempt an evaluation here; however Table 19–1 provides basic data on world rural land resources. The favorable situation of the United States in relationship to world averages is clear.

Agricultural land in decades past has generally increased at about the same rate as population growth. Whether increases can be sustained to match the projected doubling of world population is questionable. How then can world resources be made to provide for the food and fiber needs of double and even triple the present-day population? This is a serious

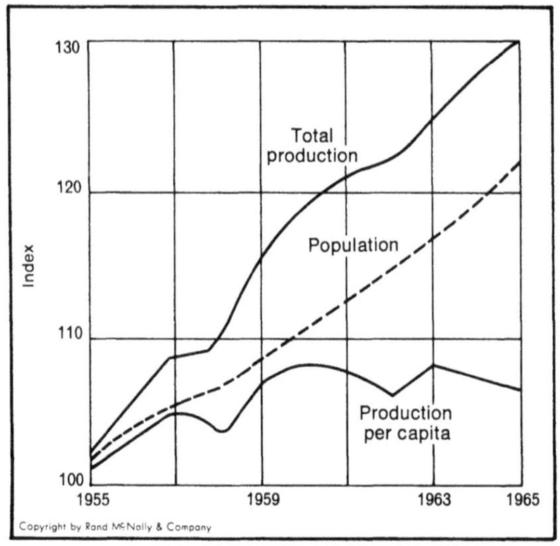

19 – 3 *Comparative indexes of world population and agricultural production, 1955–1965.* **The tendency for population increases to offset technological improvements is a problem of major concern. Moreover, it is generally agreed that large segments of world population are inadequately nourished. Thus a doubled population needs a threefold increase in food supply ("Food for Peace" 1965 Annual Report on Public Law 480 [Washington: Department of State, 1966], p. 12.)**

TABLE 19–1

WORLD RURAL LAND RESOURCES*

Area	1965 population	Arable land		Pasture land		Forest land	
	(millions)	Million acres	Acres per capita[1]	Million acres	Acres per capita[1]	Million acres	Acres per capita[1]
North America	215	561	2.6	684	3.2	1,840	8.5
United States	(194)	(457)	2.3	(633)	3.3	(746)	3.8
USSR	234	569	2.4	913	3.9	2,184	9.3
Europe	443	375	0.8	222	0.5	338	0.7
Asia	1,842	860	0.5	511	0.4	1,109	0.6
India	(482)	(402)	0.8	(27)	0.1	(140)	0.2
Latin America	248	237	0.9	1,025	4.1	2,438	9.8
Africa	211	642	2.1	1,571	5.1	1,855	6.0
Oceania	17	86	5.1	1,136	66.8	195	11.5
World	3,308	3,599	1.1	6,504	1.96	10,139	3.1

* *Production Yearbook Rome 1965* United Nations Food and Agricultural Organization, 1966. In using this comparative data it should be noted that there is considerable variation in reliability and in criteria used in world areas. No attempt has been made to adjust the UNFAO data to United States data used in other chapters. Moreover, the gross data do not reveal differentials in land productivity.

[1] Per capita acres computed using 1965 population estimates.

conservation question for many areas of the world. Some scholars believe that the hope of the world lies in greatly increased use of the humid tropics where in fact large areas such as the Amazon Basin are sparsely occupied. Others believe that the arid lands hold prospects, especially if economic technology for use of sea water is developed. Still others emphasize potentials in expanded use and management of marine resources. Most, however, believe that the major potential for food output is in increased production on lands already in use —especially through greater use of fertilizer, irrigation, pesticides, and improved seed (see Figure 19 – 4).

Whatever the ultimate answer (assuming that we accept the inevitability of population increases for the short run of time), some basic implications for rural land resource conservation seem to be crystal clear: (1) The world society cannot afford land-use practices that result in deterioration of land quality (see Figure 19 – 5). (2) Expanded research to develop technology and create economic-social conditions suitable for continuous intensive utilization of biotic potentials of the humid tropics is needed. (3) Programs that assist in implementation of improved rural land utilization ought to be encouraged and supported (see Figure 19 – 6).

INCREASED DEMANDS FOR RESOURCES. It seems inevitable that the projected population increases combined with probable higher levels of social and economic well-being will result in major realignments of demands and competition for resource that will affect the United States. In recent decades the United States, with 6 per cent of the world's population has been consuming about half the material stuff of the world (other than foods), and imports form a long list of natural resource commodities and products (see Figure 19 – 7). One of the most significant demand potentials is for commercial energy—especially the fuels such as coal, oil, and natural gas. Figure 19 – 8 reveals the major differential in per capita consumption between countries. It seems reasonable to draw the conclusion that as less developed nations advance they will be recording greater energy demands. Few scholars believe that the world, on the average, will achieve the 1960 levels of United States consumption by the year 2000, but only with continuation of present trends world demand is projected at *five times* present consumption (Table 19 – 2).

These potential demands on fuel resources are large; nevertheless most scholars believe that world fuel resources (proven and probable) are ample for at least 200 more years. Moreover, it is evident that atomic "fuels" can be-

TABLE 19–2

ENERGY CONSUMED, 1960, AND PROJECTIONS TO YEAR 2000 BY WORLD AREAS*

	Coal equivalent per capita 1960 kilograms[1]	Billions of metric tons coal equivalent 1960	2000[2] 1950–60 trend	2000 world at[2] U.S. 1960 level
World total	1,404	4.2	22.4	55.3
North America	7,785	1.55	2.68	2.61
(United States)	(7,999)	(1.45)		
Latin America	663	0.14	3.18	5.22
Western Europe	2,568	0.79	2.98	3.45
Eastern Europe & USSR	2,899	0.90	4.6	4.49
Communist Asia	600	0.40	5.0	14.4
Noncommunist Asia	406	0.24	2.9	20.0
Africa	312	0.08	1.0	5.31
Oceania	3,040	0.05	0.1	0.24

* Part of a table from Joseph L. Fisher and Neal Potter, "World Prospects for Natural Resources" (Washington: Resources for the Future, Inc., 1964), p. 47.

[1] United Nations estimates, not part of table by Fisher and Potter.

[2] Based on world population projection of 6.9 billion.

come a significant economic energy source on a commercial scale, perhaps providing 10 to 20 per cent of world consumption by the year 2000. Demand projections also have been made for other major minerals. The conclusion commonly reached is that no overall problems of adequacy exist at least through the year 2000.[3] This is not to suggest that there are no problems, for improved technology and increased costs in capital and energy will undoubtedly be required as lesser quality deposits come into use. Although definitive knowledge of potential reserves of most minerals is too little to justify quantitative comparisons of demand potentials and reserves, the great increases projected for almost all minerals make clear that the principles and practices detailed in Chapter 16 with reference to the United States' mineral resources will need to be applied throughout the world.

As a result of the probable pressures on world mineral resource deposits and the economic development of other countries, one may speculate on the future availability of mineral resource raw materials which the United States now is importing in large tonnage and variety. For example, the United States is importing most of its bauxite needs, about one-third of its iron ore needs, and 15 per cent of petroleum consumed, as well as a long list of lesser tonnage minerals. Already "developing" nations are less disposed to encourage exportation of raw materials by foreign companies, including those of the United States, and increasingly they are expecting—in some cases requiring—at least primary processing or refining as part of their national economic development programs. As a result, United States private foreign investment has tended to shift from emphasis on mining and petroleum production to processing and manufacturing. At the same time there is a strong increase in foreign investment, especially through development of multinational corporations. Since World War II United States investments abroad have increased in value from eight to fifty billion dollars. Table 19–3 indicates the very large financial investment by United States companies in foreign countries, nearly half of which is concerned with mineral resources.

NATIONAL RESOURCE DEVELOPMENT AND INTERNATIONAL CONFLICT

Development of certain resources inevitably leads to international problems. These resources

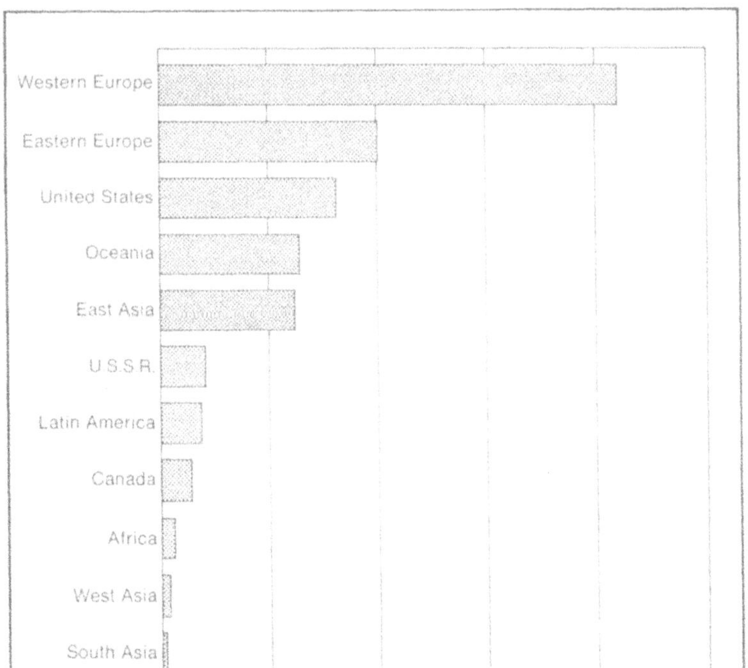

19–4 ***United States and world consumption of commercial fertilizers per cropland acre.*** **The potentials for increases in food and fiber through greater use of commercial fertilizers in less developed areas is suggested by this differential between the use levels. It should be noted, however, that fertilizer alone may not provide desired increases—water and improved seed are commonly essential cognates. ("A Graphic Summary of World Agriculture,"** ***Miscellaneous Publication 705*** **[Washington: Department of Agriculture, 1964].)**

include fish and wildlife which migrate in habitats that transcend national boundaries, international waters, and the atmosphere. Basic management concepts were developed in previous chapters; here it is intended only to refocus on a few of the problems.

How to maintain acceptable quality of atmosphere and international waters is becoming a problem as a result of advancing technology and increasing occupancy of the earth. Just as the serious deterioration of environment discussed in Chapters 11 and 17 with regard to air and water pollution in the United States cannot be adequately regulated within state boundaries, neither can it be wholly coped with by single nations. It is self-evident that economic activities which eject pollutants into the air or waters near national frontiers may lead to international controversies and the need for agreements for regulation.[4] Oil discharge from ships at sea is increasingly becoming a matter of concern to the United States as well as to other countries, especially because of potential damage to marine biota and to recreational beaches. Suggestions for use of ocean deeps as disposal "dumps" for radioactive materials is another matter of concern.

A new dimension is currently being added to the problem as a result of testing of various atomic devices. The problem is international (if not world-scale) because global air currents disperse contaminants. A recent study reported

19–5 *Severe erosion in agricultural land.* In most nations much agricultural land has suffered deterioration in quality as a result of utilization practices that have allowed accelerated physical erosion. Increased support to programs of land rehabilitation, and to education to check further losses of land from unnecessary erosion, deserves world-wide priority as one means toward assuring adequate food and fiber production. The United States has and should continue to respond generously to requests for assistance in efforts to further land conservation throughout the world and especially in countries where the programs have been initiated by and are strongly supported by local government. (Courtesy, Instituto Mexicano de Recursos Naturales Renovables, Mexico)

19 – 6 Education to implement improved rural land utilization. A United States Agency for International Development (AID) advisor in India helping diffuse knowledge and "know how" of agricultural science and technology by working with the actual operator of the land. (Courtesy Agency for International Development)

that "radioactive contamination of the environment due to tests carried out in 1961–62 had roughly doubled the total dose of radiation from radioactive debris that the world population would receive by the end of the present century."[5] The report also noted that "in Arctic regions small groups of local inhabitants showed body level of radio-caesium sometimes exceeding the world average by a factor of more than 100," believed to be the result of consumption of meat from caribou and reindeer which feed on "contaminated" vegetation. The eventual need for world-wide quality conservation of the atmosphere and waters to prevent increases of undesirable and potentially dangerous contaminants is clear.

Development of international rivers in the United States, such as the Colorado, Columbia, and St. Lawrence, to supply needs of the increased population for electricity, irrigation, navigation, and recreation inevitably requires considerations that include the well-being of neighboring peoples. The Colorado River, for example, receives essentially all of its waters from the United States, but it is the basis for the intensive commercial agriculture long established in the lower delta area—Mexico's Imperial Valley. This vested interest of Mexico led to a treaty by which a minimum of 1.5 million acre feet of water per year is guaranteed to Mexico.

TABLE 19–3

UNITED STATES DIRECT PRIVATE INVESTMENTS ABROAD BY AREA AND TYPE, 1964*

	Millions of dollars						
	Total	Mining	Petroleum	Manufacturing	Utilities	Trade	Other industries
TOTAL	44,343	3,564	14,350	16,861	2,023	3,736	3,808
Canada	13,820	1,671	3,228	6,191	467	805	1,458
Europe	12,067	56	3,086	6,547	53	1,472	854
Latin America[1]	8,932	1,098	3,142	2,340	508	951	832
Mexico	1,035	128	56	607	27	111	106
Venezuela	2,808	[2]	2,162	219	18	200	209
Other western hemisphere countries	1,386	250	569	106	49	89	263
Asia	3,062	34	2,014	535	55	238	186
Africa	1,629	356	830	225	2	93	122
Oceania	1,582	100	444	856	2	87	93
Other	1,865		1,038		827		

* U. S. Department of Commerce Survey of Current Business.

[1] Excludes investments in Cuba, 956 million dollars in 1960.

[2] Included in other industries.

The problem of ocean and anadromous fisheries is even more complicated because the resource itself migrates from national to international waters. Thus, effective management of any fishery for maximum sustained yield is impossible without cooperation. Traditionally, world nations have reserved their coastal waters within three miles for exploitation by domestic fishermen. In recent years, however, as world fishing intensity has increased, there has been a gradual extension seaward; for example Peru now reserves rights to two hundred miles. The United States in 1966 revised its limit to twelve miles[6] (see Chapter 13). It is significant to note that the extension came about in large part because of the appearance of a considerable number of Soviet fishing vessels in Pacific waters near the United States. This is a specific example of the increased competition for resource exploitation which may be expected as world population increases. Figure 19 – 9 reveals the trend of world fisheries harvest. Most fisheries scientists believe that marine food yields can be at least doubled. Yet it seems evident that management for maximum sustained yield of ocean and anadromous fisheries must in time require the cooperation of all nations.[7] Already there are a number of international treaties indicating that world nations are not unmindful of the conservation need. The following is a list of the principal international fisheries treaties to which the United States is a party:

Convention on Great Lakes Fisheries (Great Lakes Fishery Commission)
United States and Canada

Convention for the establishment of an Inter-American Tropical Tuna Commission
United States, Costa Rica, Ecuador, Mexico, Panama

International Convention for the Northwest Atlantic Fisheries (International Commission for the Northwest Atlantic Fisheries)
United States, Canada, Denmark, Norway, Spain, Italy, Portugal, Iceland, United Kingdom, France, West Germany, Poland, USSR

International Convention for the High Seas Fisheries of the North Pacific Ocean (International North Pacific Fisheries Commission)
United States, Canada, Japan

Convention for the Preservation of the Halibut Fisheries of the North Pacific Ocean and Bering Sea (International Pacific Halibut Commission)
United States and Canada

Convention for the Protection, Preservation and Extension of the Sockeye and Pink Salmon Fisheries of the Fraser River (International Pacific Salmon Commission)
United States and Canada

Interim Convention on Conservation of North Pacific Fur Seals (North Pacific Fur Seal Commission)
Canada, Japan, USSR, United States

Convention on Fishing and Conservation of Living Resources of the High Seas
Australia, Cambodia, Colombia, Dominican Republic, Finland, Haiti, Jamaica, Madagascar, Malawi, Malaysia, Mexico, Netherlands, Nigeria, Portugal, Senegal, Sierra Leone, South Africa, Switzerland, Trinidad, Uganda, United Kingdom, United States, Upper Volta, Venezuela, Yugoslavia

Convention for the Regulation of Whaling (International Whaling Commission)
Argentina, Australia, Brazil, Canada, Denmark, France, Iceland, Japan, Mexico, Netherlands, New Zealand, Norway, Panama, South Africa, USSR, United Kingdom, United States

Agreement for the establishment of the Indo-Pacific Fisheries Council

Australia, Burma, Cambodia, Ceylon, France, India, Indonesia, Japan, Korea, Malaysia, Netherlands, Pakistan, Philippines, Thailand, United Kingdom, United States, Vietnam

Agreement between the United States of America and the Union of Soviet Socialist Republics Relating to Fishing Operations in the Northeastern Pacific Ocean

United States and USSR

World Involvement of the United States

The United States, through international organizations, government agencies, private enterprise, and individual channels, has accepted broad and varied commitments to assist world-scale conservation efforts. Here only a few major features of this complicated aspect of United States involvement are included as examples.

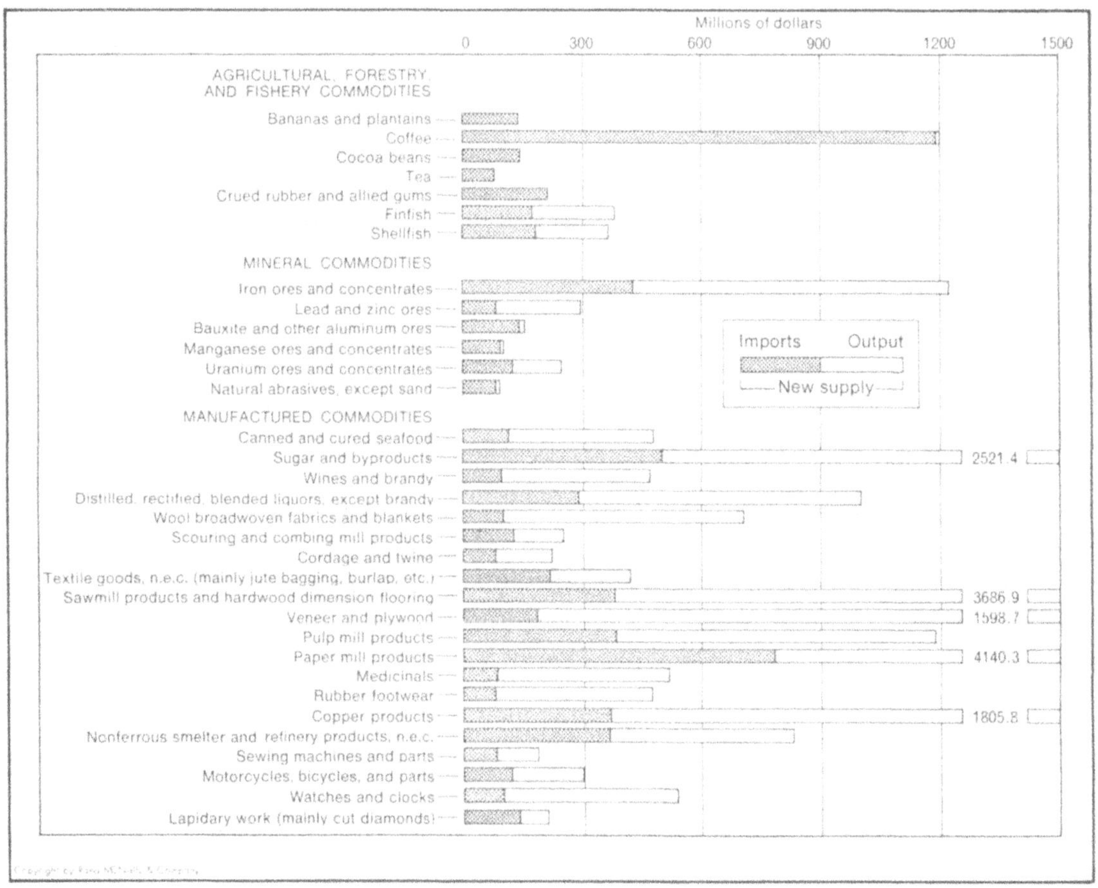

19 – 7 ***United States major imports, 1964.*** **The relationship of imports to total new supply of the thirty-three commodity groups with imports valued over fifty million dollars and at least 10 per cent of total new supply illustrates the variety and extent of U. S. demand for world resources. The United States concern for conservation problems on the world scale is implicit. (Department of Commerce)**

SUPPORT OF INTERNATIONAL ORGANIZATIONS

Financial and technical participation in the work of international agencies is one of the major facets of United States involvement in world scale conservation activities. In 1966 financial contributions totaled about 333 million dollars, accounting for an estimated 42 per cent of all member subscriptions. This estimate by the Department of Commerce does not include the cost to the United States of government employees detailed to the organizations, nor costs of international conferences, and, in general, does not include donated foods and other commodities under various aid programs. Table 19-4 lists principal international organizations and the proportion of their financial support provided by United States funds.in large part

FOOD AND AGRICULTURAL ORGANIZATION OF THE UNITED NATIONS (FAO). The FAO, organized in 1945, currently is supported by 107 member nations with the United States providing 32 per cent of the funds. In recent years the annual budget of approximately fifty million dollars supports some 1,000 technical experts in the field in about 100 nations. The Agency maintains its headquarters in Rome. FAO has made

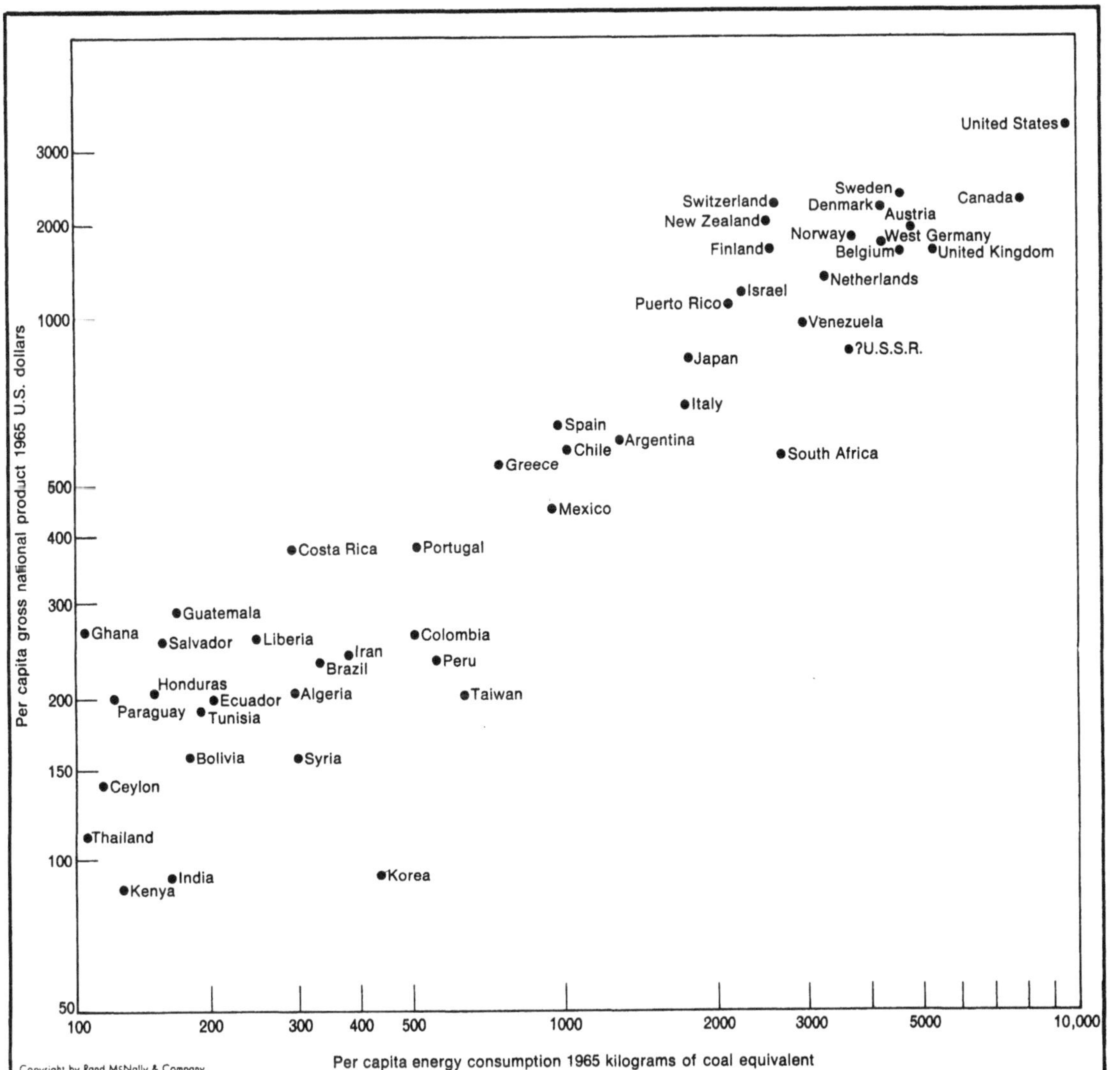

19-8 ***Relationship of per capita energy consumption and gross national product.*** **The positive correlation between energy consumption and development implies that there will be major increases in demands for energy resources as the "developing" nations raise productivity.**

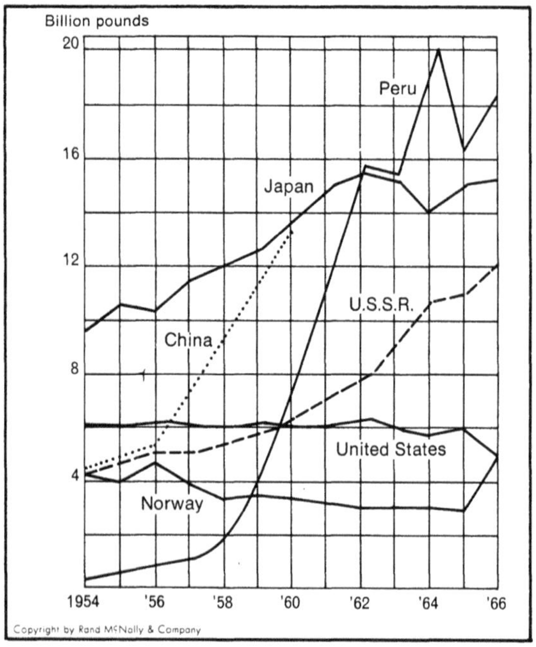

19 – 9 ***Trends in fish harvest by major nations, 1954–65.*** **Total world fish harvest in 1950 was 20.9 million tons; by 1964 tonnage had increased to 58 million tons. Increased fishing effort by vessels of the USSR and by Peru are good examples of the probable future. Potential sustained yield is uncertain, but fisheries scientists believe 120 million tons annually, or double the current harvest, is a possibility (*Yearbook of Fisheries Statistics, 1966,* [Rome: Food and Agricultural Organization of the United Nations, 1967].)**

notable contributions to the literature of resources through many technical publications and such serials as *The State of Food and Agriculture*, *The Production Yearbook*, *Yearbook of Fisheries Statistics*, *Unasylva*, and *The Yearbook of Forest Products Statistics.*

THE UNITED NATIONS EDUCATIONAL, SCIENTIFIC, AND CULTURAL ORGANIZATION (UNESCO). UNESCO, formed in 1946 with headquarters in Paris, carries on a wide variety of activities through its many divisions. A significant portion of its effort is devoted to resource problems, especially the program of arid lands research begun in 1951 and more recently expanded into a broader program of natural resources research and conferences. Its 117 members provide an annual budget of approximately 50 million dollars, of which the United States contributes 30 per cent.

WORLD HEALTH ORGANIZATION (WHO). The World Health Organization, formed in 1946, is supported by 118 member countries, with the United States contributing 31 per cent of the financial support. The annual budget of about 50 million dollars in recent years has made possible a vast program of health improvement through research, education, and operational programs for the eradication of endemic and epidemic disease. The agency maintains headquarters at Geneva, Switzerland.

WORLD METEOROLOGICAL ORGANIZATION (WMO). The World Meteorological Organization was formed in 1947 as a special agency of the United Nations. Its objectives include encouraging development of natural energy resources, especially wind and solar energy, and improving weather forecasting for agricultural needs. The organization holds a world congress every four years and maintains headquarters at Geneva, Switzerland, with a staff of about 200 scientists; it receives 24 per cent of its two million dollars financial support from the United States.

UNITED NATIONS DEVELOPMENT PROGRAM (UNDP). In 1965 the United Nations Special Fund and the expanded program of technical assistance were joined to form the Development Program to provide special assistance to "help low income nations build productive, dynamic societies based on full use of their own natural human resources." Funds from the program are derived from voluntary pledges which in recent years have totaled between one hundred and two hundred million dollars annually. The United States participates and provides 40 per cent of the funds. In particular, UNDP funds are aimed to provide for pre-investment feasibility surveys and to assist in technical training programs. By 1966 the program was supporting about 3,000 projects calling for total commitments of one and a half billion dollars, and had made it possible to send a total of about 36,000 technicians into the field to provide development assistance.

INTERNATIONAL BANK FOR RECONSTRUCTION AND DEVELOPMENT (WORLD BANK). The World Bank, formed in 1946, plays a significant role

TABLE 19–4

UNITED STATES FINANCIAL PARTICIPATION IN SELECTED INTERNATIONAL ORGANIZATIONS

	Per cent of total funds of program from United States
United Nations	
Food and Agricultural Organization	32
World Food Program	40
Educational Scientific and Cultural Organization	30
Special Fund Development Program	40
World Health Organization	31
World Meteorological Organization	24
Interamerican Institute of Agricultural Sciences	66
Interamerican Tropical Tuna Commission	92
Organization of American States	66
Pan American Health Organization	66
Pan American Institute of Geography and History	61
Colombo Plan Council	5
Organization for Economic Cooperation and Development	25
International Coffee Organization	20
International Commission for Northwest Atlantic Fisheries	7
International Cotton Advisory Committee	13
International Hydrographic Bureau	7
International Lead and Zinc Study Group	12
International North Pacific Fisheries Commission	33
International Rubber Study Group	11
International Sugar Council	12
International Whaling Commission	6
International Wheat Council	15
North Pacific Fur Seal Commission	25
Indus Basin Development	54
International Atomic Energy Agency	40

TABLE 19–5

WORLD BANK LOANS, 1964, BY PURPOSE AND AREA*

Purpose	Million dollars	Area	Million dollars
Transportation	356	Asia–Middle East	221
Electric power	239	Africa	207
Industry	68	Europe	164
Agriculture	40	Western hemisphere	103
Water supply	20	Australia	33
Education	6		

* Annual Report, International Bank for Reconstruction and Development.

through making capital available to countries which have not yet been able to accumulate sufficient national savings for development of primary resources and infrastructure. The United States has subscribed 29.9 per cent of the twenty-two billion dollars capital of the bank. Since its formation, over eight billion dollars has been loaned, and annual funding recently has amounted to 700–800 million dollars. Table 19–5 shows the 1964 loans by purpose and area. One hundred and two member countries form the bank, which has headquarters in Washington, D. C. and New York City.

INTERNATIONAL DEVELOPMENT ASSOCIATION (IDA). In 1960 the International Development Association was formed as an affiliate of the World Bank to facilitate loans to "developing" countries. The purpose was to provide for more liberal loans, which have been on a fifty-year basis free of interest with a service charge of 75 per cent. The United States subscribes 32.17 per cent of the bank's one billion dollar capital.

INTER AMERICAN DEVELOPMENT BANK. In 1959 the United States joined with nineteen Latin American Republics to form the bank with about two billion dollars capital resource. The United States has subscribed 761 million dollars. In addition, the Bank administers about 500 million dollars in Social Progress Trust Funds established in 1961 to promote social progress in Latin America as part of the Alliance for Progress program.

TABLE 19–6

UNITED STATES NET FOREIGN ASSISTANCE, 1945–1965 AND 1965*

	1945–1965 Million dollars	1965 Million dollars
TOTAL	65,509	3,373
Western Europe	23,810	1
Near East & South Asia	15,403	1,640
India	(5,150)	(849)
Pakistan	(2,593)	(348)
Far East & Pacific	14,541	655
Korea	(3,872)	(165)
Vietnam	(2,332)	(300)
Latin America	5,607	632
Africa	2,229	369
Eastern Europe	1,624	2
Other international organizations	2,295	189

* United States Department of Commerce, Foreign Grants and Credits by the United States Government.
[1] Repayments to U. S. exceed new assistance.

UNITED STATES DIRECT GOVERNMENTAL ASSISTANCE

In addition to the large indirect participation through international agencies, the United States has responded to requests and need for assistance from many areas that are struggling to improve living standards and at the same time achieve harmonious rapport with their resource environment. Table 19–6 indicates the magnitude of net financial assistance by major areas of the world.

The totality of the United States direct assistance to resource development and conservation in other countries is enormously complex. Most of the principal United States government agencies have departments oriented to international resource problems, and many participate in assisting other countries on request. For example, the United States Forest Service maintains a department devoted to international forestry problems, the Department of Agriculture maintains an International Agricultural Development Service to coordinate activities of foreign agriculture development in which the United States is involved, including the education and training of foreign nationals in agricultural sciences, and the United States Corps of Engineers maintains a special department devoted to resource inventories studies in other countries as the basis for macro-planning.

In 1961 most United States direct foreign aid was consolidated in the Department of State by the formation of the Agency for International Development, (AID) which combined the International Cooperation Administration and the Development Loan Fund. AID assumed responsibility for most nonmilitary foreign assistance programs including development loans and grants, financing of investment surveys by private enterprise (the United States government is authorized to participate in such private surveys up to 50 per cent of the cost), Alliance for Progress Activities, Food for Peace Program, and provision of capital and technical assistance to projects and programs. In 1966 AID administered about 2.6 billion dollars including 1,200 million dollars in loans, 300 million in grants, and 700 million in supporting assistance.

Conclusion

Throughout this chapter it has been implied that change, especially in degree of occupancy, will force significant realignments that in turn will require re-evaluations of United States natural resources adequacies. In addition, the projected demands on resources seem to make clear that the principles put forth in the chapters oriented to specific resources of the United States need also to be implemented by all countries and in all areas.

What the responsibility of the United States is—or ought to be—with reference to areas less fortunate in their level of economic development is not easy to define. The reality, however, is that the United States has undertaken to offer substantial aid to countries that have requested assistance and is in fact deeply and broadly involved in cooperating with other countries to help them improve living standards through development of their resource environments. There would seem to be no doubt that assistance in early stages of development is commonly needed, and thus it is desirable that the United States continue to support international organizations and generously provide technical advice along with financial assistance to countries that request aid. It would be well, however, for this nation, through its responsible agencies, to keep in mind that the best assistance is likely to be that which stimulates and aids developing countries and peoples in formulating and in bettering *their own programs.* Only assistance that leads to self-sustained and continued accomplishment will have long lasting benefit.

United States private organizations involved in resource enterprises as well as individuals also have major responsibility. Through the assignment of management and production specialists to divisions in other countries, and through training of foreign nationals, American business forms one of the significant channels by which new knowledge, improved technology, and the sense of responsibility are diffused. A similar role is played by universities, especially through cooperative programs of resources research, and education. The United States Peace Corps undoubtedly is making one of its major contributions through individual contacts which help to implement knowledge.

Finally, there is need to appreciate that the resource problems of the world are not wholly measurable in terms of physical resources divided by annual demands, nor in terms of land space divided by population. The crucial problem is how to implement and integrate the capabilities of science, technology, and societal organization for the maintenance of quality in environments, and at the same time assure continuous flow of resource commodities such that all peoples come to have a reasonable standard of living.

FOOTNOTES

[1] For a recent analysis of world population see J. Beaujeu-Garnier, *Geography of Population* (New York: St. Martins Press, 1966), 386 pp.

[2] For example see S. Mudd, ed., *Population Crisis and the Use of World Resources* (Bloomington: Indiana Univ. Press, 1964)

[3] For example see Bruce C. Netschert and Hans H. Landsberg, *The Future Supply of the Major Metals: A Reconnaissance Survey* (Baltimore: Johns Hopkins Univ. Press, 1961), 66 pp.

[4] The case of air pollution from the Trail, British Columbia Lead-Zinc Smelter which in years past devastated vegetation in the adjacent United States is a good example. It is also a good example of a major degree of control, because contaminants are now recovered and provide raw material for the fertilizer industry. The problem was adjudicated by the International Boundary Commission in 1930, and led to the early stages of the control program.

[5] Report of the United Nations Scientific Committee on the Effects of Atomic Radiation, 1964. *19th General Assembly Supplement No. 14* (New York: United Nations), 120 pp.

[6] See U. S. Senate Subcommittee on Merchant Marine and Fisheries, S2218 Hearings, May 18-20, 1966 on the Twelve Mile Fishery Zone, 89th Cong., 2nd sess., *Serial 89–65* and *Report 1280* accompanying the bill.

[7] See Francis T. Christy, Jr. and Anthony Scott, *The Common Wealth in Ocean Fisheries,* (Baltimore: Johns Hopkins Press, 1965), p. 281.

20

The Future

THE UNITED STATES TODAY possesses the highest order economy, and its people enjoy the highest standard of living of all the nations and peoples of the world. This achievement has resulted from a composite of many factors, but four may be noted as paramount. These are abundance and variety of natural resources, leadership in development and application of science and technology, comparatively small population, and a private enterprise system that stimulated individual activity.

The great wealth of resources in variety, in large quantity, and in high quality has formed a solid foundation for wide-scope commercial-industrial developments. Although some nations compare in single resource categories, none equals the composite strength of the United States in the basic resources: productive land, minerals, and water. Certainly Americans are singularly blessed in being heir to the richest estate.

The beginning of national development essentially coincided with the advent of the mechanical revolution. The westward expansion, unencumbered by older forms of settlement, progressed in a climate of constant improvements in labor-saving machinery and expanding domestic and foreign markets which stimulated individual initiative and ingenuity. By 1900 the nation was a leader in science and technology, and during this century it has had no peer in the application of these skills to the varied elements of economic production, continually advancing the efficiency of labor and output from resources. These improvements have had the paradoxical effect of creating new resources while increasing resource demand.

The citizenry of the United States totals only slightly more than 6 per cent of the world's population. As the result of the application of advanced technology to a great resource base, the populace has developed a production capacity that looms several times larger than its proportionate share of the world population. The net result is the nation's very high standard of living.

The immediate future promises even greater individual benefits. The curve of development is moving upward under a nationally supported economic policy of continued growth. This policy envisions the support of an increasing population at a higher standard of living than that enjoyed today.

There is little doubt that the nation will continue to make forward strides in science and technology. These will improve resource use and, no doubt, create additional resources. But there still remains a question of vital concern for every responsible citizen: Can the resource base of the United States support economic growth indefinitely? And if so, then how?

Despite mistakes of the past made under general economic conditions which fostered them, the country remains rich in its basic endowments. Nevertheless, as noted in preceding chapters, shortages are appearing in some resources which once were abundant, and misuse and wasteful practices continue in others. If the resource base is to meet the demands of a population of 300 to 370 million in fifty years, malpractices certainly must be minimized and science and technology implemented.

Recent Progress

Very significant forward strides have been made since 1930. On the favorable side of the conservation balance sheet many accomplishments can be noted. A land capability classification has been devised, and a detailed inventory of the nation's land resources is well under way. Experimentation and experience are proving effective methods for managing and using land commensurate with its capability. The Extension Service, Soil Conservation Service, Forest Service, Bureau of Land Management, Geologic Survey, Bureau of Mines, and other Federal and state agencies concerned with resources have developed rather strong education and assistance programs relative to conservation. Some effective legislation has been passed. River basins are now recognized as the work unit for water control and development. The technology of mining and mineral use is undergoing steady improvement. Geophysical and geochemical exploration have been greatly advanced by improved equipment. Federal and industrial research has determined much about mineral characteristics. Important advances have been made in fuel technology. Forestry is now emphasizing tree growing. Multiple-use management has been established on Federal lands, and a start in this direction has been made on private lands. Federal management agencies are planning for orderly development, and farm, ranch, and forest operators are advancing along this line. City planning is expanding and some area and regional planning has begun. Federal agencies recently have started to make long-range projections of needs.

If there were full implementation of proven technical and scientific knowledge respecting resource management and use, the nation would, indeed, be in a sound position in regard to meeting the long-range demands. Although recent progress has been noteworthy, there is no time for relaxation. Soil still is being lost in some areas; a considerable amount of quality cropland is being needlessly alienated to nonagricultural uses; much forest land is producing only a fraction of its capability; waters are still inadequately controlled in many sections of the country; and recovery in mineral exploitation is still low in many cases, and wasteful utilization is evident. Sprawl and decline in environmental quality are progressing at a higher rate than ever. In general, short-range economic considerations continue to determine the adoption and extent of conservation practices.

These adverse conditions result from some of the same factors that have resulted in the highly productive economy. Under this democracy, progress has been fostered by resource abundance, private enterprise, and the profit motive. Resource despoilment in the early years of development may be excused on the grounds of growing pains and lack of know-how. These are no longer excuses. The contemplated economic development and population increase will allow little margin for misuse. In fact, the long-range continuation of the private enterprise system, the democratic way of life, the national security, and world leadership, depend upon a prompt and full conservation adjustment.

Because there is a backlog of science and technology, and because research is continuing, it appears to the authors that the greatest immediate needs fall into the following categories: (1) allocation and acceptance of responsibilities, and (2) planning and programing future development.

Levels of Responsibility

Most Americans would agree that conservation is essential for the nation's future well-being. Few, however, recognize any responsibility for its implementation, because a major share of the population is not closely associated with the resources that actually support the economy. Yet in a democracy every citizen has a role in conservation's perpetuation. It is the contention of the authors that striving for conservation is a vital part of citizenship. Each generation has the moral obligation to pass along to the next the means for maintaining the standard of living and way of life. Every citizen has responsibility to himself and posterity to be informed on resource matters in order that he may vote intelligently and aid in bringing pressures to bear in the support of remedial and other desirable programs.

Those who are owners or managers of resources must recognize a particularly grave responsibility; they control the very basis for economic existence. They have moral obligations and are accountable to Americans for years to come. The character of their stewardship will, in a large measure, determine the potentials of the future. Thus the right of ownership carries with it a duty and trust in which conservation must pervade every decision and action respecting resource use and management.

In a democracy, government is the voice and instrument of the body politic—the organ of all of the people. It has the responsibility, rooted in constitutional grounds, of doing for the people collectively what they cannot do for themselves individually. It is the obligation of government to protect the interests of future as well as present citizens. Conservation is distinctly in the interest of promoting the general welfare. Government, therefore, has the responsibility to expedite over-all programs of resource management and development which will maintain a secure foundation and sustained yield for all facets of the nation's economy, security, and economic and social growth. The exercise of this responsibility requires positive leadership to provide the favorable administrative frameworks, conditions, policies, controls, and knowledge which foster and reward conservation. These may involve all or part of the following at a given time: economic and regulatory controls, research, inventory, education, assistance, projection of needs and development, area, regional, and national planning, and taxation. Government responsibility also involves the comprehensive management and development of lands and other resources in public proprietorship in the interests of all of of the people.

In the final analysis all levels of resource responsibility and the quality of resource stewardship are determined by the total citizenry. Desires of the majority can rule. The efficiency of resource managers and of government in conservation is a reflection of the character of public interests and pressures.

Development by Design

An assured stable growth can be achieved only by bringing economic development and resource development more fully into balance and order. This will require continuation and expansion of positive programs of action now in effect by both government and private industry. Basic aspects of such programs should include inventory, planning, and development scheduling.

Inventory in this context refers to the continual and accurate appraisal of the status of all elements of the resource base, the scientific and technological tools available related to use and maintenance, and the short- and long-range resource needs of the population. Planning refers to the application of intelligent foresight to designing methods of development that will meet future needs and goals. Scheduling refers to ordering development in such a manner to gain efficiency and minimize waste while meeting both the short- and long-range needs.

Although gains have been made along these lines, the people and their various levels of government obviously still fall short in meeting their responsibilities to future Americans. The longer the delay in facing up to the problems, the more difficult will be their solutions. Enlightenment of the public is called for now to bring about intensified conservation action on all levels of responsibility. This task presents a challenge to formal education.

It rightly should fall within the role of government to provide leadership. In the interest of the general welfare, the Federal government must improve and strengthen its responsibilities in clarifying national objectives, in directing and correlating national inventories, in devising national plans and development programs, and in providing over-all conservation leadership. State and local governments need to improve leadership on their respective levels and in the context of the national and regional interests.

It is not within the scope of this book to suggest how these improvements in governmental meeting of responsibilities should be made. It would appear, however, that they can be made within the framework of constitutionally given democratic authority through administration, regulation, taxation, research, advice, education, incentives, negotiation, and excellence of management of public resources. This may well require improvements and better coordination in the executive departments; it certainly calls for greater responsibility in the legislative branches and more cooperation between all branches and levels.

Private enterprise must clearly recognize its obligation to the larger interests of all the people, now and in the future. Industry's understandable desire for immediate profits must be tempered by the acceptance of the conservational responsibility—the inexorable obligation of private ownership. Private enterprise also should provide leadership; carry out research, plan, and program development; and cooperate in meeting regional and national goals. The degree of government intervention in actual resource management will depend on the degree to which the users accept their responsibility as resource stewards.

Conclusion

The authors are not pessimists. We believe that the nation is on the verge of a new era in conservation. The resource base remains strong, and technology, respecting its management and conversion to use, is progressing at a fantastic rate. The American people have within their own hands the tools and framework for communal action to meet the challenges and needs of the future. But they must change their attitudes from those of an exuberant, youthful society to those of an experienced, mature society.

Index

www.ingramcontent.com/pod-product-compliance
Ingram Content Group UK Ltd.
Pitfield, Milton Keynes, MK11 3LW, UK
UKHW051933150926
475560UK00005B/369

9 780313 205118